Plant Chemical Ecology

Plant Chemical Ecology

Guest Editors

Chui-Hua Kong
Yonggen Lou
Peng Wang
Guoxin Zhou

MDPI

Basel • Beijing • Wuhan • Barcelona • Belgrade • Novi Sad • Cluj • Manchester

Guest Editors
Chui-Hua Kong
College of Resources and Environmental Sciences
China Agricultural University
Beijing
China

Yonggen Lou
Institute of Insect Sciences
Zhejiang University
Hangzhou
China

Peng Wang
Institute of Applied Ecology
Chinese Academy of Sciences
Shenyang
China

Guoxin Zhou
College of Advanced Agricultural Sciences
Zhejiang A&F University
Hanghzou
China

Editorial Office
MDPI AG
Grosspeteranlage 5
4052 Basel, Switzerland

This is a reprint of the Special Issue, published open access by the journal *Plants* (ISSN 2223-7747), freely accessible at: https://www.mdpi.com/journal/plants/special_issues/JWQK1KE9U5.

For citation purposes, cite each article independently as indicated on the article page online and as indicated below:

Lastname, A.A.; Lastname, B.B. Article Title. *Journal Name* **Year**, *Volume Number*, Page Range.

ISBN 978-3-7258-7947-2 (Hbk)
ISBN 978-3-7258-7948-9 (PDF)
https://doi.org/10.3390/books978-3-7258-7948-9

Contents

plants

MDPI

Review

Chemically Mediated Plant–Plant Interactions: Allelopathy and Allelobiosis

Chui-Hua Kong [1,*], Zheng Li [1], Feng-Li Li [1], Xin-Xin Xia [1] and Peng Wang [2]

[1] College of Resources and Environmental Sciences, China Agricultural University, Beijing 100193, China; lizheng268@sina.com (Z.L.); lifengli789@163.com (F.-L.L.); 13996772329@163.com (X.-X.X.)
[2] Institute of Applied Ecology, Chinese Academy of Sciences, Shenyang 110016, China; wangpeng@iae.ac.cn
* Correspondence: kongch@cau.edu.cn

Abstract: Plant–plant interactions are a central driver for plant coexistence and community assembly. Chemically mediated plant–plant interactions are represented by allelopathy and allelobiosis. Both allelopathy and allelobiosis are achieved through specialized metabolites (allelochemicals or signaling chemicals) produced and released from neighboring plants. Allelopathy exerts mostly negative effects on the establishment and growth of neighboring plants by allelochemicals, while allelobiosis provides plant neighbor detection and identity recognition mediated by signaling chemicals. Therefore, plants can chemically affect the performance of neighboring plants through the allelopathy and allelobiosis that frequently occur in plant–plant intra-specific and inter-specific interactions. Allelopathy and allelobiosis are two probably inseparable processes that occur together in plant–plant chemical interactions. Here, we comprehensively review allelopathy and allelobiosis in plant–plant interactions, including allelopathy and allelochemicals and their application for sustainable agriculture and forestry, allelobiosis and plant identity recognition, chemically mediated root–soil interactions and plant–soil feedback, and biosynthesis and the molecular mechanisms of allelochemicals and signaling chemicals. Altogether, these efforts provide the recent advancements in the wide field of allelopathy and allelobiosis, and new insights into the chemically mediated plant–plant interactions.

Keywords: allelochemical responses; chemical signals; neighbor identity recognition; molecular mechanisms; plant-plant signaling interaction; root exudates; root–soil interactions; volatiles

1. Introduction

When two or more plants coexist, a series of intra-specific and inter-specific interactions occur either positively or negatively. Productivity and the establishment of mixed-species systems or variety mixtures (single-species) are ultimately the net result of positive and negative interactions among the component species or varieties [1,2]. The interplay and balance between positive and negative interactions are important and recurrent topics in plant ecology. Plant–plant negative interactions are mainly involved in competition, parasitism, and incompatibility, while positive interactions include heterospecific facilitation, conspecific cooperation, and complementarity. These patterns can arise through a variety of ecological mechanisms, such as resource partitioning, niche divergence, kinship strategy, and other species-specific behaviors [3–6]. The performance and outcomes of plant coexistence and community assembly depend on multiple biotic and abiotic interactions, but the central driver must be plant–plant interactions.

Plant–plant interactions are mediated physically and chemically. An increasing number of studies have shown that chemically mediated plant–plant interactions are important driving forces for plant coexistence and community assembly [7–9]. Chemically mediated plant–plant interactions can be represented by allelopathy and allelobiosis. Allelopathy means that plants produce and release allelochemicals to inhibit the establishment and growth of co-occurring plants [10]. Despite the allelochemical actions that are in some way growth-stimulating, by the mainstream perspective, allelopathy is defined as negative interactions among plant species because the term is based on the Greek terms allelo, meaning 'one another', and path, for 'suffering' [11], while allelobiosis is a relatively new term and consists of allelo and biosis, meaning 'mutualism', denoting positive or neutral interactions [12,13]. Both allelopathy and allelobiosis are ecological phenomena leading to chemical interference among plants. Allelopathy describes that plants produce and release allelochemicals to exert negative effects on neighboring plants. In contrast to negative plant–plant allelopathic interactions, allelobiosis describes the transmission of signaling chemicals in plant–plant interactions where the response of the coexisting plants affects their growth and defense strategy and is related to herbivores and their natural enemies, resulting in positive or neutral effects [14,15]. Importantly, allelobiosis is involved in chemical-mediated plant neighbor detection and identity recognition [16]. Accordingly, allelopathy and allelobiosis cover any positive or negative effects on chemically mediated interactions among plants.

The importance of allelopathy and allelobiosis in plant–plant interactions has received a great deal of attention in the past decades [17–21]. However, allelopathy and allelobiosis are usually studied separately, despite their necessary linkage in nature. There is a wealth of information on plant–plant allelopathic interactions in natural and managed ecosystems [17–20,22] but a lack of information on the role of allelobiosis in plant–plant interactions; the term allelobiosis is sparsely found in the scientific literature. In particular, most of the allelopathy research has primarily focused on plants and their allelochemical interactions but ignored the plant–plant signaling interactions. Actually, for plant coexistence, plants first may detect and potentially recognize their neighbors, and then initiate allelopathic interference to regulate inter-specific or intra-specific interactions [16,23]. Therefore, allelopathy and allelobiosis are two probably inseparable processes that occur together in plant–plant chemical interactions (Figure 1).

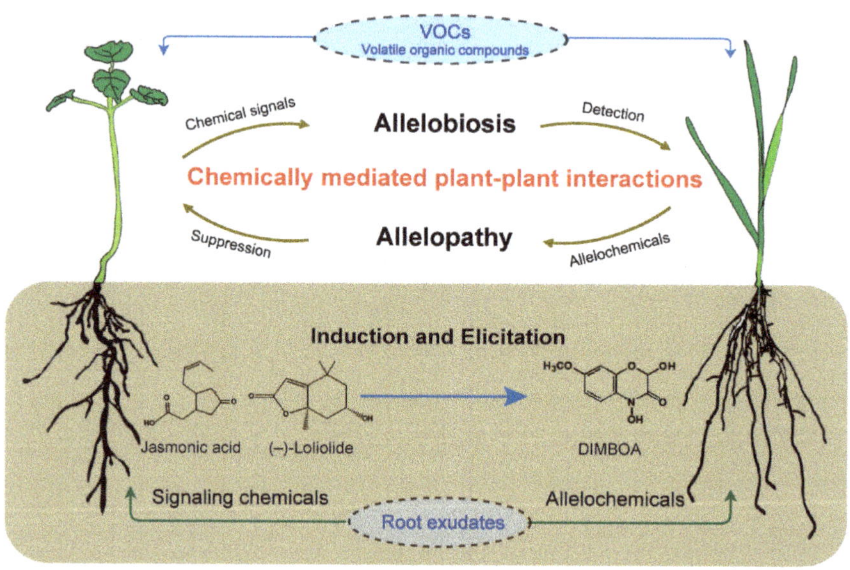

Figure 1. Allelopathy and allelobiosis in plant–plant interactions. VOCs include volatile allelochemicals, such as camphor and (–)-thujone, and volatile signaling chemicals, such as ethylene and methyl jasmonate.

Allelopathy and allelobiosis have provided fascinating insights into eco-evolutionary relationships and represent potential strategies for sustainable agriculture and forestry [9,16,20,21]. This review will focus on the recent advancements in the wide field of allelopathy and allelobiosis in plant–plant interactions. In particular, this review will extend and concretize the concept of allelobiosis from a useful definition for evaluating neighbor detection and identity recognition in plant–plant signaling interactions. Such information is critical to enhance our understanding of chemically mediated plant–plant interactions driving plant coexistence and community assembly, stimulating additional research into plant–plant interactions.

2. Allelopathy, Allelochemicals, and Their Application for Sustainable Agriculture and Forestry

2.1. Allelopathy, Competition, and Plant Invasion

Plants may affect the performance of neighboring plants through allelopathy, competition, or both. Both allelopathy and competition are naturally occurring ecological phenomena with a negative effect on neighboring plants. However, allelopathy and competition possess different mechanisms. Allelopathy means that plants produce and release allelochemicals into the environment to decrease neighbor fitness, but competition is imposed by the environment where plants draw from the same resource pools (i.e., neighboring plants deplete water and nutrients and provide shade). In addition, allelopathy is a species-specific behavior, but competition occurs invariably in plant coexistence regardless of plant species. Importantly, competition occurs only during living plant coexistence, but allelopathic potential may have an impact on successive plants by releasing allelochemicals from dead plants and their decomposing residues [11,18].

Allelopathy and competition frequently occur in conspecific and heterospecific plant–plant interactions. Intra-specific allelopathy is called autotoxicity; a plant releases allelochemicals to inhibit the growth and establishment of other conspecific plants [24]. Intra-specific competition generally must be stronger than inter-specific competition, where conspecifics interact more intensely because of their similarity [3]. Plant competition for the same resources is demonstrated by the presence of a neighbor reducing fitness, while allelopathy chemically suppresses the growth of neighboring plants, providing a competitive advantage for their own growth. Therefore, allelopathy is a plant chemical trait or behavior to increase competitive ability. Accordingly, it is not realistic to separate resource competition from allelopathy [25]. Although plants affect the performance of neighboring plants through competition and allelopathy in ways that are difficult to separate, the effects of allelopathy and competition on plant performance have been extensively investigated; however, there is a lack of quantitative synthesis [26]. A meta-analysis of 384 studies found that allelopathy reduced plant performance by 25% but the variation in allelopathy was high. In particular, allelopathy became more negative with increasing phylogenetic distance, indicating that allelopathy might contribute to the coexistence of closely related species (i.e., convergence) or dominance of a single species [27].

Allelopathy has had a checkered history. The classical cases are that black walnut (*Juglans nigra*) produced and released a 1,4-naphthoquinone (juglone) to interfere with the growth of understory plants [28] and the allelopathic interference of shrubs with grass through the release of volatile terpenes into southern California coastal grassland [29]. However, the ecological importance of allelopathy was a controversial issue because much of the early work on the preponderance of phenomenological studies was insufficiently rigorous and confirmable. In the early 2000s, allelopathy was firmly back on the center stage in a case study of invasive plants versus their new and old neighbors [30]. This case study neatly avoids the pitfalls of earlier work and provides compelling evidence that allelopathy is a mechanism for an exotic invasion of Eurasian spotted knapweed (*Centaurea maculosa*) in the western United States [31,32]. The allelopathic invasion of spotted knapweed in American grasslands resulted in the novel weapons hypothesis, as many co-occurring native plant species exhibit a lack of evolved tolerance to novel allelochemicals [33]. Subsequently, an increasing number of studies have shown that allelopathy can contribute to the success of alien plants [34–37].

Allelochemicals play a large part in the successful invasion of certain plant species. Generally, allelochemicals of invasive species have low levels when growing with their original neighbors due to long-term mutual adaptation. In the invaded habitats, however, invasive species produce and release allelochemicals at a high level to exert a strongly allelopathic interference on the native species, conferring a competitive advantage to the invasive plants [34,38]. Therefore, the allelopathy of invasive species is conditional, and there is a discrepancy between geographical sites. For example, spotted knapweed exhibited an allelopathic invasion in the western United States but not in eastern American grasslands [39]. Bohemian knotweed seems not to rely heavily on allelopathy for its persistence in invaded sites in the southwest part of Zagreb, Croatia [40].

Invasive plants apply an allelopathic impact on native species to be successful in their invaded range. It is also worthwhile to note that native species co-occurring with invaders have evolved a tolerance to allelopathy. A few studies indicated allelopathy and its coevolutionary implications between native neighbors and invasive species [35,36]. For example, co-occurring native *Juncus pallidus* may evolve tolerance to the allelochemicals induced by invasive *Cynara cardunculus*. In particular, allelochemicals induced by *C. cardunculus* exert more suppressive effects on native plants than non-native plants, linking the coevolved tolerance [36]. Although coevolution can promote the long-term coexistence of two competing species, plant invasions offer opportunities to capture the process of coevolution [35]. Therefore, allelopathy-based plant invasion may drive coexistence via coevolution, altering community structure and ecosystem functions. Not only that, allelopathy has broad ecological consequences for natural systems, and is particularly important for species distribution, conditionality of interactions, and maintenance of species diversity [10,17,20,41].

2.2. Allelochemical Responses and Interactions

Plants can modulate their morphological and biochemical traits in response to environmental conditions, particularly for the production of secondary metabolites [42–45]. Allelopathic actions depend on the production and release of allelochemicals, and thus, allelopathy is inevitably regulated by the environmental factors. Numerous studies have shown that allelopathic potential varies with plant growth conditions, and the categories and amounts of allelochemicals can be modulated by natural factors. In particular, biotic and abiotic stressors including herbivores, pathogens, competition, mechanical wounding, drought, nutrient deficiency, salt, and heavy metals can induce the production and release of allelochemicals [46–49]. Accordingly, allelopathy is not only a competitive trait, but also one induced in response to the identity of the neighboring plants. As such, allelopathy could represent the keystone in the argument that plants do differentially detect and respond to their neighbors [23,50].

Plant–plant allelopathic interactions occur in the presence of conspecific and interspecific neighbors, where co-occurring plants are prone to competition. Therefore, an allelochemical response to competitors is vital to understand allelopathic interactions. The context-dependent induction of allelopathy in plants under competition has received great attention in recent decades [23,48,51,52]. The ability to induce allelochemical production in the presence of competitors is a natural adaptive strategy so that plants can save costs of metabolite production in the absence of competitors [48,53]. Although competitor-induced allelochemical response is a general phenomenon in plant–plant allelopathic interactions, the phenomenon is density dependent. Allelopathic plant species increase allelochemical concentrations in response to competitors at low or intermediate densities but a high density of competitors may reduce the production of allelochemicals [23,54]. Plants growing with a high density of neighbors likely experience an increase in competition, resulting in a shift from allelochemical defense to growth, i.e., inducible allelopathy to trade off with constitutive growth [55].

DIMBOA (2,4-dihydroxy-7-methoxy-1,4-benzoxazin-3-one) is a dominant allelochemical in cereal (wheat, maize, and rye, but not in rice) against weeds, pathogens, and herbivores [56]. The production of DIMBOA can be induced by a diverse range of weeds [16,57].

To provide a broader context for the competitor-induced allelochemical response, a dataset of wheat DIMBOA concentration was produced in the presence of a total of 100 different plants species at varying densities. The induction of DIMBOA markedly increased in heterospecific combinations relative to DIMBOA production when competing with conspecifics. Significant increases occurred in 38/100 species at a 5:5 proportion, while occurring in 70/100 species at a 5:8 mixture [23]. Therefore, the presence of either conspecific or heterospecifc neighbors could induce the allelochemical response in a density-dependent manner.

Polyacetylenes are putative allelochemicals in invasive species such as *Solidago altissima*. A recent study indicated the inducibility of polyacetylenes in response to competition [58]. The presence of competitors induced polyacetylenes, but the inducibility was observed at an early stage of *S. altissima* invasion. In addition, induction became more prominent under nutrient depletion, where induced polyacetylenes may be effectively suppressed by the growth of competitors [58]. It appeared from the results that allelopathic plant species could respond to competition by inducing allelochemical production, but competition induced allelochemicals in a context-dependent manner, and variation in resource availability may influence plants' allelopathic capacity.

Allelopathy and allelochemical responses take place both aboveground and belowground. Aboveground allelopathy responses mediated by volatile allelochemicals are well established. This situation is primarily driven by the accessibility of aerial tissues and the availability of reliable techniques to detect and identify volatile allelochemicals [18]. However, it is extremely difficult to unambiguously demonstrate belowground allelopathy and allelochemical interactions. Unlike aboveground processes, the study of belowground allelochemical interactions is hampered by the inability to observe roots [59,60]. Actually, many allelopathic species release allelochemicals through their root systems into the soil, exerting a negative effect on cooccurring species, particularly for hindering the growth of surrounding roots [60,61].

Allelochemical interactions at the root level are commonly measured as a change in root biomass. However, a meta-analysis of 61 studies to assess the effect of allelochemicals on root morphology found that the allelochemicals significantly inhibited root length overall but had little effect on root biomass, root surface area, and root volume [62]. Furthermore, allelopathic wheat challenged with 60 other plant species was used to document the combinations of root morphology, revealing root length as a more responsive trait than root biomass in plant–plant allelopathic interactions [63]. Importantly, allelopathic plants release allelochemicals to alter the placement patterns of neighboring roots. Neighboring plants adjust root placement to avoid these allelochemicals. Allelopathic species generally proliferated roots toward the neighboring target species, while most target species showed root avoidance [60,61,63]. If neighboring species do not alter root placement, their roots should be inhibited by root-secreted allelochemicals from allelopathic species.

Allelopathic species may interfere with cooccurring plants through the induction and release of allelochemicals, resulting in neighbor suppression. However, allelopathic plants usually coexist with a broad spectrum of conspecific and inter-specific neighbors. The growth of some species is inhibited significantly by allelochemicals but that of other species is not. This issue is partly due to the allelochemicals released from the allelopathic species having relative activity on the specificity of neighbors in a dose-dependent manner, or to some cooccurring species being able to resist the allelochemicals, leading to allelochemical adaptation. Therefore, allelopathic effects may result from multiple factors rather than the allelochemicals only.

2.3. Allelopathy for Sustainable Agriculture and Forestry

Allelopathy and allelochemicals have important implications and consequences for plant coexistence and community assembly in natural ecosystems. Particularly intriguing are their potential importance and application for sustainable agriculture and forestry. Accordingly, allelopathy has received increasing attention in cropping systems [19,26,57,60,64].

Much effort has focused on allelopathic interference between crop plants and weeds, particularly for deciphering field-based evidence for crop allelopathy in weed regulation. Several practices, including the use of allelopathic plant mulches [65], incorporation into cultural management options [66], natural herbicides [67], and varietal improvement [68], have been used to improve weed management in cropping systems.

Many weeds use an allelopathic mechanism to affect crop production. A heavy infestation of allelopathic weeds in cultivated fields can result in economically destructive effects on the growth and yield of crop plants. For example, the worst weeds, *Ageratum conyzoides* and *Ambrosia trifida*, can infest and colonize quickly into cultivated fields by producing and releasing allelochemicals, resulting in adverse effects on the growth and yield of crop plants [46,69,70]. Attempts to control them by herbicides have met with limited success. However, these allelopathic weeds and their allelochemicals have been incorporated into ecological pest management or employed for biorational pesticides [67,71–73], providing an alternative weed control strategy. Allelopathic weeds have threatened crop production. Fortunately, a few crop varieties can synthesize and release their own 'herbicides', i.e., allelochemicals against weeds in fields, resulting in a natural interference of crop plants with weeds [66]. Allelopathic crop varieties hold promise for sustainable weed regulation. However, their agronomic characteristics and grain yield do not meet the commercial standards of the crop industry. Using traditional and biotechnological approaches, selected elite allelopathic crop genotypes have enabled breeding efforts to improve weed suppression traits in commercial cultivars. The successful breeding of commercially acceptable allelopathic rice cultivars with high yield and weed suppression has been incorporated into present rice production systems to minimize the amount of herbicide used [68].

Allelopathy also offers implications and applications in forestry. Forest regeneration is a critical ecological process that sustains reproduction through the establishment of saplings. However, regeneration failure usually occurs in natural and managed tree systems, particularly in monoculture tree plantations [74–76]. Autotoxicity caused by intra-specific allelopathy may be responsible for this problem. Long-term exposure to tree-derived allelochemicals from litter and root exudates may create a barrier effect on the understory-regenerated saplings, resulting in forest regeneration failure. Apart from regeneration failure, autotoxicity also is a major reason for the productivity decline of tree plantations. A tree plantation is an artificial forest where fast-growing tree species are established in a monoculture manner. However, successive rotations of plantations usually cause a replanting problem or soil disease, resulting in a decline in productivity and the loss of biodiversity in plantations [77,78].

Eucalyptus is one of the most widely planted forestry genera with the replanting problem. Allelochemicals from *Eucalyptus* litter and roots penetrate into the soil, limiting the regeneration of understory plants [79]. Moreover, intra-specific allelopathy is more crucial than resource competition in the replanting problem of *Eucalyptus* plantations [80]. *Juglans* is another widely planted forestry genus. Several species, such as *Juglans nigra* (Black walnut) and *Juglans mandshurica* (*Manchurian walnut*), are commercially planted worldwide. However, the establishment and productivity decline of replanted walnut plantations remain a significant problem. Allelopathy of black walnut through the allelochemical juglone has been observed and investigated for thousands of years [28]. Similarly, autotoxicity is a major reason for the productivity decline of monospecific *Manchurian walnut* plantations [81]. Chinese fir (*Cunninghamia lanceolata*) has been planted for the large-scale production of wood in China. However, regeneration failure and productivity decline have remained critical problems in monospecific plantations. The major driver of the problems is tree-derived allelochemicals and their autotoxicity [75,76].

To mitigate and overcome autotoxicity in tree plantations, one promising option involves mixed-species plantations instead of monocultures. The introduction of certain tree species can alleviate the autotoxicity and soil deterioration caused by allelochemicals in plantations [59,81,82]. Mixed-species plantations of *Eucalyptus* with an introduced N-fixing species *Albizia lebbeck* increase productivity and maintain soil fertility [82]. The establishment and productivity of *Manchurian walnut* can be improved in the presence of *Larix gmelini* [81]. A broadleaf species *Michelia macclurei* enhances the growth and regeneration of Chinese fir [59]. One of the explanations for the improvement is that allelochemicals released from an autotoxic tree species may be decreased in the presence of a "good" neighbor like *L. gmelini* and *M. macclurei*. Furthermore, good neighbors not only induce a microbial shift to accelerate the decomposition rates of allelochemicals but also balance the negative effects of root competition [59,83]. These successful mixed-species plantations, based on empirically established traditional practices, reveal many potential advantages and the ecological importance of allelopathy. Understanding allelopathy will be a key step in using ecological approaches to design species mixtures in tree plantations.

3. Allelobiosis and Plant Identity Recognition

3.1. Allelobiosis and Plant–Plant Signaling Interactions

Plant-to-plant signaling is a key mediator of interactions among plant species. The interpretation of plant-to-plant signaling can help plants adapt to the local environment. In particular, plants can perceive and respond to chemical cues emitted from their neighbors, altering survival and performance and impacting plant coexistence and community assembly [7,9,84]. Plant-emitted chemical cues mediate plant sensing and communication [85], acting as alarm signals to warn neighboring plants of an imminent herbivorous or pathogen attack [86], or serving as an inter- and intra-plant signal for the detection of neighbors, even kin and non-kin individuals within a species [23,87]. Much of the research into chemical-mediated plant–plant signaling interactions has dealt with volatiles induced by herbivory in initiating defensive responses from a tritrophic perspective [88–90]. Therefore, allelobiosis was first introduced by Ninkovic as a volatile signal interpretation and its effects were studied on the plant and subsequent trophic levels. The term "allelobiosis" was given three key aspects of the effect: (1) chemical communication between undamaged plants, (2) benefits for the recipient plants, and (3) the recipient plant's response influencing other trophic levels of organisms [13].

Allelobiosis originated from volatile-mediated aboveground signaling interactions that have been investigated in greater detail [7,89,90]. However, a great deal of recent attention has been paid to root exudate-mediated belowground signaling interactions among plant species [16,84,91,92]. Accordingly, allelobiosis reasonably extends to plant–plant belowground signaling interactions mediated by root-secreted chemical signals. Allelobiosis refers to the communication among plants through phytochemicals, carrying valuable information that transmits signals between adjacent plants, thereby influencing plant growth, defense, and reproduction. However, the use of the term allelobiosis is relatively limited because allelobiosis shares conceptual similarities with plant chemical communication and signal transduction. Plant chemical communication sometimes includes allelochemical interactions. A few allelochemicals, such as benzoxazinoids and momilactones, may also serve both roles of toxic inhibitors and nontoxic signals [93,94]. Whereas allelobiosis places a stronger emphasis on the role of non-toxic signaling chemicals with an informative value for co-occurring plants, highlighting a specific contribution to chemically mediated plant–plant interactions. Accordingly, allelobiosis in the context of plant–plant interactions is orchestrated by signaling chemicals, operating in both aboveground and belowground components, influencing intra-specific and inter-specific dynamics within plant populations.

Plant–plant signaling interactions involve both physical and chemical signals, including light reflection, nutrient availability, and plant-released secondary metabolites, forming a complex and intricate plant social network. These signals trigger a series of plant response strategies, such as shade avoidance, root foraging, and chemical defense [74,84,89,95]. The perception of light signals plays a crucial role in various aspects of plant growth and development, competition, and adaptation to the environment [96–98]. Plants perceive light signals through phytochromes, photoreceptors, and UVR8, and shaded plants primarily perceive light in the green and far-red spectra [99–101]. This alteration influences the synthesis and transport of auxins, triggering shade responses in plants, such as stem elongation, adjustment of leaf angle, and modulation of flowering time [100,101], and consequently, the outcomes of aboveground interactions among plants.

Plants are surrounded by a diverse array of chemical signals in their growth environment (Table 1). Plants need to discern and capture valuable information from these signals, eliciting specific physiological and biochemical responses. Ethylene was the first gaseous signal within the plants, changing the immediate environment of the producer, its neighbors, and attackers [102]. Subsequently, methyl jasmonate [103,104], methyl salicylate [105], and other volatile organic compounds (VOCs) [106–113] have been identified as chemical signals for inter- and intra-plant communication, mediated through the air. Methyl benzoate, serving as an aboveground chemical signal in plants, is frequently released into the air environment, participating in the information transmission and interaction between plants [114]. Isoprene, β-caryophyllene, trans-β-ocimene, and several terpenes [7,109] also serve as air-borne chemical signals, inducing neighboring plants to develop resistance against pests. Plant-emitted indole induces the production of defensive metabolites in neighboring plants, contributing to information transmission among plants [115]. Such signaling perception-dependent defenses enable plants to more effectively cope with external environmental challenges and enhance their survival.

Table 1. Plant-released Aboveground and Belowground Signaling Chemicals.

	Signaling Chemicals	Function/Response	References
Air-borne signals	Ethylene	Alarm and induce defense Participate in plant neighbor detection Induce shade avoidance response Alarm for roots to avoid compacted soils	[102] [116] [117] [118]
	Methyl jasmonate	Alarm and induce defense Attract natural enemies of the herbivores	[103] [104]
	(*Z*)-3-Hexenyl acetate	Induce glycosylation reaction	[106]
	(*E*)-2-Hexenal	Induce defense responses	[107]
	(*Z*)-3-Hexenol	Induce defense responses	[108]
	β-caryophyllene	Induce defense responses	[109]
	Myrcene	Induce the accumulation of MeJA	[110]
	Ocimene	Induce the accumulation of MeJA	[110]
	4,8-Dimethylnona-1,3,7-triene	Induce defense responses	[111]
	(*E*)-Nerolidol	Induce defense responses	[112]
	4,8,12-Trimethyltrideca-1,3,7,11-tetraene	Induce defense responses	[113]
	Methyl salicylate	Induce defense responses	[105]
	Indole	Induce defense responses	[115]
	Methyl benzoate	Trigger phytotoxic effects	[114]
Soil-borne signals	Jasmonic acid	Induce allelopathy Participate in plant neighbor detection	[119] [23]
	Salicylic acid	Participate in plant neighbor detection and allelopathy	[16]
	Allantoin	Participate in kin recognition in rice Activate stress response	[87] [120]
	(−)-Loliolide	Participate in plant neighbor detection Induce defense responses Mediate plant kin recognition Modulate flowering time Modify the rhizosphere microbial community	[23] [121] [122] [123] [124]
Root–microbe signals	Strigolactones	The symbiosis between roots and arbuscular mycorrhizal fungi Induce parasitic weeds to search for potential hosts	[125] [126]
	Luteolin Quercetin-3-*O*-rhamnoside	Attract rhizobia; induce the early formation of root nodules	[127]

Compared to air-borne chemical signals, soil-borne chemical signals (Table 1) have faced challenges and limitations due to the complexity, opacity, and difficulties in directly observing and collecting underground secretions imposed by the soil environment [84,92]. This hinders our understanding of the critical component for comprehensive plant–plant interactions and plant survival strategies. In recent years, there has been increasing in-depth research on the identification and functionality of signaling chemicals in the root exudates and soil environment [23,128–130]. Root-secreted jasmonic acid and salicylic acid from barnyardgrass can induce the production of rice allelochemicals [119]. Salicylic acid also stimulates glucosinolate in the root exudation of rapeseed [131]. Root-secreted (−)-loliolide, the most ubiquitous lactone, can induce the production of allelochemicals in rice and wheat [23,130]. This ubiquitous (−)-loliolide can act as a general soil-borne signaling chemical common to plants [132,133]. Co-occurring plants utilize root-secreted (−)-loliolide to convey information on neighbor detection and identity recognition. In addition, (−)-loliolide may act as a phytochemical cue to explain the fitness cost of herbicide-resistant weeds by regulating vitality and fecundity, yielding critical insights into plant–plant signaling interactions [134].

Root-secreted chemical signals involve the common mycelial networks (CMNs) of arbuscular mycorrhizal fungi (AMF) and plant roots (Table 1). CMNs can function as conduits, transferring signals between plants. Between-plant signaling mediated by CMNs has a profound impact on plant–plant interactions [135]. Strigolactones, serving as root signals, play a role in recruiting AMF to colonize plant roots [136]. However, these signals can also be intercepted by parasitic plants, facilitating successful parasitism [125,126,135]. Quercetin-3-*O*-rhamnoside attracts rhizobia, inducing the early formation of root nodules [127]. The signaling mechanisms between plants and rhizospheric microorganisms contribute to plant adaptation in complex soil environments, and the promotion of synergistic growth among plants.

3.2. Plant Neighbor Detection and Identity Recognition

Due to the sessile growth characteristics, a plant cannot select its living environment and neighbors. Therefore, plant–plant interactions are inherently local in nature and mostly occur among immediate neighbors. From a traditional perspective, plants only passively adapt to the presence of their neighbors. Recent studies have shown that plants also engage in active detection and recognition of neighboring individuals [23,84,137]. Unlike the passive adjustments that plants make to adapt to their environment, the detection of neighbors by plants represents a proactive behavior, actively engaging in communication and interaction with surrounding plants to achieve more efficient resource utilization and survival strategies [8,84]. In particular, plants actively respond to their neighbors by changing their defensive strategy based on the identity of neighbors [44,63]. Such plant neighbor detection and identity recognition are driven by chemical signals [23,84,92,138].

Chemically, plants can modify their environment by releasing secondary metabolites such as root exudates (liquid) or gaseous VOCs [84,139]. Therefore, plants may detect neighbors through plant volatiles as air-borne signals or root exudates as soil-borne signals. These chemical signals drive aboveground and belowground plant–plant signaling interactions (Figure 1). However, most studies on plant neighbor detection and identity recognition have focused on air-borne signals (Table 1). The studies on soil-borne signals are extremely difficult because the transduction of belowground chemical signals requires root–soil interactions. Nevertheless, neighbor detection and identity recognition do occur at the root level in plants [91]. Many studies have shown that allelopathic plants may detect and recognize their neighbors through the root exudates and respond by increasing the allelochemicals [23,51,57,140,141]. However, most studies did not clarify which individual constituents in the root exudates were responsible for the neighbor detection and allelochemical response.

To unambiguously demonstrate plant neighbor detection and allelochemical response by root-secreted chemical signals, three open questions must be addressed: (1) Is plant neighbor detection and allelochemical response a general phenomenon or is it restricted to the specific identity of neighbors? (2) How do researchers demonstrate that the chemical-mediated responses happen belowground rather than in aboveground signaling interactions? (3) Which signaling chemicals are responsible for neighbor detection and allelochemical response in allelopathic plant–plant interactions? It appears from an allelopathy-based wheat–neighbor system that plant neighbor detection and allelochemical response are a general phenomenon. When wheat was paired with 100 other plant species, wheat was able to detect the presence of neighbors in a density-dependent response to neighboring plants regardless of species. Further, in a segregation-based approach with and without root contact, mycorrhizal fungi and soil clearly distinguished aboveground and belowground signaling interactions in neighbor detection and allelochemical DIMBOA response, and the study showed that root exudates are sufficient to induce the response (Figure 1). Importantly, root-secreted (−)-loliolide and jasmonic acid are responsible for the plant neighbor detection and allelochemical response [23]. In particular, root-secreted (−)-loliolide as a general soil-borne signal is common to competitors, and belowground exposure to (−)-loliolide provides an effective signal-mediating response for plants against competitors, and ultimately increases plant survival and fitness.

Although there are likely to be additional signaling mechanisms that could allow species-specific responses, plants are able to commonly detect and respond to their competitors. In cropping systems, the signaling chemicals enable crucial staple food crop plants to detect their competing weeds and respond accordingly to ensure their survival [23,64,130,138]. Plant competitor detection and allelochemical elicitation play important roles in regulating crop–weed inter-specific interactions. Once a cultivated field is infested with weeds, allelopathic crop plants may detect and then enhance the release of allelochemicals to inhibit the weeds, reducing reliance on traditional herbicides for effective weed management.

Plants have evolved chemical defense mechanisms that produce and release specialized metabolites against interacting competitors. These defense mechanisms start from the recognition of intricate signaling networks while plant defensive metabolites reduce the performance of competing plants to strongly affect plant survival and fitness [19,134,142]. Although aboveground plant–plant signaling interactions and chemical defenses have substantial impacts on plant survival and fitness, relatively less is known about the signaling-mediated belowground interactions. Signals regulating underground plant–plant interactions are whispers in the dark. A key but still poorly understood mechanism for plant neighbor detection and identity recognition is through the perception and recognition of signals derived from root exudates [92].

3.3. *Relatedness-Mediated Neighbor Recognition*

Plant neighbor detection and identity recognition occur among one or more species and within a species, even individuals at a family level. Accordingly, the concept of identity recognition includes species recognition, self/non-self-recognition, kin recognition, and cultivar recognition. Species recognition, that plants can distinguish between members of the same species and those of different species, is described as above in Section 3.2. The self/non-self-recognition arises from the neighbor detection generated by the plant itself. i.e., a plant may also specifically respond to the presence of a competing neighbor by competing or not competing itself [50,143]. Therefore, self/non-self-recognition is identity recognition that allows plants to avoid within-plant competition for resources. The most of research on self/non-self-recognition has particularly focused on the root systems where apparent competition or non-competition between plants can be found [143,144]. For example, the clone plant *Buchloe dactyloides* responds to both self and non-self root systems. When exposed to its own root system, the roots become smaller and shorter. However, coexistence with a non-self root system significantly enhances root growth [144]. Similar root responses to self and non-self phenomena have also been observed in *Ambrosia dumosa* [145], *Glycine max* [146], *Phaseolus vulgaris* var. Kenya [147], *Pisum sativum* [148], and *Trifolium repens* [149].

Despite a variety of observations that have been obtained for self/non-recognition via root systems in a relatively short burst of activity, these studies have been criticized for their experimental design because these experiments simply show root growth differently from one larger twin plant. In particular, there is no convincing mechanism that could account for self/non-self-recognition [50]. Due to the flaw in experimental design and the problem of lacking a convincing mechanism, work on self/non-self-recognition has no longer been conducted in the last decade. In contrast, identity recognition has been focused on relatedness-mediated neighbor recognition.

Relatedness-mediated neighbor interactions involve kin recognition, kin discrimination, and kin selection. Kin recognition is the ability to detect the difference in level of relatedness between oneself and another individual, while kin discrimination is a relatedness-dependent cooperative behavior. Plants can distinguish their conspecific neighbors based on their genetic relatedness (i.e., kin recognition) and adjust their 'behavior' accordingly (i.e., kin discrimination) [150]. Such kin recognition and kin discrimination have been observed in individuals [151], populations, biotypes [122,152], and cultivar levels [87]. Kin recognition and kin discrimination allow behaviors toward kin groups. Cooperative behavior can be promoted through kin selection. Kin selection theory states that an altruistic trait can be favored by natural selection when the altruist and the recipient are related [153]. Importantly, kin recognition is associated with kin discrimination, but kin recognition and associated kin discrimination are not necessary for kin selection to occur [6].

Kin recognition is identity recognition at the level of siblings vs. non-siblings from the same population and species. Kin recognition in plants was proposed in 1980 [154] and first found in an annual plant, *Cakile edentula*, in 2007. The annual plant greatly expanded its root systems in the presence of unrelated plants but reduced root growth and competition with relatives, allowing greater shared access to resources [151]. Subsequent studies have shown that many plants are able to differentiate relatives from unrelated plants from the same population. Such kin recognition occurs in various types of plant species, ranging either from gymnosperms to angiosperms, or from wild species to crop plants [87,122,152,155,156]. The ability to detect relatedness would allow plants to discriminate closely from distantly related neighbors and optimize response strategies for the composition of their local neighborhood [6,150,157,158]. Plant kin recognition, without compromising individual fitness, enhances the survival chances of population offspring, playing a positive role in plant population reproduction and community assembly.

Cooperation arising from kin recognition makes for behaviors toward relatives that promote the increase of related offspring. Reducing the energy devoted to competitive organs allows for greater allocation to reproduction in kin groups [159–161]. Kin recognition and kin discrimination in crop plants can result in less intra-specific competition, maximizing stand performance and increasing yield [87,162]. For example, rice cultivars with the ability for kin recognition are capable of distinguishing between closely and distantly related cultivars and respond to them by altering biomass allocation, increasing grain yield in closely related cultivar mixtures [87]. Therefore, kin recognition and kin discrimination in crop plants may be a potential mechanism responsible for performance and yield in cultivar mixtures [5,87,151]. Different from naturally occurring species where kin represents plants sharing the same mother as either full or half siblings, artificial selection generates crop cultivars that are genetically and morphologically uniform. In the case of highly self-pollinating cultivars, the coefficient of relatedness is very close to 1, i.e., $r \approx 1$ [6,156,163]. Therefore, kinship in self-pollinated crop plants is functionally represented by closely related cultivars rather than siblings within a natural species, and kin recognition in crop plants occurs at the cultivar level [87,142].

Kin recognition and kin discrimination involve signaling interactions among intra-specific neighbors [150,161,164]. Most evidence suggests that kin recognition is mediated by a chemical mechanism and requires active secretion by roots, as recognition can take place under root segregation or before physical contact occurs [87,164]. A few studies have shown that root exudates can invoke kin and non-kin responses. However, it is not clear which root-secreted signaling chemicals are responsible for these responses [87,128,165]. A specific root-secreted allantoin has a role in kin recognition in rice lines [87]. However, the nitrogen-rich allantoin may not be the signal of relatedness but rather the effect of the underlying signal [84]. Root exudates contain signaling chemicals, such as (–)-loliolide, a common signal for neighbor detection [23,132,133]. This plant–plant signaling is reminiscent of the role of (–)-loliolide in relatedness-mediated neighbor identity recognition and discrimination. (–)-Loliolide may function as a biochemical mechanism to discriminate closely related cultivars from distantly related cultivars (that have therefore not been published).

Ecological and evolutionary approaches to improving crop mixtures have been commonly used throughout the world [2,166]. The benefits of mixed cropping systems have been attributed to niche complementarity, facilitation, and within-species diversification in inter-specific interactions [166–168] but have rarely been invoked in relatedness-mediated intra-specific interactions. Kin recognition and kin discrimination among crop plants are predicted to increase yield by reducing intra-specific competition and shifting resource allocation to reproduction [87,150,169]. Preferentially reducing competitive effects on relatives may improve our understanding of the processes at play in kin recognition to develop cultivar mixtures that increase grain yields in the limited area suitable for agriculture.

4. Chemically Mediated Root–Soil Interactions and Plant–Soil Feedback

4.1. Root–Root Interactions and Root Placement Patterns

Roots function primarily in the acquisition of nutrients and water and interacting with soil. Therefore, root systems have recognizable developmental plans that are primarily driven by resource availability. However, root–root interactions are much more sophisticated than previously thought as regards foraging for resources only. Plant roots detect and respond to complex physical and chemical cues from the soil environment [170,171]. The ability of roots to avoid obstacles has been observed for over a century. The phenomenon and its mechanisms may be explained by chemically mediated root detection and action. The roots of *Pisum sativum* can detect and avoid growth towards obstacles. Importantly, *Pisum sativum* root-secreted allelochemicals to inhibit their own roots, resulting in root navigation by self-inhibition [172]. Such obstacle avoidance by self-inhibition could optimize root systems, increasing plant performance by limiting resource allocation. Soil compaction limits the ability of roots to penetrate harder soils, reducing root growth. A recent study has shown that plant roots sense soil compaction through restricted ethylene diffusion. In a compacted soil environment, ethylene accumulated in root tissues acts as a warning signal for roots to restrict root growth towards compacted soils [118]. Importantly, roots can chemically identify and respond to roots of conspecific and inter-specific neighbors and even self from non-self roots and kin from non-kin roots within a family [40,87,143,144,151], resulting in a series of root–root interactions that determine belowground ecological processes to a large extent. Any change in root–root interactions is likely to alter the soil resource availability, and subsequently, the plant fitness.

In the interactions among roots of *Ambrosia dumosa* and *Larrea tridentata* in the Mojave Desert community, Larrea roots released allelochemicals into the soil to inhibit either Larrea or Ambrosia roots in their vicinity. However, Ambrosia roots inhibited conspecific contacted roots only, indicating their different root–root interactions are mediated by allelopathy rather than simple competition for a limiting resource. [145]. Chemically mediated root–root interactions have been extensively observed in crop–weed interactions and mixed-species plantations. The paddy weeds detect and respond to the presence of rice by altering root growth and distribution in a cultivar-dependent manner. In the presence of common rice cultivars, the root growth and distribution of paddy weeds are intrusive, unresponsive, and avoidant towards rice roots. However, the roots of paddy weeds avoid the roots of allelopathic rice cultivars [60]. In the *M. macclurei* and Chinese fir mixed-species plantation, Chinese fir root systems develop differently in the presence of their own roots than *M. macclurei* roots. In particular, Chinese fir root growth increased towards neighboring *M.*

macclurei roots. This response was attenuated by activated carbon, indicating that there were chemically mediated root–root interactions in the mixed-species plantation [59].

Roots may discriminate the roots of self/non-self and respond by either stimulating the growth of self-roots or inhibiting the intrusion of non-self-roots. Self/non-self-root–root recognition is carried out with electrical or turgor-associated signals [91,173], such as the electrical signal-mediated root communication between *A. dumosa* and *L. tridentat* [157], and root self-recognition by differential internal oscillatory signals in *Pisum sativum* and *Buchloe dactyloides* [143,144]. Nonetheless, self- and non-self-root–root interactions are likely to be mediated by root-secreted chemical signals alone or in combination with physical signaling.

Roots can detect and discern roots from close relatives and unrelated plants from the same species or family through the root exudates [165]. Rice cultivars with the ability for kin recognition can detect the presence of closely and distantly related cultivars, and respond to them by adjusting their root placements and biomass. A root-secreted allantoin may be responsible for the relatedness-mediated root–root interactions [87]. Furthermore, the intra-specific kin recognition contributes to inter-specific allelopathy in allelopathic rice interference with paddy weeds. Closely related cultivar mixtures not only reduce the production of rice allelochemicals by lowering the defense cost, but also are capable of preferential root placement towards weeds [142].

In response to neighbor roots and root–root interactions, plants actively alter root placement patterns to buffer root competition and provide potential advantages for their own growth and fitness. Generally, root–root interactions represent intrusive (approaching, overproliferation), avoidant (repelling, underproliferation), or unresponsive patterns [60,174]. However, root placement responses are reciprocal with neighboring roots. Therefore, pairwise plant–plant interactions can potentially generate nine combined patterns. A recent study has experimentally documented the occurrence of these nine root placement patterns, including intrusive-intrusive, intrusive-avoidance, intrusive-unresponsive, avoidance-intrusive, avoidance-avoidance, avoidance-unresponsive, unresponsive-intrusive, unresponsive-avoidance, and unresponsive-unresponsive patterns, in pairwise allelopathic wheat and 60 other plant species systems [63]. Importantly, the neighbor-modulated root placement patterns are mediated by both allelochemicals and signaling chemicals. Signaling (–)-loliolide from neighbor species triggered wheat allelochemicals to inhibit root growth with degrading starch grains and inducing an uneven distribution of auxin in neighbor species roots, altering root placement patterns [63].

Neighbor-modulated root–root interactions and root placement patterns driven by root-secreted allelochemicals and signaling chemicals are important mediators of belowground interactions, impacting plant fitness, reproduction, and survival [84,128,173]. Root placement patterns alter the spatial and temporal distribution of root systems in the soil, whereas the spatial and temporal distribution of root systems affect the dormant microbes of over 95% of the microbial biomass in soil. These dormant soil microbes can be activated by exudates from arriving roots. In turn, activated microbes may induce root exudation [175]. Accordingly, root placement patterns may determine the rhizosphere microbiome and soil microbial community assembly, ultimately altering root–soil interactions and belowground ecological processes.

4.2. Soil Mobility and Microbial Interactions

Plant–plant chemical interactions are mediated by allelochemicals and signaling chemicals, either volatiles or root exudates from plants into the environment, mostly into the soil. The action of allelochemicals and signaling chemicals in soil requires their presence in the vicinity of the target plants. Thus, allelochemicals and signaling chemicals endure under soil processes, such as retention, transport, and transformation, once released [176]. Of these processes, the mobility of allelochemicals and signaling chemicals, as well as their bio-transformed by soil microbes, are a determinant for belowground plant–plant interactions. The mobility and microbial degradation affect the final destination and concentration of the allelochemicals and signaling chemicals in the soil surrounding neighbors, and subsequently, their activities.

The mobility of 10 typical allelochemicals, including ferulic, p-hydroxymandelic, phydroxybenzoic and vanillic acids, vanillin, coumarin, daidzein, 1α-angeloyloxycarotol, DIMBOA, and m-tyrosine, was evaluated by using a soil thin-layer chromatography (soil TLC) combined with a bioassay approach [177]. There were significant differences in mobility factor (*Rf*) among these allelochemicals. Four phenolic acids (ferulic, p-hydroxymandelic, phydroxybenzoic, and vanillic acids) had very poor mobility (*Rf* < 0.1). In contrast to phenolic acids, phenolic aldehyde and lactone (vanillin and coumarin) showed excellent mobility (*Rf* > 0.5) in soil. Whereas daidzein (flavonoids), 1α-angeloyloxycarotol (terpenoids), DIMBOA, and m-tyrosine (non-protein amino acid) showed good mobility and their *Rf* values range from 0.24 to 0.32. Accordingly, the mobility factor is dependent on the structural specificity of allelochemicals, as polar allelochemicals could not easily be moved from the allelopathic plants to target species in the soil environment. However, binary mixtures of these allelochemicals led to an increase in mobility factors for selected combinations [177]. Similarly, the mobility of the four signaling chemicals, jasmonic acid (*Rf*, 0.21), salicylic acid (*Rf*, 0.07), (−)-loliolide (*Rf*, 0.68), and luteolin (*Rf*, 0.32), in soil was evaluated by using a soil TLC. Among them, the root-secreted signaling chemical (−)-loliolide had a higher soil mobility and could easily be moved from the rhizosphere to bulk soil [23]. Therefore, (−)-loliolide may be a soil-borne signaling chemical in plant–plant signaling interactions.

Plants release allelochemicals and signaling chemicals at rates of significance to interact with soil microbes. Allelochemicals and signaling chemicals are able to exert an effect on soil microbes. In turn, soil microbes consume and decompose allelochemicals and signaling chemicals. Allelopathic rice seedlings released 5,4′-dihydroxy-3′,5′-dimethoxy-7-*O*-β-glucopyranosylflavone into soil, where the glucopyranosylflavone rapidly hydrolyzed glucose by soil bacteria into allelochemical 5,7,4′-trihydroxy-3′,5′-dimethoxyflavone [178]. The allelochemical not only inhibited the root growth of paddy weeds but also reduced the culturable microbial population and total microbes in the soil. Therefore, allelopathic rice can modify the soil microbial community through root-secreted allelochemicals and their microbe-transformed production [179]. Similar results occur in wheat and barley allelopathy. Allelopathic wheat root-secreted DIMBOA and then microbe-transformed into 6-methoxy-benzoxazolin-2-one (MBOA) as allelochemicals against pests. There was a positive relationship between wheat DIMBOA and MBOA levels and soil microbial population. DIMBOA associated with yielding MBOA increased soil fungi, affecting the soil microbial community structure [140]. Barley produces the allelopathic alkaloids gramine and hordenine, whose biosynthesis and accumulation are preferentially located in roots [180,181]. The allelopathic alkaloids impact the community structure of field soil bacteria. Gramine caused the proliferation of many potentially beneficial strains. Furthermore, gramine biosynthesis was accompanied with the association of distinct bacterial communities in the soil [181].

Soil disease caused by a continuous monoculture of crop plants and woody trees was involved in the build-up of soil pathogens. Allelochemicals may contribute to the shift in soil microbial communities in continuously cultivated fields and plantations. The levels of allelochemicals in the rhizosphere correlate with the presence of various microbes in soil [140,182,183]. A case study indicated rhizosphere isoflavone (daidzein and genistein) levels and their relation to soil microbial community structure in a 13-year experiment of continuous soybean monocultures. As a result, soil microbial community structures were induced by both daidzein and genistein, and thereby daidzein and genistein in the soybean rhizosphere may act as allelochemicals in the interactions between the root and soil microbial community in a long-term mono-cropped soybean field [183]. Similar results have been found in 25-year-old monospecific Chinese fir plantations. There were negative relationships among allelochemical cyclic dipeptide concentration, microbial community composition, and Chinese fir root biomass in plantations. In particular, allelochemical cyclic dipeptide accumulated potentially pathogenic fungi, altering the soil microbial compositions and community structures. Allelochemical-mediated soil microbial community with a negative feedback result in a productivity decline of continuous Chinese fir monocultures [76,77]. Further clarification of the root–microbe interactions mediated by allelochemicals may contribute to solving the problem of continuous monocultures in cropping systems and tree plantations.

Extensive research has shown that soil microorganisms are an important determinant of allelopathic activity [182,184,185]. However, determining how allelochemicals affect the soil microbiome and how soil microbial changes mediate plant performance is challenging. Recent microbiome technology development makes an effort directed toward understanding the impact on soil microbial composition and the community structure of allelochemicals in the soil environment. Several studies have shown allelochemical-selected microbiomes and their effects on plant performance [186,187]. Allelochemicals affect soil microbiome structure by altering core bacterial and fungal taxa. The allelochemical-selected soil microbiome mediated plant performance across 13 plant species, mitigating allelopathic inhibition of plant productivity [186]. Rhizosphere microbial communities are vital for plant performance. A recent study indicates metagenomic insights into the responses of rhizobacteria and their alleviation role in licorice allelopathy. Allelochemical glycyrrhizin inhibited the growth and development of licorice, resulting in autotoxicity. However, specific rhizobacteria, such as the *Novo-sphingobium* genus, accounted for a relatively high proportion of the enriched taxa and appeared in metagenomic assembly genomes, such as degraded glycyrrhizin, to alleviate the allelopathic autotoxicity [187]. The results provide potential implications for resolving the problem of continuous monocultures using an allelochemical-selected rhizobacterial community.

Despite increasing knowledge of the allelochemicals involved in the soil microbial community, little is known about signaling chemicals in the rhizosphere and interactions with the soil microbial community [188]. The nitrogen-rich allantoin, as a potential root-secreted signaling chemical, significantly stimulated populations of bacteria and actinomycetes but had no effect on fungi in paddy soil [189]. Signaling allantoin increased the ratio of anaerobe:aerobe and bacteria:fungi but decreased the ratio of Gram-negative bacteria:Gram-positive bacteria, and the fungal signature lipid marker 18:2ω6,9c under flooded rice soil. The allantoin-induced shifts in microbial community composition increased microbial diversity as well as the living microbial biomass in rice soil [190].

Several studies have shown the role of root-secreted signaling (−)-loliolide in soil microbial compositions and community structures [116,117,124]. (−)-Loliolide affected the signature lipid biomarkers of soil bacteria, actinobacteria, and fungi, resulting in changes in soil microbial community structures [191]. In penoxsulam-resistant barnyardgrass-mediated root microbial communities, (−)-loliolide and jasmonic acid in the root exudates were correlated with the core microbes in the rhizosphere soil [192]. Similar correlations between (−)-loliolide and rhizosphere bacteria occurred in Arabidopsis, indicating that Arabidopsis may regulate the rhizosphere bacteria through root signaling (−)-loliolide [124]. In addition, ethylene is a typical air-borne chemical signal to participate in plant neighbor detection [116] and to induce a shade avoidance response [117]. A recent study has shown that ethylene emitted from peanut roots can reorganize the rhizosphere microbial network, and the reassembled rhizosphere microbial community can provide more available nutrients for peanut roots to promote growth and reproduction [193].

4.3. Belowground and Aboveground Interactions

Soil microbes play a crucial role in belowground ecological processes, such as nutrient cycling and antibiotic resistance [193–196]. Importantly, soil microbial communities affect aboveground performance, either through inter-specific or conspecific plants, or even kin and non-kin individuals [116,142,197]. Plants can shape soil microbiota through root-secreted allelochemicals and signaling chemicals [140,175,179,184,198]. In turn, chemically mediated soil microbial communities drive plant–soil feedback, affecting flowering and reproduction [22,199,200], and thus altering plant fitness (Figure 2).

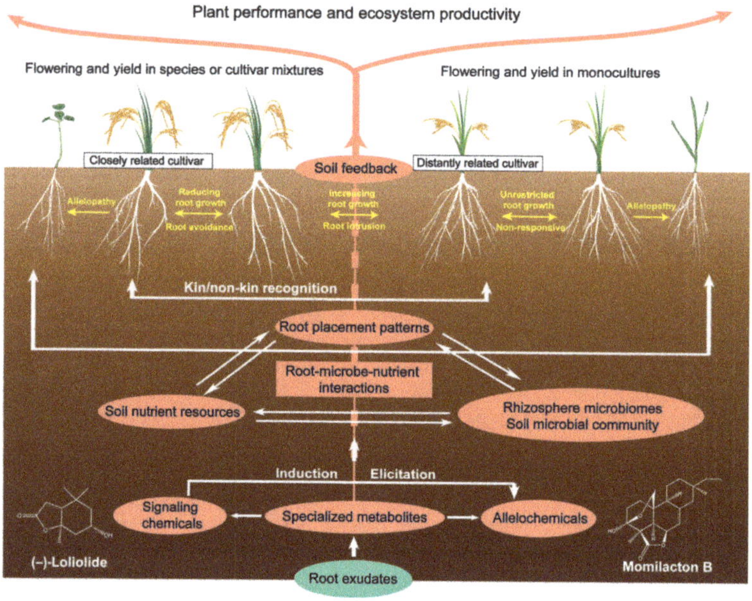

Figure 2. Chemically mediated root–soil interactions and plant–soil feedback.

Flowering is a key means for plant species to obtain maximum reproductive capacity. The flowering of plants is mostly aboveground leaves sensing environmental factors. Recent studies have shown that rhizosphere and soil microbial communities can alter the selection intensity of plant flowering phenology and flowering time [199,201–203]. In particular, root-secreted signaling chemicals can recruit specific microbes to alter the rhizosphere and soil microbial communities, impacting the flowering of plants [124,201]. A novel metabolic network in which soil microbiota influenced plant flowering time has been found in Arabidopsis, where the root exudates affected plant rhizosphere microbiota, modulating flowering time by IAA production [201]. Furthermore, the flowering time of Arabidopsis shifted with root signaling (−)-loliolide and rhizosphere bacteria. There were differential rhizosphere and soil bacterial communities between two genotypes with and without root-secreted (−)-loliolide, suggesting that the flowering time of Arabidopsis may be mediated by rhizosphere bacteria with root-secreted chemical signals [124]. Despite microbiome being a major contributor to phenotypic variation, flowering time may be determined by specific rhizobacteria rather than the whole microbiome. A recent study has shown that rhizosphere bacteria (*Cupriavidus metallidurans*, *Bacillus*, *Solibacillus*, and *Planococcus*) affect the flowering time of Arabidopsis [204].

Chemically mediated belowground ecological interactions among roots, microbes, and nutrients can feed back to the plant aboveground, affecting growth, flowering, and reproduction [200–203]. Plants release allelochemicals and signaling chemicals to alter the rhizosphere and soil microorganisms that increase and decrease nutrient bioavailability, resulting in a positive or negative plant–soil feedback (Figure 2). Plant–soil feedback has been recognized as an important mechanism behind intra- and inter-specific interactions, affecting plant performance, species diversity, and community structure [205]. Allelochemicals and signaling chemicals affect plant inputs into the soil subsystem via litter and rhizodeposits. In particular, root exudate metabolites drive plant–soil feedback on growth and defense by shaping the rhizosphere microbiota and modulating microbial succession [200,206–208]. These chemically mediated interactions may cause specific plant–soil feedback where the match between the species identity of living roots plant nutrition is evaluated [200,205].

The intricate chemical communication plays an instrumental role in regulating the balance and function of the plant–microbe–soil system. The communication network profoundly affects the growth, metabolism, and interactions of microorganisms, thereby influencing the structure and function of soil ecosystems. Different types of allelochemicals and signaling chemicals exhibit selective effects on different microorganisms, thereby governing the composition and abundance of microbial communities. The regulation of such interactions is a complex process that involves the integrated effects of multiple factors and signaling pathways. Moreover, the specific mechanisms and effects may vary in different plant species and under different environmental conditions, necessitating further research and exploration.

5. Biosynthesis and Molecular Mechanisms of Allelochemicals and Signaling Chemicals

5.1. Biosynthesis and Mechanisms of Allelochemicals

Plants biosynthesize a wide variety of allelochemicals including phenolics (simple phenolic acids, flavonoids, coumarins, and quinones), terpenoids (monoterpenes, sesquiterpenes, diterpenes, and steroids), nitrogen-containing chemicals (alkaloids, benzoxazinoids, and cyanogenic glycosides), and many other chemical families [18,19]. These allelochemicals are regulated by various genes, including those that encode transcription factors, subsequently determining allelopathic potentials [209].

The biosynthesis of phenolic allelochemicals is related to the shikimic acid and the phenylpropanoid pathways. In the shikimic acid pathway, hosphoenolpyruvate synthetase (PEPs) and erythrose-4-phosphate (E4P) are catalyzed into phenylalanine under the 3-deoxy-D-arabino-heptulosonate-7-phosphate synthase (DAHPS), 3-dehydroquinate synthase (DHQS), and shikimate dehydrogenases (SDH) [210]. The phenylpropanoid metabolism pathway begins with phenylalanine; under the catalysis of phenylalanine ammonia-lyase (PAL) and cinnamic acid 4-hydroxylase (C4H), it transforms into p-coumaric acid. Through the catalysis of enzymes, such as p-coumarate-3-hydroxylase (C3H), caffeate 3-O-methyltransferase (COMT), ferulate-5-hydroxylase (F5H), and 4-coumaroyl-CoA ligase (4CL), p-coumaric acid is further converted into caffeic acid, ferulic acid, sinapic acid, p-coumaroyl-CoA, and chlorogenic acid [210]. Finally, under the catalysis of chalcone synthase (CHS) and chalcone isomerase (CHI), flavanone 3-hydroxylase (F3H), flavanone 3′-hydroxylase (F3′H), flavonol synthase (FLS), and flavone synthase (FNS), p-coumaroyl-CoA is transformed into flavones, flavonols, and other related flavonoids [210,211]. In addition to key enzymes before, TFs such as MYB, bHLH, WD-WDR, NAC, and WRKY can coordinately regulate the expression of enzymatic genes at the transcriptional level in a spatial manner [212].

Phenolic acids are common allelochemicals and their molecular mechanisms have been revealed. In maize, the constitutive expression of ZmMyb-IF35 can promote ferulic acid and chlorogenic acid accumulation [213]. The specific expression of *StAN1* and *StMTF1* in *Solanum lycopersicum* promotes the synthesis of chlorogenic acid and caffeic acid [214]. The constitutive expression of *AtMYB12* in *Arabidopsis thaliana* increases the content of chlorogenic acid [215]. *SmMYB98* positively regulates the synthesis of salvianolic acids in *Salvia miltiorrhiza*, and *SmMYB2* significantly increases the level of salvianolic acids by upregulating the synthesis gene *CYP98A14* [216]. In rice, the transcriptional activator VP64 can upregulate *OsMYB57*, inducing the synthesis of phenolic acids and enhancing rice allelopathy against barnyardgrass [209].

Flavonoids are a crucial component of phenolic allelochemicals. In Arabidopsis, *AtMYB11*, *AtMYB12*, and *AtMYB111* can independently activate genes encoding CHS, CHI, F3H, and FLS, and then determine the content of flavonols, collectively [217]. *AtMYB4* has the ability to inhibit the expression of *AtMYB7*, and silencing *AtMYB4* can increase the biosynthesis of flavonols in *Arabidopsis thaliana* [218]. *MYBP3* and *MYBP4* can enhance the flavonol level by activating related genes in both *Arabidopsis thaliana* and *Gentiana trifloral* [219]. Furthermore, the transcription factor MYB regulates flavonol synthesis in *Vitis vinifera*, *Malus pumila*, *Fragaria ananassa*, *Glycine max*, and *Ginkgo biloba* [214,220–222].

Transient expression of *GmMYBJ3* can activate CHS8 and CHI1A, promoting the synthesis of isoflavones in soybeans [223]. Moreover, in the roots of leguminous plant *Callerya speciosa*, nine CsMYBs TFs are identified to form a regulatory network with bHLH, increasing isoflavone content by activating synthesis genes [224]. Overexpression of *CaMYB39* activates early flavonoid biosynthetic genes (EBG), including *SYNTHASE2*, which can induce the biosynthesis of flavanones, flavonols, isoflavones, and others [225].

Quinone is another important group of phenolic allelochemicals, including benzoquinones, naphthoquinones, anthraquinones, and others. Juglone is a classic naphthoquinone allelochemical derived from several pathways, including the o-succinylbenzoate pathway (OSB), the 4-hydroxybenzoic acid (4HBA)/geranyl diphosphate (GPP) pathway, the acetate-polymalonate pathway, and the homogentisate (HGA)/mevalonic acid (MVA) pathway [226]. Recent studies indicate that specialized 1,4-naphthoquinones may evolve from 1,4-dihydroxynaphthoic acid (DHNA), an intermediate in the biosynthesis of phylloquinone [227]. In particular, the naphthalenoid moiety of juglone in *Juglans nigra* originates from DHNA produced via the phylloquinone pathway [228].

Terpenoids constitute the second-largest group of allelochemicals, characterized by poor water solubility, making it challenging for rainwater to leach them into the soil [229]. The precursors of terpenoids are synthesized through two pathways: the MVA pathway that occurs in the cytoplasm and the methylerythritol phosphate (MEP) pathway that takes place in the plastids. The MVA synthesis pathway involves six enzyme-catalyzed reactions, starting from three molecules of acetyl CoA and being catalyzed by enzymes such as AACT, HMGS, HMG CoA, and others to produce isopentenyl diphosphate (IPP). The MEP synthesis pathway is completed through seven enzyme-catalyzed reactions, beginning with the condensation of D-glyceraldehyde-3-phosphate and pyruvate. Enzymes such as DXS, DXR, CMS, and HDR catalyze the formation of IPP and dimethylallyl pyrophosphate (DMAPP). Finally, under the catalysis of different enzymes, IPP and DMAPP go through various condensation reactions to produce terpenoids [230].

The synthesis genes for various terpenoids have been identified. In tea trees, genes such as *CsTPS*, *CsFPS*, *CsOCS2*, and *CsBOS1* have been identified to regulate the levels of allelochemicals in tea plants, such as (E)-nerolidol, (E)-β-ocimene, geraniol, β-myrcene, D-limonene, and (E)-β-farnesene [231,232]. A-type diterpenoid allelochemical momilactone is catalyzed by class II diterpenoid cyclases and class I diterpenoid synthetases, respectively. In rice, *CYP76M8* is responsible for the generation of momilactone [233]. In addition, TFs involved in the synthesis of plant terpenoids include AP2/ERF, bHLH, and MYB. These TFs can effectively regulate multiple enzyme genes in the terpenoid biosynthetic pathway. Therefore, achieving global regulation of the terpenoid biosynthetic pathway through transcriptional regulatory networks is more effective.

Compared to phenolics and terpenoids, nitrogen-containing allelochemicals are relatively scarce, mainly comprising alkaloids, such as gramine and hordenine [180,181] and non-protein amino acids, such as δ-hydroxynorleucine, L-canavanine, L-DOPA, L-meta-tyrosine, β-(3-isoxazolin-5-one-2-yl)-alanine, and mimosine [18,19]. Solanidine or solanine is a general term for glycosylated derivatives of steroidal alkaloids (SAs) in plants. Among them, α-chaconine and α-solanine are the most abundant, constituting 90% of the total solanidine content [234]. Chemically, solanidine is a glycosylated derivative of cholesterol and its synthesis involves a multi-gene process divided into three stages: (1) acetyl-CoA is converted into terpenoid compounds; (2) synthesis of cycloartenol; and (3) reduction of cycloartenol to cycloartanol and further synthesis into cholesterol, which is then transformed into solanidine. Finally, solanidine is glycosylated into α-solanine and α-chaconine through three different solanidine glycosyltransferases (SGT) [234].

In addition to alkaloids and non-protein amino acids, there are other nitrogen-containing allelochemicals, with the most representative being benzoxazinoids (BXs) produced by wheat and maize. Natural BXs are divided into three groups based on their N-substituents: lactams, methyl derivatives, and hydroxamic acids [235]. The BX synthesis pathway is a branch of the tryptophan biosynthetic, initiating with 1-(2-carboxyphenylamino)-ldeoxyribulose-5-phosphate that catalyzes by indole-3-glycerolphosphate synthase (IGPS) to form indole-3-glycerol phosphate, followed by the action of indole-glycerol phosphate aldolase (Bx1) and four consecutive cytochrome P450 enzymes (Bx2–Bx5) and glucosyltransferases (Bx8/Bx9) to generate DIBOA-Glc (2,4-dihydroxy-1,4-benzoxazin-3-one-glucoside) [56,236]. DIBOA-Glc is further transformed by O-methyltransferases (BX10 to BX12) into 2-hydroxy-4,7-dimethoxy-1,4-benzoxazin-3-one glucoside (HDMBOA-Glc). When DIMBOA-Glc and HDMBOA-Glc stored in vacuoles come in contact with specific glucosidases, they are converted into DIMBOA and MBOA, respectively [237].

Although the genes involved in the biosynthesis of BXs have been extensively investigated, the transcriptional regulation of BX genes is beginning. Some transcription factors have been reported to play a role in maize. Upon biotic stress, *ZmWRKY75*, *ZmMYB61*, *ZmNAC35*, and *ZmGRAS37* are upregulated in maize, leading to the upregulation of *ZmBx1* and *ZmBx13* expression and an increase in BX level [238]. Zhou et al. (2020) identified transcription factor families, such as MYB, NAC, and GRAS, through gene regulatory networks (GRNs) that can regulate BX genes in maize [239]. However, these studies were conducted using maize transcriptomics methods and did not directly test the functions of identified transcription factors. The following studies reported that in wheat, *TaMYB31* can bind to the promoters of Bx1 and Bx4, catalyzing the biosynthesis of BXs in response to biotic and abiotic stress [240].

5.2. Biosynthesis and Mechanisms of Signaling Chemicals

Plants can rely on information mediated by signaling chemicals to recognize neighbors and adjust their survival strategies. Volatile organic compounds (VOCs) emitted from plants are vital signaling chemicals, including green leaf volatiles (GLVs), terpenes, methyl jasmonate (MeJA), methyl salicylate (MeSA), ethylene, and indole. Root exudates also are key components of plant–plant signaling interactions but the root-secreted signaling chemicals have not been well-known. So far, strigolactones (SLs), (−)-loliolide, jasmonic acid (JA), and allantoin have been identified as root-secreted signaling chemicals that are responsible for belowground signaling interactions (Table 1). Once these signaling chemicals are received and recognized by neighboring plants, they may trigger a series of plant responses through signal transduction, changes in the transcriptome, proteome, and metabolome, resulting in corresponding functional alterations [84,89]. Therefore, the signaling chemicals and their biosynthetic pathways as well as regulatory mechanisms have been investigated in the past decades.

Terpenes and terpenoids are major signaling chemicals in plants. Abscisic acid, brassinosteroids, cytokinins, gibberellins, strigolactones, and (−)-loliolides are all classified as terpenoids. The biosynthesis of terpenoids involves three stages: (1) The generation of C5 precursors isopentenyl pyrophosphate (IPP) and its isomer dimethylallyl pyrophosphate (DMAPP); (2) The generation of direct precursors farnesyl diphosphate (FPP), geranyl diphosphate (GPP), geranylgeranyl diphosphate (GGPP), etc.); (3) The terpenoid formation and modification stage [241]. The synthesis of IPP and DMAPP occurs through either the mevalonic acid (MVA) pathway, which is located in the cytoplasm and utilizes acetyl-CoA as a substrate, or the deoxyxylulose-5-phosphate (DXP) or methylerythritol phosphate (MEP) pathway, which is located in the plastids and utilizes pyruvic acid and glyceraldehyde-3-phosphate as substrates [242]. The MVA pathway provides precursors (such as FPP, C15) for the biosynthesis of sesquiterpenes, triterpenes, and sterols (such as brassinosteroids). The rate-limiting enzyme in this pathway is 3-hydroxy-3-methylglutaryl-coenzyme A reductase (HMGR). Meanwhile, the MEP pathway provides precursors for the synthesis of monoterpenes (GPP, C10) and precursors for the synthesis of diterpenes, gibberellins, and carotenoids (GGPP, C20). The key enzyme in this pathway is 1-deoxyxylulose-5-phosphate synthase (DXPS). Finally, the direct precursors are catalyzed by terpene synthases to produce various terpenes [243].

The biosynthesis of terpenoids is regulated by various transcription factors. Artemisinin is a sesquiterpene lactone isolated from *Artemisia annua*, and several transcription factors associated with artemisinin biosynthesis have been identified. The sesquiterpene synthase amorpha-4,11-diene synthase (ADS) and P450 monooxygenase (CYP71AV1) are two key enzymes involved in artemisinin biosynthesis. Two JA-sensitive AP2/ERF transcription factors, AaERF1 and AaERF2, induce the promoter activity and promote the transcription of ADS and CYP71AV1. Overexpression of *AaERF1* or *AaERF2* leads to increased accumulation of artemisinin and artemisinic acid [244]. The AabHLH1 protein could bind to E-box cis-elements in the promoters of ADS and CYP71AV1 and had transcription activity in yeast. In addition, transient expression of AabHLH1 in *A. annua* leaves increased the transcription levels of genes involved in artemisinin biosynthesis (such as ADS, CYP71AV1, and HMGR). These results indicate that AabHLH1 positively regulates artemisinin biosynthesis [245]. The synthesis of allelochemical saponins in *Medicago truncatula* and *Conyza blinii* interaction is regulated by bHLH and the jasmonic acid-responsive transcription factor WRKY, respectively. The bHLH transcription factors TSAR1 and TSAR2 impact saponin biosynthesis in *M. truncatula* by regulating the expression of *HMGR1* [246], CbWRKY24 enhances the total saponin content in *C. blinii* by upregulating the expression levels of multiple MVA pathway genes [247].

Generally, volatile terpenoids participate in plant chemical communication by modulating the expression patterns of key genes, activating relevant signal transduction pathways. Monoterpene volatiles regulated the expression of Arabidopsis transcription factors or defense genes, inducing the accumulation of methyl jasmonate [110]. The release of 4,8-dimethylnona-1,3,7-triene (DMNT) from tea plants damaged by *Ectropis obliqua* induced the expression of *LOX1* and *LOX3* in neighboring plants [111]. *Fusarium oxysporum* infection in tomatoes revealed the release of different volatiles from susceptible and resistant tomato plants. In resistant tomato plants, several genes involved in volatile biosynthesis and fatty acid derivative pathways (such as *MTS1*, *TomloxC*, *TomloxD*, *AOS*, and *JAZ7*) are significantly induced by *F. oxysporum*, activating the JA pathway [248]. *Diaphorina citri* is involved in the recognition process of host plants through the regulation of genes encoding odorant-binding proteins (OBPs) and chemosensory proteins (CSPs) such as *DcitOBP3*, *DcitOBP6*, *DcitOBP8*, *DcitOBP9*, *DcitCSP1*, and *DcitCSP12* [249]. Nerolidol, as a volatile signaling chemical, participated in regulating the expression of *MAPK* and *WRKY* genes and proteins, defense-related signaling molecules, and H_2O_2 levels in tea plants [112].

Strigolactone (SLs), as a type of carotenoid-derived sesquiterpenoid, serves not only as an internal hormone but also as a signaling chemical mediating plant–root interactions in the rhizosphere [250]. SLs are synthesized through the carotenoid pathway. β-Carotenoid isomerase (D27), carotenoid cleavage dioxygenase 7 (CCD7), carotenoid cleavage dioxygenase 8 (CCD8), and cytochrome P450 monooxygenase (P450) play an important role [251]. Additionally, AtD14, MAX2 encoding an F-box protein, and the transcription factors BRC1/2 are involved in the strigolactone signaling pathway in Arabidopsis [251]. Strigolactones not only participate in stimulating the germination of parasitic plants like striga but also play a crucial role in the symbiosis between plant roots and arbuscular mycorrhizal fungi [252]. Recent studies have found that root-secreted strigolactones promote plant symbiosis with beneficial microorganisms and participate in the interaction between plants and pathogenic microorganisms. For example, SLs play a negative regulatory role in the defense of rice against the rice blast fungus (*pyricaria oryzae*). The reduction in strigolactone levels in rice enhances its defense against the rice blast fungus by activating jasmonic acid and sugar signaling pathways, as well as accumulating flavonoid phytoalexins [252].

(−)-Loliolide, a β-carotene derivative, serves both exogenous and endogenous roles in plant–plant signaling interactions [132,133,137]. (−)-Loliolide are commonly found in the root exudates and participate in plant–plant allelopathic interactions, inducing the production of allelochemicals and upregulating the expression of key genes involved in the biosynthesis of allelochemicals, such as the rice allelochemicals tricin (*CYP75B3* and *CYP75B4*) and momilacton B (*CPS4*, *KSL4*, and *MAS*) [130]. (−)-Loliolide triggered the production of wheat allelochemicals by upregulating the expression of DIMBOA biosynthetic genes, especially *Taglu*, inducing the expression of defense-related genes, and activating defense-related signaling pathways accompanied by an increase in jasmonic acid and H_2O_2 concentrations [132]. A recent study has found that (–)-loliolide participates in regulating the defense response and altering the flowering time of tobacco by modulating the expression patterns of tobacco defense and flowering-related genes [137].

Salicylic acid (SA) is an important plant defense hormone. The biosynthesis of SA begins with chorismate and proceeds through the isochorismate synthase (ICS) and phenylalanine ammonia-lyase (PAL) pathways. SA is recognized by receptor proteins such as NPR1 and NPR3/4. SA enhances its DNA binding ability by phosphorylating TGA3 and promotes the formation of the NPR1-TGA3 complex, thereby activating the expression of *PR* genes to enhance Arabidopsis resistance to diseases [253]. The salicylate hydroxylase gene *FgNahG* from *Fusarium graminearum* encodes a protein that can catalyze the conversion of SA to catechol in wheat against fusarium head blight. The expression of *FgNahG* in transgenic Arabidopsis reduces SA levels and decreases resistance to *F. graminearum* in leaves [254]. Furthermore, plant immune responses may involve crosstalk between different signaling pathways in the plants. Arabidopsis AtSR1 (a Ca^{2+}/calmodulin-binding transcription factor) is a negative regulator of plant immunity, inhibiting the expression of *EDS1* (a SA regulator) by interacting with its promoter. In addition, the binding of Ca^{2+}/calmodulin to AtSR1 is necessary for suppressing plant defense. Thus, the Ca^{2+} signal is linked to SA-mediated immune responses through calmodulin, AtSR1, and EDS1 [255]. Transcription factors play a crucial regulatory role in salicylic acid-mediated plant defense responses. Arabidopsis transcription factors TGA3 and WRKY53 interact with the CmYLCV promoter through the NPR1-dependent SA signaling pathway, enhancing its promoter activity [256].

Methyl salicylate (MeSA) plays a crucial role as a signaling chemical in regulating plant communication processes. In the natural environment, pathogen-induced volatile MeSA affects both the plant itself and neighboring plants. The expression of pathogenesis-related 1a (NbPR1a) and NbPR2, associated with disease resistance, is upregulated in response to MeSA treatment, confirming that MeSA enhances the systemic acquired resistance (SAR) capacity of tobacco. Thus, plants can perceive volatile defense chemicals, forming plant memory to enhance their SAR [257]. Indole, as a signaling chemical, is produced via the shikimate pathway via tryptophan or indole-3-pyruvate in higher plants. Indole can initiate the biosynthesis of defense-related metabolites and induce defense responses in tea trees infested with *Ectropis oblique* by enhancing Ca^{2+} signaling pathway and JA signaling [258]. Exposure of rice to exogenous indole induces the expression of its leucine-rich repeat receptor-like kinase gene (*OsLRR-RLK1*) and activates the expression of mitogen-activated protein kinases *OsMPK3* and *OsWRKY70* as well as JA biosynthesis-related genes, leading to the accumulation of JA, which regulates rice's resistance to herbivores [259].

Jasmonic acid (JA) and methyl jasmonate (MeJA) are ubiquitous signaling chemicals in a plant's response to biotic stressors. Feeding by herbivores induces the release of JA and MeJA from leaves, activating the expression of defense-related genes in neighboring healthy plants [113]. JA and MeJA are formed through the specific oxidation of C18 unsaturated fatty acids by lipoxygenase (LOX), resulting in intermediates with hydroperoxy groups at positions 9 and 13. The 13-hydroperoxide intermediate is then converted to JA through the catalysis of allene oxide synthase (AOS) and 12-oxo-phytodienoic acid reductase 3 (OPR3). JA is further transformed into methyl jasmonate (MeJA) under the catalysis of JA carboxyl methyltransferase (JMT). In plant cells, MeJA can be converted to jasmonic acid-isoleucine (JA-Ile) in an ATP-dependent manner by the action of jasmonic acid-amino synthetase (JAR1) [260].

Green leaf volatiles (GLVs) are produced in the fatty acid derivative pathway. The intermediate products of 9-hydroperoxide and 13-hydroperoxide are converted to C6 and C9 aldehydes through the action of hydroperoxide lyase (HPL). These aldehydes can then be reduced to alcohols and further transformed into esters. The resulting products are referred to as GLVs [261]. Z-3-hexenyl acetate (Z-3-HAC) in GLVs stimulates the generation of reactive oxygen species in wheat, leading to the regulation of the plant's phenylpropanoid pathway and induction of its glycosylation reactions [106]. In Arabidopsis, after pre-treatment with (E)-2-hexenal, the content of anthocyanins significantly increases upon treatment with MeJA, indicating that (E)-2-hexenal can trigger MeJA-induced plant defense responses [107]. Z-3-hexenol (Z-3-HOL) can trigger a tomato plant's defense pathways mediated by JA and SA, thereby enhancing resistance to whitefly infestation. Pre-treatment of maize with Z-3-HOL enhances the expression of the *ZmWRKY12* gene, which can regulate maize's defense against herbivorous insects [108].

Allantoin plays an important role in plant stress and plant–organism interactions [120,189]. Root-secreted allantoin is also a potential signaling chemical for kin recognition in rice lines [87]. Allantoin is an intermediate product of purine metabolism. Adenosine monophosphate (AMP) and guanosine monophosphate (GMP) undergo deamination in the cytoplasm to form xanthine. Subsequently, xanthine is catalyzed by xanthine dehydrogenase (XDH) to produce urate. Urate then enters the peroxisome and is catalyzed by urate oxidase (UOX) to generate 5-hydroxyisourate (5-HIU). The 5-HIU is further converted to allantoin through the action of allantoin synthase [262]. Plants accumulate allantoin to help in scavenging reactive oxygen species and activating stress responses. The enhanced plant stress tolerance regulated by allantoin may be achieved through the activation of abscisic acid metabolism [263]. Allantoin can also activate the JA signaling pathway regulated by MYC2 through abscisic acid [264].

Biosynthesis and molecular mechanisms of signaling chemicals involve distinct biosynthetic pathways, signaling pathways, and regulatory transcription factors. However, plants are complex organisms; the reception of a chemical signal often triggers the crosstalk of several different signaling pathways. Therefore, a systematic analysis of the signal transduction pathways induced by plant chemical communication is essential to explore the key components and molecular mechanisms underlying signal crosstalk.

5.3. *Molecular Mechanisms of Signaling-Induced Allelochemical Responses*

Plant neighbor detection and allelochemical response are driven by signaling chemicals. The signals released by neighbor plants enter the cell through stomata or the cuticle. After being recognized by receptors on the cell membrane, they enter the cell compartment (stomata or pass through the cuticular wax layer) and regulate changes in the plant cell's ion flux, triggering a burst of ROS and activating MAPK and various signaling pathways. Subsequently, the activated pathways further regulate the expression of downstream genes and promote the synthesis of secondary metabolites, including allelochemicals and phytoalexins, etc. (Figure 3).

The cuticle serves as the final barrier for the release of VOCs into the atmosphere and acts as an obstacle for VOCs to enter the plant [265]. Previous research on stomata mostly focused on their involvement in plant immunity. Recent results indicate that stomata are also one of the effective pathways for VOCs to enter neighboring plants. During the day, VOCs can enter plants through stomata. At night, due to stomatal closure, signaling chemicals primarily diffuse through the cuticle layer for propagation [266]. The signals traverse the cell wall and bind to the receptor complexes on the membrane, thus further transmitting the signal downstream. However, there is limited research on signal receptors in plants [85]. The ethylene receptor ETHYLENE RESPONSE1 (ETR1) was identified in *Arabidopsis thaliana* in the 1990s [267]. Recently, TOPLESS-like proteins (TPLs) induced by THE tpl (*TOPLESS*) and tpr (*TOPLESS-related*) genes were found to specifically bind the signal β-ionone in tobacco. Simultaneously, they interact with TFs, modulating gene expression in hormone signaling (influencing auxin signaling) and stress responses (influencing JA signaling) [268]. Additionally, plants can facilitate the passage of VOCs through the membrane via transport proteins. After silencing the ABC transporter synthesis gene *PhABCG1* in *Petunia hybrida*, this protein is involved in the active transport of VOCs between plant cells [269]. Apart from traditional signals, plant signals can also be considered damage-associated molecular patterns (DAMPs), microbe-associated molecular patterns (MAMPs), and herbivore-associated molecular patterns (HAMPs) [270]. GLVs are fatty acid derivatives that may be sensed by leucine-rich repeat-receptor-like protein kinases (LRR-RLKs) localized on the membrane [266].

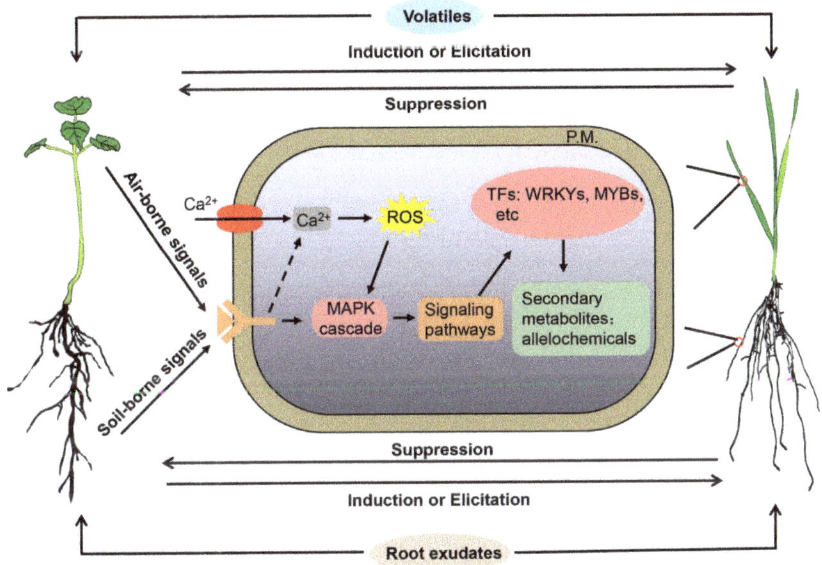

Figure 3. The signaling-induced allelochemical responses at the molecular level.

Calcium serves as a 'second messenger' involved in almost all external environmental responses and self-development signal transduction. Upon receiving external stimuli, there is a transient increase in calcium ion concentration in plant cells, leading to an influx. Under the action of Ca^{2+} sensors, such as CDPK, CaM, and CBL-CIPK, the signals are transmitted downstream [255]. Additionally, ocimene, myrcene, and pinene can also trigger cytosolic calcium influx in *Arabidopsis thaliana* [271]. ROS not only plays a role when plants are infected by pathogens but also is an indispensable part of the signal transduction network. The transcription factor *VqWRKY31* increases the accumulation of SA and ROS, facilitating the expression of defense-related genes and the accumulation of disease-resistant metabolites, including stilbenes, flavonoids, and proanthocyanidins, and enhances grapevine resistance to powdery mildew [272]. Furthermore, the production of ROS has been proven to depend on Ca^{2+} signals in many studies. After knocking out the Ca^{2+} channel genes *CNGC2* and *GLR3.3*, the ROS generation induced by MAMPs and DAMPs is significantly inhibited. In Arabidopsis, MAMP-induced ROS bursts are inhibited in *CDPK4/5/6/11* mutants, while the overexpression of *CDPK* promotes ROS generation in tomatoes [273]. A recent study has demonstrated that (–)-loliolide may mediate the content of allelochemicals based on Ca^{2+} and ROS signaling [132].

MAPK cascades are highly conserved and crucial signal transduction pathways in eukaryotes, primarily composed of three classes of protein kinases: MAPKKK/MEKK, MAPKK/MEK, and MAPK. MAPK cascades can receive upstream stimulus signals and amplify them through cascading, participating in the regulation of various processes such as plant growth, development, and responses to biotic and abiotic stresses [274]. Decreasing the expression of *MPK3* and *MPK4* can reduce the defense response induced by the volatile compound indole [259]. PD98059 is a MAPK inhibitor widely used in the research of MAPK cascades. After treating corn with PD98059, the activity of the ABA-mediated MAPK pathway was decreased and the expression of related antioxidant genes was downregulated. When grasses are cut or damaged, the released signal can rapidly activate the MAPK cascade response in neighbor plants, like *Lolium temulentum* [275]. Additionally, sesquiterpene

(E)-nerolidol and indole can directly or indirectly regulate the expression of key genes in the MAPK cascade [259]. As a central pathway in plant metabolism, the MAPK cascade not only responds to upstream signals but also participates in regulating downstream hormone signal interaction networks and activating defense genes. A well-known example is the AtMAPK3/5-AtMAPKK4/5-AtMAPK3/6 cascade reaction, which is involved in regulating the synthesis of ETH and camalexin in Arabidopsis [276]. Furthermore, the MAPK cascades are also involved in the JA and SA signal cross-talk. *SlMPK6-1* and *SlMPK6-2* in tomatoes can act as positive regulators of JA biosynthesis and signal pathways, and silencing these genes can reduce the expression of JA biosynthesis and JA-dependent defense genes [277].

After the aforementioned series of signaling cascades, the synthesis of secondary metabolites, including allelochemicals, is regulated by activating corresponding transcription factors. TFs associated with secondary metabolites include WRKYs, MYBs, and others. For example, the transcription factor NaWRKY70 mediates the synergistic regulation of scopoletin and scopolin biosynthesis by JA and ethylene signaling [278]. The receptor proteins NPR1and NPR3/4, as well as the transcription factor TGA, are involved in the SA signaling pathway. The SmNPR4-SmNPR1-SmTGA2 and SmNPR4-SmTGA5 module mediates SA regulation of phenolic biosynthesis in *Salvia miltiorrhiza* [279]. MeJA upregulates *CYP98A14* by activating *SmMYB2* and *SmMYB98*, promoting the generation of phenolic substances [216]. Additionally, MeJA can also enhance the binding of *AaWRKY1* to the *ADS* promoter in *A. annua*, thereby increasing the content of artemisinin [280].

Plant–plant chemical interactions mediated by specialized metabolites are crucial for affecting plant performance and survival. Allelopathy and allelobiosis mediated by allelochemicals and signaling chemicals are interconnected and inseparable. There is extensive research on the signaling-inducing allelochemical response, but the molecular mechanisms involved remain limited. There is a lack of molecular evidence regarding how signaling chemicals affect the entire process of plant–plant allelopathic interactions. Therefore, the content introduced in this section primarily pertains to plant defense rather than solely allelopathic interactions. Exploring the signaling-induced allelochemical responses at the molecular level is crucial for a further understanding of the chemically mediated plant–plant interactions.

6. Perspectives and Outlooks

The importance of plant–plant interactions in fitness-determining processes, such as performance, survival, resource capture, and biochemical interactions, cannot be overemphasized in natural and managed ecosystems. Recent efforts have made considerable progress toward understanding allelopathy and allelobiosis in plant–plant interactions. Nevertheless, much work is needed to clarify the details of this interesting and important area in the near future. The central focus at present is the identification and determination of new signaling chemicals, particularly for soil-borne signaling chemicals. The others are as follows:

- Identifying chemical mediation before a plant–plant interaction is considered to be allelopathy or allelobiosis.
- Developing an ecological context for allelopathy and allelobiosis, particularly for a linkage between allelopathy and allelobiosis.
- Deciphering field-based evidence for allelopathy and allelobiosis in plant coexistence and community assembly.
- Detecting and determining spatiotemporal dynamics of allelochemicals and signaling chemicals from the living plants and their environments in situ.
- Interpreting the concept of plant neighbor detection and identity recognition, particularly for relatedness-mediated neighbor detection and identity recognition and their altering of the consequences of intra-specific and inter-specific interactions in species or cultivar mixtures.
- Deciphering root–soil interactions, particularly for chemical mechanisms that structure competitive rhizosphere interactions as well as the rhizosphere microbiome.

- Evaluating allelopathy and allelobiosis under environmental variability and global changes.
- Elucidating molecular mechanisms of allelopathy and allelobiosis in plant–plant interactions from gene expressions, protein receptors, and their transportation.
- Applying ecological approaches to agriculture and forestry by allelopathy with allelochemicals and allelobiosis with signaling chemicals.

Allelopathy and allelobiosis profoundly affect the performance of plants, altering the consequences of intra-specific and inter-specific interactions in plant coexistence and community assembly. Understanding the chemically mediated plant–plant interactions is necessarily a multidisciplinary science, integrating cellular processes, gene regulation, soil microbiology, and ecological interactions. Some new technologies, such as microdialysis, stable isotope labeling (e.g., ^{13}C, ^{15}N), imageology and multi-omics, and metatranscriptomics, could shed light on identifying the role of allelopathy and allelobiosis in plant–plant interactions via the indirect way of rhizosphere microbiomes and belowground signaling interactions. In particular, recent methodological developments in high-resolution chemical detectives with molecular and imaging technologies will certainly clear the way for additional root chemical signals and their mediated interactions to be discovered.

Allelopathy and allelobiosis synergistically impact interacting plants or coexisting species. Their effects may be expressed by changes in growth, competitive outcomes, and soil resource dynamics, among others. Of course, any effects and their consequences are only part of the potential suite of interactions between species or individuals within a species. It is the overall direction and strength of the interaction that will determine community dynamics and structure. What these more traditional ecological studies bring to the conceptual table is a broader perspective on the implications of chemically mediated plant–plant interactions and the potential to incorporate these interactions into broader theories of community organization. A thorough understanding of allelopathy and allelobiosis and the mechanisms underlying the chemically mediated plant–plant interactions not only broadens our insight into the eco-evolutionary relationships but also helps us to explore ecological approaches to improve tillage management in agriculture and forestry, and enhance the predictive ability of natural and managed ecosystems to respond to environmental changes.

Author Contributions: Conceptualization, C.-H.K.; resources, Z.L., F.-L.L., X.-X.X. and P.W.; writing—original draft preparation, C.-H.K., Z.L., F.-L.L., X.-X.X. and P.W.; writing—review and editing, C.-H.K.; visualization, C.-H.K., Z.L, X.-X.X. and F.-L.L.; supervision, C.-H.K. All authors have read and agreed to the published version of the manuscript.

Funding: This research received no external funding.

Data Availability Statement: Not applicable.

Acknowledgments: The authors sincerely thank two anonymous referees for their constructive comments and suggestions.

Conflicts of Interest: The authors declare no conflicts of interest.

References

1. Zepeda, V.; Martorell, C. Seed mass equalises the strength of positive and negative plant-plant interactions in a semi-arid grassland. *Oecologia* **2019**, *190*, 287–296. [CrossRef]
2. Wuest, S.E.; Peter, R.; Niklaus, P.A. Ecological and evolutionary approaches to improving crop variety mixtures. *Nat. Ecol. Evol.* **2021**, *5*, 1068–1077. [CrossRef] [PubMed]
3. Chesson, P. Mechanisms of maintenance of species diversity. *Ann. Rev. Ecol. Evol. Syst.* **2000**, *31*, 343–366. [CrossRef]
4. Zepeda, V.; Martorell, C. Fluctuation-independent niche differentiation and relative non-linearity drive coexistence in a species-rich grassland. *Ecology* **2019**, *100*, e02726. [CrossRef] [PubMed]
5. Fréville, H.; Roumet, P.; Rode, N.O.; Rocher, A.; Latreille, M.; Muller, M.H. Preferential helping to relatives: A potential mechanism responsible for lower yield of crop variety mixtures? *Evol. Appl.* **2019**, *12*, 1837–1849. [CrossRef]
6. Mazal, L.; Fajardo, A.; Till-Bottraud, I.T.; Corenblit, D.; Fumanal, B. Kin selection, kin recognition and kin discrimination in plants revisited: A claim for considering environmental and genetic variability. *Plant Cell Environ.* **2023**, *46*, 2007–2016. [CrossRef] [PubMed]

7. Baldwin, I.T.; Halitschke, R.; Paschold, A.; von Dahl, C.C.; Preston, C.A. Volatile signaling in plant-plant interactions: "Talking Trees" in the genomics era. *Science* **2006**, *311*, 812–815. [CrossRef]
8. Pierik, R.; Mommer, L.; Voesenek, L.A.C.J. Molecular mechanisms of plant competition: Neighbour detection and response strategies. *Funct. Ecol.* **2013**, *27*, 841–853. [CrossRef]
9. Karban, R. Plant communication. *Ann. Rev. Ecol. Evol. Syst.* **2021**, *52*, 1–24. [CrossRef]
10. Meiners, S.J.; Kong, C.; Ladwig, L.M.; Pisula, N.L.; Lang, K.A. Developing an ecological context for allelopathy. *Plant Ecol.* **2012**, *213*, 1221–1227. [CrossRef]
11. Rice, E.L. *Allelopathy*; Academic Press: New York, NY, USA, 1974.
12. Pettersson, J.; Ninkovic, V.; Glinwood, R. Plant activation of barley by intercropped conspecifics and weeds: Allelobiosis. *Proc. BCPC Int. Congr. Crops Sci. Technol.* **2003**, *2*, 1135–1144.
13. Ninkovic, V.; Glinwood, R.; Pettersson, J. Communication between undamaged plants by volatiles: The role of allelobiosis. In *Communication in Plants*; Baluška, F., Mancuso, S., Volkmann, D., Eds.; Springer: Berlin/Heidelberg, Germany, 2006; pp. 421–434.
14. Ninkovic, V. Volatile interaction between undamaged plants: A short cut to coexistence. In *Plant Communication from an Ecological Perspective*; Baluska, F., Ninkovic, V., Eds.; Springer: Berlin/Heidelberg, Germany, 2010; pp. 75–86.
15. Glinwood, R.; Ninkovic, V.; Pettersson, J. Chemical interaction between undamaged plants—Effects on herbivores and natural enemies. *Phytochemistry* **2011**, *72*, 1683–1689. [CrossRef]
16. Li, Y.H.; Xia, Z.C.; Kong, C.H. Allelobiosis in the interference of allelopathic wheat with weeds. *Pest Manag. Sci.* **2016**, *72*, 2146–2153. [CrossRef] [PubMed]
17. Inderjit, S.; Wardle, D.A.; Karban, R.; Callaway, R.M. The ecosystem and evolutionary contexts of allelopathy. *Trends Ecol. Evol.* **2011**, *26*, 655–662. [CrossRef] [PubMed]
18. Kong, C.H.; Xuan, T.D.; Khanh, T.D.; Tran, H.; Trung, N.T. Allelochemicals and signaling chemicals in plants. *Molecules* **2019**, *24*, 2737. [CrossRef] [PubMed]
19. Macias, F.A.; Mecias, F.J.; Molinillo, J.M. Recent advances in allelopathy for weed control: From knowledge to applications. *Pest Manag. Sci.* **2019**, *75*, 2413–2436. [CrossRef] [PubMed]
20. Hierro, J.L.; Callaway, R.M. The ecological importance of allelopathy. *Ann. Rev. Ecol. Evol. Syst.* **2021**, *52*, 25–45. [CrossRef]
21. Han, M.; Yang, H.; Huang, H.; Du, J.; Zhang, S.; Fu, Y. Allelopathy and allelobiosis: Efficient and economical alternatives in agroecosystems. *Plant Biol.* **2024**, *26*, 11–27. [CrossRef] [PubMed]
22. Xu, Y.; Chen, X.; Ding, L.; Kong, C.H. Allelopathy and allelochemicals in grasslands and forests. *Forests* **2023**, *14*, 562. [CrossRef]
23. Kong, C.H.; Zhang, S.Z.; Li, Y.H.; Xia, Z.C.; Yang, X.F.; Meiners, S.J.; Wang, P. Plant neighbor detection and allelochemical response are driven by root-secreted signaling chemicals. *Nat. Commun.* **2018**, *9*, 3867. [CrossRef]
24. Singh, H.P.; Batish, D.R.; Kohli, R.K. Autotoxicity: Concept, organisms and ecological significance. *Crit. Rev. Plant Sci.* **1999**, *18*, 757–772. [CrossRef]
25. Inderjit; del Moral, R. Is separating resource competition from allelopathy realistic? *Bot. Rev.* **1997**, *63*, 221–230. [CrossRef]
26. Mahé, I.; Chauvel, B.; Colbach, N.; Cordeau, S.; Gfeller, A.; Reiss, A.; Moreau, D. Deciphering field-based evidences for crop allelopathy in weed regulation. A review. *Agron. Sustain. Dev.* **2022**, *42*, 50. [CrossRef]
27. Zhang, Z.J.; Liu, Y.J.; Yuan, L.; Weber, E.; van Kleunen, M. Effect of allelopathy on plant performance: A meta-analysis. *Ecol. Lett.* **2021**, *24*, 348–362. [CrossRef] [PubMed]
28. Soderquist, C.J. Juglone and allelopathy. *J. Chem. Educ.* **1973**, *50*, 782. [CrossRef]
29. Muller, C.H.; Muller, W.H.; Haines, B.L. Volatile growth inhibitors produced by aromatic shrubs. *Science* **1964**, *143*, 471–473. [CrossRef]
30. Fitter, A. Making allelopathy respectable. *Science* **2003**, *301*, 1337–1338. [CrossRef]
31. Callaway, A.; Aschehoug, E.T. Invasive plants versus their new and old neighbors: A mechanism for exotic invasion. *Science* **2000**, *290*, 521–523. [CrossRef]
32. Bais, H.P.; Vepachedu, R.; Gilroy, S.; Callaway, R.M.; Vivanco, J.M. Allelopathy and exotic plant invasion: From molecules and genes to species interactions. *Science* **2003**, *301*, 1377–1380. [CrossRef]
33. Callaway, R.M.; Ridenour, W.M. Novel weapons: Invasive success and the evolution of increased competitive ability. *Front. Ecol. Environ.* **2004**, *2*, 436–443. [CrossRef]
34. Thorpe, A.S.; Thelen, G.C.; Diaconu, A.; Callaway, R.M. Root exudate is allelopathic in invaded community but not in native community: Field evidence for the novel weapons hypothesis. *J. Ecol.* **2009**, *97*, 641–645. [CrossRef]
35. Huang, F.; Lankau, R.; Peng, S. Coexistence via coevolution driven by reduced allelochemical effects and increased tolerance to competition between invasive and native plants. *New Phytol.* **2018**, *218*, 357–369. [CrossRef]
36. Uddin, M.N.; Asaeda, T.; Shampa, S.H.; Robinson, R.W. Allelopathy and its coevolutionary implications between native and non-native neighbors of invasive *Cynara cardunculus* L. *Ecol. Evol.* **2020**, *10*, 7463–7475. [CrossRef] [PubMed]
37. Jeddi, K.; Fakhfakh, L.M.; Siddique, K.H.M.; Hessini, K.; Chaieb, M. Effect of *Acacia saligna* (Labill.) Wendl. extracts on seed germination and seedling performance of three native Mediterranean shrubs. *Bot. Lett.* **2022**, *169*, 51–60. [CrossRef]
38. Kim, Y.O.; Lee, E.J. Comparison of phenolic compounds and the effects of invasive and native species in East Asia: Support for the novel weapons hypothesis. *Ecol. Res.* **2011**, *26*, 87–94. [CrossRef]
39. Reinhart, K.O.; Rinella, M. Comparing susceptibility of eastern and western US grasslands to competition and allelopathy from spotted knapweed [*Centaurea stoebe* L. subsp. micranthos (Gugler) Hayek]. *Plant Ecol.* **2011**, *212*, 821–828. [CrossRef]

40. Levacic, D.; Perkovic, L.; Vukovic, N.; Jelaska, S.D. Bohemian Knotweed Reynoutria x bohemica Chrtek et Chrtkova seems not to rely heavily on allelopathy for its persistence in invaded sites in the Southwest Part of the Zagreb, Croatia. *Plants* **2023**, *12*, 2222. [CrossRef] [PubMed]
41. Lankau, R.A.; Strauss, S.Y. Mutual feedbacks maintain both genetic and species diversity in a plant community. *Science* **2007**, *317*, 1561–1563. [CrossRef] [PubMed]
42. Ehlers, B.K.; Thompson, J. Do co-occurring plant species adapt to one another? The response of *Bromus erectus* to the presence of different *Thymus vulgaris* chemotypes. *Oecologia* **2004**, *141*, 511–518. [CrossRef]
43. Metlen, K.L.; Aschehoug, E.T.; Callaway, R.M. Plant behavioural ecology: Dynamic plasticity in secondary metabolites. *Plant Cell Environ.* **2009**, *32*, 641–653. [CrossRef]
44. Broz, A.K.; Broeckling, C.D.; De-La-Peña, C.; Lewis, M.R.; Greene, E.; Callaway, R.M.; Vivanco, S.J.M. Plant neighbour identity influences plant biochemistry and physiology related to defence. *BMC Plant Biol.* **2010**, *10*, 115–119. [CrossRef]
45. Suzuki, R.O.; Suzuki, S.N. Morphological adaptation of a palatable plant to long-term grazing can shift interactions with an unpalatable plant from facilitative to competitive. *Plant Ecol.* **2012**, *213*, 175–183. [CrossRef]
46. Kong, C.H.; Hu, F.; Xu, X.H. Allelopathic potential and chemical constituents of volatiles from *Ageratum conyzoides* under stress. *J. Chem. Ecol.* **2002**, *28*, 1173–1182. [CrossRef] [PubMed]
47. Dayan, F.E. Factors modulating the levels of the allelochemical sorgolene in *Sorghum bicolor*. *Planta* **2006**, *224*, 339–346. [CrossRef] [PubMed]
48. Uesugi, A.; Kessler, A. Herbivore exclusion drives the evolution of plant competitiveness via increased allelopathy. *New Phytol.* **2013**, *198*, 916–924. [CrossRef]
49. Shan, Z.X.; Zhou, S.X.; Shah, A.S.; Arafat, Y.; Rizvi, S.A.H.; Shao, H. Plant allelopathy in response to biotic and abiotic factors. *Agronomy* **2023**, *13*, 2358. [CrossRef]
50. Bilas, R.D.; Bretman, A.; Bennett, T. Friends, neighbours and enemies: An overview of the communal and social biology of plants. *Plant Cell Environ.* **2021**, *44*, 997–1013. [CrossRef]
51. Kato-Noguchi, H. Barnyard grass-induced rice allelopathy and momilactone B. *J. Plant Physiol.* **2011**, *168*, 1016–1020. [CrossRef]
52. Rasher, D.B.; Hay, M.E. Competition induces allelopathy but suppresses growth and anti-herbivore defence in a chemically rich seaweed. *Proc. R. Soc. B* **2014**, *281*, 20132615. [CrossRef]
53. Lankau, R.A.; Kliebenstein, D.J. Competition, herbivory and genetics interact to determine the accumulation and fitness consequences of a defence metabolite. *J. Ecol.* **2009**, *97*, 78–88. [CrossRef]
54. Rivoal, A.; Fernandez, C.; Greff, S.; Montes, N.; Vila, B. Does competition stress decrease allelopathic potential? *Biochem. Syst. Ecol.* **2011**, *39*, 401–407. [CrossRef]
55. Wasternack, C. A plant's balance of growth and defense—Revisited. *New Phytol.* **2017**, *215*, 1291–1294. [CrossRef]
56. Niemeyer, H.M. Hydroxamic acids derived from 2-hydroxy-2H-1,4-benzoxazin-3(4H)-one: Key defense chemicals of cereals. *J. Agric. Food Chem.* **2009**, *57*, 1677–1696. [CrossRef]
57. Zhang, S.Z.; Li, Y.H.; Kong, C.H.; Xu, X.H. Interference of allelopathic wheat with different weeds. *Pest Manag. Sci.* **2016**, *72*, 172–178. [CrossRef]
58. Uesugi, A.; Johnson, R.; Kessler, A. Context-dependent induction of allelopathy in plants under competition. *Oikos* **2019**, *128*, 1492–1502. [CrossRef]
59. Xia, Z.C.; Kong, C.H.; Chen, L.C.; Wang, P.; Wang, S.L. A broadleaf species enhances an autotoxic conifers growth through belowground chemical interactions. *Ecology* **2016**, *97*, 2283–2292. [CrossRef]
60. Yang, X.F.; Kong, C.H. Interference of allelopathic rice with paddy weeds at the root level. *Plant Biol.* **2017**, *19*, 584–591. [CrossRef]
61. Asaduzzaman, M.; An, M.; Pratley, J.E.; Luckett, D.J.; Lemerle, D.; Coombes, N. The seedling root response of annual ryegrass (*Lolium rigidum*) to neighbouring seedlings of a highly-allelopathic canola (*Brassica napus*). *Flora* **2016**, *219*, 18–24. [CrossRef]
62. Yan, S.B.; Wang, P. Effects of alleolchemicals on morphological traits of roots: A meta-analysis. *Chin. J. Appl. Ecol.* **2020**, *31*, 2168–2174.
63. Wang, C.Y.; Li, L.L.; Meiners, S.J.; Kong, C.H. Root placement patterns in allelopathic plant-plant interactions. *New Phytol.* **2023**, *237*, 563–575. [CrossRef]
64. Kato-Noguchi, H.; Ino, T. The chemical-mediated allelopathic interaction between rice and barnyard grass. *Plant Soil* **2013**, *370*, 267–275. [CrossRef]
65. Tabaglio, V.; Gavazzi, C.; Schulz, M.; Marocco, A. Alternative weed control using the allelopathic effect of natural benzoxazinoids from rye mulch. *Agron. Sustain. Dev.* **2008**, *28*, 397–401. [CrossRef]
66. Kong, C.H.; Hu, F.; Wang, P.; Wu, J.L. Effect of allelopathic rice varieties combined with cultural management options on paddy field weeds. *Pest Manag. Sci.* **2008**, *64*, 276–282. [CrossRef]
67. Macias, F.A.; Chinchilla, N.; Varela, R.M.; Molinillo, J.M.G.; Marin, D.; de Siqueira, J.M. Aromatic-ring-functionalised benzoxazinones in the system *Oryza sativa–Echinochloa crus-galli* as biorational herbicide models. *Pest Manag. Sci.* **2009**, *65*, 1104–1113. [CrossRef] [PubMed]
68. Kong, C.H.; Chen, X.H.; Hu, F.; Zhang, S.Z. Breeding of commercially acceptable allelopathic rice cultivars in China. *Pest Manag. Sci.* **2011**, *67*, 1100–1106. [CrossRef] [PubMed]
69. Kong, C.H.; Wang, P.; Xu, X.H. Allelopathic interference of *Ambrosia trifida* with wheat (*Triticum aestivum*). *Agric. Ecosyst. Environ.* **2007**, *119*, 416–420. [CrossRef]

70. Williams, M.M.; Masiunas, J.B. Functional relationships between giant ragweed (*Ambrosia trifida*) interference and sweet corn yield and ear traits. *Weed Sci.* **2006**, *54*, 948–953. [CrossRef]
71. Kong, C.H.; Hu, F.; Xu, X.H.; Zhang, M.X.; Liang, W.J. Volatile allelochemicals in the *Ageratum conyzoides* intercropped citrus orchard and their effects on mites *Amblyseius newsami* and *Panonychus citri*. *J. Chem. Ecol.* **2005**, *31*, 2193–2203. [CrossRef]
72. Kong, C.H. Ecological pest management and control by using allelopathic weeds (*Ageratum conyzoides, Ambrosia trifida* and *Lantana camara*) and their allelochemicals in China. *Weed Biol. Manag.* **2010**, *10*, 73–80. [CrossRef]
73. Zhao, H.H.; Kong, C.H.; Xu, X.H. Herbicidal efficacy and ecological safety of an allelochemical-based benzothiazine derivative. *Pest Manag. Sci.* **2019**, *75*, 2690–2697. [CrossRef]
74. Fernandez, C.; Voiriot, S.; Mevy, J.P.; Vila, B.; Ormeno, E.; Dupouyet, S.; Bousquet-Melou, A. Regeneration failure of *Pinus halepensis* Mill.: The role of autotoxicity and some abiotic environmental parameters. *For. Ecol. Manag.* **2008**, *255*, 2928–2936. [CrossRef]
75. Kong, C.H.; Chen, L.C.; Xu, X.H.; Wang, P.; Wang, S.L. Allelochemicals and activities in replanted Chinese fir (*Cunninghamia lanceolata* (Lamb.) Hook.) tree ecosystem. *J. Agric. Food Chem.* **2008**, *56*, 11734–11739. [CrossRef]
76. Chen, L.C.; Wang, S.L.; Wang, P.; Kong, C.H. Autoinhibition and soil allelochemical (cyclic dipeptide) levels in replanted Chinese fir (*Cunninghamia lanceolata*) plantations. *Plant Soil* **2014**, *374*, 793–801. [CrossRef]
77. Xia, Z.C.; Kong, C.H.; Chen, L.C.; Wang, S.L. Allelochemical-mediated soil microbial community in long-term monospecific Chinese fir forest plantations. *Appl. Soil Ecol.* **2015**, *96*, 52–59. [CrossRef]
78. Williams, R.A. Mitigating biodiversity concerns in Eucalyptus plantations located in South China. *J. Biosci. Med.* **2015**, *3*, 1. [CrossRef]
79. Zhang, C.; Li, X.; Chen, Y.; Zhao, J.; Wan, S.; Lin, Y.; Fu, S. Effects of Eucalyptus litter and roots on the establishment of native tree species in Eucalyptus plantations in South China. *For. Ecol. Manag.* **2016**, *375*, 76–83. [CrossRef]
80. Qin, F.; Liu, S.; Yu, S. Effects of allelopathy and competition for water and nutrients on survival and growth of tree species in *Eucalyptus urophylla* plantations. *For. Ecol. Manag.* **2018**, *424*, 387–395. [CrossRef]
81. Yang, L.X.; Wang, P.; Kong, C.H. Effect of larch (*Larix gmelini* Rupr.) root exudates on Manchurian walnut (*Juglans mandshurica* Maxim.) growth and soil juglone in mixed-species plantation. *Plant Soil* **2010**, *329*, 249–258. [CrossRef]
82. Forrester, D.I.; Bauhus, J.; Cowie, A.L.; Vanclay, J.K. Mixed-species plantations of Eucalyptus with nitrogen-fixing trees: A review. *For. Ecol. Manag.* **2006**, *233*, 211–230. [CrossRef]
83. Bever, J.D.; Dickie, I.A.; Facelli, E.; Facelli, J.M.; Klironomos, J.; Moora, M.; Rillig, M.C.; Stock, W.D.; Tibbett, M.; Zobel, M. Rooting theories of plant community ecology in microbial interactions. *Trends Ecol. Evol.* **2010**, *25*, 468–478. [CrossRef]
84. Wang, N.Q.; Kong, C.H.; Wang, P.; Meiners, S.J. Root exudate signals in plant-plant interactions. *Plant Cell Environ.* **2021**, *44*, 1044–1058. [CrossRef]
85. Loreto, F.; D'Auria, S. How do plants sense volatiles sent by other plants? *Trends Plant Sci.* **2022**, *27*, 29–38. [CrossRef] [PubMed]
86. Berens, M.L.; Berry, H.M.; Mine, A.; Argueso, C.T.; Tsuda, K. Evolution of hormone signaling networks in plant defense. *Annu. Rev. Phytopathol.* **2017**, *55*, 401–425. [CrossRef] [PubMed]
87. Yang, X.F.; Li, L.L.; Xu, Y.; Kong, C.H. Kin recognition in rice (*Oryza sativa* L.) lines. *New Phytol.* **2018**, *220*, 567–578. [CrossRef]
88. Karban, R.; Yang, L.H.; Edwards, K.F. Volatile communication between plants that affects herbivory: A meta-analysis. *Ecol. Lett.* **2014**, *17*, 44–52. [CrossRef]
89. Ninkovic, V.; Markovic, D.; Rensing, M. Plant volatiles as cues and signals in plant communication. *Plant Cell Environ.* **2021**, *44*, 1030–1043. [CrossRef] [PubMed]
90. Brosset, A.; Blande, J.D. Volatile-mediated plant-plant interactions: Volatile organic compounds as modulators of receiver plant defence, growth, and reproduction. *J. Exp. Bot.* **2022**, *73*, 511–528. [CrossRef]
91. Chen, B.J.W.; During, H.J.; Anten, N.P.R. Detect thy neighbor: Identity recognition at the root level in plants. *Plant Sci.* **2012**, *195*, 157–167. [CrossRef]
92. Yoneyama, K.; Bennett, T. Whispers in the dark: Signals regulating underground plant–plant interactions. *Curr. Opin. Plant Biol.* **2024**, *77*, 102456. [CrossRef]
93. Maag, D.; Erb, M.; Koellner, T.G.; Gershenzon, J. Defensive weapons and defense signals in plants: Some metabolites serve both roles. *Bioassays* **2015**, *37*, 167–174. [CrossRef]
94. Rasmann, S.; Turlings, T.C.J. Root signals that mediate mutualistic interactions in the rhizosphere. *Curr. Opin. Plant Biol.* **2016**, *32*, 62–68. [CrossRef]
95. Huber, M.; Nieuwendijk, N.M.; Pantazopoulou, C.K.; Pierik, R. Light signalling shapes plant–plant interactions in dense canopies. *Plant Cell Environ.* **2021**, *44*, 1014–1029. [CrossRef]
96. Berger, K.; Machwitz, M.; Kycko, M.; Kefauver, S.C.; Van Wittenberghe, S.; Gerhards, M.; Verrelst, J.; Atzberger, C.; van der Tol, C.; Damm, A. Multi-sensor spectral synergies for crop stress detection and monitoring in the optical domain: A review. *Remote Sens. Environ.* **2022**, *280*, 113198. [CrossRef]
97. Zeng, Y.; Chen, M.; Hao, D.; Damm, A.; Badgley, G.; Rascher, U.; Johnson, J.E.; Dechant, B.; Siegmann, B.; Ryu, Y. Combining near-infrared radiance of vegetation and fluorescence spectroscopy to detect effects of abiotic changes and stresses. *Remote Sens. Environ.* **2022**, *270*, 112856. [CrossRef]
98. Eskelinen, A.; Harpole, W.S.; Jessen, M.; Virtanen, R.; Hautier, Y. Light competition drives herbivore and nutrient effects on plant diversity. *Nature* **2022**, *611*, 301–305. [CrossRef]

99. Li, H.; Burgie, E.S.; Gannam, Z.T.; Li, H.; Vierstra, R.D. Plant phytochrome B is an asymmetric dimer with unique signalling potential. *Nature* **2022**, *604*, 127–133. [CrossRef]
100. Casal, J.J. Photoreceptor signaling networks in plant responses to shade. *Ann. Rev. Plant Biol.* **2013**, *64*, 403–427. [CrossRef]
101. Keller, M.M.; Jaillais, Y.; Pedmale, U.V.; Moreno, J.E.; Chory, J.; Ballaré, C.L. Cryptochrome 1 and phytochrome B control shade-avoidance responses in Arabidopsis via partially independent hormonal cascades. *Plant J.* **2011**, *67*, 195–207. [CrossRef]
102. Farmer, E.E. Surface-to-air signals. *Nature* **2001**, *411*, 854–856. [CrossRef]
103. Farmer, E.E.; Ryan, C.A. Interplant communication: Airborne methyl jasmonate induces synthesis of proteinase inhibitors in plant leaves. *Proc. Natl. Acad. Sci. USA* **1990**, *87*, 7713–7716. [CrossRef]
104. Kessler, A.; Baldwin, I.T. Defensive function of herbivore-induced plant volatile emissions in nature. *Science* **2001**, *291*, 2141–2144. [CrossRef]
105. Shulaev, V.; Silverman, P.; Raskin, I. Airborne signalling by methyl salicylate in plant pathogen resistance. *Nature* **1997**, *385*, 718–721. [CrossRef]
106. Ameye, M.; Van Meulebroek, L.; Meuninck, B.; Vanhaecke, L.; Smagghe, G.; Haesaert, G.; Audenaert, K. Metabolomics reveal induction of ROS production and glycosylation events in wheat upon exposure to the green leaf volatile Z-3-hexenyl acetate. *Front. Plant Sci.* **2020**, *11*, 596271. [CrossRef]
107. Hirao, T.; Okazawa, A.; Harada, K.; Kobayashi, A.; Muranaka, T.; Hirata, K. Green leaf volatiles enhance methyl jasmonate response in Arabidopsis. *J. Biosci. Bioeng.* **2012**, *114*, 540–545. [CrossRef]
108. Engelberth, J.; Contreras, C.F.; Dalvi, C.; Li, T.; Engelberth, M. Early transcriptome analyses of Z-3-hexenol-treated Zea mays revealed distinct transcriptional networks and anti-herbivore defense potential of green leaf volatiles. *PLoS ONE* **2013**, *8*, e77465. [CrossRef]
109. Frank, L.; Wenig, M.; Ghirardo, A.; van der Krol, A.; Vlot, A.C.; Schnitzler, J.P.; Rosenkranz, M. Isoprene and β-caryophyllene confer plant resistance via different plant internal signalling pathways. *Plant Cell Environ.* **2021**, *44*, 1151–1164. [CrossRef]
110. Godard, K.A.; White, R.; Bohlmann, J. Monoterpene-induced molecular responses in Arabidopsis thaliana. *Phytochemistry* **2008**, *69*, 1838–1849. [CrossRef]
111. Jing, T.T.; Du, W.K.; Gao, T.; Wu, Y.; Zhang, N.; Zhao, M.Y.; Jin, J.Y.; Wang, J.M.; Schwab, W.; Wan, X.C. Herbivore-induced DMNT catalyzed by CYP82D47 plays an important role in the induction of JA-dependent herbivore resistance of neighboring tea plants. *Plant Cell Environ.* **2020**, *44*, 1178–1191. [CrossRef] [PubMed]
112. Chen, S.L.; Zhang, L.P.; Cai, X.M.; Li, X.; Bian, L.; Luo, Z.X.; Li, Z.Q.; Chen, Z.M.; Xin, Z.J. (E)-Nerolidol is a volatile signal that induces defenses against insects and pathogens in tea plants. *Hortic. Res.* **2020**, *7*, 52. [CrossRef] [PubMed]
113. Arimura, G.; Ozawa, R.; Shimoda, T.; Nishioka, T.; Boland, W.; Takabayashi, J. Herbivory-induced volatiles elicit defence genes in lima bean leaves. *Nature* **2000**, *406*, 512–515. [CrossRef]
114. Horiuchi, J.; Badri, D.V.; Kimball, B.A.; Negre, F.; Dudareva, N.; Paschke, M.W.; Vivanco, J.M. The floral volatile, methyl benzoate, from snapdragon (*Antirrhinum majus*) triggers phytotoxic effects in *Arabidopsis thaliana*. *Planta* **2007**, *226*, 1–10. [CrossRef]
115. Erb, M.; Veyrat, N.; Robert, C.A.; Xu, H.; Frey, M.; Ton, J.; Turlings, T.C. Indole is an essential herbivore-induced volatile priming signal in maize. *Nat. Commun.* **2015**, *6*, 6273. [CrossRef]
116. Pierik, R.; Visser, E.J.W.; De Kroon, H.; Voesenek, L.A.C.J. Ethylene is required in tobacco to successfully compete with proximate neighbours. *Plant Cell Environ.* **2003**, *26*, 1229–1234. [CrossRef]
117. Pierik, R.; Whitelam, G.C.; Voesenek, L.A.C.J.; De Kroon, H.; Visser, E.J.W. Canopy studies on ethylene-insensitive tobacco identify ethylene as a novel element in blue light and plant–plant signalling. *Plant J.* **2004**, *38*, 310–319. [CrossRef] [PubMed]
118. Pandey, B.K.; Huang, G.Q.; Bhosale, R.; Sjon, H.; Sturrock, C.J.; Jose, L.; Martin, O.C.; Karady, M.; Voesenek, L.A.C.J.; Ljung, K.; et al. Plant roots sense soil compaction through restricted ethylene diffusion. *Science* **2021**, *371*, 276–280. [CrossRef] [PubMed]
119. You, L.; Wang, P.; Kong, C.H. The levels of jasmonic acid and salicylic acid in a rice-barnyardgrass coexistence system and their relation to rice allelochemicals. *Biochem. Syst. Ecol.* **2011**, *39*, 491–497. [CrossRef]
120. Zhang, P.F.; Wang, X.R.; Lu, Q.; Zhang, H.W.; Chen, J.T.; Zhang, H.; Wang, Y.B.; Li, C.F. Allantoin, a purine metabolite, confers saline–alkaline tolerance to sugar beet by triggering a self-amplifying feedback loop comprising jasmonic acid and nitric oxide. *Environ. Exp. Bot.* **2023**, *206*, 105172. [CrossRef]
121. Murata, M.; Nakai, Y.; Kawazu, K.; Ishizaka, M.; Kajiwara, H.; Abe, H.; Takeuchi, K.; Ichinose, Y.; Mitsuhara, I.; Mochizuki, A.; et al. Loliolide, a carotenoid metabolite, is a potential endogenous inducer of herbivore resistance. *Plant Physiol.* **2019**, *179*, 1822–1833. [CrossRef] [PubMed]
122. Ding, L.; Zhao, H.H.; Li, H.Y.; Yang, X.F.; Kong, C.H. Kin recognition in an herbicide-resistant barnyardgrass (*Echinochloa crus-galli* L.) biotype. *Plants* **2023**, *12*, 1498. [CrossRef]
123. Li, F.L.; Chen, X.; Luo, H.M.; Meiners, S.J.; Kong, C.H. Root-secreted (−)-loliolide modulates both belowground defense and aboveground flowering in Arabidopsis and tobacco. *J. Exp. Bot.* **2023**, *74*, 964–975. [CrossRef]
124. Chen, X.; Li, F.L.; Kong, C.H. Rhizosphere bacteria mediate flowering time of two genotypes of Arabidopsis with and without root-secreted signaling (−)-loliolide. *Rhizosphere* **2023**, *27*, 100774. [CrossRef]
125. Akiyama, K.; Ogasawara, S.; Ito, S.; Hayashi, H. Structural requirements of strigolactones for hyphal branching in AM fungi. *Plant Cell Physiol.* **2010**, *51*, 1104–1117. [CrossRef]
126. Akiyama, K.; Hayashi, H. Strigolactones: Chemical signals for fungal symbionts and parasitic weeds in plant roots. *Ann. Bot.* **2006**, *97*, 925–931. [CrossRef] [PubMed]

127. Peters, N.K.; Frost, J.W.; Long, S.R. A plant flavone, luteolin, induces expression of *Rhizobium meliloti* nodulation genes. *Science* **1986**, *233*, 977–980. [CrossRef]
128. Semchenko, M.; Saar, S.; Lepik, A. Plant root exudates mediate neighbour recognition and trigger complex behavioural changes. *New Phytol.* **2014**, *204*, 631–637. [CrossRef]
129. Pickles, B.J.; Wilhelm, R.; Asay, A.K.; Hahn, A.S.; Simard, S.W.; Mohn, W.W. Transfer of ^{13}C between paired Douglas-fir seedlings reveals plant kinship effects and uptake of exudates by ectomycorrhizas. *New Phytol.* **2017**, *214*, 400–411. [CrossRef] [PubMed]
130. Li, L.L.; Zhao, H.H.; Kong, C.H. (–)-Loliolide, the most ubiquitous lactone, is involved in barnyardgrass-induced rice allelopathy. *J. Exp. Bot.* **2020**, *71*, 1540–1550. [CrossRef] [PubMed]
131. Schreiner, M.; Krumbein, A.; Knorr, D.; Smetanska, I. Enhanced glucosinolates in root exudates of *Brassica rapa* ssp. rapa mediated by salicylic acid and methyl jasmonate. *J. Agric. Food Chem.* **2011**, *59*, 1400–1405. [CrossRef]
132. Li, L.L.; Li, Z.; Lou, Y.G.; Meiners, S.J.; Kong, C.H. (–)-Loliolide is a general signal of plant stress that activates jasmonate-related responses. *New Phytol.* **2023**, *238*, 2099–2112. [CrossRef] [PubMed]
133. Frost, C.J. Information potential of an ubiquitous phytochemical cue. *New Phytol.* **2023**, *238*, 1749–1751. [CrossRef] [PubMed]
134. Li, H.Y.; Guo, Y.; Jin, B.Y.; Yang, X.F.; Kong, C.H. Phytochemical cue for the fitness cost of herbicide-resistant weeds. *Plants* **2023**, *12*, 3158. [CrossRef] [PubMed]
135. Shen, G.J.; Zhang, J.X.; Lei, Y.T.; Xu, Y.X.; Wu, J.Q. Between-plant signaling. *Annu. Rev. Plant Biol.* **2023**, *74*, 367–386. [CrossRef] [PubMed]
136. Lanfranco, L.; Fiorilli, V.; Gutjahr, C. Partner communication and role of nutrients in the arbuscular mycorrhizal symbiosis. *New Phytol.* **2018**, *220*, 1031–1046. [CrossRef] [PubMed]
137. Ninkovic, V.; Rensing, M.; Dahlin, I.; Markovic, D. Who is my neighbor? Volatile cues in plant interactions. *Plant Signal. Behav.* **2019**, *14*, 1634993. [CrossRef]
138. Patni, B.; Bhattacharyya, M.; Pokhriyal, A. The role of signaling compounds in enhancing rice allelochemicals for sustainable agriculture: An overview. *Planta* **2023**, *258*, 90. [CrossRef] [PubMed]
139. McCormick, A.C.; Effah, E.; Najar-Rodriguez, A. Ecological aspects of volatile organic compounds emitted by exotic invasive plants. *Front. Ecol. Evol.* **2023**, *11*, 1059125. [CrossRef]
140. Chen, K.J.; Zheng, Y.Q.; Kong, C.H.; Zhang, S.Z.; Li, J.; Liu, X.G. 2, 4-Dihydroxy-7- methoxy-1,4-benzoxazin-3-one (DIMBOA) and 6-methoxy-benzoxazolin-2-one (MBOA) levels in the wheat rhizosphere and their effect on soil microbial community structure. *J. Agric. Food Chem.* **2010**, *58*, 12710–12716. [CrossRef]
141. Li, B.; Li, Y.Y.; Wu, H.M.; Zhang, F.F.; Li, C.J.; Li, X.X.; Lambers, H.; Li, L. Root exudates drive interspecific facilitation by enhancing nodulation and N2 fixation. *Proc. Natl. Acad. Sci. USA* **2016**, *113*, 6496–6501. [CrossRef]
142. Xu, Y.; Cheng, H.F.; Kong, C.H.; Meiners, S.J. Intra-specific kin recognition contributes to inter-specific allelopathy: A case study of allelopathic rice interference with paddy weeds. *Plant Cell Environ.* **2021**, *44*, 3709–3721. [CrossRef]
143. Falik, O.; Reides, P.; Gersani, M.; Novoplansky, A. Self/non-self discrimination in roots. *J. Ecol.* **2003**, *91*, 525–531. [CrossRef]
144. Gruntman, M.; Novoplansky, A. Physiologically mediated self/non-self discrimination in roots. *Proc. Natl. Acad. Sci. USA* **2004**, *101*, 3863–3867. [CrossRef]
145. Mahall, B.E.; Callaway, R.M. Root communication mechanisms and intracommunity distributions of 2 Mojave desert shrubs. *Ecology* **1992**, *73*, 2145–2151. [CrossRef]
146. Gersani, M.; Brown, J.S.; O'Brien, E.E.; Maina, G.M.; Abramsky, Z. Tragedy of the commons as a result of root competition. *J. Ecol.* **2001**, *89*, 660–669. [CrossRef]
147. Maina, G.G.; Brown, J.S.; Gersani, M. Intra-plant versus inter-plant root competition in beans: Avoidance, resource matching or tragedy of the commons. *Plant Ecol.* **2002**, *160*, 235–247. [CrossRef]
148. O'Brien, E.E.; Gersani, M.; Brown, J.S. Root proliferation and seed yield in response to spatial heterogeneity of below-ground competition. *New Phytol.* **2005**, *168*, 401–412. [CrossRef]
149. Falik, O.; de Kroon, H.; Novoplansky, A. Physiologically-mediated self/non-self root discrimination in Trifolium repens has mixed effects on plant performance. *Plant Signal. Behav.* **2006**, *1*, 116–121. [CrossRef] [PubMed]
150. Anten, N.P.R.; Chen, B.J.W. Detect thy family: Mechanisms, ecology, and agricultural aspects of kin recognition in plants. *Plant Cell Environ.* **2021**, *44*, 1059–1071. [CrossRef] [PubMed]
151. Dudley, S.A.; File, A.L. Kin recognition in an annual plant. *Biol. Lett.* **2007**, *3*, 435–438. [CrossRef]
152. Torices, R.; DeSoto, L.; Narbona, E.; Gómez, J.M.; Pannell, J.R. Effects of the relatedness of neighbours on floral colour. *Front. Ecol. Evol.* **2021**, *9*, 589781. [CrossRef]
153. Lehmann, L.; Perrin, N. Altruism, dispersal, and phenotype-matching kin recognition. *Am. Nat.* **2002**, *159*, 451–468. [CrossRef]
154. Nakamura, R.R. Plant kin selection. *Evol. Theory* **1980**, *5*, 113–117.
155. Lepik, A.; Abakumova, M.; Zobel, K.; Semchenko, M. Kin recognition is density-dependent and uncommon among temperate grassland plants. *Funct. Ecol.* **2012**, *26*, 1214–1220. [CrossRef]
156. Pezzola, E.; Pandolfi, C.; Mancuso, S. Resource availability affects kin selection in two cultivars of *Pisum sativum*. *Plant Growth Regul.* **2020**, *90*, 321–329. [CrossRef]
157. Dudley, S.A.; Murphy, G.P.; File, A.L. Kin recognition and competition in plants. *Funct. Ecol.* **2013**, *27*, 898–906. [CrossRef]
158. Murphy, G.P.; Swanton, C.J.; van Acker, R.C.; Dudley, S.A. Kin recognition, multilevel selection and altruism in crop sustainability. *J. Ecol.* **2017**, *105*, 930–934. [CrossRef]

159. Bhatt, V.M.; Khandelwal, A.; Dudley, S.A. Kin recognition, not competitive interactions predicts root allocation in young *Cakile edentula* seeding pairs. *New Phytol.* **2011**, *189*, 1135–1142. [CrossRef]
160. Belter, P.R.; Cahill, J.F., Jr. Disentangling root system responses to neighbours: Identification of novel root behavioural strategies. *AoB Plants* **2015**, *7*, plv059. [CrossRef]
161. Biedrzycki, M.L.; Bais, H.P. Kin recognition in plants: Did we learn anything from roots? *Front. Ecol. Evol.* **2022**, *9*, 785019. [CrossRef]
162. Pereira, L.; Sadras, V.O.; Batista, W.; Casal, J.J.; Hall, A.J. Light mediated self-organization of sunflower stands increases oil yield in the field. *Proc. Natl. Acad. Sci. USA* **2017**, *114*, 7975–7980. [CrossRef] [PubMed]
163. Simonsen, A.K.; Chow, T.; Stinchcombe, J.R. Reduced plant competition among kin can be explained by Jensen's inequality. *Ecol. Evol.* **2014**, *4*, 4454–4466. [CrossRef] [PubMed]
164. Goddard, E.L.; Varga, S.; John, E.A.; Soulsbury, C.D. Evidence for root kin recognition in the clonal plant species *Glechoma hederacea*. *Front. Ecol. Evol.* **2020**, *8*, 578141. [CrossRef]
165. Biedrzycki, M.L.; Jilany, T.A.; Dudley, S.A.; Bais, H.P. Root exudates mediate kin recognition in plants. *Commun. Integr. Biol.* **2010**, *3*, 28–35. [CrossRef] [PubMed]
166. Yang, L.N.; Pan, Z.C.; Zhu, W.; Wu, E.J.; He, D.C.; Yuan, X.; Qin, Y.Y.; Wang, Y.; Chen, R.S.; Thrall, P.H.; et al. Enhanced agricultural sustainability through within-species diversification. *Nat. Sustain.* **2019**, *2*, 46–52. [CrossRef]
167. Ryan, M.R. Crops better when grown together. *Nat. Sustain.* **2021**, *4*, 926–927. [CrossRef]
168. Yu, R.P.; Lambers, H.; Callaway, R.M.; Wright, A.J.; Li, L. Belowground facilitation and trait matching: Two or three to tango? *Trends Plant Sci.* **2021**, *26*, 1227–1235. [CrossRef]
169. Bawa, K.S. Kin selection and the evolution of plant reproductive traits. *Proc. R. Soc. B* **2016**, *283*, 20160789. [CrossRef]
170. Callaway, R.M.; Li, L. Decisions, decisions, decisions: Plant roots detect and respond to complex environmental cues. *New Phytol.* **2020**, *226*, 11–12. [CrossRef]
171. Hodge, A. Root decisions. *Plant Cell Environ.* **2009**, *32*, 628–640. [CrossRef]
172. Falik, O.; Reides, P.; Gersani, M.; Novoplansky, A. Root navigation by self inhibition. *Plant Cell Environ.* **2005**, *28*, 562–569. [CrossRef]
173. Novoplansky, A. What plant roots know? *Semin. Cell Dev. Biol.* **2019**, *92*, 126–133. [CrossRef]
174. Semchenko, M.; John, E.A.; Hutchings, M.J. Effects of physical connection and genetic identity of neighbouring ramets on root placement patterns in two clonal species. *New Phytol.* **2007**, *176*, 644–654. [CrossRef]
175. Korenblum, E.; Dong, Y.H.; Szymanski, J.; Panda, S.; Jozwiak, A.; Massalha, H.; Meir, S.; Rogachev, I.; Aharoni, A. Rhizosphere microbiome mediates systemic root metabolite exudation by root-to-root signaling. *Proc. Natl. Acad. Sci. USA* **2020**, *117*, 3874–3883. [CrossRef] [PubMed]
176. Kong, C.H.; Wang, P.; Gu, Y.; Xu, X.H.; Wang, M.L. The fate and impact on microorganisms of rice allelochemicals in paddy soil. *J. Agric. Food Chem.* **2008**, *56*, 5043–5049. [CrossRef] [PubMed]
177. Li, X.J.; Xia, Z.C.; Kong, C.H.; Xu, X.H. Mobility and microbial activity of allelochemicals in soil. *J. Agric. Food Chem.* **2013**, *61*, 5072–5079. [CrossRef] [PubMed]
178. Kong, C.H.; Zhao, H.; Xu, X.H.; Wang, P.; Gu, Y. Activity and allelopathy of soil of flavone O-glycosides from rice. *J. Agric. Food Chem.* **2007**, *55*, 6007–6012. [CrossRef] [PubMed]
179. Kong, C.H.; Wang, P.; Zhao, H.; Xu, X.H.; Zhu, Y.D. Impact of allelochemical exuded from allelopathic rice on soil microbial community. *Soil Biol. Biochem.* **2008**, *40*, 1862–1869. [CrossRef]
180. Maver, M.; Trevisan, F.; Miras-Moreno, B.; Lucini, L.; Trevisan, M.; Cesco, S.; Mimmo, T. The interplay between nitrogenated allelochemicals, mineral nutrition and metabolic profile in barley roots. *Plant Soil* **2022**, *479*, 715–730. [CrossRef]
181. Schütz, V.; Frindte, K.; Cui, J.X.; Zhang, P.F.; Hacquard, S.; Schulze-Lefert, P.; Knief, C.; Schulz, M.; Dörmann, P. Differential impact of plant secondary metabolites on the soil microbiota. *Front. Microbiol.* **2021**, *12*, 666010. [CrossRef]
182. Dong, H.Y.; Kong, C.H.; Wang, P.; Huang, Q.L. Temporal variation of soil friedelin and microbial community under different land uses in a long-term agroecosystem. *Soil Biol. Biochem.* **2014**, *69*, 275–281. [CrossRef]
183. Guo, Z.Y.; Kong, C.H.; Wang, J.G.; Wang, Y.F. Rhizosphere isoflavones (daidzein and genistein) levels and their relation to the microbial community structure of mono-cropped soybean soil in field and controlled conditions. *Soil Biol. Biochem.* **2011**, *43*, 2257–2264. [CrossRef]
184. Shi, K.; Shao, H. Changes in the soil fungal community mediated by a *Peganum harmala* allelochemical. *Front. Microbiol.* **2022**, *13*, 911836. [CrossRef]
185. Inderjit. Soil microorganisms: An important determinant of allelopathic activity. *Plant Soil* **2005**, *274*, 227–236. [CrossRef]
186. Revillini, D.; David, A.S.; Reyes, A.L.; Knecht, L.D.; Vigo, C.; Allen, P.; Searcy, C.A.; Afkhami, M.E. Allelopathy-selected microbiomes mitigate chemical inhibition of plant performance. *New Phytol.* **2023**, *240*, 2007–2019. [CrossRef]
187. Liu, Y.; Wang, H.; Qian, X.; Gu, J.; Chen, W.M.; Shen, X.H.; Tao, S.H.; Jiao, S.; Wei, G.H. Metagenomics insights into responses of rhizobacteria and their alleviation role in licorice allelopathy. *Microbiome* **2023**, *11*, 109. [CrossRef] [PubMed]
188. Korenblum, E.; Massalha, H.; Aharoni, A. Plant–microbe interactions in the rhizosphere via a circular metabolic economy. *Plant Cell* **2022**, *34*, 3168–3182. [CrossRef] [PubMed]
189. Wang, P.; Kong, C.H.; Hu, F.; Xu, X.H. Allantoin involved in species interactions with rice and other organisms in paddy soil. *Plant Soil* **2007**, *296*, 43–51. [CrossRef]

190. Wang, P.; Kong, C.H.; Sun, B.; Xu, X.H. Allantoin-induced change of microbial diversity and community in rice soil. *Plant Soil* **2010**, *332*, 357–368. [CrossRef]
191. Zhou, B.; Kong, C.H.; Li, Y.H.; Wang, P.; Xu, X.H. Crabgrass (*Digitaria sanguinalis*) allelochemicals that interfere with crop growth and the soil microbial community. *J. Agric. Food Chem.* **2013**, *61*, 5310–5317. [CrossRef]
192. Zhao, H.H.; Li, H.Y.; Kong, C.H. Penoxsulam-resistant barnyardgrass-mediated root microbial communities affect the growth of rice. *Pest Manag. Sci.* **2023**, *79*, 2664–2674. [CrossRef] [PubMed]
193. Chen, Y.; Bonkowski, M.; Shen, Y.; Griffiths, B.S.; Jiang, Y.J.; Wang, X.Y.; Sun, B. Root ethylene mediates rhizosphere microbial community reconstruction when chemically detecting cyanide produced by neighbouring plants. *Microbiome* **2020**, *8*, 4. [CrossRef]
194. Yin, C.T.; Vargas, J.M.C.; Schlatter, D.C.; Hagerty, C.H.; Hulbert, S.H.; Paulitz, T.C. Rhizosphere community selection reveals bacteria associated with reduced root disease. *Microbiome* **2021**, *9*, 86. [CrossRef]
195. Elhaissoufi, W.; Ghoulam, C.; Barakat, A.; Zeroual, Y.; Bargaz, A. Phosphate bacterial solubilization: A key rhizosphere driving force enabling higher P use efficiency and crop productivity. *J. Adv. Res.* **2022**, *38*, 13–28. [CrossRef]
196. Hao, C.K.; Dungait, J.A.J.; Wei, X.M.; Ge, T.D.; Kuzyakov, Y.; Cui, Z.L.; Tian, J.; Zhang, F.S. Maize root exudate composition alters rhizosphere bacterial community to control hotspots of hydrolase activity in response to nitrogen supply. *Soil Biol. Biochem.* **2022**, *170*, 108717. [CrossRef]
197. Ulbrich, T.C.; Rivas-Ubach, A.; Tiemann, L.K.; Friesen, M.L.; Evans, S.E. Plant root exudates and rhizosphere bacterial communities shift with neighbor context. *Soil Biol. Biochem.* **2022**, *172*, 108753. [CrossRef]
198. Singh, K.; Chandra, R.; Purchase, D. Unraveling the secrets of rhizobacteria signaling in rhizosphere. *Rhizosphere* **2022**, *21*, 100484. [CrossRef]
199. Chaney, L.; Baucom, R.S. The soil microbial community alters patterns of selection on flowering time and fitness-related traits in ipomoea purpurea. *Am. J. Bot.* **2020**, *107*, 186–194. [CrossRef]
200. Hu, L.; Robert, C.A.M.; Cadot, S.; Zhang, X.; Ye, M.; Li, B.; Manzo, D.; Chervet, N.; Steinger, T.; van der Heijden, M.G.A. Root exudate metabolites drive plant–soil feedbacks on growth and defense by shaping the rhizosphere microbiota. *Nat. Commun.* **2018**, *9*, 2738. [CrossRef] [PubMed]
201. Lu, T.; Ke, M.J.; Lavoie, M.; Jin, Y.J.; Fan, X.J.; Zhang, Z.Y.; Fu, Z.W.; Sun, L.W.; Gillings, M.; Penuelas, J.; et al. Rhizosphere microorganisms can influence the timing of plant flowering. *Microbiome* **2018**, *6*, 231. [CrossRef]
202. Panke-Buisse, K.; Poole, A.C.; Goodrich, J.K.; Ley, R.E.; Kao-Kniffin, J. Selection on soil microbiomes reveals reproducible impacts on plant function. *ISME J.* **2015**, *9*, 980–989. [CrossRef]
203. Wagner, M.R.; Lundberg, D.S.; Coleman-Derr, D.; Tringe, S.G.; Dangl, J.L.; Mitchell-Olds, T. Natural soil microbes alter flowering phenology and the intensity of selection on flowering time in a wild Arabidopsis relative. *Ecol. Lett.* **2014**, *17*, 717–726. [CrossRef]
204. Ramandi, A.; Nourashrafeddin, S.M.; Marashi, S.H.; Seifi, A. Microbiome contributes to phenotypic plasticity in saffron crocus. *World J. Microbiol. Biotechol.* **2023**, *39*, 9. [CrossRef] [PubMed]
205. Bennett, J.A.; Klironomos, J. Mechanisms of plant-soil feedback: Interactions among biotic and abiotic drivers. *New Phytol.* **2019**, *222*, 91–96. [CrossRef] [PubMed]
206. Doornbos, R.F.; van Loon, L.C.; Bakker, P.A.H.M. Impact of root exudates and plant defense signaling on bacterial communities in the rhizosphere. A review. *Agron. Sustain. Dev.* **2012**, *32*, 227–243. [CrossRef]
207. Sun, B.; Wang, P.; Kong, C.H. Plant-soil feedback in the interference of allelopathic rice with barnyardgrass. *Plant Soil* **2014**, *377*, 309–321. [CrossRef]
208. Massalha, H.; Korenblum, E.; Tholl, D.; Aharoni, A. Small molecules below-ground: The role of specialized metabolites in the rhizosphere. *Plant J.* **2017**, *90*, 788–807. [CrossRef] [PubMed]
209. Fang, C.X.; Yang, L.K.; Chen, W.S.; Li, L.L.; Zhang, P.L.; Li, Y.Z.; He, H.B.; Lin, W.X. MYB57 transcriptionally regulates MAPK11 to interact with PAL2;3 and modulate rice allelopathy. *J. Exp. Bot.* **2020**, *71*, 2127–2141. [CrossRef]
210. Fraser, C.M.; Chapple, C. The phenylpropanoid pathway in arabidopsis. *Arab. Book* **2011**, *9*, e0152. [CrossRef] [PubMed]
211. Muro-Villanueva, F.; Mao, X.Y.; Chapple, C. Linking phenylpropanoid metabolism, lignin deposition, and plant growth inhibition. *Curr. Opin. Biotech.* **2019**, *56*, 202–208. [CrossRef]
212. Li, C.N.; Ng, C.; Fan, L.M. MYB transcription factors, active players in abiotic stress signaling. *Environ. Exp. Bot.* **2015**, *114*, 80–91. [CrossRef]
213. Dias, A.P.; Grotewold, E. Manipulating the accumulation of phenolics in maize cultured cells using transcription factors. *Biochem. Eng. J.* **2003**, *14*, 207–216. [CrossRef]
214. Rommens, C.M.; Richael, C.M.; Yan, H.; Navarre, D.A.; Ye, J.S.; Krucker, M.; Swords, K. Engineered native pathways for high kaempferol and caffeoylquinate production in potato. *Plant Biotech. J.* **2008**, *6*, 870–886. [CrossRef] [PubMed]
215. Luo, J.; Butelli, E.; Hill, L.; Parr, A.; Niggeweg, R.; Bailey, P.; Weisshaar, B.; Martin, C. AtMYB12 regulates caffeoyl quinic acid and flavonol synthesis in tomato: Expression in fruit results in very high levels of both types of polyphenol. *Plant J.* **2008**, *56*, 316–326. [CrossRef] [PubMed]
216. Deng, C.P.; Wang, Y.; Huang, F.F.; Lu, S.J.; Zhao, L.M.; Ma, X.Y.; Kai, G.Y. SmMYB2 promotes salvianolic acid biosynthesis in the medicinal herb *Salvia miltiorrhiza*. *J. Integr. Plant Biol.* **2020**, *62*, 1688–1702. [CrossRef] [PubMed]

217. Pandey, A.; Misra, P.; Khan, M.P.; Swarnkar, G.; Tewari, M.C.; Bhambhani, S.; Trivedi, R.; Chattopadhyay, N.; Trivedi, P.K. Co-expression of Arabidopsis transcription factor, AtMYB12, and soybean isoflavone synthase, GmIFS1, genes in tobacco leads to enhanced biosynthesis of isoflavones and flavonols resulting in osteoprotective activity. *Plant Biotech. J.* **2014**, *12*, 69–80. [CrossRef] [PubMed]
218. Fornalé, S.; Lopez, E.; Salazar-Henao, J.E.; Fernández-Nohales, P.; Rigau, J.; Caparros-Ruiz, D. AtMYB7, a new player in the regulation of uv-sunscreens in *Arabidopsis thaliana*. *Plant Cell Physiol.* **2014**, *55*, 507–516. [CrossRef]
219. Nakatsuka, T.; Saito, M.; Yamada, E.; Fujita, K.; Kakizaki, Y.; Nishihara, M. Isolation and characterization of GtMYBP3 and GtMYBP4, orthologues of R2R3-MYB transcription factors that regulate early flavonoid biosynthesis, in gentian flowers. *J. Exp. Bot.* **2012**, *63*, 6505–6517. [CrossRef]
220. Vimolmangkang, S.; Han, Y.P.; Wei, G.C.; Korban, S.S. An apple MYB transcription factor, MdMYB3, is involved in regulation of anthocyanin biosynthesis and flower development. *BMC Plant Biol.* **2013**, *13*, 176. [CrossRef]
221. Xu, F.; Ning, Y.J.; Zhang, W.W.; Liao, Y.L.; Li, L.L.; Cheng, H.; Cheng, S.Y. An R2R3-MYB transcription factor as a negative regulator of the flavonoid biosynthesis pathway in Ginkgo biloba. *Funct. Integr. Genom.* **2014**, *14*, 177–189. [CrossRef]
222. Yan, J.H.; Wang, B.; Zhong, Y.P.; Yao, L.M.; Cheng, L.J.; Wu, T.L. The soybean R2R3 MYB transcription factor GmMYB100 negatively regulates plant flavonoid biosynthesis. *Plant Mol. Biol.* **2015**, *89*, 35–48. [CrossRef]
223. Zhao, M.Z.; Wang, T.L.; Wu, P.; Guo, W.Y.; Su, L.T.; Wang, Y.; Liu, Y.J.; Yan, F.; Wang, Q.Y. Isolation and characterization of GmMYBJ3, an R2R3-MYB transcription factor that affects isoflavonoids biosynthesis in soybean. *PLoS ONE* **2017**, *12*, e0179990. [CrossRef]
224. Huang, D.; Yu, L.C.; Ming, R.H.; Tan, X.M.; Li, L.B.; Huang, R.S.; Tan, Y.; Yao, S.C. A chromosome-level genome assembly of *Callerya speciosa* sheds new light on the biosynthesis of root-specific isoflavonoids. *Ind. Crop. Prod.* **2023**, *200*, 116877. [CrossRef]
225. Saxena, S.; Pal, L.; Naik, J.; Singh, Y.; Verma, P.K.; Chattopadhyay, D.; Pandey, A. The R2R3-MYB-SG7 transcription factor CaMYB39 orchestrates surface phenylpropanoid metabolism and pathogen resistance in chickpea. *New Phytol.* **2023**, *238*, 798–816. [CrossRef] [PubMed]
226. Widhalm, J.R.; Rhodes, D. Biosynthesis and molecular actions of specialized 1,4-naphthoquinone natural products produced by horticultural plants. *Hortic. Res.* **2016**, *3*, 16046. [CrossRef] [PubMed]
227. Hao, X.L.; Pu, Z.Q.; Cao, G.; You, D.W.; Zhou, Y.; Deng, C.P.; Shi, M.; Nile, S.H.; Wang, Y.; Zhou, W.; et al. Tanshinone and salvianolic acid biosynthesis are regulated by SmMYB98 in *Salvia miltiorrhiza* hairy roots. *J. Adv. Res.* **2020**, *23*, 1–12. [CrossRef]
228. McCoy, R.M.; Utturkar, S.M.; Crook, J.W.; Thimmapuram, J.; Widhalm, J.R. The origin and biosynthesis of the naphthalenoid moiety of juglone in black walnut. *Hortic. Res.* **2018**, *5*, 67. [CrossRef] [PubMed]
229. Wei, J.C.; Yang, Y.; Peng, Y.; Wang, S.Y.; Zhang, J.; Liu, X.B.; Liu, J.J.; Wen, B.B.; Li, M.F. Biosynthesis and the transcriptional regulation of terpenoids in tea plants (*Camellia sinensis*). *Int. J. Mol. Sci.* **2023**, *24*, 6937. [CrossRef] [PubMed]
230. Karunanithi, P.S.; Zerbe, P. Terpene synthases as metabolic gatekeepers in the evolution of plant terpenoid chemical diversity. *Front. Plant Sci.* **2019**, *10*, 1166. [CrossRef]
231. Chen, S.; Xie, P.F.; Li, Y.Y.; Wang, X.X.; Liu, H.H.; Wang, S.S.; Han, W.B.; Wu, R.M.; Li, X.L.; Guan, Y.F.; et al. New insights into stress-induced β-Ocimene biosynthesis in tea (*Camellia sinensis*) leaves during oolong tea processing. *J. Agric. Food Chem.* **2021**, *69*, 11656–11664. [CrossRef]
232. Xu, Q.S.; He, Y.X.; Yan, X.M.; Zhao, S.Q.; Zhu, J.Y.; Wei, C.L. Unraveling a crosstalk regulatory network of temporal aroma accumulation in tea plant (*Camellia sinensis*) leaves by integration of metabolomics and transcriptomics. *Environ. Exp. Bot.* **2018**, *149*, 81–94. [CrossRef]
233. Kitaoka, N.; Zhang, J.; Oyagbenro, R.K.; Brown, B.; Wu, Y.S.; Yang, B.; Li, Z.H.; Peters, R.J. Interdependent evolution of biosynthetic gene clusters for momilactone production in rice. *Plant Cell* **2021**, *33*, 290–305. [CrossRef]
234. Cárdenas, P.D.; Sonawane, P.D.; Heinig, U.; Bocobza, S.E.; Burdman, S.; Aharoni, A. The bitter side of the nightshades: Genomics drives discovery in Solanaceae steroidal alkaloid metabolism. *Phytochemistry* **2015**, *113*, 24–32. [CrossRef] [PubMed]
235. Cambier, V.; Hance, T.; de Hoffmann, E. Variation of DIMBOA and related compounds content in relation to the age and plant organ in maize. *Phytochemistry* **2000**, *53*, 223–229. [CrossRef] [PubMed]
236. Richter, A.; Powell, A.F.; Mirzaei, M.; Wang, L.J.; Movahed, N.; Miller, J.K.; Piñeros, M.A.; Jander, G. Indole-3-glycerolphosphate synthase, a branchpoint for the biosynthesis of tryptophan, indole, and benzoxazinoids in maize. *Plant J.* **2021**, *106*, 245–257. [CrossRef] [PubMed]
237. Czjzek, M.; Cicek, M.; Zamboni, V.; Bevan, D.R.; Henrissat, B.; Esen, A. The mechanism of substrate (aglycone) specificity in β-glucosidases is revealed by crystal structures of mutant maize β-glucosidase-DIMBOA, -DIMBOAGlc, and -dhurrin complexes. *Proc. Natl. Acad. Sci. USA* **2000**, *97*, 13555–13560. [CrossRef] [PubMed]
238. Song, J.; Liu, H.; Zhuang, H.F.; Zhao, C.X.; Xu, Y.X.; Wu, S.B.; Qi, J.F.; Li, J.; Hettenhausen, C.; Wu, J.Q. Transcriptomics and alternative splicing analyses reveal large differences between maize lines b73 and mo17 in response to aphid *Rhopalosiphum padi* infestation. *Front. Plant Sci.* **2017**, *8*, 1738. [CrossRef] [PubMed]
239. Zhou, P.; Li, Z.; Magnusson, E.; Cano, F.G.; Crisp, P.A.; Noshay, J.M.; Grotewold, E.; Hirsch, C.N.; Briggs, S.P.; Springer, N.M. Meta gene regulatory networks in maize highlight functionally relevant regulatory interactions. *Plant Cell* **2020**, *32*, 1377–1396. [CrossRef]
240. Batyrshina, Z.S.; Shavit, R.; Yaakov, B.; Bocobza, S.; Tzin, V. The transcription factor TaMYB31 regulates the benzoxazinoid biosynthetic pathway in wheat. *J. Exp. Bot.* **2022**, *73*, 5634–5649. [CrossRef]

241. Dudareva, N.; Pichersky, E.; Gershenzon, J. Biochemistry of plant volatiles. *Plant Physiol.* **2004**, *135*, 1893–1902. [CrossRef]
242. Rodríguez-Concepción, M.; Boronat, A. Elucidation of the methylerythritol phosphate pathway for isoprenoid biosynthesis in bacteria and plastids.: A metabolic milestone achieved through genomics. *Plant Physiol.* **2002**, *130*, 1079–1089. [CrossRef]
243. Chen, F.; Tholl, D.; Bohlmann, J.; Pichersky, E. The family of terpene synthases in plants: A mid-size family of genes for specialized metabolism that is highly diversified throughout the kingdom. *Plant J.* **2011**, *66*, 212–229. [CrossRef]
244. Yu, Z.X.; Li, J.X.; Yang, C.Q.; Hu, W.L.; Wang, L.J.; Chen, X.Y. The jasmonate-responsive AP2/ERF transcription factors AaERF1 and AaERF2 positively regulate artemisinin biosynthesis in *Artemisia annua* L. *Mol. Plant* **2012**, *5*, 353–365. [CrossRef]
245. Ji, Y.P.; Xiao, J.W.; Shen, Y.L.; Ma, D.M.; Li, Z.Q.; Pu, G.B.; Li, X.; Huang, L.L.; Liu, B.Y.; Ye, H.C.; et al. Cloning and characterization of AabHLH1, a bHLH transcription factor that positively regulates artemisinin biosynthesis in *Artemisia annua*. *Plant Cell Physiol.* **2014**, *55*, 1592–1604. [CrossRef]
246. Mertens, J.; Pollier, J.; Vanden Bossche, R.; Lopez-Vidriero, I.; Franco-Zorrilla, J.M.; Goossens, A. The bHLH transcription factors TSAR1 and TSAR2 regulate triterpene saponin biosynthesis in *Medicago truncatula*. *Plant Physiol.* **2016**, *170*, 194–210. [CrossRef]
247. Sun, W.J.; Zhan, J.Y.; Zheng, T.R.; Sun, R.; Wang, T.; Tang, Z.Z.; Bu, T.L.; Li, C.L.; Wu, Q.; Chen, H. The jasmonate-responsive transcription factor CbWRKY24 regulates terpenoid biosynthetic genes to promote saponin biosynthesis in *Conyza blinii* H. Lév. *J. Genet.* **2018**, *97*, 1379–1388. [CrossRef]
248. Hernández-Aparicio, F.; Lisón, P.; Rodrigo, I.; Bellés, J.M.; López-Gresa, M.P. Signaling in the tomato immunity against *Fusarium oxysporum*. *Molecules* **2021**, *26*, 1818. [CrossRef]
249. Zhang, H.; Chen, J.L.; Lin, J.H.; Lin, J.T.; Wu, Z.Z. Odorant-binding proteins and chemosensory proteins potentially involved in host plant recognition in the Asian citrus psyllid, *Diaphorina citri*. *Pest Manag. Sci.* **2020**, *76*, 2609–2618. [CrossRef]
250. Clark, J.; Bennett, T. Cracking the enigma: Understanding strigolactone signalling in the rhizosphere. *J. Exp. Bot.* **2024**, *75*, 1159–1173. [CrossRef]
251. Jia, K.P.; Baz, L.; Al-Babili, S. From carotenoids to strigolactones. *J. Exp. Bot.* **2018**, *69*, 2189–2204. [CrossRef]
252. Lahari, Z.; van Boerdonk, S.; Omoboye, O.O.; Reichelt, M.; Höfte, M.; Gershenzon, J.; Gheysen, G.; Ullah, C. Strigolactone deficiency induces jasmonate, sugar and flavonoid phytoalexin accumulation enhancing rice defense against the blast fungus *Pyricularia oryzae*. *New Phytol.* **2024**, *241*, 827–844. [CrossRef]
253. Han, Z.; Zhang, C.; Zhang, H.; Duan, Y.; Zou, Z.; Zhou, L.; Zhu, X.; Fang, W.; Ma, Y. CsMYB transcription factors participate in jasmonic acid signal transduction in response to cold stress in tea plant (*Camellia sinensis*). *Plants* **2022**, *11*, 2869. [CrossRef]
254. Qi, P.F.; Zhang, Y.Z.; Liu, C.H.; Chen, Q.; Guo, Z.R.; Wang, Y.; Xu, B.J.; Jiang, Y.F.; Zheng, T.; Gong, X.; et al. Functional analysis of FgNahG clarifies the contribution of salicylic acid to wheat (*Triticum aestivum*) resistance against fusarium head blight. *Toxins* **2019**, *11*, 59. [CrossRef]
255. Du, L.Q.; Ali, G.S.; Simons, K.A.; Hou, J.G.; Yang, T.B.; Reddy, A.S.N.; Poovaiah, B.W. Ca^{2+}/calmodulin regulates salicylic-acid-mediated plant immunity. *Nature* **2009**, *457*, 1154–1158. [CrossRef]
256. Sarkar, S.; Das, A.; Khandagale, P.; Maiti, I.B.; Chattopadhyay, S.; Dey, N. Interaction of Arabidopsis TGA3 and WRKY53 transcription factors on Cestrum yellow leaf curling virus (CmYLCV) promoter mediates salicylic acid-dependent gene expression in plants. *Planta* **2018**, *247*, 181–199. [CrossRef]
257. Song, G.C.; Ryu, C.M. Evidence for volatile memory in plants: Boosting defence priming through the recurrent application of plant volatiles. *Mol. Cells* **2018**, *41*, 724–732.
258. Ye, M.; Liu, M.M.; Erb, M.; Glauser, G.; Zhang, J.; Li, X.W.; Sun, X.L. Indole primes defence signalling and increases herbivore resistance in tea plants. *Plant Cell Environ.* **2021**, *44*, 1165–1177. [CrossRef]
259. Ye, M.; Glauser, G.; Lou, Y.G.; Erb, M.; Hu, L.F. Molecular dissection of early defense signaling underlying volatile-mediated defense regulation and herbivore resistance in rice. *Plant Cell* **2019**, *31*, 687–698. [CrossRef]
260. Stenzel, I.; Hause, B.; Miersch, O.; Kurz, T.; Maucher, H.; Weichert, H.; Ziegler, J.; Feussner, I.; Wasternack, C. Jasmonate biosynthesis and the allene oxide cyclase family of *Arabidopsis thaliana*. *Plant Mol. Biol.* **2003**, *51*, 895–911. [CrossRef]
261. Matsui, K. Green leaf volatiles: Hydroperoxide lyase pathway of oxylipin metabolism. *Curr. Opin. Plant Biol.* **2006**, *9*, 274–280. [CrossRef]
262. Werner, A.K.; Witte, C.P. The biochemistry of nitrogen mobilization: Purine ring catabolism. *Trends Plant Sci.* **2011**, *16*, 381–387. [CrossRef]
263. Watanabe, S.; Matsumoto, M.; Hakomori, Y.; Takagi, H.; Shimada, H.; Sakamoto, A. The purine metabolite allantoin enhances abiotic stress tolerance through synergistic activation of abscisic acid metabolism. *Plant Cell Environ.* **2014**, *37*, 1022–1036. [CrossRef]
264. Takagi, H.; Ishiga, Y.; Watanabe, S.; Konishi, T.; Egusa, M.; Akiyoshi, N.; Matsuura, T.; Mori, I.C.; Hirayama, T.; Kaminaka, H.; et al. Allantoin, a stress-related purine metabolite, can activate jasmonate signaling in a MYC2-regulated and abscisic acid-dependent manner. *J. Exp. Bot.* **2016**, *67*, 2519–2532. [CrossRef]
265. Liao, P.; Ray, S.; Boachon, B.; Lynch, J.H.; Deshpande, A.; McAdam, S.; Morgan, J.A.; Dudareva, N. Cuticle thickness affects dynamics of volatile emission from petunia flowers. *Nat. Chem. Biol.* **2021**, *17*, 138–145. [CrossRef]
266. Wang, L.; Erb, M. Volatile uptake, transport, perception, and signaling shape a plant's nose? *Essays Biochem.* **2022**, *66*, 695–702.
267. Chang, C.; Kwok, S.F.; Bleecker, A.B.; Meyerowitz, E.M. Arabidopsis ethylene-response gene etr1—Similarity of product to 2-component regulators. *Science* **1993**, *262*, 539–544. [CrossRef]

268. Nagashima, A.; Higaki, T.; Koeduka, T.; Ishigami, K.; Hosokawa, S.; Watanabe, H.; Matsui, K.; Hasezawa, S.; Touhara, K. Transcriptional regulators involved in responses to volatile organic compounds in plants. *J. Biol. Chem.* **2019**, *294*, 2256–2266. [CrossRef]
269. Adebesin, F.; Widhalm, J.R.; Boachon, B.; Lefèvre, F.; Pierman, B.; Lynch, J.H.; Alam, I.; Junqueira, B.; Benke, R.; Ray, S.; et al. Emission of volatile organic compounds from petunia flowers is facilitated by an ABC transporter. *Science* **2017**, *356*, 1386–1388. [CrossRef]
270. Meents, A.K.; Mithöfer, A. Plant-plant communication: Is there a role for volatile damage-associated molecular patterns? *Front. Plant Sci.* **2020**, *11*, 583275. [CrossRef]
271. Asai, N.; Nishioka, T.; Takabayashi, J.; Furuichi, T. Plant volatiles regulate the activities of Ca^{2+}-permeable channels and promote cytoplasmic calcium transients in Arabidopsis leaf cells. *Plant Signal. Behav.* **2009**, *4*, 294–300. [CrossRef]
272. Yin, W.C.; Wang, X.H.; Liu, H.; Wang, Y.; van Nocker, S.; Tu, M.X.; Fang, J.H.; Guo, J.Q.; Li, Z.; Wang, X.P. Overexpression of VqWRKY31 enhances powdery mildew resistance in grapevine by promoting salicylic acid signaling and specific metabolite synthesis. *Hortic. Res.* **2022**, *9*, uhab064. [CrossRef]
273. Boudsocq, M.; Willmann, M.R.; McCormack, M.; Lee, H.; Shan, L.B.; He, P.; Bush, J.; Cheng, S.H.; Sheen, J. Differential innate immune signalling via Ca^{2+} sensor protein kinases. *Nature* **2010**, *464*, 418–422. [CrossRef]
274. Manna, M.; Rengasamy, B.; Sinha, A.K. Revisiting the role of MAPK signalling pathway in plants and its manipulation for crop improvement. *Plant Cell Environ.* **2023**, *46*, 2277–2295. [CrossRef]
275. Dombrowski, J.E.; Kronmiller, B.A.; Hollenbeck, V.G.; Rhodes, A.C.; Henning, J.A.; Martin, R.C. Transcriptome analysis of the model grass Lolium temulentum exposed to green leaf volatiles. *BMC Plant Biol.* **2019**, *19*, 222. [CrossRef]
276. Zhang, M.M.; Su, J.B.; Zhang, Y.; Xu, J.; Zhang, S.Q. Conveying endogenous and exogenous signals: MAPK cascades in plant growth and defense. *Curr. Opin. Plant Biol.* **2018**, *45*, 1–10. [CrossRef]
277. Kandoth, P.K.; Ranf, S.; Pancholi, S.S.; Jayanty, S.; Walla, M.D.; Miller, W.; Howe, G.A.; Lincoln, D.E.; Stratmann, J.W. Tomato MAPKs LeMPK1, LeMPK2, and LeMPK3 function in the systemin-mediated defense response against herbivorous insects. *Proc. Natl. Acad. Sci. USA* **2007**, *104*, 12205–12210. [CrossRef]
278. Song, N.; Wu, J.S. Synergistic induction of phytoalexins in *Nicotiana attenuata* by jasmonate and ethylene signaling mediated by NaWRKY70. *J. Exp. Bot.* **2024**, *75*, 1063–1080. [CrossRef]
279. Ding, M.L.; Xie, Y.F.; Zhang, Y.H.; Cai, X.N.; Zhang, B.; Ma, P.D.; Dong, J.E. Salicylic acid regulates phenolic acid biosynthesis via SmNPR1-SmTGA2/SmNPR4 modules in *Salvia miltiorrhiza*. *J. Exp. Bot.* **2023**, *74*, 5736–5751. [CrossRef]
280. Jiang, W.M.; Fu, X.Q.; Pan, Q.F.; Tang, Y.L.; Shen, Q.; Lv, Z.Y.; Yan, T.X.; Shi, P.; Li, L.; Zhang, L.D.; et al. Overexpression of AaWRKY1 leads to an enhanced content of artemisinin in *Artemisia annua*. *Biomed. Res. Int.* **2016**, *971*, 7314971.

plants

MDPI

Article

Isolation and Identification of Allelopathic Substances from *Forsythia suspensa* Leaves, and Their Metabolism and Activity

Hisashi Kato-Noguchi [1,*], Yuga Takahashi [1], Shunya Tojo [2] and Toshiaki Teruya [3]

[1] Department of Applied Biological Science, Faculty of Agriculture, Kagawa University, Miki 761-0795, Kagawa, Japan; s20a068@kagawa-u.ac.jp
[2] Graduate School of Engineering and Science, University of the Ryukyus, 1 Senbaru, Nishihara 903-0213, Okinawa, Japan; k23803@eve.u-ryukyu.ac.jp
[3] Faculty of Education, University of the Ryukyus, 1 Senbaru, Nishihara 903-0213, Okinawa, Japan; t-teruya@edu.u-ryukyu.ac.jp
* Correspondence: kato.hisashi@kagawa-u.ac.jp

Abstract: The fruit of *Forsythia suspensa* (Thunb.) Vahl has been used in traditional Chinese medicine as "*Forsythiae fructus*". The species is also grown in parks and gardens, and on streets and building lots, as an ornamental plant, but it requires pruning. In this study, the allelopathic activity and allelopathic substances in the leaves of pruned branches of *F. suspensa* were investigated to determine any potential application. The leaf extracts of *F. suspensa* showed growth inhibitory activity against three weed species; *Echinochloa crus-galli*, *Lolium multiflorum*, and *Vulpia myuros.* Two allelopathic substances in the extracts were isolated through the bioassay-guided purification process, and identified as (-)-matairesinol and (-)-arctigenin. (-)-Matairesinol and (-)-arctigenin, which showed significant growth inhibitory activity at concentrations greater than 0.3 mM in vitro. The inhibitory activity of (-)-arctigenin was greater than that of (-)-matairesinol. However, both compounds were more active than (+)-pinolesinol which is their precursor in the biosynthetic pathway. The investigation suggests that *F. suspensa* leaves are allelopathic, and (-)-matairesinol and (-)-arctigenin may contribute to the growth inhibitory activities. Therefore, the leaves of the pruned branches can be applied as a weed management strategy in some agricultural practices such as using the leaf extracts in a foliar spray and the leaves in a soil mixture, thereby reducing the dependency on synthetic herbicides in the crop cultivation and contributing to developing eco-friendly agriculture.

Keywords: arctigenin; eco-friendly agriculture; lignan; matairesinol; soil additive; pinolesinol; pruned waste

1. Introduction

Forsythia suspensa (Thunb.) Vahl belongs to the Oleaceae family. The genus of *Forsythia* consists of about 10 species, which are native to eastern Asia except one species (*Forsythia europaea*) that is native to southern Europe. *F. suspensa* is native to China, is a deciduous shrub with 1 m–3 m high. The opposite leaves are usually simple, but occasionally three-foliolate. Leaf blades are ovate-oblong, 6 cm–10 cm long and 1.5 cm–5 cm wide with a rounded to cuneate base and an acute apex. The petioles are 0.8 cm–1.5 cm long. The species bears flowers at leaf axils with a 5 mm–6 mm long pedicel. The corollas are yellow and divided into four lobes, and the lobes are 1.2 cm–2 cm long and 0.6 cm–1 cm wide. Pistils are 5 mm–7 mm long with 3 mm–5 mm long stamens. The fruits (capsules) are ovoid to ellipsoid, 1.2 cm–2.5 cm long and 0.6 cm–1.2 cm wide with 0.7 cm–1.5 cm long stalks [1–3]. Its chromosome number is 2n = 48 [4,5]. The species grows in grassy and thicket areas in valleys and on slopes at 300 m–2200 m above sea level in the warm temperature zones [1,6]. Some of the *F. suspensa* population showed cold- and/or drought tolerance [6].

The dried fruits of the species have been used in traditional Chinese medicine as "*Forsythiae fructus*". *Forsythiae fructus* has a bitter flavor, and has been traditionally applied for thousands of years as an effective remedy for the treatments of carbuncle, scrofula, gall tumor, malignant ulcer, gonorrhea, nephritis, and erysipelas. This medicinal herb is effective in clearing heat channel and releasing the heart of stomach and spleen according to traditional medicine [1,7,8]. Recent investigations have shown that *Forsythiae fructus* has several pharmacological activities such as anti-inflammatory, antibacterial, anti-allergic, anti-cancer, anti-HIV, antioxidant, and neuroprotective activities in vivo and in vitro [1,7–9]. More than 300 compounds in many chemical classes such as lignans, phenylethanoid glycosides, flavonoids, terpenoids, steroids and alkaloids have been identified in *Forsythiae fructus*, and the pharmacological activity of some of these compounds has been demonstrated [1,7,9].

F. suspensa is also cultivated widely in East Asia, Europe and North America in parks and gardens, and on streets and building lots as an ornamental plant because of its beautiful flowers [10–13]. The species blooms in early spring before the leaves appear, and the whole tree looks to be golden, which is attractive to people as an ornamental plant species (Figure 1). However, the plant grows vigorously and its branches straggle pendulously. Therefore, pruning is necessary to maintain the species as an ornamental plant (Figure 2) [10–13]. Pruning produces a large amount of waste because the plant is fast-growing. Consequently, developing a beneficial application of the pruned waste to help reduce the environmental and economic concerns is necessary.

Figure 1. Leaves and flowers of *F. suspensa*.

Figure 2. Before and after the pruning of *F. suspensa*.

Many medicinal plants have been shown to possess relatively high allelopathic activity [14,15]. Allelopathy is the chemical interaction between donor plants and receiver plants through certain plant secondary metabolites defined as allelochemicals [16–19]. The allelochemicals can suppress the germination and growth of other plants including weed plant species [20,21]. Several allelopathic plant species are effective in weed management practices in agricultural fields [22–25]. Therefore, allelopathic plants can be useful as a weed management option to reduce the application of commercial herbicides. *F. suspensa* may also contain allelochemicals because it is used as a medicinal plant and contains many secondary metabolites, so its pruned waste may be effective in weed management practices. However, no information on the allelopathy and allelochemicals of *F. suspensa* is available. The objective of this research was to evaluate the allelopathic activity of *F. suspensa* leaves against three weed species; *Echinochloa crus-galli* (L.) P.Beauv., *Lolium multiflorum* Lam., and *Vulpia myuros* (L.) C.C.Gmel., and to isolate and characterize the allelochemicals in the leaves of the pruned branches to develop a beneficial application of the pruned waste.

2. Results

2.1. Allelopathic Activity of the F. suspensa Leaves

The leaf extracts of *F. suspensa* significantly inhibited the growth of the roots and coleoptiles of *E. crus-galli*, *L. multiflorum* and *V. myuros* at concentrations higher than 100 mg leaf extract per mL (Figure 3). The concentrations of the extracts causing 50% growth inhibition (IC_{50} values) of the extracts for the root growth were 3.8 mg, 39.4 mg and 12.4 mg leaf equivalent extract per mL for *E. crus-galli*, *L. multiflorum* and *V. myuros*, respectively, and the IC_{50} values for the coleoptiles were 43.9 mg, 46.1 mg and 18.1 mg leaf extract per mL for *E. crus-galli*, *L. multiflorum* and *V. myuros*, respectively. The extracts also inhibited the roots and hypocotyls of cress (*Lepidum sativum* L.) at concentrations greater than 100 mg and 30 mg leaf extract per mL for roots and hypocotyls, respectively (Figure 4). The IC_{50} values of the extracts against the cress roots and hypocotyls were 91.5 mg and 40.1 mg leaf equivalent extract per mL, respectively.

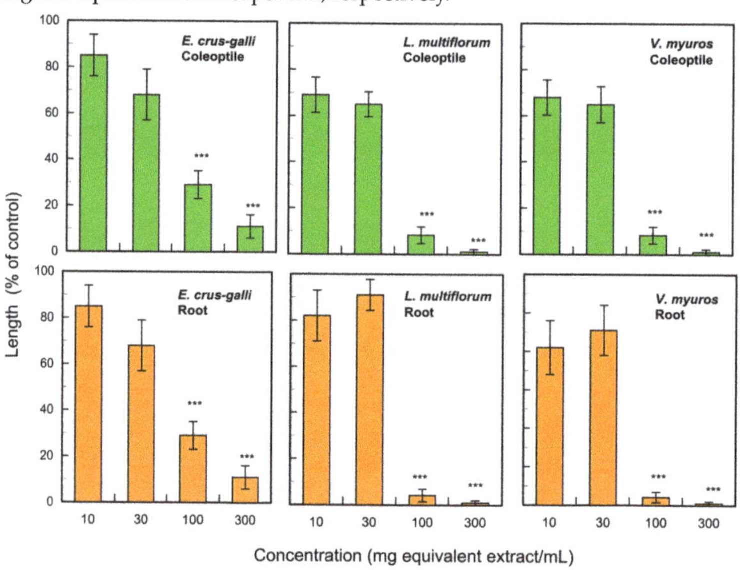

Figure 3. Effects of the leaf extracts of *F. suspensa* on the growth of the roots and coleoptiles of *E. crus-galli*, *L. multiflorum* and *V. myuros*. Concentration (mg leaf equivalent extract/mL) corresponds to the amount of the extracts obtained from 10, 30, 100 or 300 mg leaves per mL. Mean ± SE from 4 independent experiments with 10 seedlings for each determination are shown. Asterisks indicate significant differences between control and treatment: ***, $p < 0.001$.

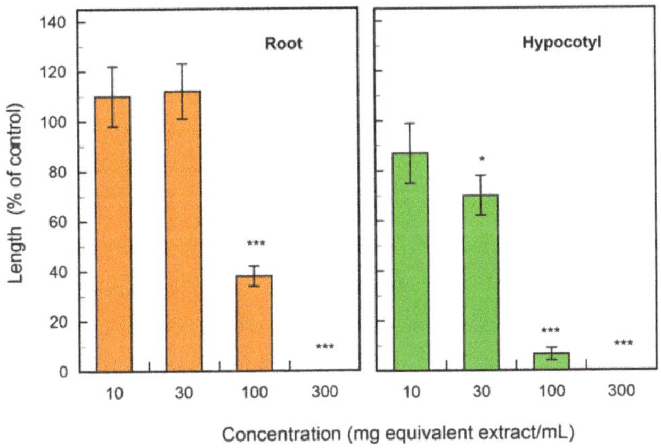

Figure 4. Effects of the *F. suspensa* leaf extracts on the growth of the roots and coleoptiles of cress. Other conditions were as described in Figure 3. Asterisks indicate significant differences between control and treatment: *, $p < 0.05$, ***, $p < 0.001$.

2.2. Purification and Identification of the Allelochemicals in the Leaves

The extract of the *F. suspensa* leaves was purified through the bioassay-guided purification process outlined in Figure 5. The ethyl acetate fraction was separated through silica gel chromatography, and the activity of all the separated fractions was determined using a cress bioassay. Allelopathic activity was found in fractions 5 and 9 (Figure 6). However, the fraction 9 significantly inhibited the growth of cress roots but not the hypocotyls. Fraction 5 inhibited the growth of the cress roots and hypocotyls to 35% and 41% of the control roots and hypocotyls, respectively. Therefore, the fraction 5 was further purified using a Sephadex LH-20 and ODS cartridge (Figure 5), and two compounds, 1 and 2, showing allelopathic activity were isolated using HPLC (Figure 7).

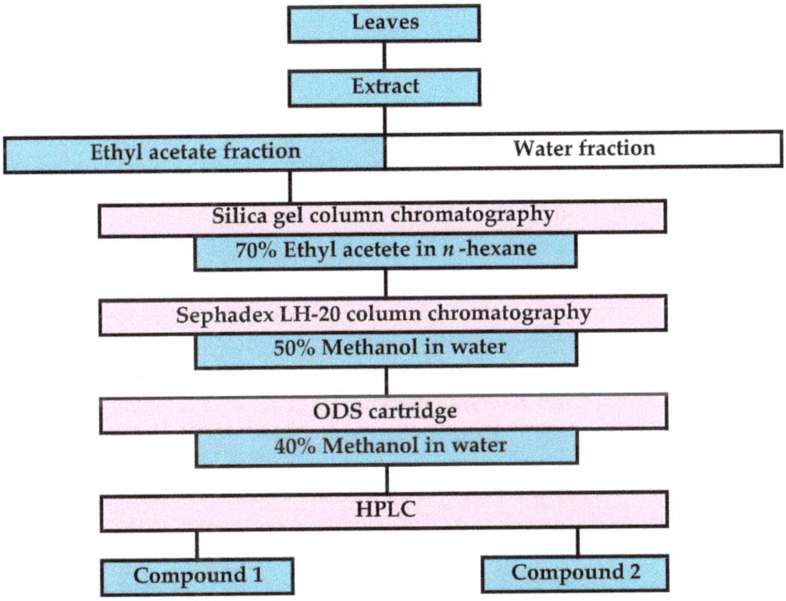

Figure 5. Purification process of compounds 1 and 2 from the leaf extract of *F. suspensa*.

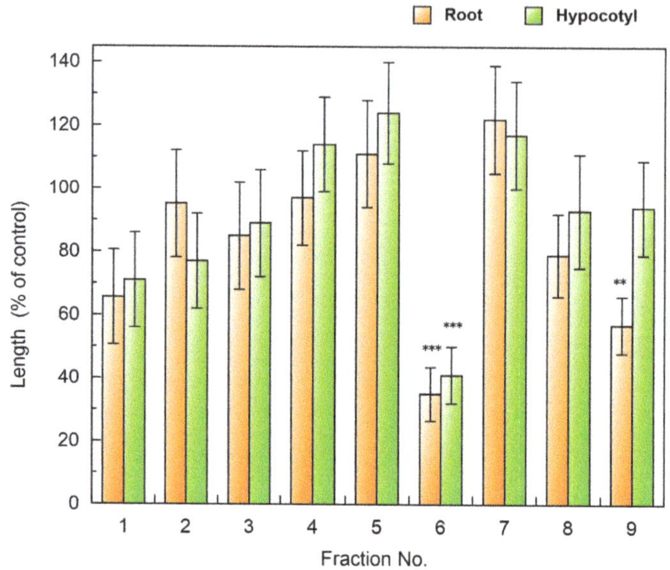

Figure 6. Effects of the fractions separated by a silica gel column chromatography on the root and hypocotyl growth of cress. Concentration of tested samples corresponded to the extract obtained from 0.3 g of *F. suspensa* per mL. Other conditions were as described in Figure 3. Asterisks indicate significant differences between control and treatment: **, $p < 0.01$, ***, $p < 0.001$.

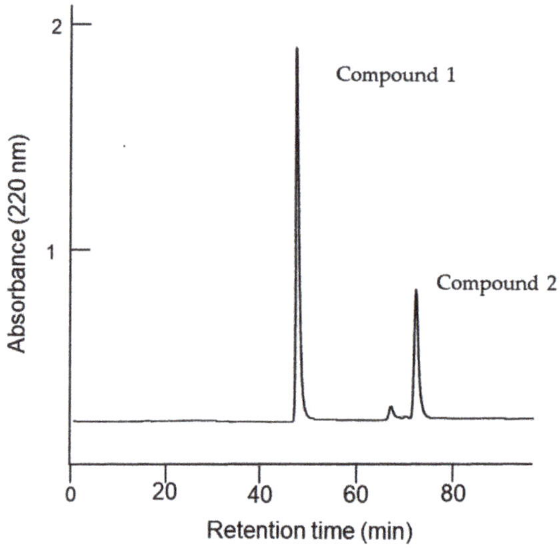

Figure 7. HPLC chromatogram and chemical structures of compounds 1 and 2. HPLC condition; column: reverse-phase, Inertsil ODS-3, 4.6 mm i.d. × 250 mm in length (YMC Ltd., Kyoto, Japan), solvent: 70% aqueous methanol (flow rate, 0.8 mL), detection: 220 nm.

The molecular formula of compound 1 was determined as $C_{20}H_{22}O_6$ by HRESI-MS. The ^{1}H NMR spectrum showed the presence of six aromatic proton signals at δ_H 6.81 (1H, d, J = 7.9), 6.80 (1H, d, J = 8.0), 6.61 (1H, d, J = 1.7), 6.60 (1H, dd, J = 7.9, 1.7), 6.51 (1H, dd, J = 8.0, 1.7) and 6.41 (1H, d, J = 1.7), two methyl proton signals at δ_H 3.82 (3H, s) and

3.81 (3H, s); and two methine proton and six methylene proton signals at δ_H 4.15 (1H, dd, J = 9.1, 7.4), 3.89 (1H, dd, J = 9.1, 7.3), 2.95 (1H, dd, J = 14.1, 5.3), 2.88 (1H, dd, J = 14.1, 7.0), 2.61 (1H, dd, J = 13.6, 6.7) and 2.58−2.42 (3H, m). The optical rotation was $[\alpha]^{27}{}_D = -38.6$ (c 0.07, acetone). Based on the spectrum data and published data in the literature [26,27], the chemical structure of compound 1 was characterized as (-)-matairesinol (Figure 8).

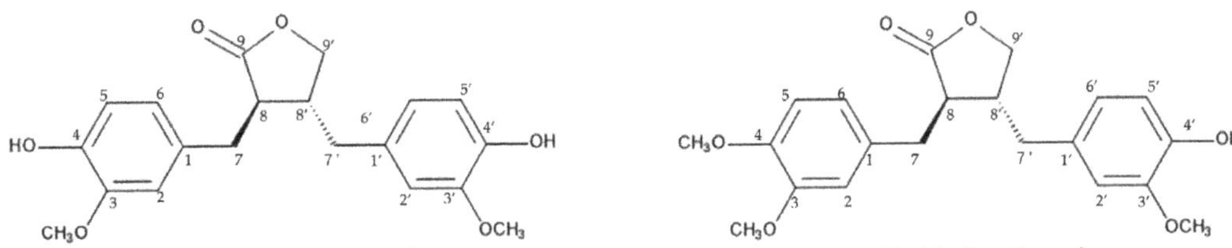

1: (-)-Matairesinol **2:** (-)-Arctigenin

Figure 8. Chemical structures of (-)-matairesinol and (-)-arctigenin.

The molecular formula of compound 2 was found to be $C_{21}H_{24}O_6$ by HRESI-MS. The ^{1}H NMR spectrum showed the presence of six aromatic proton signals at δ_H 6.82 (1H, d, J = 8.0), 6.75 (1H, d, J = 8.2), 6.64 (1H, d, J = 1.2), 6.61 (1H, dd, J = 8.0, 1.2), 6.55 (1H, dd, J = 8.2, 1.4) and 6.46 (1H, d, J = 1.4), three methyl proton signals at δ_H 3.85 (3H, s), 3.82 (3H, s) and 3.81 (3H, s); and two methine proton and six methylene proton signals at δ_H 4.14 (1H, dd, J = 8.9, 7.5), 3.89 (1H, dd, J = 8.9, 7.8), 2.95 (1H, dd, J = 14.1, 5.3), 2.90 (1H, dd, J = 14.1, 6.7), 2.64 (1H, dd, J = 13.4, 6.1) and 2.59−2.44 (3H, m). The optical rotation was $[\alpha]^{27}{}_D = -20.6$ (c 0.13, MeOH). Compound 2 was identified as (-)-arctigenin (Figure 8) through these spectrum data and published data in the literature [27].

2.3. Allelopathic Activity of the Isolated Compounds and Pinolesinol

The allelopathic activity of the isolated compounds, (-)-matairesinol and (-)-arctigenin, and their biosynthetic precursor, (+)-pinolesinol was determined. (-)-Matairesinol significantly inhibited the growth of the cress roots and hypocotyls at concentrations greater than 1 mM, and the growth of the *L. multiflorum* roots and coleoptiles at concentrations greater than 0.3 mM (Figure 9). The IC_{50} values of (-)-matairesinol against the cress roots and hypocotyls were 1.1 mM and 2.1 mM, respectively, and those for the *L. multiflorum* roots and coleoptiles were 0.93 mM and 2.2 mM, respectively (Table 1). (-)-Arctigenin also showed the growth inhibitory activity against the cress roots and hypocotyls, and the *L. multiflorum* roots and coleoptiles at concentrations greater than 0.3 mM (Figure 10). The IC_{50} values of (-)-arctigenin against the cress roots and hypocotyls were 0.79 mM and 1.3 mM, respectively, and those for the *L. multiflorum* roots and coleoptiles were 0.85 mM and 1.3 mM, respectively (Table 1). (+)-Pinolesinol was active at concentrations greater than 1 mM against the cress roots and hypocotyls, and *L. multiflorum* roots and coleoptiles (Figure 11). The IC_{50} values of (+)-pinolesinol for the cress roots and hypocotyls were 2.1 mM and 2.3 mM, respectively, and those for the *L. multiflorum* roots and coleoptiles were 1.1 mM and 2.5 mM, respectively (Table 1).

Table 1. IC_{50} values (mM) of (-)-matairesinol, (-)-arctigenin and (+)-pinoresinol against the growth of the roots and hypocotyls/coleoptiles of cress and *L. multiflorum*. Different letters on in the columns (same organ) indicate significant differences at the $p < 0.05$ level.

	Cress		*L. multiflorum*	
Compound	**Root**	**Hypocotyl**	**Root**	**Coleoptile**
(-)-Matairesinol	1.1 b	2.1 b	0.93 b	2.2 b
(-)-Arctigenin	0.79 a	1.3 a	0.85 a	1.3 a
(+)-Pinoresinol	2.1 c	2.3 b	1.1 c	2.5 b

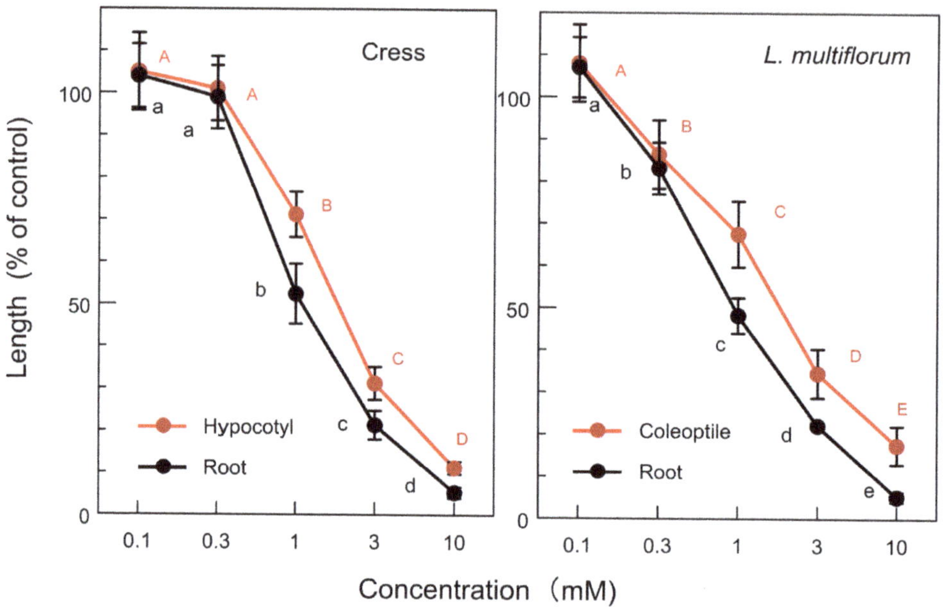

Figure 9. Effects of (-)-matairesinol on the growth of the roots and hypocotyls/coleoptiles of cress and *L. multiflorum*. Means ± SE from 4 independent experiments with 10 seedlings for each determination are shown. Different letters on the symbols in the same panels indicate significant differences at the $p < 0.05$ level.

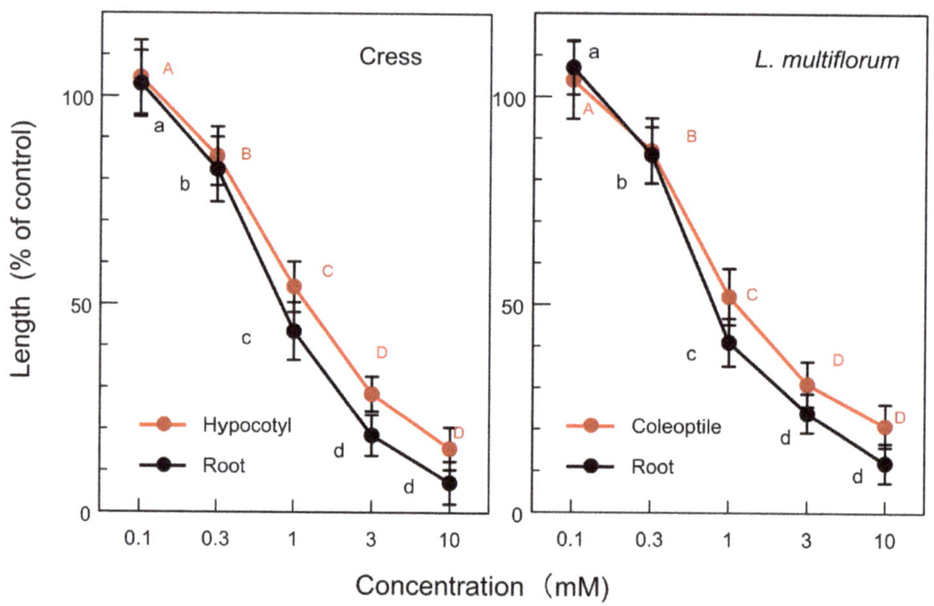

Figure 10. Effects of (-)-arctigenin on the growth of the roots and hypocotyls/coleoptiles of cress and *L. multiflorum*. Other conditions were as described in Figure 9.

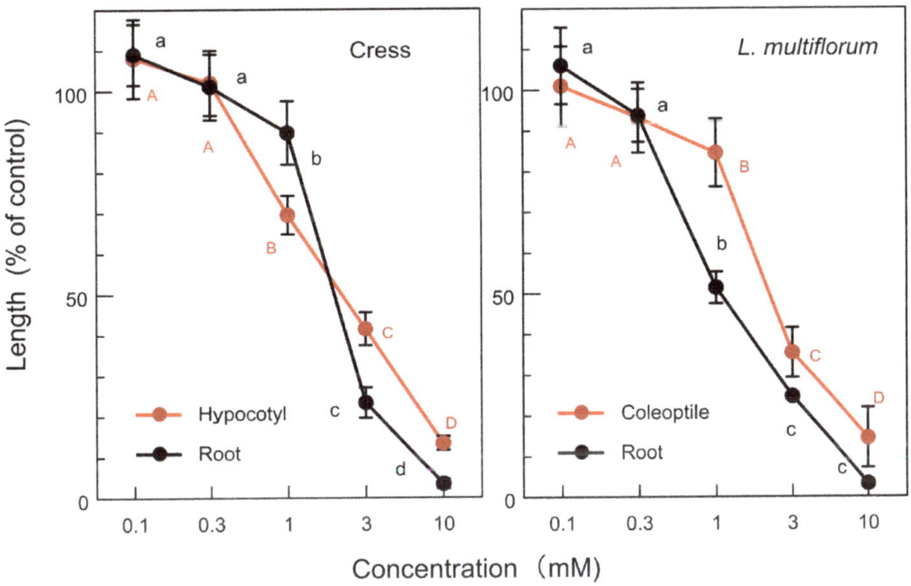

Figure 11. Effects of (+)-pinoresinol on the growth of the roots and hypocotyls/coleoptiles against of cress and *L. multiflorum*. Other conditions were as described in Figure 9.

3. Discussion

As an ornamental tree, *F. suspensa* needs adequate pruning treatments because it is fast-growing [11–13], and developing a beneficial application of the pruned waste to help alleviate environmental and economic concerns is necessary. Accordingly, we determined the allelopathic activity of the leaves of *F. suspensa*. The leaf extracts inhibited the growth of the roots and coleoptiles of three weed species (Figure 3): the *E. crus-galli* roots were most sensitive to the extracts, and *L. multiflorum* coleoptiles were the least sensitive of the three weed species. Based on the growth inhibitory activity of the leaf extracts of *F. suspensa*, the leaves may contain certain allelochemicals. Allelochemicals are synthesized and accumulated in some plant tissues including leaves, and released into the neighboring environment including the rhizosphere soil through volatilization, root exudation, and decomposition processes of the donor plant residues in the soil [20–23]. Deciduous tree plants can also release allelochemicals into their rhizosphere soil over many years through the decomposition process of their fallen leaves [28–30]. *F. suspensa* is a deciduous plants species [1–3]. The compounds liberated from the fallen leaves may accumulate in the soil, and inhibit the germination and growth of other nearby plant species [31–34]. Therefore, allelochemicals in *F. suspensa* leaves may also be released into the neighboring environments and/or the rhizosphere soil.

The *F. suspensa* leaf extract was then purified through the bioassay-guided purification process (Figure 5), and two active compounds 1 and 2 were isolated (Figure 7). During the purification process, the allelopathic activity of all fractions obtained after each chromatography step was evaluated using a cress bioassay, and the fraction showed the highest activity was subjected to the next chromatography process as described by [35]. The germination rate of *E. crus-galli*, *L. multiflorum*, and *V. myuros* was 20–40% at 48 h after sowing, and was not stable during the experiments, while the germination rate of cress was 90% at 48 h after the sowing and stable during the experiments. Therefore, selecting germinated seeds for the cress bioassay is not necessary because of its high and stable germination behavior. Comparing the IC_{50} values of the extracts of the *F. suspensa* leaves against cress with IC_{50} values of the extracts against the three weed species showed that the sensitivity of the cress roots was low. However, the sensitivity of the cress hypocotyls was not high and not low. Therefore, cress was chosen as a bioassay species because of the convenience, and the growth of cress was monitored during extract purification.

Compounds 1 and 2 were characterized as (-)-matairesinol and (-)-arctigenin, respectively, by the spectrum analysis of HRESI-MS, ^{1}H-NMR and optical rotation (Figure 8). (-)-Matairesinol and (-)-arctigenin are classified as lignans. Lignans are a structurally diverse group of plant natural products originating from phenylpropanoids, have important physiological functions in plants, and exhibit pharmacological activity [36]. *Forsythia* species accumulate a considerable amount of lignans such as (+)-pinoresinol, (-)-matairesinol, (-)-arctigenin and their respective glycosides in the entire plant body [37]. The biosynthetic pathway of these lignans in *Forsythia species* has been reported (Figure 12). (+)-Pinoresinol is synthesized by radical coupling of two coniferyl alcohol molecules [38–42], and then converted to (-)-secoisolariciresinol via (+)-lariciresinol by pinoresinol/lariciresinol reductase (PLR).

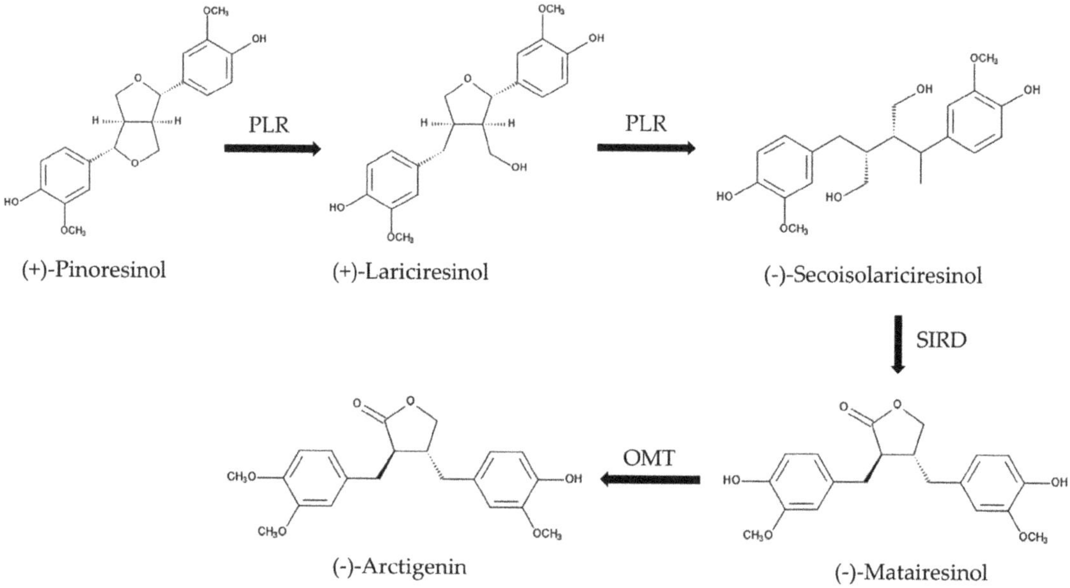

Figure 12. The biosynthetic pathway of (-)-matairesinol and (-)-arctigenin in the *Forsythia* species. PLR: pinoresinol/lariciresinol reductase, SIRD: secoisolariciresinol dehydrogenase, OMT: *O*-methyltransferases.

(-)-Secoisolariciresinol is metabolized to (-)-matairesinol by secoisolariciresinol dehydrogenase (SIRD), and (-)-matairesinol is metabolized to (-)-arctigenin by *O*-methyltransferases (OMT). The majority of those lignans are glycosylated at their hydroxyl groups by a superfamily of enzymes; the uridine diphosphate (UDP)-sugar dependent glycosyltransferases (UGT). UGT transfers a sugar moiety from an activated UDP-sugar to an accepting substrate [37–42]. (-)-Matairesinol, (-)-arctigenin and (+)-pinoresinol have been reported to show several pharmacological activities such as anti-tumor, anti-HIV, anti-angiogenic, and estrogenic activity [36,43,44]. (+)-Pinoresinol isolated from Oleaceae species has also shown the growth inhibitory activity against several plant species as an allelopathic agent [35,45]. However, the allelopathic activity of (-)-matairesinol and (-)-arctigenin has not yet been reported.

Cress was selected as for a bioassay species for the isolated compounds because its growth could be monitored during extract purification. *L. multiflorum* was also selected as a bioassay species because its sensitivity to the *F. suspensa* extract was not high among the three weed species, *E. crus-galli*, *L. multiflorum* and *V. myuros* (Figure 3). (-)-Matairesinol and (-)-arctigenin suppressed the growth of the cress roots and hypocotyls and *L. multiflorum* roots and coleoptiles (Figures 9 and 10). The active threshold of the growth inhibition of (-)-matairesinol was 1 mM and 0.3 mM for cress and *L. multiflorum*, respectively (Figure 9), and that of (-)-arctigenin was 0.3 mM for cress and *L. multiflorum* (Figure 10). The IC_{50} values of (-)-matairesinol and (-)-arctigenin against the growth of the hypocotyls/coleoptiles and roots of cress and *L. multiflorum* were 0.93 mM–2.2 mM and 0.79 mM 1.3 mM, respectively (Table 1). Therefore, the IC_{50} values and active threshold of the compounds show that the growth inhibitory activity of (-)-arctigenin was greater than that of (-)-matairesinol, which suggests that the methoxy group (CH_3O-) at the C-4 position of (-)-arctigenin may increase the growth inhibitory activity compared with the hydroxy group (HO-) at the C-4 position of (-)-matairesinol (Figure 8). The IC_{50} values of their precursor in the biosynthetic pathway (Figure 12), (+)-pinolesinol were 1.1 mM–2.5 mM (Table 1). The active threshold of the growth inhibition of (+)-pinolesinol was 1 mM for cress and *L. multiflorum* (Figure 11). Comparing those IC_{50} values and the active threshold revealed that the activities of (-)-arctigenin and (-)-matairesinol were greater than (+)-pinolesinol. In addition, we isolated 26 mg of (-)-matairesinol and 35 mg of (-)-arctigenin from 100 g of *F. suspensa* leaves, which indicates the leaves may contain a relatively large amount of these compounds. Because (+)-lariciresinol and (-)-secoisolariciresinol (Figure 12) were not isolated from the extracts of the *F. suspensa* leaves through the bioassay-guided purification process (Figure 5), the growth inhibitory activity of both compounds may not be high compared with that of (-)-matairesinol and (-)-arctigenin. There have also been no reports on the plant growth inhibitory activity of these compounds. In addition, modifying the structure to increase allelopathic activity is possible [46].

When the leaf extracts and their soaking water obtained from some plant species such as *Lantana camara* L., *Imperata cylindrica* (L.) Beauv., *Solidago canadensis* L, *Tithonia diversifolia* (Hemsl.) A. Gray, *Bidens pilosa* L., and *Leucaena leucocephala* (Lam.) de Wit. were applied as foliar spray and/or irrigation water, the emergence and growth of several weed species were significantly inhibited under field, greenhouse and laboratory conditions [47–55]. When the leaves of *L. camara*, *I. cylindrica*, and *L. leucocephala* were mixed with soil, the emergence and growth of several weed species were also suppressed under field and greenhouse conditions [56–60]. These observations suggest that the leaf extracts and the soaking water of these plant species, and their residues incorporated into soil showed allelopathic activity against the emergence and growth of the weed species, and they may contain allelochemicals. Some of these allelochemicals in the residues may be liberated into the soil during the decomposition of the leaves [16–23]. Therefore, the leaves of allelopathic plants can be applied for the weed management purposes in eco-friendly agriculture. In addition, the allelochemicals of several plant extracts and residues also disturb this mutualism of weed plant species with rhizobia and/or arbuscular mycorrhizal fungi [31,36, 60]. Rhizobium nodulation enhances the performance of legume plants through nitrogen supply [60,61]. Arbuscular mycorrhizal fungi enhance the performance of most territorial plants through increasing the absorption of nutrients, photosynthesis, and their defense functions [61–65]. The disturbance of this mutualism by allelochemicals may reduce the competitive ability of these weed species, and may relatively increase the competitive ability of crop plants.

The present investigation suggests that the leaves of *F. suspensa* exhibit allelopathic activity against the growth of *E. crus-galli*, *L. multiflorum* and *V. myuros*, and contain the allelopathic substances, (-)-matairesinol and (-)-arctigenin. When the leaves of *F. suspensa* were incorporated into the soil as a soil mixture, certain allelochemicals in the leaves may be liberated into the field soil during the decomposition, and the liberated allelochemicals may suppress the emergence and growth of some weed species. The leaf extracts and its soaking water can be applied as a foliar spray to inhibit the emergence and growth of some weed plant species. The fruit of *F. suspensa* has been used in traditional Chinese medicine [1,7–9], and the leaves from the pruned branches of *F. suspensa* may also be useful for weed management.

4. Materials and Methods

4.1. Plant Material

F. suspensa leaves were obtained at the Kagawa University campus in April 2023. Three weed species; *Echinochloa crus-galli* (L.) P.Beauv., *Lolium multiflorum* Lam., and *Vulpia myuros* (L.) C.C.Gmel., were used to evaluate allelopathic activity. Cress (*Lepidum sativum* L.) was used during the purification process of the allelochemicals as a bioassay plant because of its convenient handling procedure and stable germination rate.

4.2. Extraction and Determination of the Allelopathic Activity of *F. suspensa*

F. suspensa leaves (100 g fresh weight) were extracted by soaking in a mixture of water and methanol (20%: 80%, *v*/*v*, 1.5 L; Nacalai, Kyoto, Japan) for 48 h, and filtered using filter paper (No. 2; Advantec Toyo, Tokyo, Japan). The residue obtained from the first filtration was extracted again by soaking in methanol (1.5 L) for 48 h and filtering. The two filtrates were mixed and concentrated under reduced pressure at 40 °C.

The leaf extract was dissolved in methanol, and the dissolved extract was added to filter paper in a Petri dishes (2.8 cm i.d.). After the methanol in the Petri dishes was completely evaporated in a fume hood, 0.6 mL solution of Tween 20 (0.05% *w*/*v*, Nacalai, Kyoto, Japan) was applied to the filter paper. The seeds of *E. crus-galli*, *L. multiflorum* and *V. myuros* were germinated on another filter paper moistened with distilled water in the dark at 25 °C for 48 h. Ten germinated seeds of each species were then arranged on the filter paper in the Petri dishes. The length of the roots and coleoptiles of *E. crus-galli*, *L. multiflorum* and *V. myuros* was determined after the incubation of 48 h in the dark at 25 °C. The percentage length of the roots and coleoptiles of these weed species was calculated against the length of the roots and coleoptiles of the control plants. The control plants were treated using same procedure, with the only difference being not adding the leaf extracts to the filter paper in the Petri dishes. The allelopathic activity of the extracts was also determined with cress bioassay. The procedure of the cress bioassay was the same as for the weed species, but the seeds (not germinated) were arranged in the filter paper on the Petri dishes.

4.3. Separation of the *F. suspensa* Extract

The leaf extract of *F. suspensa* obtained as described in the above section was adjusted to pH 7.0 with a buffer of 1 M phosphate, and partitioned with ethyl acetate to separate the ethyl acetate and aqueous fractions (Figure 5). The ethyl acetate fraction was then applied to silica gel chromatography (silica gel 60, 70–230 mesh; Merck, Darmstadt, Germany). The allelopathic activity of all the separated fractions by the silica gel chromatography was determined using a cress bioassay as described above. The most active fraction 5 (Figure 6) was further subjected to a Sephadex LH-20 chromatography (Sigma-Aldrich, Burlington, VT, USA) and an ODS cartridge (YMC-Dispo SPE; YMC Ltd., Kyoto, Japan). Active compounds 1 and 2 were then isolated using a reverse-phase HPLC (Figure 7). The chemical structures of the compounds were determined by the spectrum analyses of HRESI-MS (Orbitrap Exploris 240, Thermo-Fisher K.K., Tokyo, Japan) and ^{1}H-NMR (500 MHz, $CDCl_3$; Bruker Avance III 500, Bruker Co., Yokohama, Japan), and optical rotation (P-1010 polarimeter, Jasco Co., Tokyo, Japan). All other reagents were purchased from Nacalai at a purity > 99.8.

4.4. Allelopathic Activity of the Isolated Compounds and Pinoresinol

(-)-Matairesinol and (-)-arctigenin were isolated from the *F. suspensa* leaves as described above. (+)-Pinoresinol was isolated from the leaves of *Osmanthus* × *fortunei* Carrière as described by [35]. A given quantity of the methanol solution of each compound was applied onto a filter paper in the Petri dish. After complete evaporation of the methanol solvent in the Petri dishes, 0.6 mL of a Tween 20 solution was added onto the filter paper. Then, 10 cress seeds and 10 germinated *L. multiflorum* seeds, were separately put on the filter paper and grown, and the allelopathic activity of the compounds was determined as described in the Section 4.2. The concentrations of the applied compounds were 0.1, 0.3, 1, 3 and 10 mM.

4.5. Statistical Analysis

The bioassays were conducted with four replications using a completely randomized design with 10 plants for each determination. Significant differences between control and treatment were analyzed using Welch's *t*-test (Figures 3, 4 and 6). The bioassay data for (-)-matairesinol, (-)-arctigenin and (+)-pinoresinol were subjected to a one-way analysis of ANOVA (SPSS, version 16.0, SPSS Inc., Chicago, IL, USA) and subsequent post-hoc analysis using Tukey's HSD test at the $p < 0.05$ level (Figures 9–11, Table 1). The IC_{50} values were obtained using GraphPad Prism 6.0 (Software of GraphPad, Inc., La Jolla, CA, USA).

5. Conclusions

The extracts of the *F. suspensa* leaves were allelopathic and significantly inhibited the growth of three weed species; *E. crus-galli*, *L. multiflorum* and *V. myuros* in vitro. (-)-Matairesinol and (-)-arctigenin were isolated and identified from the extracts as allelopathic agents. Both compounds inhibited the growth of *L. multiflorum* and cress in vitro, and the activity of (-)-arctigenin was greater than that of (-)-matairesinol. Moreover, (-)-arctigenin and (-)-matairesinol were more active than (+)-pinolesinol, which is their precursor in the biosynthetic pathway. The leaves may contain a large amount of (-)-matairesinol and (-)-arctigenin. When the leaves of *F. suspensa* were incorporated into the soil as a soil mixture, certain allelochemicals in the leaves may be liberated into the field soil during the decomposition process. The liberated allelochemicals may suppress the emergence and growth of some weed species. The leaf extracts and its soaking water can be applied as a foliar spray to inhibit the emergence and growth of some weed plant species. Therefore, because of its allelopathic activity, the leaves and extracts of *F. suspensa* are potentially useful for weed manegement options to reduce the application of synthetic herbicides in crop production and to develop to eco-friendly agriculture. However, evaluating the inhibitory activity of the leaf extracts of *F. suspensa* as a foliar spray, and the leaves in a soil mixture is necessary under field conditions.

Author Contributions: Conceptualization, H.K.-N.; methodology, Y.T. and S.T.; software Y.T.: validation, T.T. and H.K.-N.; formal analysis, T.T. and H.K.-N.; investigation, Y.T. and S.T.; data curation, T.T. and H.K.-N.; writing H.K.-N.; supervision, H.K.-N. All authors have read and agreed to the published version of the manuscript.

Funding: This research received no external funding.

Institutional Review Board Statement: Not applicable.

Informed Consent Statement: Not applicable.

Data Availability Statement: No supporting data in this study.

Acknowledgments: We thank Dennis Murphy, The United Graduate School of Agricultural Sciences, Ehime University, Japan, for editing the English of the manuscript.

Conflicts of Interest: The authors declare no conflicts of interest.

References

1. Wang, Z.; Xia, Q.; Liu, X.; Liu, W.; Huang, W.; Mei, X.; Luo, J.; Shan, M.; Lin, R.; Zou, D.; et al. Phytochemistry, pharmacology, quality control and future research of *Forsythia suspensa* (Thunb.) Vahl: A review. *J. Ethnopharmacol.* **2018**, *210*, 318–339. [CrossRef] [PubMed]
2. Makino, T. Observation on the flora of Japan. *Bot. Mag.* **1912**, *28*, 105–107. [CrossRef] [PubMed]
3. Leshcheniuk, O.; Koniakin, S. Morphological and anatomical parameters of *Forsythia suspensa* (Thunb.) Vahl leaves in the urban environment. *Plant Introd.* **2022**, *95/96*, 75–84. [CrossRef]
4. Iwatsubo, Y.; Sato, K.; Miura, N.; Ohno, M.; Fujino, H.; Tatsuo, Y.; Sugiyama, R.; Kurosaki, F. Chromosome numbers of the genus *Forsythia* (Oleaceae). *J. Phytogeogr. Taxon.* **2008**, *56*, 95–97.

5. Shen, J.S.; Xie, Y.J.; Pan, H.T.; Zhang, Q.X. Karyotype analysis of twelve *Forsythia* species and cultivars. *J. Plant Genet. Resour.* **2015**, *16*, 178–184.
6. Yang, J.; Miao, C.Y.; Mao, R.L.; Li, Y. Landscape population genomics of Forsythia (*Forsythia suspensa*) reveal that ecological habitats determine the adaptive evolution of species. *Front. Plant Sci.* **2017**, *8*, 481. [CrossRef] [PubMed]
7. Peng, H.; Lu, P.; Liu, Y.; Liu, C.; Gao, Y.; Nwafor, E.O.; Zhang, Y.; Liu, Z. Research progress on chemical composition, pharmacological effects of *Forsythia suspensa* (Thunb.) Vahl and predictive analysis on Q-marker. *TMR Mod. Herb. Med.* **2020**, *3*, 86–112. [CrossRef]
8. Bailly, C. Forsythosides as essential components of Forsythia-based traditional Chinese medicines used to treat inflammatory diseases and COVID-19. *World J. Traditi. Chin. Med.* **2022**, *8*, 1–20. [CrossRef]
9. Gong, L.; Wang, C.; Zhou, H.; Ma, C.; Zhang, Y.; Peng, C.; Li, Y. A review of pharmacological and pharmacokinetic properties of Forsythiaside A. *Pharmacolo. Res.* **2021**, *169*, 105690. [CrossRef]
10. Plants of the World. Royal Botanical Gardens—Kew, *Mimosa pigra* L. Available online: https://powo.science.kew.org/taxon/urn:lsid:ipni.org:names:608900-1 (accessed on 17 December 2023).
11. Gardener's Path. *Forsythia suspensa*. Available online: https://gardenerspath.com/plants/ornamentals/grow-weeping-forsythia/ (accessed on 17 December 2023).
12. Hortipedia. *Forsythia suspensa*. Available online: https://en.hortipedia.com/Forsythia_suspensa (accessed on 17 December 2023).
13. Plant for a Furture. *Forsythia suspensa*. Available online: https://pfaf.org/User/Plant.aspx?LatinName=Forsythia+suspensa (accessed on 17 December 2023).
14. Fujii, Y.; Shibuya, T.; Yasuda, T. Survey of Japanese weed and crops for the detection of water-extractable allelopathic chemicals using Richards' function fitted to lettuce germination test. *Weed Res. Jpn.* **1990**, *35*, 362–370.
15. Fujii, Y.; Parvez, S.S.; Parvez, M.M.; Ohmae, Y.; Iida, O. Screening of 239 medicinal plant species for allelopathic activity using the sandwich method. *Weed Biol. Manag.* **2003**, *3*, 233–241. [CrossRef]
16. Rice, E.L. *Allelopathy*, 2nd ed.; Academic Press: Orlando, FL, USA, 1984; pp. 1–422.
17. Bais, H.P.; Weir, T.L.; Perry, L.G.; Gilroy, S.; Vivanco, J.M. The role of root exudates in rhizosphere interactions with plants and other organisms. *Annu. Rev. Plant Biol.* **2006**, *57*, 233–266. [CrossRef]
18. Bonanomi, G.; Sicurezza, M.G.; Caporaso, S.; Esposito, A.; Mazzoleni, S. Phytotoxicity dynamics of decaying plant materials. *New Phytol.* **2006**, *169*, 571–578. [CrossRef] [PubMed]
19. Kato-Noguchi, H. Defensive molecules momilactones A and B: Function, biosynthesis, induction and occurrence. *Toxins* **2023**, *15*, 241. [CrossRef] [PubMed]
20. Belz, R.G. Allelopathy in crop/weed interactions—An update. *Pest Manag. Sci.* **2007**, *63*, 308–326. [CrossRef] [PubMed]
21. Macías, F.A.; Molinillo, J.M.G.; Varela, R.M.; Galindo, J.G.G. Allelopathy—A natural alternative for weed control. *Pest Manag. Sci.* **2007**, *63*, 327–348. [CrossRef] [PubMed]
22. Narwal, S.S. Allelopathy in weed management. In *Allelopathy Update. Basic and Applied Aspects*; Narwal, S.S., Ed.; Science Publishers Inc.: Enfield, NH, USA, 1999; Volume 2, pp. 203–254.
23. Duke, S.O.; Dayan, F.E.; Romagni, J.G.; Rimando, A.M. Natural products as sources of herbicide, current status and future trends. *Weed Res.* **2000**, *40*, 99–111. [CrossRef]
24. Kato-Noguchi, H. Phytotoxic substances involved in teak allelopathy and agroforestry. *Appl. Sci.* **2021**, *11*, 3314. [CrossRef]
25. Kato-Noguchi, H.; Kato, M. Evolution of the secondary metabolites in invasive plant species *Chromolaena odorata* for the defense and allelopathic functions. *Plants* **2023**, *12*, 521. [CrossRef]
26. Niemeyer, H.B.; Honig, D.; Lange-Böhmer, A.; Jacobs, E.; Kulling, S.E.; Metzler, M. Oxidative metabolites of the mammalian lignans enterodiol and enterolactone in rat bile and urine. *J. Agric. Food Chem.* **2000**, *48*, 2910–2919. [CrossRef]
27. Fischer, J.; Reynolds, A.J.; Sharp, L.A.; Sherburn, M.S. Radical carboxyarylation approach to lignans. Total synthesis of (−)-arctigenin, (−)-matairesinol, and related natural products. *Org. Lett.* **2004**, *6*, 1345–1348. [CrossRef] [PubMed]
28. Mallik, A.U. Allelopathy and competition in coniferous forests. *Environ. For. Sci.* **1998**, *54*, 309–315.
29. Kato-Noguchi, H.; Kimura, F.; Ohno, O.; Suenaga, K. Involvement of allelopathy in inhibition of understory growth in red pine forests. *J. Plant Physiol.* **2017**, *218*, 66–73. [CrossRef]
30. Matuda, Y.; Iwasaki, A.; Suenaga, K.; Kato-Noguchi, H. Allelopathy and allelopathic substances of fossil tree species *Metasequoia glyptostroboides*. *Agronomy* **2022**, *12*, 83. [CrossRef]
31. Kato-Noguchi, H.; Kurniadie, D. Allelopathy and allelochemicals of *Leucaena leucocephala* as an invasive plant species. *Plants* **2022**, *11*, 1672. [CrossRef] [PubMed]
32. Kato-Noguchi, H.; Kato, M. Allelopathy and allelochemicals of *Solidago canadensis* L. and *S. altissima* L. for their naturalization. *Plants* **2022**, *11*, 3235. [CrossRef] [PubMed]
33. Kato-Noguchi, H. Involvement of allelopathy in the invasive potential of *Tithonia diversifolia*. *Plants* **2020**, *9*, 766. [CrossRef]
34. Kato-Noguchi, H. Allelopathy and allelochemicals of *Imperata cylindrica* as an invasive plant species. *Plants* **2022**, *11*, 2551. [CrossRef]
35. Kato-Noguchi, H.; Hamada, Y.; Kojima, M.; Kumagai, S.; Iwasaki, A.; Suenaga, K. Allelopathic substances of *Osmanthus* spp. for developing sustainable agriculture. *Plants* **2023**, *12*, 376. [CrossRef]

36. Lewis, N.G.; Davin, L.B. Lignans: Biosynthesis and function. In *Comprehensive Natural Products Chemistry, I*; Barton, S.D., Nakanishi, K., Meth-Cohn, O., Eds.; Elsevier: Amsterdam, The Netherlands, 1999; pp. 639–712.
37. Ono, E.; Kim, H.J.; Murata, J.; Morimoto, K.; Okazawa, A.; Kobayashi, A.; Umezawa, T.; Satake, H. Molecular and functional characterization of novel furofuran-class lignan glucosyltransferases from *Forsythia*. *Plant Botechnol.* **2010**, *27*, 317–324. [CrossRef]
38. Schmitt, J.; Petersen, M. Pinoresinol and matairesinol accumulation in a *Forsythia* × intermedia cell suspension culture. *Plant Cell Tissue Organ Cult.* **2002**, *68*, 91–98. [CrossRef]
39. Ward, R.S. Recent advances in the chemistry of lignans. In *Studies in Natural Products Chemistry*; Attaur, R., Ed.; Elsevier: Amsterdam, The Netherlands, 2000; pp. 739–798.
40. Davin, L.B.; Lewis, N.G. A historical perspective on lignan biosynthesis: Monolignol, allylphenol and hydroxycinnamic acid coupling and downstream metabolism. *Phytochem. Rev.* **2003**, *2*, 257–288. [CrossRef]
41. Ferrer, J.L.; Austin, M.B.; Stewart, C., Jr.; Noel, J.P. Structure and function of enzymes involved in the biosynthesis of phenylpropanoids. *Plant Physiol. Biochem.* **2008**, *46*, 356–370. [CrossRef] [PubMed]
42. Patyra, A.; Kołtun-Jasion, M.; Jakubiak, O.; Kiss, A.K. Extraction techniques and analytical methods for isolation and characterization of lignans. *Plants* **2022**, *11*, 2323. [CrossRef] [PubMed]
43. Yokoyama, T.; Okano, M.; Noshita, T.; Funayama, S.; Ohtsuki, K. Characterization of (−)-matairesinol as a potent inhibitor of casein kinase I in vitro. *Biol. Pharm. Bull.* **2003**, *26*, 371–374. [CrossRef] [PubMed]
44. Lee, B.; Kim, K.H.; Jung, H.J.; Kwon, H.J. Matairesinol inhibits angiogenesis via suppression of mitochondrial reactive oxygen species. *Biochem. Biophys. Res. Commun.* **2012**, *421*, 76–80. [CrossRef] [PubMed]
45. Fuentes-Gandaraa, F.; Torresb, A.; Fernández-Poncec, M.T.; Casasc, L.; Mantellc, C.; Varela, R.; de la Ossa-Fernándezc, E.J.M.; Macías, F.A. Selective fractionation and isolation of allelopathic compounds from *Helianthus annuus* L. leaves by means of high-pressure techniques. *J. Supercriti. Fluids* **2019**, *143*, 32–41. [CrossRef]
46. Zorrilla, J.G.; Cárdenas, D.M.; Rial, C.; Molinillo, J.M.; Varela, R.M.; Masi, M.; Macías, F.A. Bioprospection of phytotoxic plant-derived eudesmanolides and guaianolides for the control of *Amaranthus viridis*, *Echinochloa crus-galli*, and *Lolium perenne* weeds. *J. Agric. Food Chem.* **2024**, *72*, 1797–1810. [CrossRef]
47. Heděnec, P.; Novotný, D.; Ust'ak, S.; Honzík, R.; Kovářová, M.; Šimáčková, H.; Frouz, J. Allelopathic effect of new introduced biofuel crops on the soil biota: A comparative study. *Eur. J. Soil Biol.* **2014**, *63*, 14–20. [CrossRef]
48. Inderjit, S.; Dakshini, K.M.M. Investigations on some aspects of chemical ecology of cogongrass, *Imperata cylindrica* (L.) Beauv. *J. Chem. Ecol.* **1991**, *17*, 343–352. [CrossRef]
49. Ishak, M.S.; Sahid, I. Allelopathic effects of the aqueous extract of the leaf and seed of *Leucaena leucocephala* on three selected weed species. *AIP Conf. Proc.* **2014**, *1614*, 659–664.
50. Oke, S.O.; Awowoyin, A.V.; Osein, S.R.; Adediwura, E.L. Effects of aqueous shoot extract of *Tithonia diversifolia* on the growth of seedlings of *Monodora tenuifolia* (Benth.), *Dialium guineense* (Willd.) and *Hildegardia barteri* (Mast.) Kosterm. *Not. Sci. Biol.* **2011**, *3*, 64–70. [CrossRef]
51. Zheng, H.; Wei, N.; Wang, L.; He, P. Effects of *Lantana camara* leaf extract on the activity of superoxide dismutase and accumulation of H_2O_2 in water hyacinth leaf. *J. Plant Physiol. Mol. Biol.* **2006**, *32*, 189–194.
52. Poonpaiboonpipat, T.; Poolkum, S. Utilization of *Bidens pilosa* var. radiata (Sch. Bip.) Sherff integrated with water irrigation for paddy weed control and rice yield production. *Weed Biol. Manag.* **2019**, *19*, 31–38. [CrossRef]
53. Kato-Noguchi, H.; Kurniadie, D. Allelopathy of *Lantana camara* as an invasive plant. *Plants* **2021**, *10*, 1028. [CrossRef] [PubMed]
54. Kato-Noguchi, H. Invasive mechanisms of one of the world's worst alien plant species *Mimosa pigra* and its management. *Plants* **2023**, *12*, 1960. [CrossRef] [PubMed]
55. Kato-Noguchi, H. The impact and invasive mechanisms of *Pueraria montana* var. lobata, one of the world's worst alien species. *Plants* **2023**, *12*, 3066. [CrossRef]
56. Tominaga, T.; Watanabe, O. Weed growth suppression by cogongrass (*Imperata cylindrica*) leaves. *Weed Sci. Tech.* **1997**, *42*, 289–293. [CrossRef]
57. Koger, C.H.; Bryson, C.T. Effect of cogongrass (*Imperata cylindrica*) extracts on germination and seedling growth of selected grass and broadleaf species. *Weed Technol.* **2004**, *18*, 236–242. [CrossRef]
58. Chou, C.H.; Kuo, Y.L. Allelopathic research of subtropical vegetation in Taiwan. *J. Chem. Ecol.* **1986**, *12*, 1431–1448. [CrossRef]
59. Wang, R.; Kang, X.; Quan, G.; Zhang, J. Influence of *Lantana camara* on soil II. Effects of *Lantana camara* leaf litter on plants and soil properties. *Allelopath. J.* **2015**, *35*, 207–216.
60. Kato-Noguchi, H. Allelopathy of knotweeds as invasive plants. *Plants* **2022**, *11*, 3. [CrossRef] [PubMed]
61. Tsyganova, A.V.; Brewin, N.J.; Tsyganov, V.E. Structure and development of the legume-rhizobial symbiotic interface in infection threads. *Cells* **2021**, *10*, 1050. [CrossRef]
62. Mathesius, U. Are legumes different? Origins and consequences of evolving nitrogen fixing symbioses. *J. Plant Physiol.* **2022**, *276*, 153765. [CrossRef] [PubMed]
63. Smith, S.E.; Read, D.J. *Mycorrhizal Symbiosis*, 3rd ed.; Academic Press: London, UK, 2008; pp. 1–815.

64. Diagne, N.; Ngom, M.; Djighaly, P.I.; Fall, D.; Hocher, V.; Svistoonoff, S. Roles of arbuscular mycorrhizal fungi on plant growth and performance: Importance in biotic and abiotic stressed regulation. *Diversity* **2020**, *12*, 370. [CrossRef]
65. Tang, H.; Hassan, M.U.; Feng, L.; Nawaz, M.; Shah, A.N.; Qari, S.H.; Liu, Y.; Miao, J. The critical role of arbuscular mycorrhizal fungi to improve drought tolerance and nitrogen use efficiency in crops. *Front. Plant Sci.* **2022**, *13*, 919166. [CrossRef]

plants

MDPI

Article

Effects of Seven Plant Essential Oils on the Growth, Development and Feeding Behavior of the Wingless *Aphis gossypii* Glover

Xinhang Wang [1], Ying Zhang [2], Haibin Yuan [1,*] and Yanhui Lu [2,3,*]

1 Department of Plant Protection, Jilin Agricultural University, Changchun 130118, China; wangxinhang0302@163.com
2 State Key Laboratory for Biology of Plant Diseases and Insect Pests, Institute of Plant Protection, Chinese Academy of Agricultural Sciences, Beijing 100193, China; yingzhang6565@163.com
3 Western Agricultural Research Center, Chinese Academy of Agricultural Sciences, Changji 831100, China
* Correspondence: yuanhaibin@jlau.edu.cn (H.Y.); luyanhui@caas.cn (Y.L.)

Abstract: Cotton aphid *Aphis gossypii* Glover damages plants such as cotton directly by feeding on leaves and indirectly by transmitting viruses and excreting honeydew, which interferes with photosynthesis. The control of *A. gossypii* is still dominated by the frequent use of insecticides, which leads to a gradual increase in pesticide resistance in *A. gossypii*. Research is therefore needed on non-pesticide controls. In this study, seven plant essential oils (EOs) of *Ocimum sanctum* L., *Ocimum basilicum* L., *Ocimum gratissimum* L., *Mentha piperita* L., *Mentha arvensis* L., *Tagetes erecta* L., and *Lavandula angustifolia* Mill. were examined as potential controls for *A. gossypii*. We used life tables and electrical penetration graphs (EPG) to explore the effects of these EOs on the growth, development, and feeding behavior of *A. gossypii*, followed by a study of effects of the EOs on honeydew secretion by *A. gossypii* as a measure of their antifeedant activity. We found that the EOs of *O. sanctum*, *M. piperita*, *M. arvensis* and *T. erecta* significantly extended the pre-adult developmental period. Also, adult longevity, number of oviposition days, and total fecundity of *A. gossypii* treated with the EOs of *M. arvensis* or *T. erecta* were all significantly reduced. *Aphids* treated with the EOs of *O. sanctum*, *M. piperita*, or *L. angustifolia* showed significant reductions in their net reproductive rate (R_0), intrinsic rate of increase (r_m), and finite rate of increase (λ), and significant increases in mean generation time (T). In terms of their effects on the feeding behavior of *A. gossypii*, all seven EOs significantly reduced the total duration of phloem feeding (E2 waves), the number of phloem-feeding bouts, and the proportion of time spent in secretion of saliva into phloem sieve elements (E1 waves) and phloem feeding (E2). The total duration and number of E1 waves (saliva secretion) were significantly reduced by the EOs of *O. sanctum*, *O. gratissimum*, and *M. arvensis*. For C waves (probing in non-vascular tissues), the total duration spent in this behavior was significantly increased by the EOs of *O. gratissimum*, *M. piperita*, and *L. angustifolia*, but the number of such probing events was increased only by *L. angustifolia* EO. The EOs of *O. basilicum*, *M. arvensis*, and *T. erecta* significantly increased the total duration of ingestion of xylem sap (G waves), while the total time of mechanical difficulty in stylet penetration (F waves) was increased by *M. arvensis*. The total duration and number of the non-probing events (Np waves) were significantly increased by EOs of *O. sanctum* and *O. basilicum*. After treatment with all seven of these EOs, the area covered by honeydew was significantly reduced compared with the control. Studies have analyzed that EOs of *O. sanctum*, *M. piperita*, and *T. erecta* were most effective, followed by the EOs of *M. arvensis* and *L. angustifolia*, and finally the EOs of *O. basilicum* and *O. gratissimum*. In the present study, the EOs of *O. sanctum*, *M. piperita*, and *T. erecta* were found to have potential for the development as antifeedants of *A. gossypii*, and these data provide a basis for future research on non-pesticide chemical control of *A. gossypii*.

Keywords: essential oil; *Aphis gossypii* Glover; antifeedant activity; growth and development

1. Introduction

Cotton aphid *Aphis gossypii* Glover (Hemiptera: Aphididae) has a wide range of hosts and is harmful to cotton, potato, and melons, among others [1]. Its damage includes direct feeding injury to foliage and stems and mission transmission of plant viruses [2], including cotton leafroll dwarf virus [3,4]. Wingless aphids develop faster than winged aphids. Wingless aphids have adaptations that maximize fecundity, making them more difficult to control [5]. Insecticides are still the main method for control of *A. gossypii* in most agricultural settings. However, past frequent use of insecticides has resulted in *A. gossypii* populations resistant to all the major insecticide groups [6]. Nauen et al. showed that field populations of *A. gossypii* and *Myzus persicae* Sulzer in several European countries were highly resistant to pirimicarb and oxydemeton-methyl [7]. Patima et al. demonstrated that *A. gossypii* populations in cotton fields in several regions of Xinjiang, China, have developed very high levels of resistance to deltamethrin, cypermethrin, and omethoate [8]. Insecticide use may also harm populations of important natural enemies of pest insect [9], contributing in this case to resurgence of aphid populations. Therefore, research and development of non-chemical pesticides for cotton aphid control are needed.

Plant essential oils (EOs) are volatile, naturally occurring, complex compounds found in aromatic plants, where they act as secondary metabolites. EOs intended for commercial uses can be collected by distillation, mechanical pressure, or extraction [10]. EOs have potential as alternatives to commercial synthetic pesticides for use in green agriculture. Karamanoli demonstrated that EOs of spearmint and peppermint in soil declined about 90% within 30 days [11]. Because of such rapid degradation of EOs in nature, they do not accumulate in the environment and thus leave no polluting residues. Some botanical insecticides such as pyrethrum and neem are already well established as commercial pest control products [12], and are recommended for control of insects such as aphids, mirid bugs and whiteflies on field crops, as well as pests of glasshouse crops, house plants, and stored products [13]. Today, more than 100 commercial neem formulations are used worldwide [14]. Insecticides from the oils of citrus peel, containing limonene, have recently had moderate commercial success in North America and Europe [15]. Following neem oil, orange oil was the most frequently used botanical insecticide in California in 2016, with 7.0 tons of active ingredient being applied [16]. Additionally, the EOs of *Chenopodium ambrosioides* L. are used for insect pest management [17], with an average of 8000 kg sold per year in California, United States [15].

Plant EOs have great potential in the control of aphids. A number of articles have been published both at home and abroad reporting the fumigant and contact activities of EOs in the Lamiaceae and Asteraceae families against aphids. The EOs of *Mentha piperita* L., *Mentha pulegium* L., and *Ocimum basilicum* L. showed excellent efficacy in both contact and fumigation tests against *Lipaphis pseudobrassicae* Davis, *M. persicae*, and *A. gossypii* [18]. *Lavandula angustifolia* Bubani oils had fumigant activity against *Acyrthosiphon pisum* Harris [19]. *Mentha longifolia* L. oils had fumigant and contact activities against *Aphis craccivora* Koch, and affected the development, survival, and reproduction of *A. craccivora* [20]. *Artemisia dracunculus* L. had high fumigant toxicity to *A. gossypii* [21]. The EOs *Santolina chamaecyparissus* L. and *Tagetes patula* L. showed contact activity against *Myzus persicae* Sulzer and *Rhopalosiphum padi* L., while sublethal concentrations of the two essential oils reduced aphid fecundity [22].

At present, the EOs of Lamiaceae and Asteraceae have been reported to have antifeedant activity against a variety of aphids. Abualfia et al. discovered that the EOs of *Salvia officinalis* L., *Lavandula spica* L. and *Mentha spica* L. caused different degrees of interference with aphid feeding [23]. Zhang et al. found that the ethanol extract of *Xanthium strumarium* L. had stronger antifeedant activity against aphids feeding on *Lycium barbarum* L., with a plant rejection rate of more than 82% [24]. Zhou demonstrated that *Flaveria bidentis* L. oil had different levels of antifeedant activity on *Lipaphis erysimi* Kaltenbach and *Rhopalosiphum maidis* Fitch [25]. However, research on the antifeedant activity of plant EOs on aphids is relatively scarce. We sought to determine whether the EOs from Lamiaceae and Asteraceae plants exhibit antifeedant effects on cotton aphids, which will provide a theoretical basis for the development of subsequent antifeedant.

In this study, we examined the effects of EOs from seven plants (*Ocimum sanctum* L., *Ocimum basilicum* L., *Ocimum gratissimum* L., *Mentha piperita* L., *Mentha arvensis* L., *Tagetes erecta* L., and *Lavandula angustifolia* Mill.) on the growth, development and fecundity of *A. gossypii*. In addition, we analyzed the feeding behaviors of *A. gossypii* using the EPG technique. Lastly, we measured honeydew excretion by aphids feeding on cotton leaves treated with EOs of seven plant species, using honeydew excretion as a measure of aphid feeding.

2. Results

2.1. Effect of EOs on Cotton Aphid Growth, Development, and Fecundity

The developmental durations of all nymphal instars of *A. gossypii*, except the third instar, were affected by exposure to plant EOs (Table 1). Duration of the first instar was significantly prolonged by three plants: *O. basilicum* ($p = 0.002$), *M. arvensis* ($p = 0.005$), and *T. erecta* ($p = 0.004$). Similarly, the stage duration of second instars was affected by three different plant EOs: *O. sanctum* ($p = 0.004$), *M. piperita* ($p < 0.001$), and *L. angustifolia* ($p = 0.012$). For fourth instars, stage duration was affected by two plants—*O. basilicum* ($p = 0.002$) and *O. gratissimum* ($p = 0.024$)—while for all instars as a group, four plants' EOs had significant effects: *O. sanctum* ($p = 0.003$), *M. piperita* ($p = 0.001$), *M. arvensis* ($p = 0.003$), and *T. erecta* ($p = 0.042$).

Table 1. Effects of essential oils of seven plants on the growth and development of *Aphis gossypii*.

EOs	Developmental Times (d) of Life Stages				Pre-Adults
	1st	2nd	3rd	4th	
Control	1.18 ± 0.06 cd	1.09 ± 0.04 d	1.09 ± 0.05 ab	1.44 ± 0.07 ab	4.80 ± 0.09 d
O. sanctum	1.29 ± 0.08 bc	1.39 ± 0.07 ab	1.12 ± 0.05 ab	1.40 ± 0.07 abc	5.19 ± 0.07 ab
O. basilicum	1.49 ± 0.07 a	1.18 ± 0.06 bcd	1.09 ± 0.04 ab	1.14 ± 0.05 d	4.89 ± 0.09 cd
O. gratissimum	1.36 ± 0.07 abc	1.20 ± 0.05 bcd	1.10 ± 0.04 ab	1.22 ± 0.06 cd	4.88 ± 0.11 cd
M. piperita	1.34 ± 0.07 abc	1.51 ± 0.14 a	1.21 ± 0.06 a	1.38 ± 0.07 abc	5.23 ± 0.06 a
M. arvensis	1.46 ± 0.08 ab	1.16 ± 0.05 cd	1.18 ± 0.05 a	1.38 ± 0.07 abc	5.18 ± 0.11 ab
T. erecta	1.53 ± 0.08 ab	1.15 ± 0.05 d	1.10 ± 0.04 ab	1.33 ± 0.07 bc	5.10 ± 0.10 abc
L. angustifolia	1.06 ± 0.03 d	1.35 ± 0.07 abc	1.00 ± 0.00 b	1.54 ± 0.07 a	4.96 ± 0.05 bcd
	$F_{7, 385} = 4.760$	$F_{7, 375} = 3.954$	$F_{7, 373} = 1.885$	$F_{7, 371} = 3.332$	$F_{7, 371} = 3.339$
	$p < 0.001$	$p < 0.001$	$p = 0.071$	$p = 0.002$	$p = 0.002$

Note: Values are means ± SE. Data in the same column followed by different lowercase letters differed significantly ($p < 0.05$).

The longevity and fecundity of *A. gossypii* treated with different plant EOs were generally shortened by plant EOs (Table 2). The EOs of all seven plants significantly reduced adult longevity ($p < 0.001$). Aphid fecundity (as number of offspring produced) was reduced by the EOs of five plants: *O. sanctum* ($p < 0.001$), *M. piperita* ($p < 0.001$), *M. arvensis* ($p = 0.011$), *T. erecta* ($p = 0.001$), and *L. angustifolia* ($p < 0.001$). The number of reproductive days for adult aphids was shortened by the EOs of four plants: *O. basilicum* ($p = 0.001$), *O. gratissimum* ($p < 0.001$), *M. arvensis* ($p < 0.001$), and *T. erecta* ($p < 0.001$).

Compared with the control group, *A. gossypii* treated with EOs of three plants (*O. sanctum, M. piperita,* and *L. angustifolia*) showed significant reductions in net reproductive rate (R_0), intrinsic rate of increase (r_m), and finite rate of increase (λ), and significant increases in mean generation time (T) (Table 3).

2.2. Effect of Plant EOs on Aphid Feeding Behavior

Plant EOs showed effects on various non-phloem feeding behaviors in *A. gossypii* (Table 4). Except for the total duration of probing inside cells (Pd waves), which should not be affected, the seven EOs not only lengthened the total duration of probing in non-vascular tissues (C waves), ingestion of xylem sap (G waves), and non-probing activity (Np waves).

EO exposure also increased the number of probing events in non-vascular tissues (C waves) and of non-probing events (Np waves). For the behavior of probing in non-vascular tissues (C waves), the total duration was significantly increased by *O. gratissimum* ($p = 0.019$), *M. piperita* ($p = 0.013$), and *L. angustifolia* ($p = 0.004$). For *L. angustifolia*, the number of such events, but not their total duration, was increased ($p = 0.002$). The EOs of *O. basilicum* ($p = 0.021$), *M. arvensis* ($p = 0.001$), and *T. erecta* ($p < 0.001$) significantly increased the total duration of xylem feeding (G waves). The EO of *M. arvensis* also increased the total duration of mechanical difficulty in stylet penetration (F waves) ($p = 0.011$). The total duration of the non-probing behavior (Np waves) was significantly increased by *O. sanctum* ($p = 0.002$), *O. basilicum* ($p = 0.002$), and *O. gratissimum* ($p = 0.025$). The number of non-probing behavior events (Np waves) was significantly increased by *O. sanctum* ($p = 0.001$) and *O. basilicum* ($p = 0.010$).

Table 2. Effects of essential oils of seven plants on the longevity and fecundity of *Aphis gossypii*.

EOs	Adult Longevity (d)	Reproductive Days (d)	Fecundity
Control	29.67 ± 0.67 a	14.04 ± 0.32 a	53.18 ± 0.99 a
O. sanctum	26.25 ± 1.12 b	14.21 ± 0.78 a	41.69 ± 2.44 cde
O. basilicum	26.36 ± 0.89 b	11.30 ± 0.36 b	49.55 ± 1.88 ab
O. gratissimum	26.76 ± 0.75 b	11.08 ± 0.34 b	49.24 ± 1.27 ab
M. piperita	25.32 ± 1.16 bc	13.47 ± 0.82 a	39.53 ± 2.6 de
M. arvensis	25.80 ± 0.98 b	10.81 ± 0.36 b	46.20 ± 1.95 bc
T. erecta	22.92 ± 0.85 c	11.25 ± 0.44 b	43.79 ± 1.46 cd
L. angustifolia	24.25 ± 0.99 bc	13.55 ± 0.66 a	36.63 ± 2.12 e
	$F_{7,371} = 4.27$ $p < 0.001$	$F_{7,367} = 7.219$ $p < 0.001$	$F_{7,371} = 8.390$ $p < 0.001$

Note: Values are means ± SE. Data in the same column followed by different lowercase letters differed significantly ($p < 0.05$).

Table 3. Effects of essential oils of seven plants on the life table parameters of *Aphis gossypii*.

EOs	R_0	T (d)	r_m	λ
Control	47.86 ± 2.41 ab	8.99 ± 0.16 c	0.4304 ± 0.0099 a	1.5379 ± 0.0152 a
O. sanctum	40.02 ± 2.59 cde	10.24 ± 0.15 a	0.3605 ± 0.0072 c	1.4340 ± 0.0104 c
O. basilicum	43.60 ± 2.80 abcd	9.27 ± 0.13 bc	0.4071 ± 0.0096 ab	1.5024 ± 0.0143 ab
O. gratissimum	48.26 ± 1.57 a	9.23 ± 0.15 bc	0.4199 ± 0.0073 ab	1.5218 ± 0.0111 ab
M. piperita	37.16 ± 2.77 de	10.04 ± 0.13 a	0.3601 ± 0.0080 c	1.4335 ± 0.0115 c
M. arvensis	46.20 ± 1.92 abc	9.55 ± 0.16 b	0.4014 ± 0.0073 b	1.4940 ± 0.0109 b
T. erecta	42.02 ± 1.84 bcd	9.25 ± 0.16 bc	0.4043 ± 0.0078 b	1.4982 ± 0.0117 ab
L. angustifolia	35.14 ± 2.26 e	9.51 ± 0.13 b	0.3742 ± 0.0068 c	1.4538 ± 0.0099 c

Note: Values are means ± SE. Data in the same column followed by different lowercase letters differed significantly ($p < 0.05$).

Plant EOs also, importantly, affected the phloem-feeding behaviors of *A. gossypii* (Table 5). All seven EOs significantly reduced the total duration of phloem ingestion (E2 waves) ($p = 0.001$), the number of phloem-feeding events ($p = 0.045$), and the proportion of time occupied by saliva secretion and phloem feeding combined (E1 and E2 waves) ($p < 0.001$). However, for the other behaviors, each essential oil had different effects. The total duration of salivation (E1 waves) was significantly reduced by *O. sanctum* ($p = 0.002$), *O. basilicum* ($p = 0.025$), *O. gratissimum* ($p = 0.017$), *M. arvensis* ($p = 0.001$), and *L. angustifolia* ($p = 0.004$). Also, the number of salivation events (E1 events) was significantly reduced by *O. sanctum* ($p = 0.025$), *O. gratissimum* ($p = 0.029$), and *M. arvensis* ($p = 0.004$). The EOs of *O. sanctum* ($p = 0.026$) and *T. erecta* ($p = 0.022$) also significantly prolonged the time to the first salivation event (E1), but had no effect on the time to the first phloem feeding (E2).

Table 4. Effects of essential oils of seven plants on non-phloem phase feeding behaviors of *Aphis gossypii*, as total duration of behaviors (min) for C, F, G, Np, and Pd behaviors, as well as total number of behavioral events: C-events and Np-events.

Treatment (*n*)	C	F	G	Np	Pd	C Events	Np Events
Control (10)	205.97 ± 18.369 b	6.97 ± 5.416 b	13.00 ± 7.365 d	0.18 ± 0.175 c	13.44 ± 1.443 ab	186.00 ± 17.578 bc	0.20 ± 0.200 c
O. sanctum (11)	240.90 ± 16.574 ab	20.69 ± 10.040 ab	58.22 ± 14.374 bcd	34.89 ± 8.350 a	12.58 ± 1.459 b	186.91 ± 20.866 bc	9.64 ± 2.180 a
O. basilicum (11)	237.38 ± 25.908 ab	4.52 ± 3.902 b	73.22 ± 21.420 abc	33.46 ± 13.111 a	14.54 ± 1.823 ab	230.82 ± 27.775 ab	7.64 ± 2.636 ab
O. gratissimum (11)	281.85 ± 26.821 a	9.55 ± 7.996 b	55.79 ± 14.204 bcd	24.55 ± 7.615 ab	13.19 ± 1.400 ab	172.82 ± 12.156 c	4.00 ± 1.279 bc
M. piperita (10)	288.70 ± 26.326 a	14.21 ± 11.502 ab	39.03 ± 11.134 cd	6.32 ± 4.161 bc	15.35 ± 0.971 ab	217.40 ± 12.199 bc	1.40 ± 0.884 c
M. arvensis (12)	247.52 ± 23.507 ab	40.96 ± 14.030 a	100.21 ± 24.490 ab	18.06 ± 5.136 abc	14.62 ± 1.635 ab	210.00 ± 22.016 bc	5.58 ± 1.960 abc
T. erecta (12)	240.65 ± 10.570 ab	1.77 ± 1.766 b	112.25 ± 21.620 a	14.67 ± 6.030 abc	13.90 ± 1.071 ab	207.25 ± 12.283 bc	3.50 ± 1.222 bc
L. angustifolia (12)	297.43 ± 21.609 a	23.36 ± 10.341 ab	21.26 ± 13.919 d	10.67 ± 6.310 bc	17.08 ± 1.337 a	272.08 ± 19.719 a	3.75 ± 2.937 bc
	$F_{7, 81} = 2.004$ $p = 0.064$	$F_{7, 81} = 2.093$ $p = 0.054$	$F_{7, 81} = 4.114$ $p = 0.001$	$F_{7, 81} = 2.783$ $p = 0.012$	$F_{7, 81} = 1.018$ $p = 0.425$	$F_{7, 81} = 2.786$ $p = 0.012$	$F_{7, 81} = 2.445$ $p = 0.025$

Note: Values are means ± SE. Data in the same column followed by different lowercase letters differed significantly ($p < 0.05$). C = Probing in non-vascular tissues, F = mechanical difficulty in stylet penetration, G = ingestion of xylem sap, Np = non-probing, Pd = probing inside cells.

Table 5. Effects of essential oils of seven plants on phloem phase of *Aphis gossypii*, as total duration of behaviors (min) for E1, and E2 behaviors, as well as total number of behavioral events: as E1-events and E2-events.

Treatment (*n*)	Time to First E1 (min)	E1	E1 Events	Time to First E2 (min)	E2	E2 Events	Percentage of E1 + E2	Total Number of Single E1
Control (10)	54.58 ± 7.751 b	77.76 ± 15.591 a	15.30 ± 1.633 a	112.76 ± 42.052 ab	162.69 ± 24.491 a	9.10 ± 1.560 a	0.50 ± 0.039 a	2.60 ± 0.562 b
O. sanctum (11)	175.03 ± 46.451 a	27.57 ± 6.986 c	8.54 ± 1.951 bc	126.78 ± 41.123 ab	85.15 ± 21.611 b	3.82 ± 1.381 b	0.23 ± 0.047 b	2.91 ± 1.282 b
O. basilicum (11)	102.84 ± 26.238 ab	41.86 ± 8.664 bc	12.91 ± 1.781 ab	182.44 ± 30.427 a	75.02 ± 16.794 b	4.27 ± 1.121 b	0.24 ± 0.040 b	6.27 ± 1.602 a
O. gratissimum (11)	159.31 ± 40.401 ab	39.22 ± 9.034 bc	8.73 ± 2.067 bc	197.40 ± 53.812 a	55.76 ± 24.627 b	3.73 ± 1.630 b	0.20 ± 0.067 b	3.45 ± 0.857 ab
M. piperita (10)	135.14 ± 49.783 ab	50.62 ± 10.683 abc	10.30 ± 2.271 abc	32.75 ± 12.343 b	65.57 ± 28.194 b	3.30 ± 1.257 b	0.24 ± 0.067 b	4.90 ± 1.159 ab
M. arvensis (12)	115.89 ± 31.654 ab	25.10 ± 5.532 c	6.75 ± 1.286 c	161.45 ± 38.100 a	33.53 ± 14.907 b	2.33 ± 0.752 b	0.12 ± 0.032 b	3.83 ± 1.086 ab
T. erecta (12)	176.36 ± 39.023 a	64.48 ± 17.068 ab	12.00 ± 2.336 abc	154.60 ± 42.219 a	32.15±11.671 b	4.00 ± 1.219 b	0.20 ± 0.052 b	6.33 ± 1.356 a
L. angustifolia (12)	150.97 ± 32.706 ab	31.95 ± 7.718 bc	13.67 ± 2.524 ab	144.79 ± 38.275 ab	79.96 ± 21.324 b	5.67 ± 1.734 ab	0.23 ± 0.053 b	5.00 ± 1.030 ab
	$F_{7, 81} = 1.240$ $p = 0.291$	$F_{7, 81} = 2.880$ $p = 0.010$	$F_{7, 81} = 2.074$ $p = 0.056$	$F_{7, 81} = 1.560$ $p = 0.159$	$F_{7, 81} = 3.791$ $p = 0.001$	$F_{7, 81} = 2.177$ $p = 0.045$	$F_{7, 81} = 4.392$ $p < 0.001$	$F_{7, 81} = 1.509$ $p = 0.176$

Note: Values are means ± SE. Data in the same column followed by different lowercase letters differed significantly ($p < 0.05$). E1 = watery salivation into sieve elements, E2 = phloem sap ingestion.

2.3. Effect of Plant EOs on Daily Secretion of Aphid Honeydew

In the experiment, the honeydew distribution was scattered and no large spots were formed, so the error caused by stacking could not be considered. Exposure to plant EOs significantly reduced the level of honeydew excretion by *A. gossypii* compared with the control ($p < 0.001$) (Table 6). The levels of antifeedant impact (based on the reduction in honeydew by a treated aphid relative to a control aphid) due to EOs of *O. basilicum*, *O. gratissimum*, *M. piperita*, *M. arvensis*, and *T. erecta* were 64.6%, 51.1%, 50.7%, 65.2%, and 59.2%.

Table 6. Effect of essential oils of seven plants on honeydew excretion by *Aphis gossypii*.

EOs	*A. gossypii*	
	Proportion Honeydew/mm^2	**Antifeedant Rate%**
Control	11.53 ± 0.809 a	—
O. sanctum	7.61 ± 0.853 b	33.40
O. basilicum	4.08 ± 0.833 c	64.61
O. gratissimum	5.64 ± 0.870 bc	51.08
M. piperita	5.69 ± 0.878 bc	50.65
M. arvensis	4.01 ± 0.686 c	65.22
T. erecta	4.70 ± 0.533 c	59.24
L. angustifolia	6.17 ± 0.821 bc	46.49
	$F_{7,136} = 9.669$ $p < 0.001$	

Note: Values are means ± SE. Data in the same column followed by different lowercase letters differed significantly ($p < 0.05$).

3. Discussion

We examined the effects of seven plant EOs (*O. sanctum*, *O. basilicum*, *O. gratissimum*, *M. piperita*, *M. arvensis*, *T. erecta*, and *L. angustifolia*) on the growth, development, feeding behavior, and honeydew excretion of *A. gossypii*. All seven EOs had varying effects on *A. gossypii*, with the most effective from the point of view of aphid control being the EOs of *O. sanctum*, *M. piperita*, and *T. erecta*.

In general, it is the case that plant EOs can adversely affect the growth and development of aphids. For example, the EO of *Tagetes minuta* L. significantly reduced the fecundity of *Acyrthosiphon pisum* Harris, *M. persicae*, and *Aulacorthum solani* Kaltenbach [26]. The EOs of *Origanum majorana* L., *Mentha pulegium* L., and *Melissa officinalis* L. significantly reduced the longevity and fecundity of *M. persicae* [27]. We showed that the adult longevity of *A. gossypii* was significantly shortened and its fecundity significantly reduced by the EOs of six of our seven test plants (*O. sanctum*, *O. gratissimum*, *M. piperita*, *M. arvensis*, *T. erecta*, and *L. angustifolia*). Also, the net reproductive rate (R_0), intrinsic rate of increase (r_m), and finite rate of increase (λ) of *A. gossypii* were all significantly reduced, and the mean generation time (T) was significantly increased after treatment with EOs of *O. sanctum*, *M. piperita* and *L. angustifolia*. The growth, development, and fecundity of *A. gossypii* were all diminished, possibly due to a reduction in time spent in phloem and xylem feeding, reducing the level of nutrition available for growth [28].

Insects assess whether a plant is suitable for feeding by sampling plant parts for their nutrients [29]. Aphid probing and prolonged feeding are closely associated with the level of plant tissue damage and plant virus transmission [30]. *A. gossypii* obtains the nutrients it needs for growth and development mainly by feeding on the phloem sap [31]. Aphids promote the success of such feeding by secreting water-soluble saliva, which can prevent plant phloem proteins from clogging sieve tubes [32]. In our experiment, we observed that secretion of saliva (E1) and phloem feeding (E2 waves) by *A. gossypii* were both reduced after exposure to EOs, leading to less successful feeding by *A. gossypii*. Similarly, for pea aphids (*A. pisum*) treating its host plants with kaempferol led to an increase in the time spent in non-probing behaviors [33]. When plants infested with *A. gossypii* were treated with imidacloprid, *A. gossypii* spent significantly longer time on intercellular apoplastic stylet pathway activities, which are non-production behaviors taking time away from feeding [34]. For *Diaphorina citri* Kuwayama non-probing behaviors (Np) were prolonged after exposure to the EOs of guava (*Psidium guajava* L.) [35]. Similarly, in our study, *A. gossypii* spent more time in non-probing behaviors after exposure to the plant EOs examined here. Also, the duration and number of events of non-vascular probing (C waves) increased in *A. gossypii* following exposure to our test EOs, indicating that more probing occurred between the plant epidermis and the microtubule bundles, requiring more time to find a suitable feeding site. Zhou et al. observed plants' aqueous extracts of *Nerium indicum* Mill., *Cinnamomum camphora* L., *Ginkgo biloba* L., and *Lycopersicon esculentum* Mill. significantly prolonged the duration of non-vascular probing in *A. gossypii* [36]. In contrast to phloem ingestion, xylem ingestion supplies water but not nutrition, and increased time spent in xylem feeding may be linked to a decrease in phloem feeding [37]. Aphids spend more time xylem feeding when under stress [38]. We found the same behavior in our experiment, in which the xylem feeding was prolonged after treatment with EOs, presumably to maintain the internal water equilibrium [39].

After aphids feed on phloem, excess water, which still contains some sugar, is excreted as honeydew to concentrate sugars in the fluids retained for ingestion [40]. In contrast, xylem feeding results in little or no honeydew production because most ingested xylem is retained to maintain water equilibrium [41]. Honeydew excretion is, therefore, closely tied to the level of phloem feeding by an aphid [42]. Honeydew often covers plant surfaces and affects photosynthesis and respiration [43]. When an aphid's level of phloem feeding is reduced, its rate of growth and development declines, and it excretes less honeydew. By quantifying the volume of honeydew excreted by cotton aphids, we were able to determine the degree to which plant essential oils had influenced the feeding of cotton aphids. Extracts of *Syngonium podophyllum* Schott, *Xanthium sibiricum* Petr.et Widd, and *Tephrosia vogelli* Hook have been shown to reduce honeydew excretion by *M. persicae* and *Lipaphis erysimi* Kaltenbach [44]. In our study, the daily rate of honeydew excretion by *A. gossypii* declined significantly after treatment with plant EOs. This decline in honeydew production indicates that plant EOs have significant antifeedant effects on *A. gossypii*, but the EOs of different plants varied in the levels of their antifeedant effects on *A. gossypii*. We observed that the honeydew of *A. gossypii* was dispersed on the filter paper, likely indicating that the EO-treated leaves were not acceptable for aphid feeding, and aphids kept moving and searching for better sites. This suggests that the honeydew production of *A. gossypii* was reduced and no suitable sites were located, given that there was no overlapping of honeydew to from blotches on the filter paper.

From an integrated pest management point of view, when considering the effects of EOs on pest control, we also need to examine their effects on natural enemies. Chemical control not only leads to an increase in pest resistance but also kills large numbers of natural enemies, weakening their contribution to pest control, which may lead to pest outbreaks. It has, however, been demonstrated that many EOs are much less damaging to natural enemies than to the target pests. Papadimitriou et al. compared the biological activity of the EO of *M. pulegium* against *A. gossypii*, *Tetranychus urticae* Koch, and the mirid *Nesidiocoris tenuis* Reuter, and found that the plant's EO was highly lethal to both *A. gossypii* and *T. urticae* at a concentration of 1000 μL/L, but had no toxic effects on *N. tenuis* [45]. In another study, the LC_{50} values of the EOs of *M. piperita*, *Mentha longifolia* L., *Salvia officinalis* L., and *Salvia rosmarinus* Spenn. were about four times higher against the coccinellid *Coccinella undecimpunctata* L. than to the pest, *Aphis punicae* Passerini [46]. Ebadollahi et al. observed that the EO of *Satureja intermedia* C.A.Mey showed significant contact activity against *Aphis nerii* Kaltenbach, but was safe against its predator *Coccinella septempunctata* L. [47]. These findings suggest that insect natural enemies may be more tolerant to a wide range of plant EOs than are target pests.

Although this paper showed that the EOs of *O. sanctum*, *O. basilicum*, *O. gratissimum*, *M. piperita*, *M. arvensis*, *T. erecta*, and *L. angustifolia* all had different antifeedant effects on *A. gossypii*, in-depth studies are needed to determine their individual antifeedant mechanisms.

We will identify the active ingredients in plant essential oils in subsequent research and further evaluate which components in the plant essential oils have antifeedant effects on cotton aphids. *A. gossypii*, as a major pest of cotton, has developed a high level of resistance to conventional chemical control, and botanical antifeedants are a novel means for its control. Antifeedants appear to both reduce colonization of plants by reducing their attraction for pests and reduce feeding of any pest individuals that do begin to attack the host plant, slowing their population growth. EOs protect the host plant and also avoid harming natural enemies and other non-target insects [48]. EOs can, therefore, be used in combination with biological control, natural enemy conservation, and other such techniques to better promote sustainable management of *A. gossypii* in cotton fields.

Currently, botanical insect antifeedants are not yet used on a large scale due to costs in their mass production [49]. Plant EOs are also limited by their relatively high volatility, which limits the duration of their efficacy. Technologies that promote the slow, controlled release of EOs by means of microencapsulation [50] using polymer films have potential to increase the persistence of EOs, allowing for and less frequent applications.

4. Materials and Methods

4.1. Sources of Essential Oils and Test Insects

Essential oils were purchased from Poli Aroma Pharmaceutical Technology Co, Shanghai and the EOs were diluted with acetone to 1 μg/μL for testing. The seven essential oils were extracted from plants using the distillation method. The entire plants of *O. sanctum*, *O. basilicum*, and *O. gratissimum* were subjected to extraction. For *M. piperita* and *M. arvensis*, the extraction was performed on the flowering tops and leaves. For *T. erecta* and *L. angustifolia*, the flowers were the parts used for extraction.

We initiated a colony of *A. gossypii* by collecting wingless aphids from experimental cotton fields (CAAS, 39.53° N, 116.70° E) of the Lang Fang Experiment Station of the Chinese Academy of Agricultural Sciences (Hebei Province, China). The aphid colony was reared in 12-well plates, with each well two-thirds filled with 2% agar on which a cotton leaf disk excised with a leaf punch from uninfested leaves of cotton plants grown in the greenhouse. The colony was held at 26 ± 1 °C, 50 ± 5% rh, a 16 L:8 D photoperiod.

4.2. Growth, Development, and Fecundity

Based on Chen's experimental method [51], fresh round cotton leaves were cut using a hole punch (23 mm diameter), and were then placed in an inverted orientation on each cell of a 12-well cell culture plate on a bed of 2% agar to which we applied 50 μL of plant essential oil at a concentration of 1 μg/μL. According to Liu and Wang 's experimental method, we used 12-well cell culture plates for the experiment [52,53]. Following Izakmehri and Castilhos [54,55], we used acetone as the solvent for dissolving the essential oils. The acetone was allowed to evaporate naturally. For observation of nymphal and adult development, adult aphids were transferred onto the leaves. One adult aphid was placed in each cell. After twenty-four hours, one newborn nymph was retained, and the adults and other nymphs removed with a brush. Plates were covered with nylon gauze to prevent aphid escape, but still allow the EO to evaporate naturally. The leaves were replaced every 2 days and the essential oil renewed at the same level. Observations were made once a day to record the molting time of nymphs, the number of nymphs surviving, the daily production of adult aphids and the number of adult aphids surviving until all adult aphids were dead. Each treatment had 50 replicates, 1 insect being a replicate. The experiments were conducted in a constant temperature and light incubator at 26 ± 1 °C, 50 ± 5% rh, and a 16 L:8 D photoperiod.

4.3. Effect of EOs on Aphid Feeding Behaviors

To assess the effect of EOs on aphid feeding behavior, we used a Giga-8 DC-EPG device (EPG-systems Wageningen, Netherlands) with a signal converter card (Di 710, EPG-systems, Wageningen, Netherlands), and the Giga-8 DC-EPG was placed in a Faraday cage

(80 cm × 60 cm × 60 cm). Vigorous *A. gossypii* adults were selected and starved for 1 h before testing. A wooden stick was inserted vertically into each cotton seedling nutrient bowl (specifications: upper diameter of 12 cm, lower diameter of 9 cm, height of 10 cm). A piece of cardboard larger than the leaf was mounted on the stick, with the height in line with the cotton leaf. The cotton leaf was turned and affixed to the cardboard. The cotton leaf was coated with 1 μg/μL of an EO with an artist's brush, while the control leaf was coated with acetone. The volume of essential oil used was 500 μL on each leaf. Acetone did not damage the plant tissue. Aphids were gently placed on the lower surface of individual cotton leaf (choosing the third true leaf) after being attached to the gold-wire electrode with conductive silver paint. One aphid was placed on each cotton plant for testing. Recordings continued for 8 h for each recording session, and at the end of the recording, the cotton leaf and aphid were replaced with new ones. The experiment was run in a greenhouse at 26 °C, 75% rh, and continuous light. Each EO treatment and control was replicated more than 10 times [56].

Following Liu and Sandanayaka et al. [28,57], we recorded the following aphid feeding behaviors: (1) non-probing (Np waves), (2) probing in non-vascular tissues (C waves), (3) probing inside cells (Pd waves), (4) secretion of saliva into phloem sieve elements (E1 waves), (5) phloem ingestion (E2 waves), (6) ingestion of xylem sap (G waves), and (7) mechanical difficulty in stylet penetration (F waves).

4.4. Honeydew Excretion

To measure aphid feeding, we recorded the amount of honeydew produced as a surrogate parameter using the honeydew color spot area method of Zhang et al. [58], in which bromophenol green reacts with the amino acids in the honeydew to produce a visible spot, whose area is determined. To do so, the qualitative filter paper (9 cm dia) was first soaked with 0.1% bromocresol green solution, an average of 100 mL of solution is soaked in 10 qualitative filter papers, and then dried in an oven at 75 °C. The filter paper was then bright orange. Agar (2%) was then poured into plastic cups (6.5 cm dia), and fresh cotton leaves spread onto the agar to preserve freshness. To each leaf, we applied 500 μL of the plant essential oil solution, at a concentration of 1 μg/μL. Five third instar aphids were placed in each cup (after being starved for 1 h before the test). The plastic cups were then inverted over filter paper so that most of honeydew secreted by *A. gossypii* would fall onto the filter paper, which turned peacock blue on contact with honeydew. After 24 h, the filter paper was removed, and the blue area (mm^2) was measured with a transparent standard calculation paper. The experiment was conducted in a greenhouse at 26 ± 1 °C, 50 ± 5% rh, and under natural light. Treatments and controls were each replicated 18 times.

4.5. Statistical Analyses

Aphid development, reproduction and population parameters were analyzed according to the age-stage hermaphroditic life table theory using the TWOSEX-MS Chart program (Version 2021.10.30) (http://140.120.197.173/Ecology/prod02.htm) (accessed on 1 December 2021) [59,60]. Standard errors for nymphal developmental duration, adult aphid longevity, reproduction, and population parameters were calculated using the bootstrap method within the software, with a bootstrap count of 100,000, and differences between treatments were analyzed using the software's paired bootstrap test ($p < 0.05$).

The EPG test data were analyzed using Stylet+a software (Version 1.30) for each waveform, Excel was used for simple statistics on the time and numbers of each waveform, and then SPSS 21.0 software was used to assess differences between treatments with one-way analysis of variance (ANOVA). Differences between treatments were compared using the LSD multiple comparisons method ($p < 0.05$).

Evaluation of *A. gossypii* antifeedant activity using antifeedant rate:

$$\text{Antifeedant rate} = \frac{\text{Honeydew area of control aphids} - \text{Treatment groups}}{\text{Honeydew area of control aphids}}$$

Antifeedant rates were simply statistically analyzed using Excel and then SPSS 21.0 software was used to assess differences between treatments with one-way analysis of variance (ANOVA). Differences between treatments were compared using the LSD multiple comparisons method ($p < 0.05$).

5. Conclusions

In conclusion, we have demonstrated that the EOs of all seven test plants (*O. sanctum*, *O. basilicum*, *O. gratissimum*, *M. piperita*, *M. arvensis*, *T. erecta*, and *L. angustifolia*) affected the growth, development, and reproductive ability of *A. gossypii*, inhibited aphid feeding behavior, and reduced the amount of honeydew excreted daily by *A. gossypii*. The EOs of *O. sanctum*, *M. piperita*, and *T. erecta* had the greatest effect, followed by EOs of *M. arvensis* and *L. angustifolia*, and finally the EOs of *O. basilicum* and *O. gratissimum*. Thus, among the seven EOs studied, the EOs of *O. sanctum*, *M. piperita*, and *T. erecta* have potential to be developed as antifeedants for the non-pesticide chemical control of *A. gossypii*.

Author Contributions: X.W.: data curation and analysis, investigation, writing—original draft; Y.Z.: data curation and analysis, investigation, writing—original draft; H.Y.: conceptualization, writing—review and editing; Y.L.: conceptualization, data curation and analysis, funding acquisition, methodology, project administration, supervision, validation, writing—review and editing. All authors have read and agreed to the published version of the manuscript.

Funding: This work was supported by the National Key R&D Program of China (2022YFD1400300), and the China Agriculture Research System of MOF and MARA (CARS-15-19).

Data Availability Statement: Data are contained within the article.

Acknowledgments: We thank Yu Gao from the Institute of Plant Protection, Chinese Academy of Agricultural Sciences, for her help in aphid rearing.

Conflicts of Interest: The authors declare no conflicts of interest. The funders had no role in the design of the study; in the collection, analyses, or interpretation of data; in the writing of the manuscript; or in the decision to publish the results.

References

1. Li, R.; Cheng, S.; Chen, Z.; Guo, T.; Liang, P.; Zhen, C.; Wang, J.; Zhang, L.; Liang, P.; Gao, X. Establishment of toxicity and susceptibility baseline of broflanilide for *Aphis gossypii* Glover. *Insects* **2022**, *13*, 1033. [CrossRef] [PubMed]
2. Zhao, Q.; Zhang, S.; Luo, J.; Wang, C.; Lv, L.; Dong, S.; Cui, J. Ecological adaption analysis of the cotton aphid (*Aphis gossypii*) in different phenotypes by transcriptome comparison. *PLoS ONE* **2013**, *8*, e83180. [CrossRef]
3. Heilsnis, B.; Mahas, J.B.; Conner, K.; Pandey, S.; Clark, W.; Koebernick, J.; Srinivasan, R.; Martin, K.; Jacobson, A.L.; Burrack, H. Characterizing the vector competence of *Aphis gossypii*, *Myzus persicae* and *Aphis craccivora* (Hemiptera: Aphididae) to transmit cotton leafroll dwarf virus to cotton in the United States. *J. Econ. Entomol.* **2023**, *116*, 719–725. [CrossRef] [PubMed]
4. O'Hara, F.M.; Davis, J.A.; Swale, D.R. Profile of commercialized aphicides on the survivorship and feeding behavior of the cotton aphid, *Aphis gossypii*. *Pestic. Biochem. Physiol.* **2022**, *186*, 105174. [CrossRef] [PubMed]
5. Braendle, C.; Davis, G.K.; Brisson, J.A.; Stern, D.L. Wing dimorphism in aphids. *Heredity* **2006**, *97*, 192–199. [CrossRef]
6. Herron, G.A.; Powis, K.; Rophail, J. Insecticide resistance in *Aphis gossypii* Glover (Hemiptera: Aphididae), a serious threat to Australian cotton. *Aust. J. Entomol.* **2001**, *40*, 85–91. [CrossRef]
7. Nauen, R.; Elbert, A. European monitoring of resistance to insecticides in *Myzus persicae* and *Aphis gossypii* (Hemiptera: Aphididae) with special reference to imidacloprid. *Bull. Entomol. Res.* **2007**, *93*, 47–54. [CrossRef] [PubMed]
8. Patima, W.; Guo, P.; Ma, S.; Gao, X.; Zhang, L.; Zhang, S.; Ma, D. Resistance of different field populations of *Aphid gossypii* to ten insecticides in Xinjiang. *Plant Prot.* **2019**, *45*, 273–278. [CrossRef]
9. Fernandes, M.E.S.; Alves, F.M.; Pereira, R.C.; Aquino, L.A.; Fernandes, F.L.; Zanuncio, J.C. Lethal and sublethal effects of seven insecticides on three beneficial insects in laboratory assays and field trials. *Chemosphere* **2016**, *156*, 45–55. [CrossRef]
10. Bakkali, F.; Averbeck, S.; Averbeck, D.; Idaomar, M. Biological effects of essential oils-a review. *Food Chem. Toxicol.* **2008**, *46*, 446–475. [CrossRef]
11. Karamanoli, K.; Ainalidou, A.; Bouzoukla, F.; Vokou, D. Decomposition profiles of leaf essential oils in the soil environment. *Ind. Crops Prod.* **2018**, *124*, 397–401. [CrossRef]
12. Isman, M.B. Botanical insecticides, deterrents, and repellents in modern agriculture and an increasingly regulated world. *Annu. Rev. Entomol.* **2006**, *51*, 45–66. [CrossRef] [PubMed]
13. Dayan, F.E.; Cantrell, C.L.; Duke, S.O. Natural products in crop protection. *Bioorg. Med. Chem.* **2009**, *17*, 4022–4034. [CrossRef]

14. Grdiša, M.; Gršić, K. Botanical insecticides in plant protection. *Agric. Conspec. Sci.* **2013**, *78*, 85–93.
15. Isman, M.B. Botanical insecticides in the twenty-first century-fulfilling their promise? *Annu. Rev. Entomol.* **2020**, *65*, 233–249. [CrossRef] [PubMed]
16. Isman, M.B. Commercial development of plant essential oils and their constituents as active ingredients in bioinsecticides. *Phytochem. Rev.* **2019**, *19*, 235–241. [CrossRef]
17. Chiasson, H.; Vincent, C.; Bostanian, N. Insecticidal properties of a *Chenopodium*-based botanical. *J. Econ. Entomol.* **2004**, *97*, 1378–1383. [CrossRef]
18. Ikbal, C.; Pavela, R. Essential oils as active ingredients of botanical insecticides against aphids. *J. Pest Sci.* **2019**, *92*, 971–986. [CrossRef]
19. Attia, S.; Lognay, G.; Heuskin, S.; Hance, T. Insecticidal activity of *Lavandula angustifolia* Mill. against the pea aphid *Acyrtosiphum pisum*. *J. Entomol. Zool. Stud.* **2016**, *4*, 118–122.
20. Saıfı, R.; Saıfı, H.; Akca, İ.; Benabadelkader, M.; Askın, A.K.; Belghoul, M. Insecticidal and repellent effects of *Mentha longifolia* L. essential oil against *Aphis craccivora* Koch (Hemiptera: Aphididae). *Chem. Biol. Technol. Agric.* **2023**, *10*, 18. [CrossRef]
21. Mousavi, M.; Valizadegan, O. Insecticidal effects of *Artemisia dracunculus* L. (Asteraceae) essential oil on adult of *Aphis gossypii* Glover (Hemiptera: Aphididae) under laboratory conditions. *Arch. Phytopathol.* **2014**, *47*, 1737–1745. [CrossRef]
22. Czerniewicz, P.; Chrzanowski, G. The effect of *Santolina chamaecyparissus* and *Tagetes patula* essential oils on biochemical markers of oxidative stress in aphids. *Insects* **2021**, *12*, 360. [CrossRef]
23. Abualfia, R.; Samara, R. Antifeedants impact of plant essential oil on green peach aphid on potato crops. *J. Ecol. Eng.* **2021**, *23*, 274–285. [CrossRef]
24. Zhang, J.; Yang, G.; Yu, Y.; Yang, L. Antifeeding and insecticidal activity of ethanol extracts from 11 plants against *Aphis* sp. of *Lycium barbarum*. *J. Agric. Sci.* **2007**, *1*, 21–23. [CrossRef]
25. Zhou, W. Preliminary tests on the biological activity of extracts of *Flaveria bidentis* against aphids and *Ostrinia nubilalis*. *Jiang Su Agric. Sci.* **2010**, *6*, 198–199. [CrossRef]
26. Tomova, B.S.; Waterhouse, J.S.; Doberski, J. The effect of fractionated *Tagetes* oil volatiles on aphid reproduction. *Entomol. Exp. Appl.* **2005**, *115*, 153–159. [CrossRef]
27. Petrakis, E.A.; Kimbaris, A.C.; Perdikis, D.C.; Lykouressis, D.P.; Tarantilis, P.A.; Polissiou, M.G. Responses of *Myzus persicae* (Sulzer) to three Lamiaceae essential oils obtained by microwave-assisted and conventional hydrodistillation. *Ind. Crops Prod.* **2014**, *62*, 272–279. [CrossRef]
28. Liu, J.; Wang, C.; Li, H.; Gao, Y.; Yang, Y.; Lu, Y. Bottom-up effects of drought-stressed cotton plants on performance and feeding behavior of *Aphis gossypii*. *Plants* **2023**, *12*, 2886. [CrossRef]
29. Sauge, M.H.; Lacroze, J.P.; Poëssel, J.L.; Pascal, T.; Kervella, J. Induced resistance by *Myzus persicae* in the peach cultivar 'Rubira'. *Entomol. Exp. Appl.* **2002**, *102*, 29–37. [CrossRef]
30. Fereres, A.; Moreno, A. Behavioural aspects influencing plant virus transmission by homopteran insects. *Virus Res.* **2009**, *141*, 158–168. [CrossRef]
31. Powell, G.; Tosh, C.R. Host plant selection by aphids: Behavioral, evolutionary, and applied perspectives. *Annu. Rev. Entomol.* **2006**, *51*, 309–330. [CrossRef]
32. Tjallingii, W.F. Salivary secretions by aphids interacting with proteins of phloem wound responses. *J. Exp. Bot.* **2006**, *57*, 739–745. [CrossRef] [PubMed]
33. Stec, K.; Kordan, B.; Gabryś, B. Effect of soy leaf flavonoids on pea aphid probing behavior. *Insects* **2021**, *12*, 756. [CrossRef] [PubMed]
34. Wang, L.; Wang, Q.; Wang, Q.; Rui, C.; Cui, L. The feeding behavior and life history changes in imidacloprid-resistant *Aphis gossypii* Glover (Homoptera: Aphididae). *Pest Manag. Sci.* **2019**, *76*, 1402–1412. [CrossRef] [PubMed]
35. Zhu, H.; Zeng, X.; Zaka, M.S.; Cen, Y. Influence of the essential oil of guava leaves on the feeding behavior of the Asian citrus psyllid. *Environ. Entomol.* **2010**, *32*, 483–487. [CrossRef]
36. Zhou, T.; Chen, J.; Zhang, P.; Wang, Y. The influence of four kinds of plant aqueous extracts on the feeding behaviors of *Aphis gossypii*. *J. Plant Prot. Res.* **2004**, *3*, 252–258. [CrossRef]
37. Pettersson, J.; Tjallingii, W.F.; Hardie, J. Host-plant selection and feeding. In *Aphis as Crop Pests;* CABI: Wallingford, UK, 2017; pp. 173–195. [CrossRef]
38. Ramírez, C.C.; Niemeyer, H.M. The influence of previous experience and starvation on aphid feeding behavior. *J. Insect Behav.* **2000**, *13*, 699–709. [CrossRef]
39. Pompon, J.; Quiring, D.; Goyer, C.; Giordanengo, P.; Pelletier, Y. A phloem-sap feeder mixes phloem and xylem sap to regulate osmotic potential. *J. Insect Physiol.* **2011**, *57*, 1317–1322. [CrossRef]
40. Calvo-Agudo, M.; Tooker, J.F.; Dicke, M.; Tena, A. Insecticide-contaminated honeydew: Risks for beneficial insects. *Biol. Rev.* **2021**, *97*, 664–678. [CrossRef]
41. Seo, B.Y.; Kwon, Y.-H.; Jung, J.K.; Kim, G.-H. Electrical penetration graphic waveforms in relation to the actual positions of the stylet tips of *Nilaparvata lugens* in rice tissue. *J. Asia Pac. Entomol.* **2009**, *12*, 89–95. [CrossRef]
42. He, Y.; Zhao, J.; Zheng, Y.; Weng, Q.; Biondi, A.; Desneux, N.; Wu, K. Assessment of potential sublethal effects of various insecticides on key biological traits of the tobacco whitefly, *Bemisia tabaci*. *Int. J. Biol. Sci.* **2013**, *9*, 246–255. [CrossRef] [PubMed]

43. Joschinski, J.; Krauss, J. Food colouring as a new possibility to study diet ingestion and honeydew excretion by aphids. *Entomol. Exp. Appl.* **2017**, *164*, 141–149. [CrossRef]
44. Zhou, Q.; Liang, W.; Zeng, L. Study on antifeedant effect of plant extracts on *Myzus persicae* (Sulzer) and *Lipaphis erysimi* (Kaltenbach). *Nat. Prod. Res. Dev.* **2004**, *16*, 521–524. [CrossRef]
45. Papadimitriou, D.M.; Petrakis, E.A.; Arvaniti, K.A.; Kimbaris, A.C.; Polissiou, M.G.; Perdikis, D.C. Comparative bioactivity of essential oils from two *Mentha pulegium* (Lamiaceae) chemotypes against *Aphis gossypii, Aphis spiraecola, Tetranychus urticae* and the generalist predator *Nesidiocoris tenuis*. *Phytoparasitica* **2019**, *47*, 683–692. [CrossRef]
46. Sayed, S.; Soliman, M.M.; Al-Otaibi, S.; Hassan, M.M.; Elarrnaouty, S.A.; Abozeid, S.M.; El-Shehawi, A.M. Toxicity, deterrent and repellent activities of four essential oils on *Aphis punicae* (Hemiptera: Aphididae). *Plants* **2022**, *11*, 463. [CrossRef] [PubMed]
47. Ebadollahi, A.; Setzer, W.N. Evaluation of the toxicity of *Satureja intermedia* C. A. Mey essential oil to storage and greenhouse insect pests and a predator ladybird. *Foods* **2020**, *9*, 712. [CrossRef] [PubMed]
48. Mensah, R.; Moore, C.; Watts, N.; Deseo, M.A.; Glennie, P.; Pitt, A. Discovery and development of a new semiochemical biopesticide for cotton pest management: Assessment of extract effects on the cotton pest *Helicoverpa* spp. *Entomol. Exp. Appl.* **2014**, *152*, 1–15. [CrossRef]
49. Koul, O. Phytochemicals and insect control: An antifeedant approach. *Crit. Rev. Plant Sci.* **2008**, *27*, 1–24. [CrossRef]
50. Chen, M.; Hu, Y.; Zhou, J.; Xie, Y.; Wu, H.; Yuan, T.; Yang, Z. Facile fabrication of tea tree oil-loaded antibacterial microcapsules by complex coacervation of sodium alginate/quaternary ammonium salt of chitosan. *RSC Adv.* **2016**, *6*, 13032–13039. [CrossRef]
51. Chen, X.; Fan, Y.; Zhang, W.; Tian, Z.; Liu, J.; Zhao, K. Soybean aphid, *Aphis glycines* (Hemiptera: Aphididae), developmental and reproductive capacity on white clover, *Trifolium repens* (Rosales: Leguminosae), in northeast China. *Appl. Entomol. Zool.* **2017**, *52*, 491–495. [CrossRef]
52. Liu, J.; Wang, C.; Desneux, N.; Lu, Y. Impact of temperature on survival rate, fecundity, and feeding behavior of two aphids, *Aphis gossypii* and *Acyrthosiphon gossypii*, when reared on cotton. *Insects* **2021**, *12*, 565. [CrossRef] [PubMed]
53. Wang, W.; Huang, Q.; Liu, X.; Liang, G. Differences in the sublethal effects of sulfoxaflor and acetamiprid on the *Aphis gossypii* Glover (Homoptera: Aphididae) are related to its basic sensitivity level. *Insects* **2022**, *13*, 498. [CrossRef] [PubMed]
54. Izakmehri, K.; Saber, M.; Mehrvar, A.; Hassanpouraghdam, M.B.; Vojoudi, S. Lethal and sublethal effects of essential oils from *Eucalyptus camaldulensis* and *Heracleum persicum* against the adults of *Callosobruchus maculatus*. *J. Insect Sci.* **2013**, *13*, 152. [CrossRef] [PubMed]
55. Castilhos, R.V.; Grützmacher, A.D.; Coats, J.R. Acute toxicity and sublethal effects of terpenoids and essential oils on the predator *Chrysoperla externa* (Neuroptera: Chrysopidae). *Neotrop. Entomol.* **2018**, *47*, 311–317. [CrossRef] [PubMed]
56. Goławska, S.; Sprawka, I.; Łukasik, I.; Goławski, A. Are naringenin and quercetin useful chemicals in pest-management strategies? *J. Pest Sci.* **2014**, *87*, 173–180. [CrossRef] [PubMed]
57. Sandanayaka, W.R.M.; Hale, C.N. Electronically monitored stylet penetration pathway of woolly apple aphid, *Eriosoma lanigerum* (Homoptera: Aphididae), on apple (*Malus domestica*). *N. Z. J. Crop Hortic. Sci.* **2003**, *31*, 107–113. [CrossRef]
58. Zhang, P.; Zhang, Z.; He, H.; Li, J.; Bai, S.; Yan, F. Review of insect honeydew research methods. *Chin. J. Appl. Entomol.* **2020**, *57*, 466–473. [CrossRef]
59. Chi, H. Life-table analysis incorporating both sexes and variable development rates among individuals. *Environ. Entomol.* **1988**, *17*, 26–34. [CrossRef]
60. Chi, H.; Liu, H. Two new methods for the study of insect population ecology. *Bull. Inst. Zool. Acad. Sin* **1985**, *24*, 225–240.

plants

MDPI

Article

First Clarification of the Involvement of Glycosyltransferase MdUGT73CG22 in the Detoxification Metabolism of Nicosulfuron in Apple

Yuefeng Zhang [1,†], Aijuan Zhao [1,†], Lijun Mu [1], Xiao Teng [2], Yingxin Ma [1], Ru Li [1], Kang Lei [1], Lusha Ji [1], Xuekun Wang [1,*] and Pan Li [1,*]

1 State Key Laboratory for Macromolecule Drugs and Large-Scale Manufacturing, School of Pharmaceutical Sciences, Liaocheng University, Liaocheng 252059, China; 13562058239@163.com (Y.Z.); zaj15105403245@163.com (A.Z.); mlj15215354741@163.com (L.M.); m13021553065@163.com (Y.M.); liru0024@163.com (R.L.); leikang@lcu.edu.cn (K.L.); jilusha@lcu.edu.cn (L.J.)

2 Rizhao Research Institute of Agricultural Science, Rizhao 276500, China; teng-xiao-rzaas@hotmail.com

* Correspondence: xuekunwang0610@126.com (X.W.); lipan202301@163.com (P.L.)

† These authors contributed equally to this work.

Abstract: Nicosulfuron, an acetolactate synthase (ALS) inhibitor herbicide, is a broad-spectrum and highly effective post-emergence herbicide. Glycosyltransferases (GTs) are widely found in organisms and transfer sugar molecules from donors to acceptors to form glycosides or sugar esters, thereby altering the physicochemical properties of the acceptor molecule, such as participating in detoxification. In this study, nine glycosyltransferases in group D of the apple glycosyltransferase family I were predicted to possibly be involved in the detoxification metabolism of ALS-inhibiting herbicides based on gene chip data published online. In order to confirm this, we analysed whether the expression of the nine glycosyltransferase genes in group D was induced by the previously reported ALS-inhibiting herbicides by real-time PCR (polymerase chain reaction). It was found that the ALS-inhibiting herbicide nicosulfuron significantly increased the expression of the *MdUGT73CG22* gene in group D. Further investigation of the mechanism of action revealed that the apple glycosyltransferase MdUGT73CG22 glycosylated and modified nicosulfuron both in vivo and ex vivo to form nicosulfuron glycosides, which were involved in detoxification metabolism. In conclusion, a new glycosyltransferase, MdUGT73CG22, was identified for the first time in this study, which can glycosylate modifications of the ALS-inhibiting herbicide nicosulfuron and may be involved in the detoxification process in plants, which can help to further improve the knowledge of the non-targeted mechanism of herbicides.

Keywords: glycosyltransferase; MdUGT73CG22; acetolactate synthase inhibitor; nicosulfuron; herbicide; detoxification mechanism

1. Introduction

There are many types of herbicides, and they are classified in various ways. Based on the mode of action, they can be classified into selective and inactivating herbicides, with the former having different levels of resistance to different species of plants, such as gaynon, bendazone, and dinocarb [1], and the latter being toxic to all plants, as long as the green parts are in contact [2]. Herbicides can be classified according to their transmission type; the contact herbicides only kill the part of plants that come into contact with herbicides, whereas the systemic herbicides can be transported to other parts of the plant after uptake, thus poisoning the whole plant [3].

There are currently 54 ALS inhibitors used in agriculture, which belong to five subclasses: sulfonylureas, imidazolinones, triazolopyrimidines, pyrimidinylthiobenzoates, and sulfonylamino carbonyltriazolinones [4]. Among them, triazolopyrimidine sulfonamides are a new class of acetolactate synthetase (ALS) inhibitors that have been developed and rapidly applied since the 1990s. By modifying sulfonylureas, they have retained the advantage of ultra-high efficiency while solving the obvious disadvantage that some varieties have a long residual time in the soil, which can cause damage to the subsequent crop, and have thus become the first choice among the five subclasses of herbicides [5].

Nicosulfuron is a systemic herbicide that can be absorbed by weed stems, leaves, and roots and, subsequently, be conducted through the plant, causing growth stagnation, greening of stems and leaves, and gradual dieback of sensitive plants, typically within 20–25 days [6]. Plants are sensitive to sulfonylurea herbicides such as nicosulfuron. Nicosulfuron is a sulfonylurea-type endopeptic herbicide that can be quickly absorbed by the stems, leaves, and roots of plants. It inhibits the activity of acetyllactate synthase in the plant to prevent branched-chain amino acid synthesis, thereby preventing cell division and stopping the growth of sensitive plants [7]. Even if only small amounts of such herbicide residues are present in the soil, they can affect plant growth and cause a lot of harm to beneficial non-target organisms in agricultural production as well as to the environment [8]. Therefore, it is of great significance to study the reproduction and genotoxicity of herbicides for selecting safe species and protecting ecosystems.

Glycosyltransferases (GTs) are a class of highly divergent multisource gene families present in almost all organisms that catalyse biological transformation reactions, transferring sugar groups from activated donor molecules to acceptor molecules [9]. In plants, UDP-glucose is the more abundant sugar donor, in addition to UDP-galactose, UDP-rhamnose, UDP-xylose, and UDP-glucuronic acid [10]; whereas glycosyl acceptors include non-carbohydrates, such as proteins, lipids, antibiotics, steroids, phenolics, terpenes, cyanohydrols, plant hormones, and other substances, in addition to monosaccharides, oligosaccharides, polysaccharides, cyanoalcohols, phytohormones, alkaloids, phytotoxins, and exogenous substances (e.g., herbicides and insecticides), etc. The site of glycosylation is mostly at the O-atom of the receptor molecule and, to a lesser extent, at the N-, S-, and C-atoms, generating the corresponding glycosides or glycolides [11], which alter the physicochemical properties of the receptor molecule, such as their hydrophilicity, chemical stability, biologically activity, and subcellular localisation, contributing to their intracellular and intraorganismal transport and storage, etc. [12].

Taylor et al. showed that when wheat (*Triticum aestivum* L.) plants were spray-treated with three safener chemistries, namely cloquintocet mexyl, mefenpyr diethyl, and fenchlorazole ethyl, a large number of glycosyltransferase genes were induced to be expressed [13]. Furthermore, it has been claimed that several *Arabidopsis* (*Arabidopsis thaliana* (L.) *Heynh*) UGTs are involved in the metabolic detoxification of the exogenous substances 2,4,5-trichlorophenol and 3,4-dichloroaniline (DCA) [14,15]. In addition, *Arabidopsis* (*Arabidopsis thaliana* (L.) *Heynh*) plants have been shown to have enhanced detoxification of 2,4-dihydroxy-1,4-benzoxazin-3-one (DIBOA) and 2,4-dihydroxy-7-methoxy-1,4-benzoxazin-3-one (DIMBOA) by overexpression of either the *GT BX8* or *GT BX9* genes [16]. Su et al. identified and cloned an enzyme with O-glycosyltransferase activity, UGT73A17, which was responsible for the biosynthesis of several flavonoid glycosides, and it was also involved in the thermal response and quality of tea (*Camellia sinensis* (L.) *O. Ktze*) [17]. Li et al. identified 229 members of family 1 through a genome-wide analysis of the apple UGTs, which were clustered into 18 groups, from A to R [18]. However, there are fewer studies on the direct involvement of glycosyltransferases in the detoxification metabolism of plant pesticides, and the involvement in the detoxification metabolism mechanism of the ALS-inhibiting herbicide nicosulfuron has not been reported. Therefore, it is important to study the involvement of glycosyltransferases in the mechanism of pesticide detoxification metabolism in plants. In this study, we identified a new apple (*Malus × domestica Borkh.*) glycosyltransferase MdUGT73CG22 by molecular biology techniques, which can glycosylatively modify the ALS-inhibiting herbicide nicosulfuron and may be involved in the detoxification process in plants, which can help to further improve the knowledge on the mechanism of non-targeted action of herbicides.

2. Results

2.1. Candidate Gene MdUGT73CG22 Identified

Based on the gene chip database published on NCBI (https://www.ncbi.nlm.nih.gov/, accessed on 18 June 2023) and the apple (*Malus* × *domestica Borkh.*) glycosyltransferase phylogenetic tree [18], nine glycosyltransferase genes *MdUGT73B36* (*MD17G1100000*), *MdUGT73B37* (*MD17G1100300*), *MdUGT73B38* (*MD05G1086300*), *MdUGT73B39* (*MD05G-1085800*), *MdUGT73B40* (*MD05G1085700*), *MdUGT73CG21* (*MD12G1104300*), *MdUGT73CG22* (*MD12G1104800*), *MdUGT73CP3* (*MD05G1085400*), and *MdUGT73AC7* (*MD05G1085600*) were predicted to possibly be involved in the detoxification metabolism of ALS-inhibiting herbicides. To validate their functional predictions, we treated wild-type apple (*Malus* × *domestica Borkh.*) seedlings with herbicides, including bensulfuron methyl, nicosulfuron, thifensulfuron, chlorsulfuron, metsulfuron, triasulfuron, cinosulfuron, chlorimuron-ethyl, sulfometuron, pyrazosulfuron, rimsulfuron, and ethametsulfuron, and real-time PCR was performed to detect the mRNA expression levels of these nine glycosyltransferase genes, respectively. The results are shown in Figure 1 and Table S1; compared with other ALS-inhibiting herbicides and glycosyltransferase genes, nicosulfuron significantly increased the mRNA expression level of glycosyltransferase gene *MdUGT73CG22* ($p < 0.01$). For this reason, we chose the apple (*Malus* × *domestica Borkh.*) glycosyltransferase gene *MdUGT73CG22* as a candidate gene in this study.

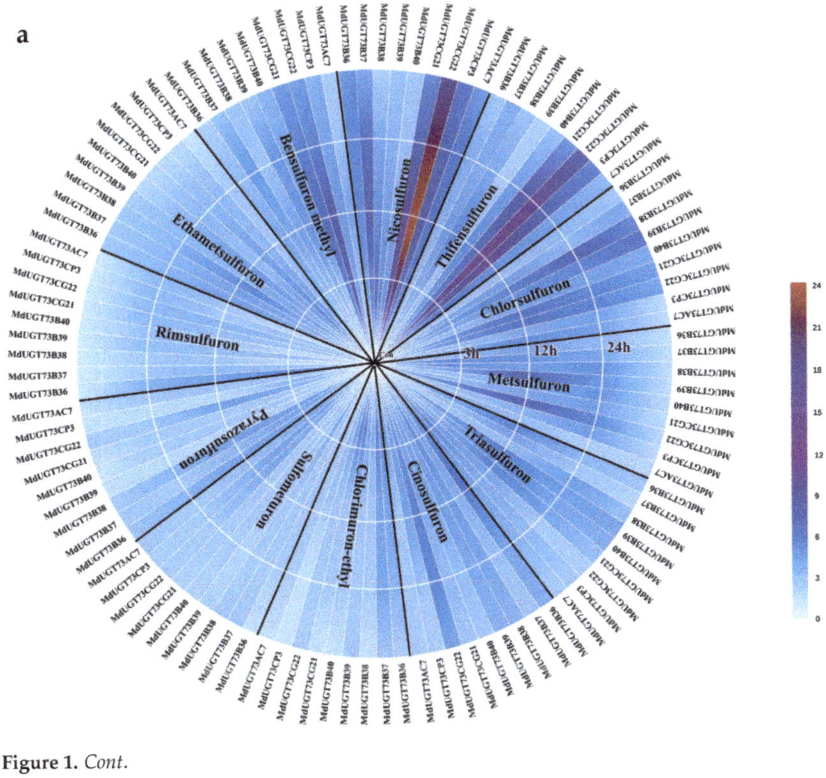

Figure 1. *Cont.*

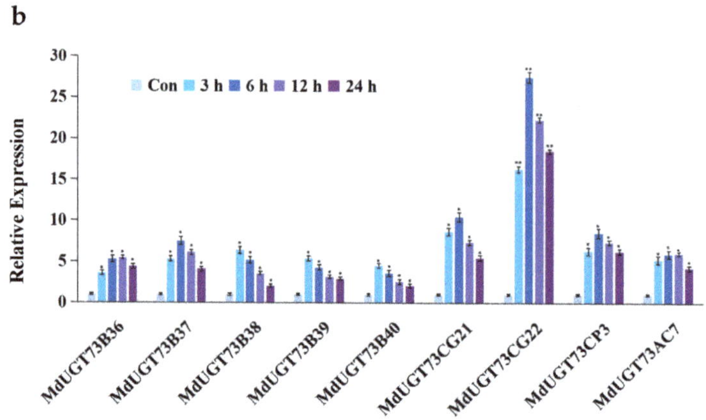

Figure 1. Relative expression of nine glycosyltransferase genes induced by 12 types of acetolactate synthase (ALS) inhibitor herbicides: (**a**) real-time PCR detection of glycosyltransferase gene *MdUGT73B40* (*MD05G1085700*), *MdUGT73CG21* (*MD12G1104300*), *MdUGT73CG22* (*MD12G1104800*), *MdUGT73CP3* (*MD05G1085400*), and *MdUGT73AC7* (*MD05G1085600*) are affected by acetolactate synthase (ALS) inhibitor herbicides: bensulfuron methyl, nicosulfuron, thifensulfuron, chlorsulfuron, metsulfuron, triasulfuron, cinosulfuron, chlorimuron-ethyl, sulfometuron, pyrazosulfuron, rimsulfuron, and ethametsulfuron induced expression at 0, 3, 6, 12, and 24 h; (**b**) real-time PCR detection of glycosyltransferase gene *MdUGT73CG22* (*MD12G1104800*) subjected to expression induced by the acetolactate synthase (ALS) inhibitor-like herbicide nicosulfuron at 0, 3, 6, 12, and 24 h. Note: * $p < 0.05$, ** $p < 0.01$, $n = 3$.

2.2. *Analysis of Apple Glycosyltransferase MdUGT73CG22 Tissue Expression Pattern*

In order to explore the expression of gene *MdUGT73CG22* in different developmental periods of apple (*Malus* × *domestica Borkh.*), RNA was extracted from fresh materials of different growth and developmental periods of apple (*Malus* × *domestica Borkh.*) and reversed transcribed into cDNA; then, the mRNA expression level of *MdUGT73CG22* was detected by real-time PCR. The results are shown in Figure 2; the gene *MdUGT73CG22* was expressed at the highest level in apple (*Malus* × *domestica Borkh.*) roots ($p < 0.05$) and the lowest level in flowers ($p < 0.05$). This result suggests that the apple glycosyltransferase gene *MdUGT73CG22* may play a primary role in the root system.

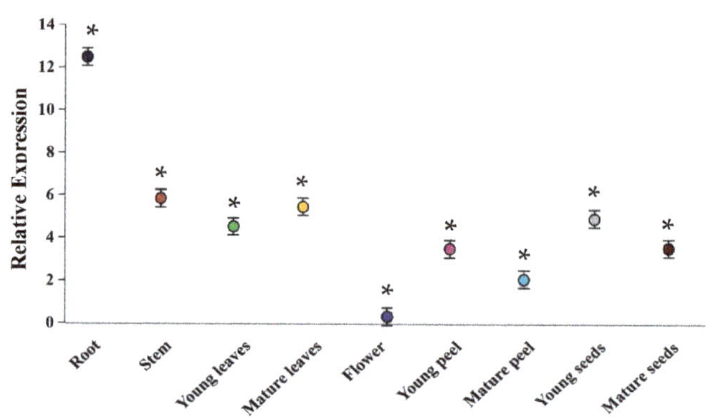

Figure 2. Real-time PCR detection of apple glycosyltransferase MdUGT73CG22 tissue expression pattern, including root, stem, young leaves, mature leaves, flower, young peel, mature peel, young seeds, and mature seeds. Note: * $p < 0.05$, $n = 3$.

2.3. Identification of Substrates for Glycosylation by Apple Glycosyltransferase MdUGT73CG22

Glycosyltransferase is a kind of modified enzyme; in order to further test its activity, we induced and purified the MdUGT73CG22 enzyme protein in vitro and analyzed the total protein and purified protein after induction by SDS-PAGE, which showed that the amount of the total protein after induction was significantly increased at 81.0 kDa compared with the amount before induction, and the purified protein band appeared was close to 81.0 kDa (the target band carrying the GST tag, which was 26.0 kDa) (Figure 3a), indicating that the MdUGT73CG22 protein was successfully expressed and purified.

In order to detect the activity of MdUGT73CG22, we used HPLC (high performance liquid chromatography) to detect the reaction products of MdUGT73CG22 protein. The results are shown in Figure 3b,c. MdUGT73CG22 catalysed nicosulfuron to nicosulfuron glycoside within 30 min; however, in the control experiments, no nicosulfuron glycoside production was found. This shows that MdUGT73CG22 can use glucose as a glycosyl donor to glycosylate nicosulfuron to produce nicosulfuron glycoside in vitro.

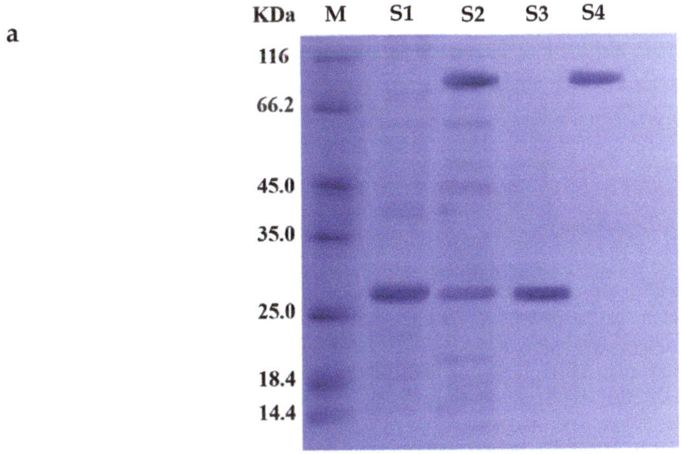

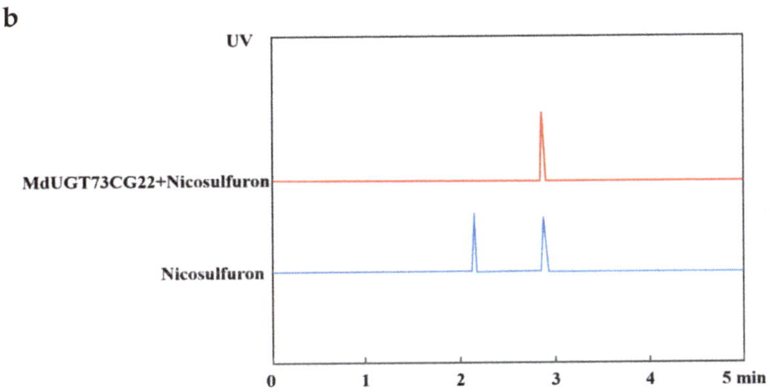

Figure 3. *Cont.*

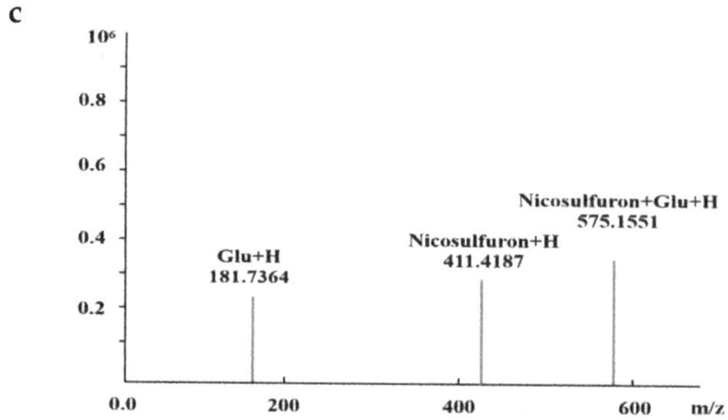

Figure 3. HPLC identification of glycosylation substrates for the apple glycosyltransferase MdUGT73CG22: (**a**) lane M is Mark; lane S1 is GST tag protein extracted from bacterial solution; lane S2 is MdUGT73CG22-GST fusion protein extracted from bacterial solution; lane S3 is purified GST tag protein; lane S4 is purified MdUGT73CG22 protein. Induced expression of apple glycosyltransferase MdUGT73CG22 protein GST tag: 26.0 KDa; MdUGT73CG22-GST: 81.0 KDa; (**b**) HPLC assay revealed that apple glycosyltransferase MdUGT73CG22 could modify nicosulfuron by glycosylation in vitro, with a peak time of 2.9 min for nicosulfuron and 2.3 min for nicosulfuron glycosides; (**c**) determination of nicosulfuron glycosides, the product of apple glycosyltransferase MdUGT73CG22, by molecular weight by HPLC-MS.

2.4. Acquisition of the MdUGT73CG22 Transgenic Apple Line and Identification of the Detoxification Function

In the above study, we found that the MdUGT73CG22 enzyme protein could catalyse nicosulfuron in vitro. Could MdUGT73CG22 also catalyse nicosulfuron in apple (*Malus* × *domestica Borkh.*)? We extracted the (*Malus* × *domestica Borkh.*) apple RNA, reverse transcribed it into cDNA, successfully cloned the *MdUGT73CG22* gene by PCR amplification, constructed the plant overexpression vector of *MdUGT73CG22*, and successfully obtained the *MdUGT73CG22* overexpression strains *OE14* and *OE26* through the mediation of *R. rhizogenes* strain K599 (Figure 4a). Then, we treated each line with an appropriate amount of nicosulfuron and extracted the total glycosides; the results showed that the content of nicosulfuron glycosides in the *MdUGT73CG22* overexpression lines were significantly increased compared with the control (Figure 4b). These results suggest that the MdUGT73CG22 enzyme protein can also catalyse nicosulfuron in vivo and convert it to nicosulfuron glycosides. Taken together, the results show that the apple (*Malus* × *domestica Borkh.*) glycosyltransferase MdUGT73CG22 can modify the ALS-inhibiting herbicide nicosulfuron by glycosylation, which may be involved in the detoxification process of plants and help to further improve the knowledge of the mechanism of non-target action of herbicides.

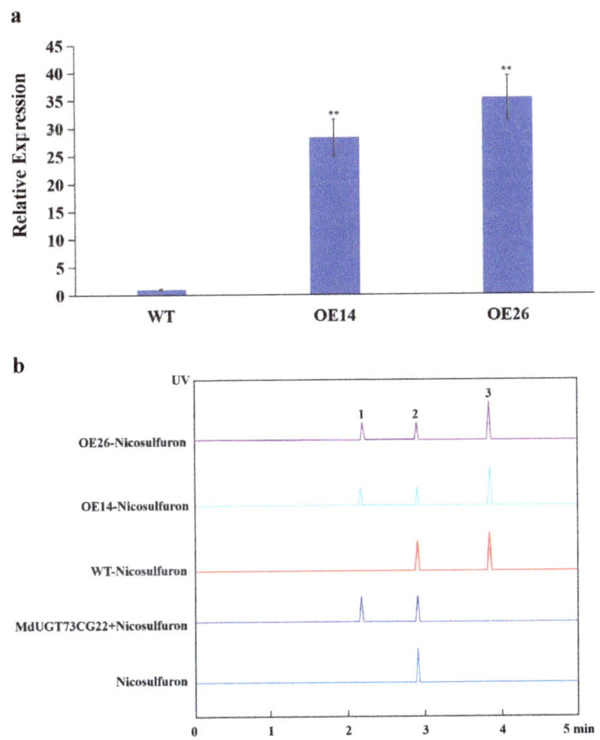

Figure 4. HPLC detection of nicosulfuron glycosides in apple glycosyltransferase gene *MdUGT73CG22* overexpression lines *OE14* and *OE26*: (**a**) expression level of gene *MdUGT73CG22* in overexpression lines detected by real-time PCR; (**b**) HPLC detection of nicosulfuron glycosides in *MdUGT73CG22* overexpression lines *OE14* and *OE26*. Note: 1, nicosulfuron glycosides, 2.3 min; 2, nicosulfuron, 2.9 min; 3, internal reference, 3.8 min. Note: ** $p < 0.01$, $n = 3$.

3. Discussion

Pesticides are effective in safeguarding agricultural production, controlling pests, diseases, and weeds, and achieving the goal of increasing crop production and generating income. However, excessive use of pesticides causes serious pollution and damage to the environment and ecology, which in turn threatens the entire food chain and human health [19]. Therefore, it is of great theoretical significance and practical value to study the metabolism and detoxification mechanism of pesticides in crops, reduce the accumulation of pesticides in crops, and ensure the safety and quality of agricultural products.

The degradation of pesticides usually involves hydrolysis, photolysis, and enzymatic degradation [20,21]. Pesticides can be biodegraded in plants through various metabolic and transformation pathways [22]. Studying its mechanism of action will help to clarify the composition of pesticide residues in agricultural commodities and to assess the dietary risk of pesticides. Meanwhile, it also can provide an important basis for the creation of new pesticides and the safe use of pesticides. The results of the related research can be an important scientific basis in the fields of crop breeding, transgenic plant engineering, and phytoremediation of environmental pollution.

The metabolic transformation of pesticides in plants is mainly divided into three stages. Phase I metabolism is the enzyme-catalysed reaction and abiotic degradation process, including oxidation, hydrolysis, and reduction [23]. In general, this process converts pesticides into more water-soluble compounds with lower phytotoxicity, while in some special cases, more phytotoxic substances may be generated, such as the metabolic activation of some herbicides in plants [24]. Phase II metabolism involves the conjugation of pesticides and their metabolites with plant endogenous substances including carbohydrates, amino acids, and glutathione [25]. The plant enzymes involved in phase II metabolism include glutathione S-transferases (GSTs), GTs, and malonyltransferases (MTs), among others [26]. Enzymes have substrate specificity; therefore, GTs can selectively search for their own substrate and combine with it. The conjugates formed in phase II metabolism are usually more water-soluble than the parent compound, have lower phytotoxicity, and can be stored in organelles, such as vesicles [27]. Phase III metabolism actively removes conjugates and associated catabolic metabolites formed by the phase II metabolism from the cytosol, mainly through adenosine triphosphate (ATP)-dependent transporter proteins, and deposits them into vesicles for temporary storage or further catabolic metabolism [28]; some conjugates are transported to other regions of plant cells, such as aggregation into the cell walls [27]. There are also some reports that classify some of the processes of phase III metabolism as phase IV metabolism, including the re-transport of vesicular products of phase III metabolism back into the cytosol for mineralisation and the incorporation of metabolic intermediates into the conjugated residues, among others [29–32]. However, at present, the definition of phase IV metabolism is not clear, and opinions are not consistent. In addition, the same pesticide can undergo different pathways of metabolism simultaneously in the same plant.

Numerous studies have shown that GTs play an important role in pesticide detoxification and plant selectivity to herbicides [33–36]. Zhang et al. investigated the metabolism of atrazine after overexpression of the *ARGT1* gene in rice (*Oryza* × *sativa*) and found that the glycosyltransferase 1 (ARGT1) played an important role in the metabolism and detoxification of atrazine in transgenic rice (*Oryza* × *sativa*). GTs were involved in the detoxification process of atrazine in *alfalfa Medicago sativa*, which was treated with atrazine, showed increased activities of GTs and GSTs (Glutathione S-transferase), and detected two O-glycosylated metabolites and homoglutathione (hGSH) conjugates [33]. Huang et al. found that *Arabidopsis thaliana* AtUGT91C1 played an important role in herbicide detoxification by glycosylating sulforaphane [34]. Yu et al. found that glutathione could reduce chlorothalonil (CHT) residues in tomato (*Lycopersicon esculentum*) by inducing the expression of the UDP-glycosyltransferase (UGT) gene [35]. Brazier- Hicks et al. reported that 44 GTs from the model plant *Arabidopsis thaliana* catalysed the O-glycosylation of the pesticide synthesis intermediate chlorophenols, with UGT72B1 also showing N-glycosylation activity towards chloranilides [36].

Individual genotypes of the cultivated apple tree (*Malus domestica Borkh.*) differ considerably in their fruit and vegetative characteristics. The results of apple genomic studies confirmed that cultivated apple varieties originate mainly from *Malus sieversii M. Roem* found in Central Asia. Other species such as the European apple (*Malus sylvestris Mill.*), the small-fruited *Malus baccata* (L.) *Borkh.* from Northeast Asia, and the Eastern Caucasian apple *Malus orientalis Uglitz.* also contributed to the creation of today's cultivated varieties. This heterogeneity can manifest itself when working with seedlings created by generative propagation, which is part of the methodical procedure. Therefore, the role of the glycosyltransferase MdUGT73CG22 identified in this study in the detoxification metabolism of nicosulfuron in apples (*Malus* × *domestica Borkh.*) may have species preference. In the following research, we will try to expand the sample size and elucidate the detoxification function of glycosyltransferase MdUGT73CG22 in different apple varieties.

In this paper, the metabolic mechanism of the apple (*Malus* × *domestica Borkh.*) glycosyltransferase gene regulating the detoxification of the acetolactate synthase (ALS) inhibitor herbicide nicosulfuron was investigated, using nicosulfuron as a test pesticide and molecular biology techniques, which resulted in the first discovery of a novel apple (*Malus* × *domestica Borkh.*) glycosyltransferase, MdUGT73CG22, which can glycosylate ALS-inhibiting herbicide nicosulfuron and may be involved in the plant detoxification process. This helps us to better understand and utilize herbicides, which are convenient and do not cause pollution to the environment or damage the plants themselves. This study will lay the foundation for subsequent biological detoxification research.

4. Materials and Methods

4.1. Plant Material

We selected the roots, stems, leaves, flowers, and fruits of the 10-year-old *Malus* × *domestic Borkh.* apple tree planted by the School of Life Sciences, Shandong Agricultural University. Twenty samples were randomly collected from each plant, with a total of 10 strains collected. Please refer to "Apple Germplasm" in the "Resource Description Specification and Data Standards" [37] for details. The samples were frozen in liquid nitrogen and stored at −80 °C for expression pattern analysis of the glycosyltransferase gene *MdUGT73CG22*.

Transgenic apple (*Malus* × *domestica Borkh.*) lines (*MdUGT73CG22-OE14* and *MdUGT73CG22-OE26*) were obtained by transforming the leaves of Gala using tissue culture seedlings of the clone apple (*Malus* × *domestica Borkh.*) "GL-3". Tissue culture seedlings were grown at 22 °C, 60 $\mu mol \cdot m^{-2} \cdot s^{-1}$, and 14/10 h photoperiod. After 30 d of succession culture and 45 d of rooting medium culture, the seedlings were incubated for 1 week, and then the tissue-cultured seedlings were transferred to plastic pots (8.5 × 8.5 × 7.5 cm^3) and incubated in a light incubator.

4.2. Treatment of Apple Seedlings with ALS-Inhibiting Herbicides

The stratified apple (*Malus* × *domestica Borkh.*) seeds were placed in a light incubator at 25 °C with 14 h of light and 10 h of darkness until the root length reached 3 cm; they were then transplanted into a 1:1 soil of nutrient soil and vermiculite and cultured until the six-leaf stage. A certain amount of acetolactate synthase (ALS) inhibitor herbicides (Table 1) was weighed separately, dissolved in DMSO, and prepared to a final concentration of 10 μmol/L by dilution with ddH_2O water. Then, 50 mL of each compound solution was taken and placed in a petri dish, the roots of the 6-leaf stage apple (*Malus* × *domestica Borkh.*) seedlings were placed in the solution, the plant materials were taken at five time points, 0 h, 3 h, 6 h, 12 h and 24 h, respectively, and they were wrapped in tin foil and labelled to be placed in a refrigerator at −80 °C for subsequent RNA extraction.

Table 1. Specific information on the acetolactate synthase (ALS) inhibitor class of herbicides.

Chemical Compound	Structure Name	Structural Formula	Manufacturer	CAS
Bensulfuron methyl	[3-(4,6-dimethoxypyrimidin-2-yl)-1-(2-methoxycarbonylbenzyl)sulfonylurea]	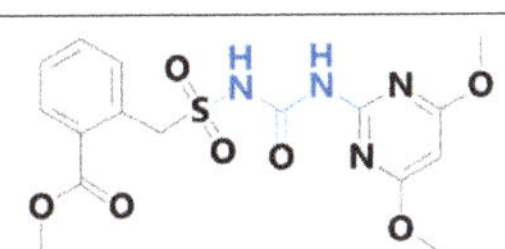	Bermuda Biotechnology Co. (Nanjing, China)	83055-99-6
Nicosulfuron	[2-(4,6-dimethoxypyrimidin-2-ylcarbamoylsulfamoyl)-N,N-dimethylnicotinamide]	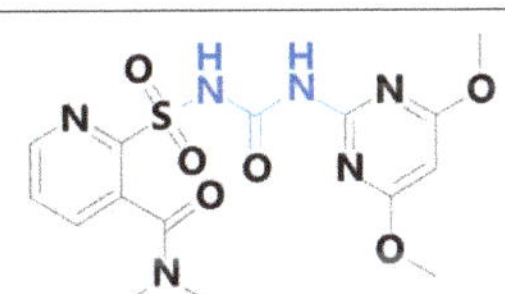	BioLabs Technology Ltd. (Beijing, China)	111991-09-4
Thifensulfuron	[3-(4-methoxy-6-methyl-1,3,5-triazin-2-ylaminocarbonylsulfamoyl)-thiophene-2-carboxylic acid]	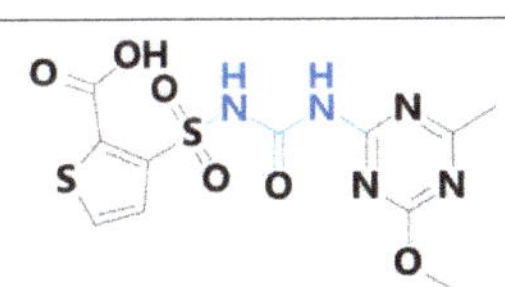	Wittig Chemical Technology Co. (Zhengzhou, China)	79277-67-1

Table 1. *Cont.*

Chemical Compound	Structure Name	Structural Formula	Manufacturer	CAS
Chlorsulfuron	[1-(2-Chlorophenylsulfonyl)3-(4-methoxy-6-methyl-1,3,5-triazin-2-yl)urea]	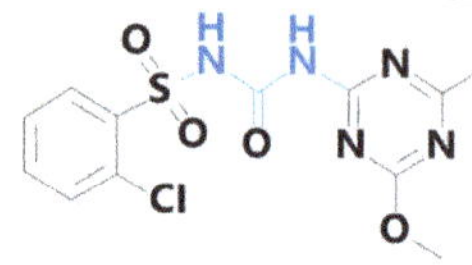	BioLabs Technology Ltd. (Beijing, China)	64902-72-3
Metsulfuron	[3-(4-methoxy-6-methyl-1,3,5-triazin-2-yl)-1-(2-formylphenyl)sulfonylurea]	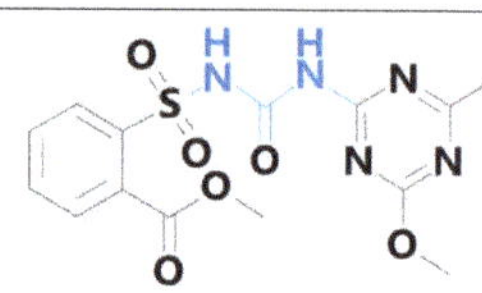	BioLabs Technology Ltd. (Beijing, China)	142469-14-5
Triasulfuron	[1-[2-(2-chloroethoxy)phenylsulfonyl]-3-(4-methoxy-6-methyl-1,3,5-triazin-2-yl)urea]	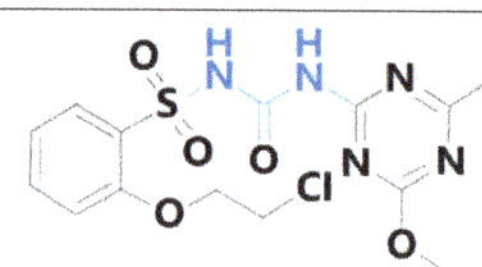	Yuanye Biotechnology Co. (Shanghai, China)	82097-50-5
Cinosulfuron	[3-(4,6-dimethoxy-1,3,5-triazin-2-yl)-1-[2-(2-methoxyethoxy)phenyl]sulfonylurea]	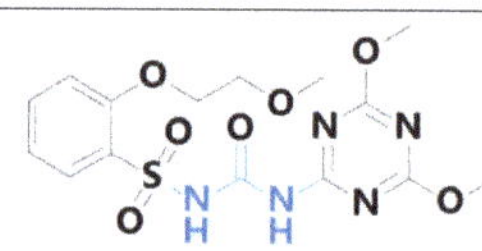	BioLabs Technology Ltd. (Beijing, China)	94593-91-6
Chlorimuron-ethyl	[2-(4-chloro-6-methoxypyrimidin-2-ylcarbamoylsulfamoyl)benzoic acid]	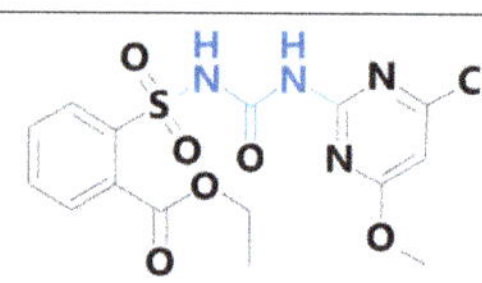	BioLabs Technology Ltd. (Beijing, China)	99283-00-8
Sulfometuron	[2-(3-(4,6-dimethylpyrimidin-2-yl)ureasulfonyl)benzoic acid]	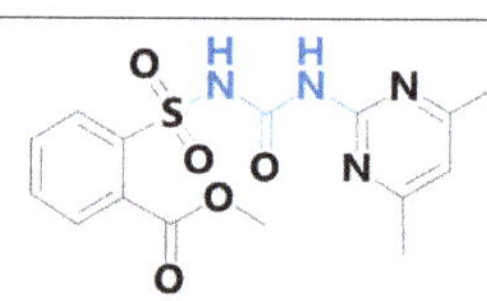	Eimage Technology Co. (Wuhan, China)	74222-97-2
Pyrazosulfuron	[5-(4,6-dimethoxypyrimidin-2-ylcarbamoylsulfamoyl)-1-methylpyrazole-4-carboxylic acid]	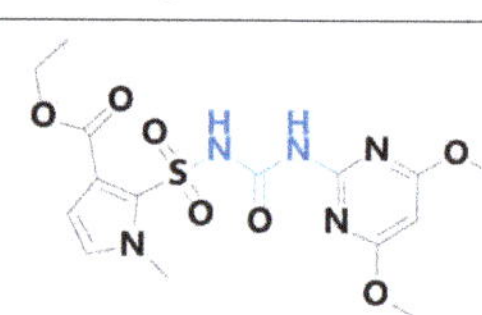	Guyan Industry Co. (Shanghai, China)	98389-04-9
Rimsulfuron	[1-(4,6-dimethoxy-2-pyrimidinyl)-3-[3-(ethylsulfonyl)-2-pyridinesulfonyl]urea]	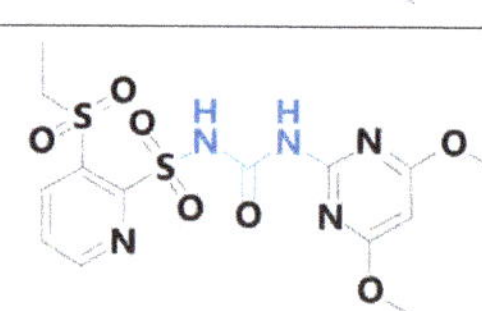	Yuanye Biotechnology Co. (Shanghai, China)	122931-48-0
Ethametsulfuron	[Methyl2-[(4-ethoxy-6-methylamino-1,3,5-triazin-2-yl)carbamoyl sulfamoyl]benzoate]	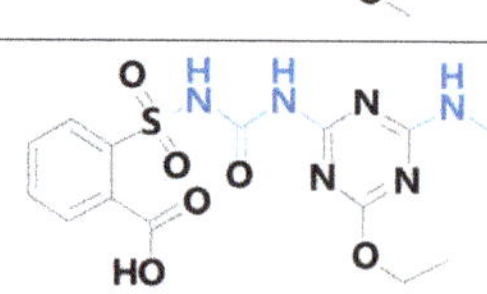	Yuanye Biotechnology Co. (Shanghai, China)	111353-84-5

4.3. Real-Time PCR

For RNA extraction, we referred to the instructions of the plant RNA kit polysaccharide polyphenol/complex plant RNA rapid extraction kit, as SPARKscript II All-in-one RT SuperMix (AC0307, Sparkjade, Jinan, China).

For cDNA reverse transcription, we referred to the instructions of the reverse tran scription kit, as 2 × Spark Taq PCR Master Mix (AG0305, Sparkjade, Jinan, China).

RT-qPCR was performed using SYBR Green PCR Master Mix kit (Q111-02, VazymE, Nanjing, China) by Bio-Rad Thermal Cycling System (CFX Connect, Hercules, CA, USA). The expression level of the reference gene β-actin was standardised using the ΔΔCt method to normalise the expression levels; the primer sequences are shown in Table S2, and the nucleotide and amino acid sequences of MdUGT73CG22 are shown in Table S3.

4.4. Expression Vector Construction

Pfu DNA Ploymerase (PC1300, Solepol Biotechnology Co., Shanghai, China) was used to amplify the target fragment from the apple (*Malus × domestica Borkh.*) seedling cDNA; then, the target fragment was recovered using an agarose gel DNA recovery kit (ZY511-100, Runye Biotechnology Co., Shanghai, China). The gel-recovered product was ligated into a pMD18-T vector, transformed into *E. coli* DH5α, and screened for recombinant plasmids by the blue–white spot method. The recombinant plasmid was double digested to identify the positive clones and sent to BGI Genomics Co., Ltd. (Shenzhen, China) for sequencing of the target fragments. Here, 100% of the sequenced fragments were correctly sequenced; then, the target gene *MdUGT73CG22* was recombinantly transferred into the prokaryotic expression vector pGEX-2T and the plant expression vector PBI121, respectively. After bacteriophage PCR and double enzyme digestion verification, the positive plasmids *MdUGT73CG22*-pGEX-2T (Figure S1a) and *MdUGT73CG22*-PBI121 (Figure S1b) were transfected into the protein expression strains *E. coli* BL-21 and *R. rhizogenes* strain K599, respectively, for the subsequent protein purification and the obtainment of transgenic strains, respectively.

4.5. Protein Purification and In Vitro Enzymatic Reactions

Protein purification: single bacteria containing *MdUGT73CG22*-pGEX-2T were inoculated into LB liquid medium at 200 rpm/min and incubated at 37 °C overnight. The activated strain was inoculated in 100 mL LB medium containing kanamycin (50 μg/mL) at a ratio of 1:50, and when the OD_{600} value of the bacterial solution was about 0.6, the bacterial solution was inoculated with a final concentration of 0.4 mmol/L of Isopropyl-beta-D-thiogalactopyranoside (IPTG), at a final concentration of 0.4 mmol/L. Soluble recombinant protein expression was induced with Isopropyl-beta-D-thiogalactopyranoside (IPTG) and incubated at 20 °C, 200 rpm/min, for 24 h. The induced bacterial fluids were subjected to target protein purification according to the instructions of the Beyotime GST-tag Protein Purification Kit (P2262, Beyotime, Shanghai, China).

In vitro enzymatic reaction: the protein obtained by purification as described above was subjected to an in vitro enzymatic reaction with the target compounds in a specific system, consisting of 10 μL of 0.5 M Tris-HCl (pH 8.0), 5 μL of 50 mM $MgSO_4$, 5 μL of 200 mM KCl, 2.5 μL of 0.1 M UDP-glucose, 1 μL of 10% β-mercaptoethanol, 1 μL of 100 mM nicosulfuron, 2 μL MdUGT73CG22 protein, and ddH_2O, and the volume was made up to 100 μL by adding ddH_2O.

4.6. Transgenic Apple Seedling Screening

The test material used for apple genetic transformation was the tissue culture seedlings of clone "GL-3", with strong regenerative ability and high transformation efficiency, obtained from the "Gala" apple. The leaves of "GL-3" were transformed with *R. rhizogenes* strain K599, which carries the sequence *MdUGT73CG22*-PBI121, and 50 mg·L^{-1} kanamycin and 250 mg·L^{-1} sodium cefotaxime were added to the MS medium for the selection of regenerating shoots. During the selection process, slow-growing and yellowing regeneration shoots were discarded, and normal-growing regeneration shoots continued to be cultured on the selection medium until they developed into robust individuals with stem and leaf tissues.

DNA was extracted from "GL-3" and transgenic apple plants using the PureLink™ Genomic Plant DNA Purification Kit (K183001, Thermo Fisher Scientific, Waltham, MA, USA). The partial sequence of the CaMV 35S promoter on the vector was used as the upstream primer, the downstream primer was designed with the sequence of *MdUGT73CG22* gene for PCR amplification, and electrophoresis was carried out to detect whether it was the target gene or not.

4.7. Endogenous Glycoside Extraction from Transgenic Apple Seedlings

Five- to six-week-old wild-type and successful transgenic apple (*Malus* × *domestica Borkh.*) seedlings treated with the target compounds for 12 h at a final concentration of 10 µmol/L were frozen in liquid nitrogen, powdered, and homogenised in 10 mL of 90% (*v*/*v*) methanol with the internal standard. After centrifugation at 10,000× *g* for 10 min, the supernatant was transferred to a new centrifuge tube, the remaining precipitate was re-extracted with 100% (*v*/*v*) methanol and centrifuged at 10,000× *g* for 10 min, the supernatants were combined into the supernatant centrifuge tube from the previous step, and the combined supernatant was dried by heating in a fast vacuum (60 °C). The residue was resuspended in 500 µL of 20% (*v*/*v*) trichloroacetic acid and sonicated for 10 min; then, it was extracted with 2 volumes of ethyl acetate-cyclopentane-isopropanol (50:50:1). The organic phase containing fumetsulfuron was then dried by heating in a rapid vacuum (60 °C). The residue was dissolved in 100 µL of aqueous solution containing 10% MeOH (*v*/*v*) and 0.1% TFA (*v*/*v*) and vortexed for 1 min. The fraction can be used for HPLC analysis.

4.8. High Performance Liquid Chromatography (HPLC) Analysis of Target Products

The HPLC detection conditions were as follows: Agilent 1290-6470A ultra performance liquid chromatography–tandem mass spectrometry, Agilent (G4261B, Agilent, Santa Clara, CA, USA); Poroshelll20EC-C18 column (2.1 mm × 50 mm, 1.9 µm); column temperature: room temperature; detection wavelength: 241 nm; mobile phase A: 0.2% formic acid in water; mobile phase B: acetonitrile. The flow rate was 0.3 mL/min; the injection volume was 5 µL.

The gradient elution conditions were as follows: 0.00 min, phase A 95%, phase B 5%; 0.50 min, phase A 95%, phase B 5%; 1.60 min, phase A 20%, phase B 80%; 3.00 min, phase A 10%, phase B 90%; 4.00 min, phase A 10%, phase B 90%; 4.10 min, phase A 95%, phase B 5%; 4.50 min, phase A 95%, phase B 5%.

The mass spectrometry conditions were as follows: electrospray ion source, positive ion ionisation mode (ESI+), drying gas temperature 325 °C, drying gas flow rate 10 L/min. nebulising gas pressure 50 psi, sheath gas temperature 400 °C, sheath gas flow rate 12 L/min, capillary voltage 4000 V, nozzle voltage 500 V; data acquisition was performed in multiple reaction monitoring mode.

Mass Spectrometer: retention time 2.09 min; parent ion 411.1; daughter ion 214 and 182, cone voltage 106 VL; collision energy 14 and 20 eV.

4.9. Data Statistics and Analyses

All experiments were performed with at least three independent biological replicates, and each measurement was repeated three times. Data are expressed as means SD. Data were statistically analysed using Student's *t*-test. * indicates significant differences (* $p < 0.05$, ** $p < 0.01$) relative to wild-type or controls in different experimental treatments.

Supplementary Materials: The following supporting information can be downloaded at: https://www.mdpi.com/article/10.3390/plants13091171/s1, Figure S1: Induced expression of apple glycosyltransferase MdUGT73CG22 protein GST tag: 26.0 KDa; MdUGT73CG22-GST: 81.0 KDa;

Table S1: mRNA expression of nine glycosyltransferase genes induced by class 12 acetolactate synthase (ALS) inhibitor herbicides; Table S2: Gene primers; Table S3: Nucleotide and amino acid sequences of MdUGT73CG22.

Author Contributions: All authors contributed to the study's conception and design. Conceptualization, Y.Z. and A.Z.; methodology, L.M.; software, X.T. and Y.M.; validation, R.L. and Y.Z.; formal analysis, R.L.; investigation, P.L.; resources, K.L. and L.J.; data curation, A.Z.; writing—original draft preparation, L.M.; writing—review and editing, Y.Z.; visualization, Y.M.; supervision, L.M.; project administration, A.Z.; funding acquisition, P.L., L.J. and X.W. All authors have read and agreed to the published version of the manuscript.

Funding: This research was funded by the Shandong Provincial Natural Science Foundation of China, grant number No. ZR202102180037, and the Guangyue Young Scholar Innovation Team of Liaocheng University, grant number No. LCUGYTD2022-04.

Data Availability Statement: Data will be made available on request.

Acknowledgments: We are grateful to the College of Life Sciences of Shandong Agricultural University for providing the plant materials.

Conflicts of Interest: The authors declare no conflicts of interest.

References

1. Choudri, B.S.; Charabi, Y.; Al-Nasiri, N. Pesticides and herbicides. *Water Environ. Res.* **2020**, *92*, 1425–1432. [CrossRef] [PubMed]
2. Gaines, T.A.; Duke, S.O.; Morran, S.; Rigon, C.A.G.; Tranel, P.J.; Küpper, A.; Dayan, F.E. Mechanisms of evolved herbicide resistance. *J. Biol. Chem.* **2020**, *295*, 10307–10330. [CrossRef] [PubMed]
3. Milesi, M.M.; Lorenz, V.; Durando, M.; Rossetti, M.F.; Varayoud, J. Glyphosate Herbicide: Reproductive Outcomes and Multigenerational Effects. *Front. Endocrinol.* **2021**, *12*, 672532. [CrossRef] [PubMed]
4. Widianto, R.; Kurniadie, D.; Widayat, D.; Umiyati, U.; Nasahi, C.; Sari, S.; Juraimi, A.S.; Kato-Noguchi, H. Acetolactate Synthase-Inhibitor Resistance in *Monochoria vaginalis* (Burm. f.) C. Presl from Indonesia. *Plants* **2022**, *11*, 400. [CrossRef] [PubMed]
5. Low, Y.S.; Garcia, M.D.; Lonhienne, T.; Fraser, J.A.; Schenk, G.; Guddat, L.W. Triazolopyrimidine herbicides are potent inhibitors of *Aspergillus fumigatus* acetohydroxyacid synthase and potential antifungal drug leads. *Sci. Rep.* **2021**, *11*, 21055. [CrossRef] [PubMed]
6. Cheron, M.; Costantini, D.; Brischoux, F. Nicosulfuron, a sulfonylurea herbicide, alters embryonic development and oxidative status of hatchlings at environmental concentrations in an amphibian species. *Ecotoxicol. Environ. Saf.* **2022**, *232*, 113277. [CrossRef] [PubMed]
7. Petric, I.; Karpouzas, D.G.; Bru, D.; Udikovic-Kolic, N.; Kandeler, E.; Djuric, S.; Martin-Laurent, F. Nicosulfuron application in agricultural soils drives the selection towards NS-tolerant microorganisms harboring various levels of sensitivity to nicosulfuron. *Environ. Sci. Pollut. Res. Int.* **2016**, *23*, 4320–4333. [CrossRef] [PubMed]
8. Hassan, S.; Ganai, B.A. Deciphering the recent trends in pesticide bioremediation using genome editing and multi-omics approaches: A review. *World J. Microbiol. Biotechnol.* **2023**, *39*, 151. [CrossRef]
9. Márquez-López, R.E.; Loyola-Vargas, V.M.; Santiago-García, P.A. Interaction between fructan metabolism and plant growth regulators. *Planta* **2022**, *255*, 49. [CrossRef]
10. Biswas, A.; Thattai, M. Promiscuity and specificity of eukaryotic glycosyltransferases. *Biochem. Soc. Trans.* **2020**, *48*, 891–900. [CrossRef]
11. Dimunová, D.; Matoušková, P.; Podlipná, R.; Boušová, I.; Skálová, L. The role of UDP-glycosyltransferases in xenobioticresistance. *Drug Metab. Rev.* **2022**, *54*, 282–298. [CrossRef] [PubMed]
12. Kurze, E.; Wüst, M.; Liao, J.; McGraphery, K.; Hoffmann, T.; Song, C.; Schwab, W. Structure-function relationship of terpenoid glycosyltransferases from plants. *Nat. Prod. Rep.* **2022**, *39*, 389–409. [CrossRef] [PubMed]
13. Taylor, V.L.; Cummins, I.; Brazier-Hicks, M.; Edwards, R. Protective responses induced by herbicide safeners in wheat. *Environ. Exp. Bot.* **2013**, *88*, 93–99. [CrossRef]
14. Messner, B.; Thulke, O.; Schäffner, A.R. Arabidopsis glucosyltransferases with activities toward both endogenous and xenobiotic substrates. *Planta* **2003**, *217*, 138–146. [CrossRef] [PubMed]
15. Loutre, C.; Dixon, D.P.; Brazier, M.; Slater, M.; Cole, D.J.; Edwards, R. Isolation of a glucosyltransferase from Arabidopsis thaliana active in the metabolism of the persistent pollutant 3,4-dichloroaniline. *Plant J.* **2003**, *34*, 485–493. [CrossRef] [PubMed]
16. von Rad, U.; Hüttl, R.; Lottspeich, F.; Gierl, A.; Frey, M. Two glucosyltransferases are involved in detoxification of benzoxazinoids in maize. *Plant J.* **2001**, *28*, 633–642. [CrossRef]
17. Su, X.; Wang, W.; Xia, T.; Gao, L.; Shen, G.; Pang, Y. Characterization of a heat responsive UDP: Flavonoid glucosyltransferase gene in tea plant (*Camellia sinensis*). *PLoS ONE* **2018**, *13*, e0207212. [CrossRef]

18. Li, Y.; Li, P.; Zhang, L.; Shu, J.; Court, M.H.; Sun, Z.; Jiang, L.; Zheng, C.; Shu, H.; Ji, L.; et al. Genome-wide analysis of the apple family 1 glycosyltransferases identified a flavonoid-modifying UGT, MdUGT83L3, which is targeted by MdMYB88 and contributes to stress adaptation. *Plant Sci.* **2022**, *321*, 111314. [CrossRef]
19. Singh, Y.; Saxena, M.K. Insights into the recent advances in nano-bioremediation of pesticides from the contaminated soil. *Front. Microbiol.* **2022**, *13*, 982611. [CrossRef]
20. Gangola, S.; Bhatt, P.; Kumar, A.J.; Bhandari, G.; Joshi, S.; Punetha, A.; Bhatt, K.; Rene, E.R. Biotechnological tools to elucidate the mechanism of pesticide degradation in the environment. *Chemosphere* **2022**, *296*, 133916. [CrossRef]
21. Mörtl, M.; Vehovszky, Á.; Klátyik, S.; Takács, E.; Győri, J.; Székács, A. Neonicotinoids: Spreading, Translocation and Aquatic Toxicity. *Int. J. Environ. Res. Public Health* **2020**, *17*, 2006. [CrossRef] [PubMed]
22. Guerrero Ramírez, J.R.; Ibarra Muñoz, L.A.; Balagurusamy, N.; Frías Ramírez, J.E.; Alfaro Hernández, L.; Carrillo Campos, J. Microbiology and Biochemistry of Pesticides Biodegradation. *Int. J. Mol. Sci.* **2023**, *24*, 15969. [CrossRef] [PubMed]
23. Qiu, J.; Zhang, T.; Zhu, F.; Ouyang, G. In vivo monitoring and exposure potency assessment of phase I metabolism of fenthion in vegetables. *J. Hazard. Mater.* **2020**, *399*, 123013. [CrossRef] [PubMed]
24. Zhang, R.; Zhang, Z.; Li, R.; Tan, Y.; Lv, S.; McClements, D.J. Impact of pesticide polarity and lipid phase dimensions on the bioaccessibility of pesticides in agricultural produce consumed with model fatty foods. *Food Funct.* **2020**, *11*, 6028–6037. [CrossRef]
25. Singh, S.; Kumar, V.; Gill, J.P.K.; Datta, S.; Singh, S.; Dhaka, V.; Kapoor, D.; Wani, A.B.; Dhanjal, D.S.; Kumar, M.; et al. Herbicide Glyphosate: Toxicity and Microbial Degradation. *Int. J. Environ. Res. Public Health* **2020**, *17*, 7519. [CrossRef] [PubMed]
26. Sirivarasai, J.; Chanprasertyothin, S.; Kongtip, P.; Woskie, S. Genetic Polymorphisms of Pesticide-Metabolizing Enzymes and Transporters in Agricultural Workers and Thyroid Hormone Levels. *Risk Manag. Healthc. Policy* **2021**, *14*, 3435–3451. [CrossRef] [PubMed]
27. Li, R.; Wang, S.; Chang, J.; Pan, X.; Dong, F.; Li, Z.; Zheng, Y.; Li, Y. Insight into the uptake and metabolism of a new insecticide cyetpyrafen in plants. *Environ. Int.* **2022**, *169*, 107522. [CrossRef] [PubMed]
28. Tungu, P.K.; Rowland, M.W.; Messenger, L.A.; Small, G.J.; Bradley, J.; Snetselaar, J.; Kirby, M.J.; Mbewe, N.J. Large-scale (Phase III) evaluation of broflanilide 50WP (VECTRON T500) for indoor residual spraying for malaria vector control in Northeast Tanzania: Study protocol for a two-arm, non-inferiority, cluster-randomised community trial. *BMC Infect. Dis.* **2022**, *22*, 171. [CrossRef] [PubMed]
29. Kawahigashi, H. Transgenic plants for phytoremediation of herbicides. *Curr. Opin. Biotechnol.* **2009**, *20*, 225–230. [CrossRef]
30. Schwartz, M.D.; Raulli, R.; Laney, J.W.; Coley, W.; Walker, R.; O'Rourke, A.W.; Raine, K.; Horwith, G.; Gao, Y.; Eisnor, D.L.; et al. Systemic Bioavailability of Sublingual Atropine Ophthalmic Solution: A Phase I Study in Healthy Volunteers with Implications for Use as a Contingency Medical Countermeasure. *J. Med. Toxicol.* **2022**, *18*, 187–197. [CrossRef]
31. Qiao, Y.; Zhang, A.P.; Ma, L.Y.; Zhang, N.; Liu, J.; Yang, H. An ABCG-type transporter intensifies ametryn catabolism by Phase III reaction mechanism in rice. *J. Hazard. Mater.* **2023**, *457*, 131804. [CrossRef] [PubMed]
32. Jie, C.Z.; Yan, Z.X.; Liu, J.; Zhang, N.; Yang, H. Detoxification and catabolism of mesotrione and fomesafen facilitated by a Phase II reaction acetyltransferase in rice. *J. Adv. Res.* **2023**, *51*, 1–11.
33. Zhang, J.J.; Gao, S.; Xu, J.Y.; Lu, Y.C.; Lu, F.F.; Ma, L.Y.; Su, X.N.; Yang, H. Degrading and Phytoextracting Atrazine Residues in Rice (*Oryza sativa*) and Growth Media Intensified by a Phase II Mechanism Modulator. *Environ. Sci. Technol.* **2017**, *51*, 11258–11268. [CrossRef] [PubMed]
34. Huang, X.X.; Zhao, S.M.; Zhang, Y.Y.; Li, Y.J.; Shen, H.N.; Li, X.; Hou, B.K. A novel UDP-glycosyltransferase 91C1 confers specific herbicide resistance through detoxification reaction in Arabidopsis. *Plant Physiol. Biochem.* **2021**, *159*, 226–233. [CrossRef] [PubMed]
35. Yu, G.; Chen, Q.; Chen, F.; Liu, H.; Lin, J.; Chen, R.; Ren, C.; Wei, J.; Zhang, Y.; Yang, F.; et al. Glutathione Promotes Degradation and Metabolism of Residual Fungicides by Inducing UDP-Glycosyltransferase Genes in Tomato. *Front. Plant Sci.* **2022**, *13*, 893508. [CrossRef]
36. Brazier-Hicks, M.; Edwards, R. Functional importance of the family 1 glucosyltransferase UGT72B1 in the metabolism of xenobiotics in Arabidopsis thaliana. *Plant J.* **2005**, *42*, 556–566. [CrossRef]
37. Wang, K.; Liu, F.Z.; Cao, Y.F. *Apple Germplasm Resources Description Specification and Data Criteria;* China Agricultural Press: Beijing, China, 2005.

plants

MDPI

Article

Allelopathic Effects of Corn Straw and Its Water Extracts on Four Weed Species and Foxtail Millet

Shuqi Dong [1,†], Jiaxin Dong [1,†], Peiyao Li [1], Bo Cao [1], Mengyao Liu [1], Zhenyu Guo [2], Xie Song [1], Yongqing Ma [3], Chunyan Hu [2,*] and Xiangyang Yuan [1,*]

1 College of Agriculture, Shanxi Agricultural University, Jinzhong 030800, China; dongshuqi@sxau.edu.cn (S.D.); djx199907@163.com (J.D.); peiyaoli666@163.com (P.L.); caobo000315@163.com (B.C.); 15903556943@163.com (M.L.); sxndsxe@163.com (X.S.)
2 College of Plant Protection, Shanxi Agricultural University, Jinzhong 030800, China; gzy2816@163.com
3 Institute of Soil and Water Conservation, CAS & MWR, Yangling 712100, China; mayongqing@ms.iswc.ac.cn
* Correspondence: hucy4216@163.com (C.H.); yuanxiangyang200@163.com (X.Y.)
† These authors have contributed equally to this work.

Abstract: Straw covering is a protective tillage measure in agricultural production, but there is relatively little research on the allelopathic effects of corn straw on weeds and foxtail millet. This experiment studied the allelopathic effects of corn straw on four weeds (*Chenopodium album, Setaria viridis, Echinochloa crus-galli* and *Amaranthus retroflexus*) in foxtail millet fields, and also measured the growth indicators of foxtail millet. The study consisted of Petri dish and field experiments. Five treatments were used in the Petri dish experiment: clear water as control (0 g/L, TCK) and four types of corn straw water extracts. They were, respectively, the stock solution (100 g/L, T1), 10 X dilution (10 g/L, T2), 50 X dilution (2 g/L, T3), and 100 X dilution (1 g/L, T4) of corn straw water extracts. Additionally, seven treatments were set up in the field experiment, consisting of three corn straw covering treatments, with covering amounts of 3000 (Z1), 6000 (Z2) and 12,000 kg/ha (Z3), and four control treatments—one treatment with no corn straw cover (CK) and three treatments involving the use of a black film to create the same shading area as the corn straw covered area, with black film coverage areas of 50% (PZ1), 70% (PZ2), and 100% (PZ3), respectively. The results showed that the corn straw water extract reduced the germination rate of the seeds of the four weeds. The T1 treatment resulted in the allelopathic promotion of *C. album* growth but the inhibition of *S. viridis, E. crus-galli*, and *A. retroflexus* growth. Treatments T2, T3, and T4 all induced the allelopathic promotion of the growth of the four weeds. The order of the effects of the corn straw water extracts on the comprehensive allelopathy index of the four weed seeds was as follows: *C. album* > *S. viridis* > *A. retroflexus* > *E. crus-galli*. With an increase in the corn straw mulching amount, the density and total coverage of the four weeds showed a gradual downward trend, whereas the plant control effect and fresh weight control effect showed a gradual upward trend. All indices showed the best results under 12,000 kg/ha of mulching and returning to the field. Overall, corn straw coverage significantly impacted the net photosynthetic rate and transpiration rate of foxtail millet and increased the yield of foxtail millet. Under coverages of 6000 and 12,000 kg/ha, the growth of foxtail millet is better. Based on our findings, we recommend a corn straw coverage of 12,000 kg/ha for the allelopathic control of weeds in foxtail millet fields.

Keywords: corn straw; water extract; allelopathy; weeds control

1. Introduction

Foxtail millet (*Setaria italica* (L.) Beauv) is an annual gramineous plant that originated in China [1,2] and has a short growth period, rich nutritional value [3–5], and strong drought and salt stress resistance. It is the main miscellaneous grain crop in dry farming ecological agriculture [6,7] and can be used as a substitute crop to supplement other economically valuable crops [8]. After years of development, the foxtail millet industry has made continuous progress under the premise of green and sustainable ecological development. In recent years, with the support of the government and promotion of the market, foxtail millet industrialization has accelerated. Foxtail millet is widely cultivated across northern China [9]. With the structural adjustment of the agricultural industry and the increased demand for miscellaneous grains, the yield of foxtail millet has shown a gradual upward trend. However, many types of weeds are found in foxtail millet fields that have strong stress resistance and are difficult to control, and few herbicides are suitable for foxtail millet fields [10]. Currently, only seven herbicides are registered with the China Pesticide Information Network. Foxtail millet is very sensitive to chemical herbicides [11]. Therefore, manual weeding remains the main method of weed control during the growth period of foxtail millet, which is laborious.

Straw is an important byproduct of crop production and a valuable renewable energy source. After straw covering, chemical substances are released through pathways such as root secretion, stem and leaf volatilization, rain and fog leaching, and plant residue decomposition, which affect the surrounding plants or microorganisms [12–14]. These substances are called allelopathic substances and are secondary metabolites of diverse types with a small molecular weight and simple structure. They affect the growth of neighboring crops, target crops, and weeds through allelopathic effects [15,16]. The majority of discovered allelochemicals are terpenes, phenols, and organic acids. These allelopathic substances can exhibit direct or indirect inhibitory or promoting effects on different receptor plants or microorganisms [17].

Straw covering can effectively reduce soil moisture evaporation and regulate soil temperature [18–20]. Microbial decomposition and rainwater leaching covered with straw can improve soil fertility [21] and physicochemical properties [22–24], can promote soil water fertilizer interaction, and are widely used in conservation tillage [25,26]. Cai et al. [27] revealed that the soil water storage capacity of corn straw covered within a soil depth of 200 cm increased. Additionally, grain yield showed an increase with the increase in straw coverage, improving water use efficiency. Valvi et al. [28] showed that the water extract of teak leaves had an inhibitory effect on rice germination and other growth parameters, which was proportional to the concentration of the extract; the higher the concentration, the stronger the inhibitory effect. Akhtar et al. [29] showed that corn straw mulching reduces the demand for nitrogen fertilizer for wheat by regulating soil moisture and temperature and improves the nitrogen use efficiency, yield, and water use efficiency of wheat. Crawford et al. [30] confirmed that the straw mulching of cereal crops inhibited the growth of weed seedlings in edamame fields and reduced the emergence rate of weeds in edamame fields by 20% whilst not interfering with the growth of soybean. Akhtar et al. [31] found that wheat straw mulching significantly improved photosynthesis, SPAD value, growth, biomass, and the seed yield of soybeans. Zhu et al. [32] found that the biomass of *E. crus-galli* covered with 1, 3, 5, 7, and 9 cm rapeseed straw decreased by 65.74%, 80.18%, 81.15%, 70.99%, 55.65%, and 27.22%, respectively. Mehra et al. [33] illustrated that the severity of wheat diseases varies depending on residual coverage rate. Disease severity increased nonlinearly with the residual level increase, and residual coverage had a significant effect on disease severity in all treatments.

Corn is the predominant crop grown in China. However, there are few reports on the allelopathic effects of corn straw on weeds in foxtail millet fields. In this study, using a combination of Petri dish and field experiments, the biological indices of four weeds in foxtail millet fields were measured in a Petri dish experiment and after corn straw mulching, and the number of weeds was investigated in a field experiment. Additionally, we analyzed the growth and physiological indicators of foxtail millet and the allelopathic effects of corn straw on foxtail millet and its associated weeds. These findings provide a scientific reference for the application of corn straw in foxtail millet field mulching.

2. Results

2.1. Effects of Corn Straw Water Extracts on Four Weed Species

All the concentrations of corn straw water extract reduced the germination rate of the seeds of the four weeds. Compared with that of TCK, the T4 treatment significantly reduced the germination rate of *C. album* and *A. retroflexus*. The seed germination rates of *C. album*, *S. viridis*, and *E. crus-galli* was lowest under the T1 treatment, which significantly decreased by only 25.55%, 23.33% and 30.00%, respectively, compared with that of TCK, whilst the seed germination rates of all four weeds were significantly reduced compared with that of TCK. The T2 treatment significantly reduced the germination rates of *E. crus-galli* and *A. retroflexus* by 11.11% and 18.89%, respectively, compared with that of TCK. There was no significant difference in the germination rates of the four weed seeds between the T3 treatment and TCK groups (Figure 1).

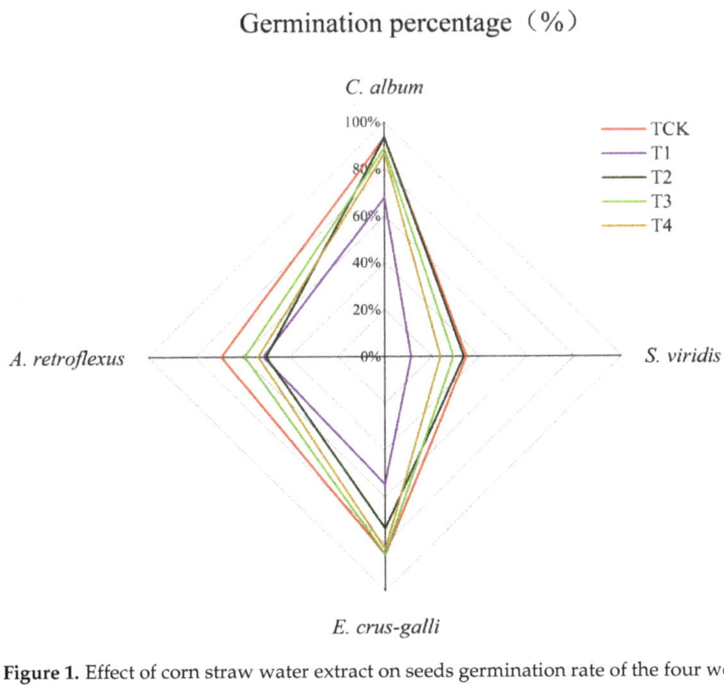

Figure 1. Effect of corn straw water extract on seeds germination rate of the four weeds (Petri dish experiment). TCK—Clear water control; T1—the stock solution of corn straw water extracts; T2—10× dilution of corn straw water extracts; T3—50× dilution of corn straw water extracts; T4—100× dilution of corn straw water extracts.

Different concentrations of corn straw water extract had different allelopathic effects on the germination rates of the seeds of the four weeds, but all the treatments showed the allelopathic inhibition of germination. Under the T1 treatment, the allelopathic index of the germination rate of the four weeds was the largest and the allelopathic inhibition effect was the strongest, compared to that of the other treatments. The order of the inhibition effect of T1 on the seed germination rates of the four weeds from strong to weak is *S. viridis* > *E. crus-galli* > *C. album* > *A. retroflexus* (Figure 2).

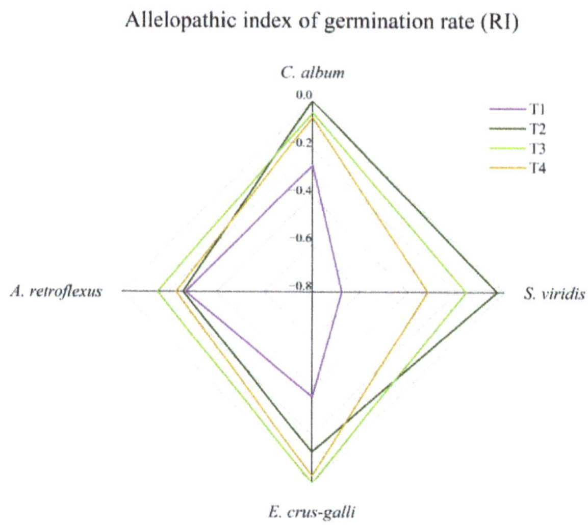

Figure 2. Effect of corn straw water extract on the allelopathy index of seeds germination rate (Petri dish experiment).

The T1 treatment allelopathically inhibited the root lengths of *S. viridis* and *E. crus-galli* by 79.48% and 67.06%, respectively, and the shoot lengths by 0.85% and 9.74%, respectively, compared with that of TCK. Thus, the inhibition effect on the root length was stronger than the effect on the shoot length (Figure 3). However, the T1 treatment increased the root and shoot length of *C. album* through allelopathy. The root length of *A. retroflexus* seeds was inhibited by allelopathy, and the shoot length of *A. retroflexus* seeds was promoted by allelopathy. The T2, T3, and T4 treatments promoted the root and shoot lengths of the four weeds, and the allelopathic-induced reduction in the shoot length by the T4 treatment was significantly different from that of the TCK treatments. With an increase in the concentration of the corn straw water extract, the increase in root length gradually decreased, while shoot length gradually increased (Figure 3).

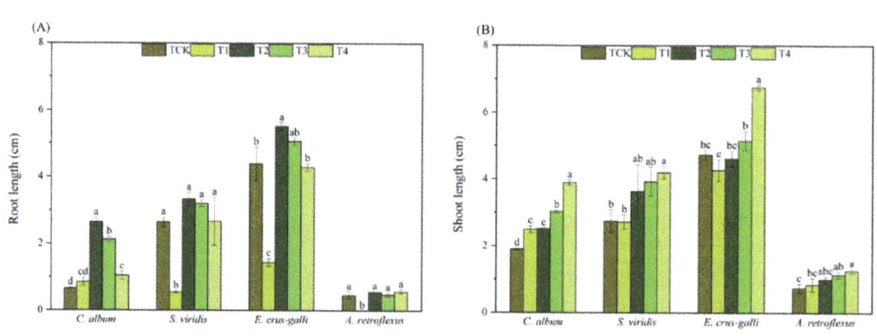

Figure 3. Effects of corn straw water extract on root length (**A**) and shoot length (**B**) (Petri dish experiment). Error bars indicate the standard error of means. Note: Comparison between treatments of different concentrations on the same weed species, with lowercase letters representing a significant difference ($p < 0.05$).

The different corn straw water extract treatments had different allelopathic effects on the root and shoot lengths of the four weeds. The T1 treatment induced allelopathic reductions in the root and shoot lengths of *S. viridis* and *E. crus-galli* and had the strongest allelopathic effect on the root lengths of *A. retroflexus*. Under the T2, T3, and T4 treatments, allelopathy resulted in a gradual decrease in root length with increased dilution ratios of corn straw water extract, whilst shoot length gradually increased with increased dilution ratios of corn straw water extract. The T4 treatment had the greatest effect on the allelopathy index of the shoot lengths of the four weeds, which were 0.51, 0.35, 0.30, and 0.41 (Figure 4).

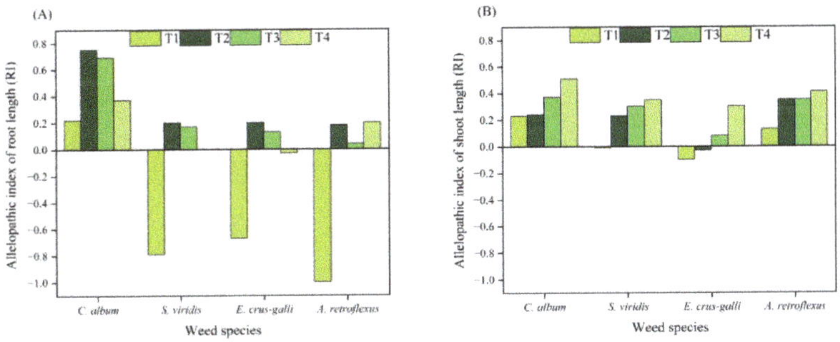

Figure 4. Effects of corn straw water extract on allelopathic index of root (**A**) and shoot (**B**) length (Petri dish experiment).

The comprehensive allelopathy index of the corn straw water extracts for the four weeds significantly differed, and the comprehensive allelopathic effects on *C. album* were the greatest. The T1 treatment resulted in the allelopathic promotion of *C. album* and allelopathic inhibition of *S. viridis*, *E. crus-galli* and *A. retroflexus*. In contrast, the T2, T3, and T4 treatments all induced the allelopathic promotion of the growth of the four weeds. With the increasing dilution ratios of the corn straw water extract, the comprehensive allelopathic effect of the extracts on *C. album* and *S. viridis* gradually decreased, whilst the comprehensive allelopathy effect of the extracts on *E. crus-galli* and *A. retroflexus* gradually increased. The order of the effects of the corn straw water extracts on the comprehensive allelopathy index of the four weeds from strong to weak was *C. album* > *S. viridis* > *A. retroflexus* > *E. crus-galli* (Figure 5).

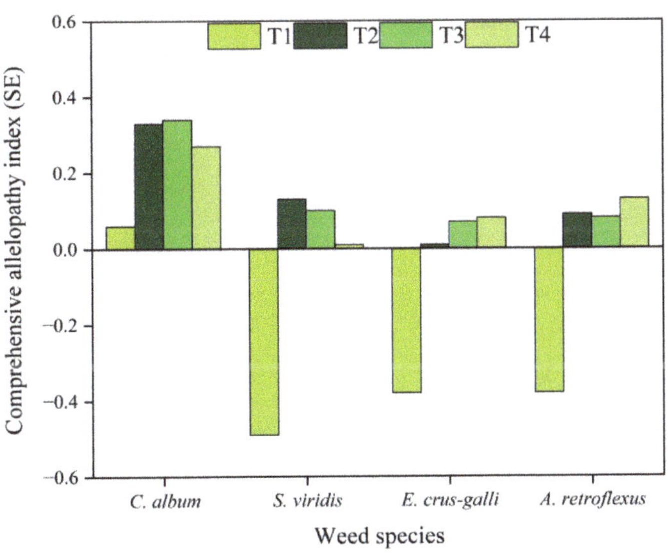

Figure 5. Effect of straw water extract on the comprehensive allelopathic index of the four weeds seeds (Petri dish experiment).

2.2. Effects of Corn Straw Mulching on the Four Weeds

Under different corn straw mulching amounts, the number of weeds under the Z3 treatment was the lowest after rainfall. With the increase in corn straw mulching, the density of weeds gradually decreased, and with increasing incidences of rainfall, the reduction in weed density due to corn straw mulching gradually decreased. Under the Z2, and Z3 treatments, after each rainfall event, the densities of *C. album* and *E. crus-galli* were significantly lower than those of CK, which has an obvious allelopathy inhibition effect. After the third rainfall, the density of *A. retroflexus* under Z3 treatment was significantly lower than that under CK treatment (Table 1).

Table 1. Weed density of four foxtail millet fields under different corn straw mulching (field experiments). "/" indicates that no samples are found in the community.

Rain Frequency	Treatment	*C. album*	*S. viridis*	*E. crus-galli*	*A. retroflexus*
First rain	CK	65.33 ± 7.07 a	26.67 ± 6.11 a	26.67 ± 6.11 a	12.00 ± 2.00 a
	PZ1	41.33 ± 8.86 b	16.00 ± 4.00 ab	16.00 ± 4.00 bc	2.67 ± 0.62 bc
	PZ2	37.33 ± 5.33 b	14.67 ± 2.31 b	21.33 ± 2.31 ab	4.00 ± 1.93 b
	PZ3	/	/	/	/
	Z1	41.33 ± 6.22 b	17.33 ± 6.11 ab	17.33 ± 6.11 b	2.67 ± 0.62 bc
	Z2	29.33 ± 2.31 c	12.00 ± 4.58 b	14.67 ± 3.58 bc	1.33 ± 0.51 c
	Z3	14.67 ± 6.86 d	8.00 ± 2.93 c	9.33 ± 6.93 d	/
Second rain	CK	65.33 ± 8.33 a	10.67 ± 1.11 b	26.67 ± 2.31 a	4.00 ± 2.00 a
	PZ1	29.33 ± 3.24 c	6.67 ± 2.31 c	18.67 ± 2.31 b	2.66 ± 0.61 ab
	PZ2	20.00 ± 2.00 d	8.00 ± 2.00 bc	16.00 ± 4.00 b	1.33 ± 2.30 b
	PZ3	/	/	/	/
	Z1	40.00 ± 4.42 b	24.00 ± 4.00 a	17.33 ± 6.11 b	2.66 ± 0.61 ab
	Z2	38.67 ± 6.07 b	8.00 ± 2.93 bc	16.00 ± 6.93 b	2.66 ± 2.30 ab
	Z3	30.67 ± 4.13 c	/	8.00 ± 6.93 c	1.33 ± 0.50 b
Third rain	CK	80.00 ± 4.33 a	14.67 ± 8.33 b	40.00 ± 6.93 a	4.00 ± 2.00 ab
	PZ1	21.33 ± 6.11 c	9.33 ± 2.31 c	13.33 ± 2.86 d	2.66 ± 0.61 b
	PZ2	20.00 ± 6.93 c	5.33 ± 2.31 d	10.67 ± 4.33 d	2.66 ± 0.61 b
	PZ3	/	/	/	/
	Z1	48.00 ± 6.93 b	26.67 ± 5.73 a	38.67 ± 2.13 a	6.66 ± 2.11 a
	Z2	46.67 ± 4.62 b	17.33 ± 4.04 b	30.67 ± 2.31 b	2.66 ± 0.61 b
	Z3	41.33 ± 8.33 b	16.00 ± 2.05 b	17.33 ± 6.11 c	1.33 ± 0.50 c

Note: Comparison between treatments of different mulching on the same days, with lowercase letters representing a significant difference ($p < 0.05$). The treatments in the table are as follows: no corn straw cover (CK); shading area without corn straw black film, 50% (PZ1), 70% (PZ2), and 100% (PZ3); and corn straw coverage, 3000 kg/ha (Z1), 6000 kg/ha (Z2), and 12,000 kg/ha (Z3).

Five days after the first rainfall event, the densities of the four weeds under the Z1, Z2, and Z3 treatments were lower than that under CK, but there was no significant difference between the Z1 and PZ1 treatments. Under the cover of Z2 straw, *C. album* and *A. retroflexus* were significantly different from the PZ2, but *S. viridis* and *E. crus-galli* were not significantly different from the PZ2. Therefore, 5 days after the first rainfall, the reduction in weed density in the Z1 and Z2 treatments primarily depended on the physical coverage provided by the straw (Table 1).

Five days after the second rainfall event, the densities of *C. album* and *E. crus-galli* under the Z1, Z2, and Z3 treatments were significantly lower than those under CK. However, the density of *S. viridis* under the Z1 treatment was significantly higher than that under CK. The density of the four weeds under the Z2 treatment were lower than those under CK. *E. crus-galli* and *A. retroflexus* showed no significant difference between the Z1 and PZ1 treatments. *S. viridis, E. crus-galli* and *A. retroflexus* showed no significant difference between the Z2 and PZ2 treatments. Therefore, the decrease in the prevalence of the four weeds under the Z1 and Z2 treatments primarily depended on the physical covering effect of the straw 5 days after the second rainfall (Table 1).

Five days after the third rainfall event, the densities of the four weeds were significantly different between the Z1 and PZ1 treatments or between the Z2 and PZ2 treatments except for *A. retroflexus*. The Z3 treatment effectively reduced the prevalence of the four weeds, which was the result of the physical covering effect of the straw and allelochemicals (Table 1).

The Z1, Z2, and Z3 treatments significantly reduced the total coverage of weeds in foxtail millet fields compared with the coverage in the CK group, and with increased corn straw coverage, the total coverage of weeds gradually decreased. Total weed coverage under the Z3 treatment was consistently lower than that under the Z1 and Z2 treatments (Figure 6).

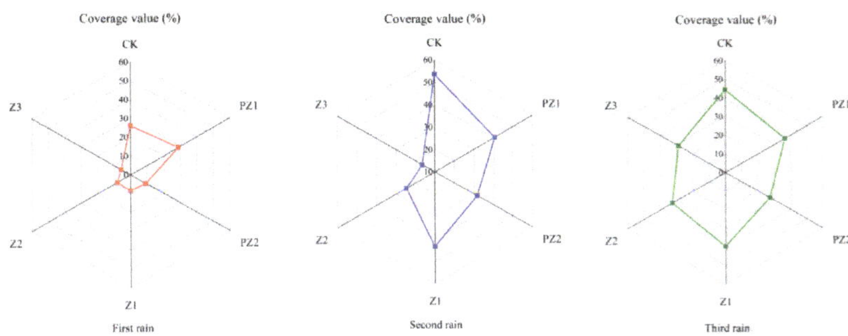

Figure 6. Total weed coverage value under different corn straw mulching (field experiments). No weed growth under PZ3 treatment.

Five days after the first rainfall, the total weed coverage under the Z1, Z2, and Z3 treatments was significantly lower than that of CK, and total weed coverage was lowest under the Z3 treatment. There was no significant difference in weed coverage between the Z1 and PZ1 treatments five days after the second and third rainfall events. Therefore, the reduction in weed coverage under the Z1 treatment five days after the first rainfall was the result of both the physical covering effect of the straw and allelochemicals, and the reduction in weed coverage five days after the second and third rainfall events primarily depended on the physical covering effect of the straw. There was no significant difference in weed coverage between the Z2 and PZ2 treatments five days after the first and second rainfall events. Total weed coverage under the Z2 treatment was significantly higher than that under the PZ2 treatment five days after the third rainfall event. Therefore, the decrease in total weed coverage under the Z2 treatment five days after the first and second rainfall events was due to the combined effect of the physical covering of the corn straw, whilst the decrease in total weed coverage five days after the third rainfall event primarily depended on the physical covering effect of corn straw and allelochemicals (Figure 6).

With increasing corn straw mulching amounts, the number and fresh weight of weeds gradually decreased, while the plant control effect and fresh weight control effect gradually increased. With increased rainfall incidences, the number and fresh weight of weeds gradually increased, while the plant control effect of and fresh weight control effect of weeds gradually decreased (Table 2).

Five days after the first rainfall, the plant control effect and fresh weight control effect under the Z3 treatment were the highest, reaching 74.38% and 61.03%, and the number and fresh weight of weeds under the Z1, Z2, and Z3 treatments were significantly lower than those of CK. Compared with that of PZ2, the control effect under the Z2 treatment increased by 14.63%. Five days after the second rainfall, the number and fresh weight of weeds treated with Z3 were significantly lower than those treated with CK, whilst plant control effect and fresh weight control effect were significantly higher than those under the Z1 and Z2 treatments. Under the Z1 and Z2 treatments, the plant control effect and fresh weight control effect were lower than those under PZ1 and PZ2 treatments. Five days after the third rainfall event, the number and fresh weight of weeds under the Z2 and Z3 treatments were significantly different from those of CK, while the plant control effect and fresh weight control effect of Z3 treatment were significantly higher than those of Z1 and Z2 (Table 2).

Table 2. Weed control effect under different corn straw mulching (field experiments). "/" indicates that no samples are found in the community.

Rain Frequency	Treatment	Weed Number (Plant)	Plant Control Effect (%)	Fresh Weight (g)	Fresh Weight Control Effect (%)
First rain	CK	130.67 a		125.67 a	
	PZ1	77.33 b	39.49 c	70.00 c	44.21 b
	PZ2	76.00 b	39.29 c	60.67 d	51.49 b
	PZ3	/	/	/	/
	Z1	77.33 b	39.49 c	99.00 b	20.71 d
	Z2	58.67 c	53.92 b	78.00 c	37.99 c
	Z3	32.00 d	74.38 a	49.33 e	61.03 a
Second rain	CK	105.33 a		530.00 a	
	PZ1	45.33 d	56.56 a	363.00 c	32.38 c
	PZ2	58.67 c	44.06 b	299.00 d	44.4 b
	PZ3	/	/	/	/
	Z1	82.67 b	21.56 c	480.67 b	12.30 d
	Z2	62.67 c	40.50 b	354.67 c	33.44 c
	Z3	38.67 d	62.76 a	211.00 e	60.11 a
Third rain	CK	134.67 a		827.00 a	
	PZ1	44.00 e	64.46 ab	797.33 a	3.37 c
	PZ2	36.00 f	71.22 a	752.67 b	11.28 a
	PZ3	/	/	/	/
	Z1	94.67 b	23.44 d	800.67 a	3.18 c
	Z2	81.33 c	36.43 c	759.00 b	8.22 b
	Z3	54.67 d	60.09 b	733.33 c	8.87 b

Note: Comparison between treatments of different mulching on the same days, with lowercase letters representing a significant difference ($p < 0.05$).

2.3. *Effects of Corn Straw Mulching on Foxtail Millet*

2.3.1. Effects of Corn Straw Mulching on Plant Height, Leaf Area and SPAD of Foxtail Millet

The plant height, leaf area, and SPAD value of foxtail millet all improved under the varying coverages of corn straw. Specifically, as corn straw coverage increased, the plant height and leaf area of foxtail millet gradually increased, while the SPAD value first increased and then decreased. Following the first rainfall that lasted for 5 days, the plant height, leaf area, and SPAD value of Z1 treatment increased by 4.67 cm, 1.36 cm^2, and 2.1, respectively, compared with CK; however, this difference was not significant. Compared with CK, the plant height, leaf area, and SPAD value of Z2 treatment increased by 9 cm, 7.12 cm^2, and 6.23, respectively, with this difference not being significant. Compared with CK, the plant height and leaf area of foxtail millet treated with Z3 increased by 12.33 cm and 9.65 cm^2, respectively, with these differences being significant. Additionally, the SPAD value increased by 5.07; however, this increase was not significant. The plant height, leaf area, and SPAD value of foxtail millet all increased under different corn straw mulching amounts compared with the corresponding area of black film mulching. The plant height, leaf area, and SPAD value of foxtail millet in Z1 treatment increased by 3.4 cm, 0.82 cm^2, and 0.8, compared with PZ1 treatment, although the differences were not significant. Compared with PZ2 treatment, the plant height of foxtail millet in Z2 treatment increased by 7.7 cm; however, the difference was not significant. The leaf area and SPAD value increased by 5.99 cm^2 and 5.5, with significant differences. Compared with PZ3 treatment, the plant height and leaf surface integral of foxtail millet significantly increased by 9.1 cm and 7.08 cm^2, while SPAD value increased by 4.3; however, this difference was not significant.

After receiving rain for the second time for 5 days, the plant height and SPAD value of foxtail millet in Z2 treatment, as well as the plant height and leaf area of foxtail millet in Z3 treatment were significantly higher than those in CK and the corresponding black film mulching treatments of PZ2 and PZ3. After receiving rain for the third time for 5 days, the plant height, leaf area, and SPAD value of foxtail millet in Z1, Z2, and Z3 treatments were higher than CK and those treated with black film covering PZ1, PZ2, and PZ3 in the corresponding areas; however, the leaf area of foxtail millet treated with Z3 was significantly higher than that of CK (Figures 7–9).

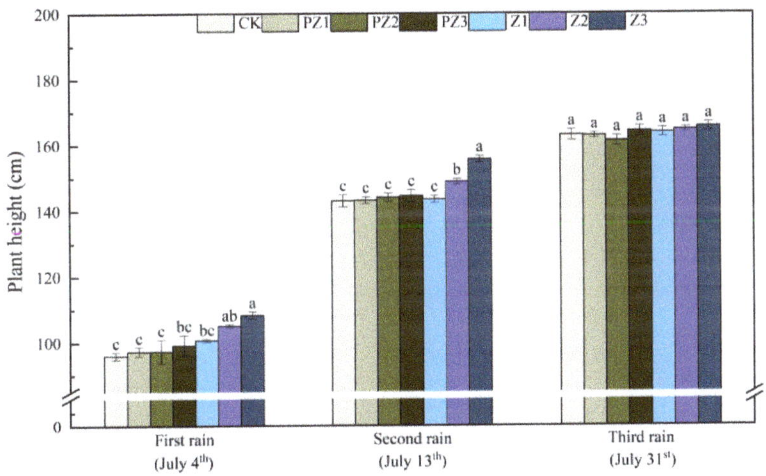

Figure 7. Effect of corn straw mulching on foxtail millet plant height (field experiments). Note: Comparison between treatments of different mulching on the same days, with lowercase letters representing a significant difference ($p < 0.05$).

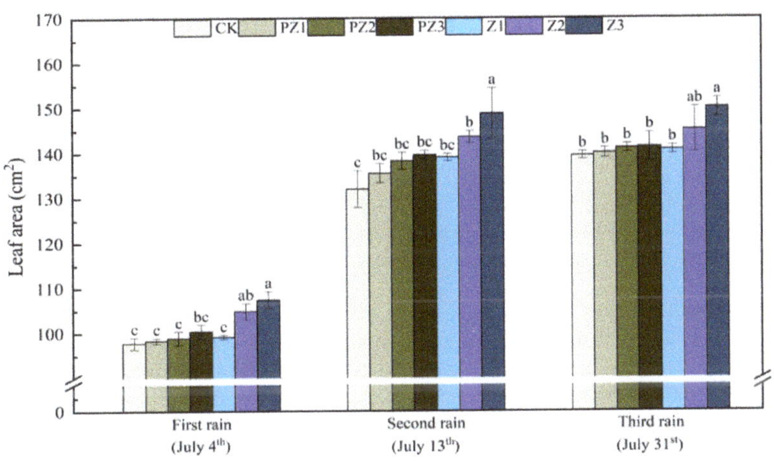

Figure 8. Effect of corn straw mulching on foxtail millet leaf area (field experiments). Note: Comparison between treatments of different mulching on the same days, with lowercase letters representing a significant difference ($p < 0.05$).

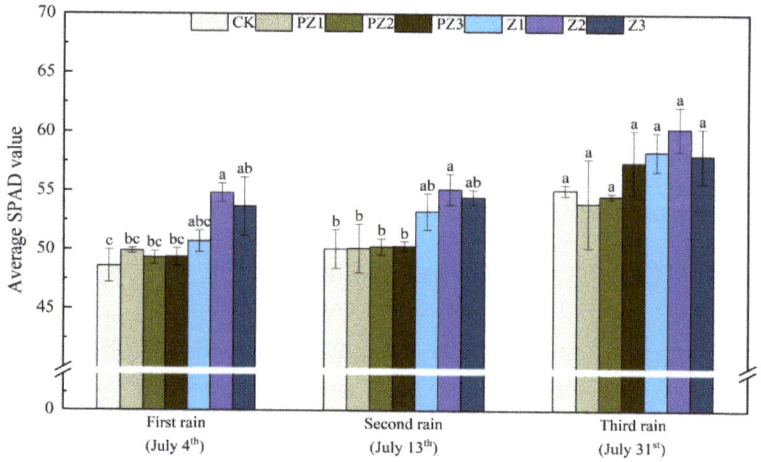

Figure 9. Effect of corn straw mulching on soil plant analysis development SPAD value of foxtail millet (field experiments). Note: Comparison between treatments of different mulching on the same days, with lowercase letters representing a significant difference ($p < 0.05$).

2.3.2. Effects of Corn Straw Mulching on Photosynthesis of Foxtail Millet Leaves

Adding corn straw to the field can improve the photosynthetic characteristics of foxtail millet. Specifically, the net photosynthetic rate (*Pn*), transpiration rate (*Tr*), and stomatal conductance (*Gs*) of foxtail millet leaves under different mulching amounts were higher than in CK and in the treatments of black film mulching with PZ1, PZ2, and PZ3 in the corresponding areas. However, the intercellular CO_2 concentration (*Ci*) was lower than in CK and in the treatments of black film mulching with PZ1, PZ2, and PZ3 in the corresponding areas. With the increase in corn straw mulching, the *Pn*, *Tr*, and *Gs* of foxtail millet leaves showed a trend of first increasing and then decreasing, while the *Ci* first decreased and increased. Following 5 days of three spells of rain, the photosynthetic efficiency of foxtail millet leaves was the highest in Z2 treatment. After the first spell of rain for 5 days, compared with CK, the Pn of foxtail millet leaves in Z1 and Z3 treatments was significantly different, which increased by 23.72% and 27.76%, respectively. Compared with PZ1 and PZ3, the *Pn*, *Tr*, *Gs*, and *Ci* of foxtail millet leaves were not significantly different. Compared with CK, the *Pn* and *Tr* of foxtail millet leaves in Z2 treatment were significantly different, increasing by 35.44% and 25.77%, respectively. Compared with PZ2 treatment, the *Pn*, *Tr*, *Gs*, and *Ci* of foxtail millet leaves were not significantly different. After the second spell of rain for 5 days, the transpiration rate of millet leaves treated with Z2 was significantly different from CK and PZ2, and that of foxtail millet leaves treated with Z3 was significantly different from CK, but not significantly different from PZ3. After the 5 days of the third spell of rain, no significant difference was observed in *Pn*, *Tr*, *Gs*, and *Ci* between treatments Z1, Z2, and Z3, as well as CK, and treatments PZ1, PZ2 and PZ3 covered with black film in corresponding areas. Therefore, our findings imply that the application of corn straw mulching and its incorporation into the field primarily impacts the *Pn* and *Tr* of foxtail millet leaves (Figures 10–13).

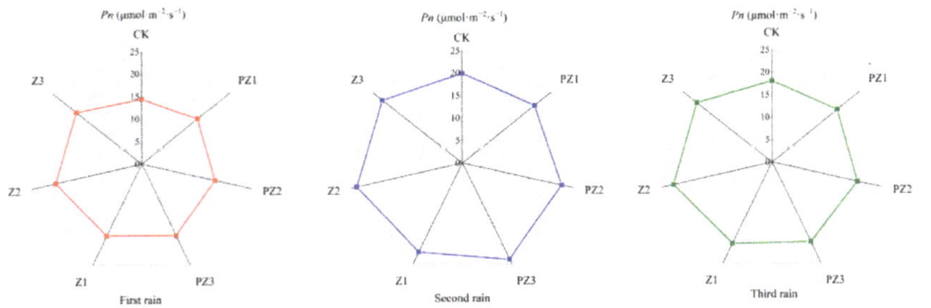

Figure 10. Effect of corn straw mulching on *Pn* of foxtail millet leaves (field experiments).

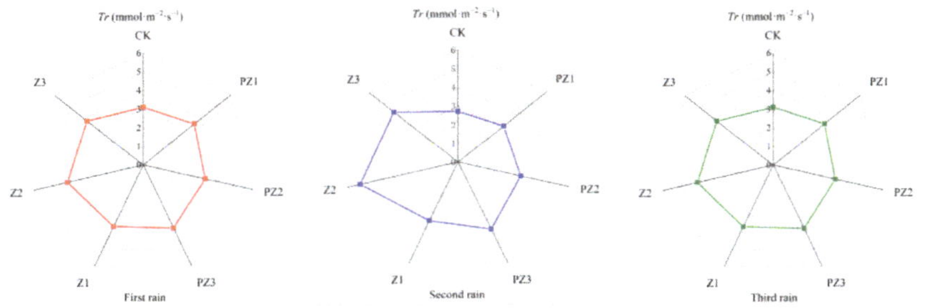

Figure 11. Effect of corn straw mulching on *Tr* of foxtail millet leaves (field experiments).

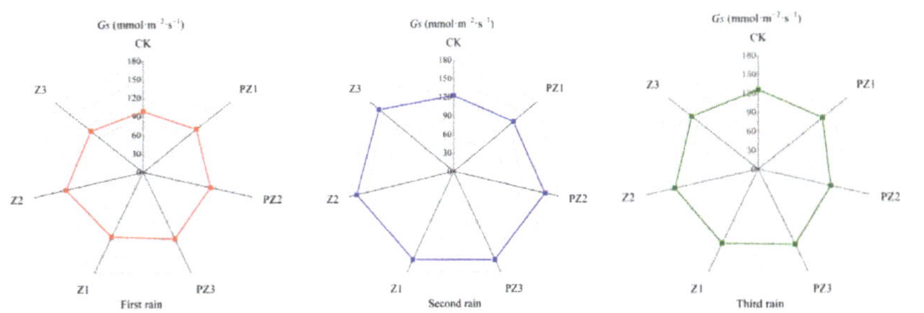

Figure 12. Effect of corn straw mulching on *Gs* of foxtail millet leaves (field experiments).

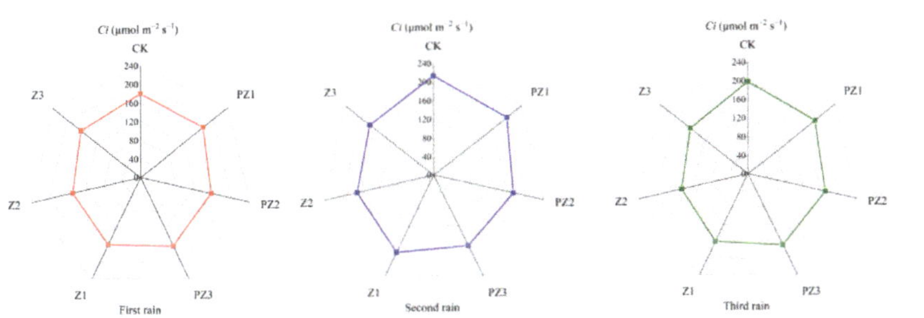

Figure 13. Effect of corn straw mulching on *Ci* of foxtail millet leaves (field experiments).

2.3.3. Effects of Corn Straw Mulching on Foxtail Millet Yield

The yield components of foxtail millet were improved by the application of different mulching amounts of corn straw. Specifically, the panicle weight, panicle length, panicle grains weight and the 1000-grain weight of foxtail millet were significantly higher under Z2 and Z3 treatments compared to CK. Compared with the black film mulching treatment in the corresponding area, the panicle length and 1000-grain weight of foxtail millet under Z2 and Z3 treatments were significantly higher than those of foxtail millet under PZ2 and PZ3 treatments, and the nutrients and allelochemicals in corn straw mainly affected the panicle length and 1000-grain weight of foxtail millet. As mulch increased, the yield of foxtail millet first increased and then decreased. The highest yield of foxtail millet under Z2 treatment was 5781.51 kg/ha, which was 10.32% higher than that of CK and 3.96% higher than that under PZ2 treatment. The physical covering effect of corn straw mulching plays a role in saving water and preserving soil moisture. Simultaneously, nutrients and allelochemicals in the straw enter the soil and are absorbed and utilized by foxtail millet, thus increasing the yield of foxtail millet, as indicated by the increasing panicle length and 1000-grain weight of foxtail millet (Table 3).

Table 3. The yield components of foxtail millet (field experiments).

Treatments	Panicle Weight (g)	Panicle Length (cm)	Panicle Grains Weight (g)	1000-Grain Weight (g)	Yield (kg/ha)
CK	18.57 ± 1.002 c	26.10 ± 2.61 c	11.51 ± 0.64 c	2.96 ± 0.01 b	5240.88 ± 159.95 c
PZ1	18.66 ± 0.50 c	27.23 ± 0.47 bc	11.99 ± 1.33 bc	2.96 ± 0.04 b	5325.65 ± 226.36 c
PZ2	19.23 ± 1.36 bc	28.07 ± 0.78 bc	12.18 ± 0.98 bc	2.97 ± 0.03 b	5561.12 ± 157.80 b
PZ3	20.81 ± 0.40 ab	28.63 ± 0.55 b	13.75 ± 0.70 bc	2.98 ± 0.04 b	5666.78 ± 136.96 ab
Z1	18.78 ± 1.73 c	29.23 ± 1.06 b	12.56 ± 2.73 bc	3.02 ± 0.01 ab	5681.40 ± 209.57 ab
Z2	22.68 ± 1.76 a	34.03 ± 0.59 a	15.48 ± 1.22 a	3.07 ± 0.05 a	5781.51 ± 333.98 a
Z3	21.29 ± 0.05 a	32.23 ± 1.36 a	14.46 ± 1.24 ab	3.07 ± 0.02 a	5711.02 ± 336.76 a

Note: Comparison between treatments of different mulching on the same days, with lowercase letters representing a significant difference ($p < 0.05$).

3. Discussion

3.1. Allelopathic Effects of Straw Water Extract on Weeds

Crop straw affects seed germination and the seedling growth of plants through allelopathy, but the sensitivity of different plants to allelochemicals varies. Allelochemicals produced by different crop straws have different effects on different plant seeds [34,35], and the intensity of the liquefaction sensitivity of straw extraction in different crops, parts, and concentrations also differs. The water extracts of wheat straw significantly reduced the germination rate, germination index, and all morphological growth parameters of Bermuda grass seedlings [36]. Crude corn straw extract had a greater inhibitory effect on seedling germination and growth than the fermented corn straw extract. The corn straw extract had remarkably inhibited the germination and seedling growth of *C. album* and the seedling growth of *A. theophrasti* [37]. A 40% water extract concentration of *Gaillardia pulchella* can inhibit the germination of *Brassica napus*. Different concentrations of water extracts of *G. pulchella* inhibited the growth of *B. napus* roots but promoted the elongation of shoots [38]. This is similar to the results of this study using corn straw water extract treatments. In this study, the T1 treatment inhibited the germination of the four weeds; reduced the root length, shoot length, and allelopathy index of *S. viridis* and *E. crus-galli*. However, T1 treatment increased the root length, shoot length, and allelopathy index of *C. album*; and reduced the root length of *A. retroflexus*, but increased shoot length. The order of the effects of corn straw water extract from strong to weak on the comprehensive allelopathy index of the four weed seeds was as follows: *C. album* > *S. viridis* > *A. retroflexus* > *E. crus-galli*.

3.2. Allelopathic Effects of Corn Straw Mulching on Weeds

Currently, weed control in fields remains an urgent problem, and the growth of weeds affects the growth, yield, and quality of crops. Various herbicides are used on agricultural lands. Long-term use of pesticides causes resistance in weeds and adverse effects, such as soil pollution [39,40]. Black film covering is a common cultivation method in agricultural production, which can increase crop yield, affect crop growth and soil environment [41,42]. Black film covering can weaken the heat exchange between the soil and the outside world, increasing the temperature of the cultivated layer of the soil [43,44]. At the same time, covering with black film can reduce the light under the film, so as to inhibit weed growth. The PZ1, PZ2 and PZ3 treatments in this study only covered the ridges; the plants inside the rows were exposed outside, so the temperature under the film was not too high. Rainfall can seep into the soil from the rows without affecting plant growth and soil moisture loss. Straw covering is a green and environmentally friendly cultivation measure that reduces the occurrence of weeds and environmental degradation, as well as effectively utilizing a renewable resource and promoting the development of green and sustainable agriculture [45]. Corn straw mulching can maintain soil temperature and moisture, it also has an allelopathic effect (promoting and inhibiting) on the germination and growth of plants. Corn straw mulching can increase the root lengths of the deep roots of crops, while the average root length density increases by 11.7% and 15.8% at depths of 30 and 40 cm, respectively [46]. The combination of corn straw mulching and no-tillage maintains soil moisture, initiates the release of available nitrogen, and enhances nitrogen uptake in plants, thereby improving grain yield, water use efficiency, and nitrogen use efficiency [47]. The degradation of straw mulching in soil after corn harvesting can improve the physiochemical properties of the soil, leading to increased water absorption, yield, and water use efficiency in the subsequent generation of corn crops [48]. In this study, the physical covering effect of corn straw mulching saved water, preserved soil moisture, and regulated the soil temperature. At the same time, the allelochemicals in the corn straw had different degrees of allelopathic effects on the weeds and foxtail millet after entering the soil. The results of this experiment showed that the germination and growth of the four weeds were inhibited by the physical covering effect and allelopathy of corn straw mulching. With increased corn straw mulching duration and rainfall, corn straw gradually decomposed and the allelochemicals in the corn straw were gradually diluted, which promoted the germination and growth of weeds, thus gradually reducing their influence on weed density and coverage. The density of the four weeds decreased with an increase in the straw mulching amount and increased with increasing rainfall under different straw mulching amounts.

In this study, the allelopathy of corn straw and its water extracts on four representative weeds in foxtail millet fields was investigated, and the weed species and allelopathy indexes studied were not comprehensive enough.

3.3. *Effect of Corn Straw Mulching on Foxtail Millet*

Straw mulching can effectively utilize soil resources and enhance the vitality of crop roots, thereby promoting the absorption of nutrients and water. This cascade of events enhances the physiological functions of plants and consequently increases crop yield. Intercropping wheat and corn with straw mulching can improve the overcompensation effect of late-maturing corn, ultimately increasing the total yield of the intercropping system [49]. Combining black film with wheat straw mulching increases the leaf area, leaf photosynthetic rate, and the 1000-grain dry weight of summer maize, thereby resulting in an increase in maize yields [50]. Straw mulching alleviated the impact of drought stress on sesame, indicative of the potential of this approach to support sesame yields in arid areas [51]. This study indicates that corn straw mulching improves the plant height, leaf area, SPAD value, leaf photosynthetic efficiency, and yield of foxtail millet. With the increase in straw mulching, the plant height and leaf area of foxtail millet gradually increased, whereas the SPAD value first increased and then decreased. The *Pn*, *Tr*, and *Gs* of foxtail millet leaves show a trend of first increasing and then decreasing with increasing coverage, while the intercellular CO_2 (*Ci*) concentration first decreased and increased. After covering with corn straw, the panicle length and 1000-grain weight of foxtail millet were increased, resulting in an increased foxtail millet yield.

In this study, the physical covering effect of straw influenced water retention and moisture retention. Additionally, nutrients and allelopathic substances in straw permeated the soil and were absorbed and utilized by foxtail millet, which is also an important driver of crop yield. In future studies, elucidating other indicators is of utmost importance. Meanwhile, further research investigating the effects of corn straw and its water extracts on other weeds and foxtail millet is necessary.

4. Materials and Methods

4.1. Experimental Materials

Weeds: seeds of four common weeds found in foxtail millet fields were collected from the farming stations: *Chenopodium album* L. (*C. album*), *Setaria viridis* L. (*S. viridis*), *Echinochloa crus-galli* L. (*E. crus-galli*) and *Amaranthus retroflexus* L. (*A. retroflexus*). These weed seeds were collected in the field of foxtail millet at the Agricultural Station of Shanxi Agricultural University in Jinzhong City, Shanxi Province.

Corn straw: corn straw grown in the countryside around Shanxi Agricultural University in Taigu District, Jinzhong City, Shanxi Province was harvested at the maturity stage, dried, and cut into 3 cm pieces, which were stored in a cool and ventilated place for subsequent use.

Foxtail millet variety tested in the field: Zhangzagu 10 (cultivated by the Zhangjiakou Academy of Agricultural Sciences).

4.2. Experimental Design

The study was divided into Petri dish and field experiments. All experiments adopted a completely randomized experimental design.

The Petri dish experiment consisted of five treatments: water control (TCK) and four corn straw water extract treatment groups (T1, T2, T3, and T4). The concentrations of T1–T4 corn straw water extracts were 100, 10, 2, and 1 g/L, respectively.

The field experiment consisted of seven treatments: four control treatments, namely, no corn straw (CK), 50% black film (PZ1), 70% black film (PZ2), and 100% black film (PZ3) covering treatments, and three corn straw covering treatments, with coverage amounts of 3000, 6000, and 12,000 kg/ha (Z1, Z2, and Z3, respectively).

4.2.1. Petri Dish Experiment

(1) Preparation of corn straw water extracts:

Air-dried corn straw was crushed using a high-speed crusher and filtered through an 80 mesh sieve to obtain corn straw powder. The corn straw powder was mixed with distilled water in a beaker at a mass ratio of 1:10, placed in an ultrasonic instrument (JP-100S, Jiemeng cleaning equipment Co., LTD., Shenzhen, China) for ultrasonic extraction for 30 min, with the temperature of water at 30 °C. It was centrifuged for 3 min at 6400 rpm, and the supernatant was obtained, which was used as the stock solution of the corn straw water extract (100 g/L). The stock solution of the corn straw extract was diluted with distilled water to obtain 10× (10 g/L), 50× (2 g/L), and 100× (1 g/L) dilutions, which were placed in a refrigerator at 4 °C for subsequent use.

(2) Petri dish weed experiment:

Petri dishes with a diameter of 9 cm were sterilized. After cooling, two layers of sterilized filter paper were placed in each Petri dish, and 5 mL of corn straw water extract stock solution, 10× diluent, 50× diluent, and 100× diluent were added; distilled water was used as a control. Thirty weed seeds were evenly seeded in each Petri dish and placed in a constant temperature incubator at 27 °C and a relative humidity of 50%. After 6 days, the root and shoot lengths of the weed seeds were measured, and each measurement was repeated three times for each treatment. The germination rate, allelopathy index, and comprehensive allelopathy index of the weed seeds were calculated according to the number of germinating seeds observed daily [17]. The calculation formulas used are as follows:

Germination rate = (number of germinations on the 6th day/total number of weed seeds per Petri dish) × 100.

Allelopathy index (RI) = 1-TCK/T or T/TCK-1.

(Where: RI is the allelopathy index, T represents the treatment group data and TCK represents the control group data. When T $\geq$ TCK, RI = 1-TCK/T, when T < TCK, RI = T/TCK-1)

Comprehensive allelopathy index = the average value of the RI of germination rate, root length and shoot length of weed seeds.

4.2.2. Field Experiment

The experiment was conducted in May 2020 in the experimental base of Shanxi Agricultural University Farming Station. The experiment used a randomized block design (Table 4). Seven treatments were set up in this experiment, each treatment was repeated three times, and the area of each treatment was 30 m^2 (5 m × 6 m). Soil nutrient content was measured before sowing foxtail millet (Table 5). After the sprouting of foxtail millet, the relevant amounts of corn straw and black film were evenly spread between the rows of foxtail millet in each plot (Figure 14). The black film used in our experiment refers to a black polyethylene plastic film with a thickness of 0.01 mm, provided by Shanxi Yingtai Plastic Co., Ltd. (Yuncheng, China).

Table 4. Treatment conditions of field experiment.

Treatment	CK	PZ1	PZ2	PZ3	Z1	Z2	Z3
Corn straw mulching amount (kg/ha)	0	0	0	0	3000	6000	12,000
Black film covering area (%)	0	50%	70%	100%	0	0	0

Table 5. Nutrient content of foxtail millet field soil.

Year	Soil Depth (cm)	pH	Available K (mg/kg)	Available P (mg/kg)	Available N (mg/kg)	Total N (g/kg)	Total P (g/kg)	Total K (g/kg)	Organic (g/kg)
2019	0~5	8.21	484.20	43.70	64.20	1.22	0.76	19.36	8.52
	5~10	8.14	471.90	26.79	42.80	1.27	0.76	18.08	8.52
	10~15	8.17	319.60	21.28	49.93	1.11	0.77	18.53	7.99
2020	0~5	8.11	377.40	21.06	48.20	1.11	0.91	21.91	8.09
	5~10	8.19	297.80	18.93	56.55	1.04	0.81	23.33	8.52
	10~15	8.10	303.90	24.10	51.77	1.07	0.87	21.76	8.22

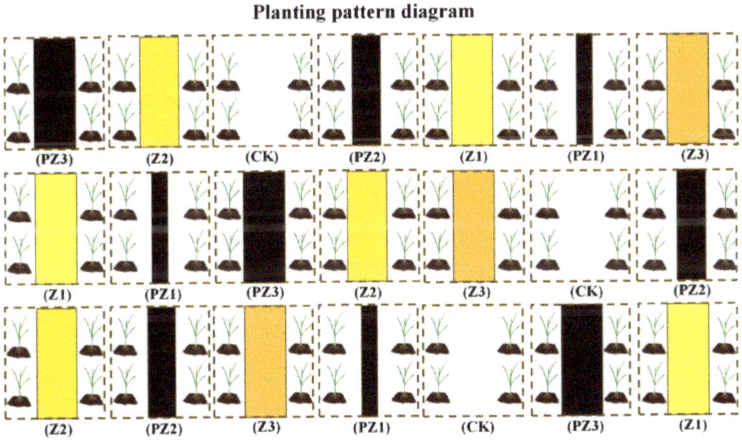

Figure 14. Planting pattern diagram of foxtail millet field.

Field weed investigation:

In 2020, random quadrat samples were conducted 5 days (29 June, 8 July and 26 July) after each of three rainfall events (Table 6). These samples were conducted in each plot and included weed density, total weed coverage value and weed control effect. The area of the quadrat was 0.25 m^2 (0.5 m × 0.5 m) and three quadrat areas were investigated for each treatment [11]. The indices were calculated using the following formulas:

Table 6. Rain events in 2020.

Rainfall Frequency	Rainfall Date	Rainfall Time	Precipitation (mm)	Measurement Date
First rain	29 June	8:40 a.m. to 11:30 a.m.	18	4 July
Second rain	8 July	6:00 a.m. to 2:30 p.m.	52.3	13 July
Third rain	26 July	11:30 a.m. to 5:20 p.m.	43.9	31 July

Field density (plant/m^2) = number of weed plants in each survey sample/area.

Coverage (%) = (green vegetation pixels in the sample/total pixels in the survey sample) × 100.

Plant control effect (%) = (number of weeds in the control area − number of weeds in the treatment area/number of weeds in the control area) × 100.

Fresh weight control effect (%) = (weed fresh weight in control area − weed fresh weight in treatment area/weed fresh weight in control area) × 100.

Determining the agronomic traits of foxtail millet:

Five days after three rains in the same year, representative and uniformly growing plants were selected for each treatment. The height of the foxtail millet, as well as the length and width of the second leaf of the plant from top to bottom, were measured using a tape measure [11]. Specifically, the leaf area was determined as follows:

Leaf area (cm^2) = leaf length × leaf width × 0.75.

Determining the photosynthetic characteristics of foxtail millet:

The SPAD-502 portable chlorophyll analyzer (Minolta Camera Co. Ltd., Osaka, Japan) was used to measure the SPAD values of the second leaf from top to bottom.

On a clear and windless morning (9:00–11:00), photosynthetic parameters including the net photosynthetic rate (*Pn*), transpiration rate (*Tr*), stomatal conductance (*Gs*), intercellular CO_2 concentration (*Ci*) of the second leaf from top to bottom were measured using a portable handheld photosynthetic instrument CI-340 (CID Bio-Science, Inc., Washington, DC, USA).

Foxtail millet yield:

Before harvesting foxtail millet, we randomly selected panicles to measure phenotypic outcomes such as panicle length, panicle weight, panicle grain weight, and 1000-grain weight. Using the diagonal sampling method, each treatment is 1 m^2 (1 × 1 m), and then converted into yield.

4.3. Data Processing

Data processing was performed using Microsoft Excel 2021 (Microsoft, Redmond, WA, USA) and IBM SPSS Statistics 25 software (SPSS Inc., Chicago, IL, USA). All data analysis was conducted using Duncan's new multiple range method, including variance and multiple comparisons, with significance at $p < 0.05$. Data were visualized using Origin 2021 software (Originlab, Northampton, MA, USA).

5. Conclusions

The seed germination rate, root length, and shoot length of four weeds, *C. album*, *S. viridis*, *E. crus-galli* and *A. retroflexus*, in foxtail millet fields showed obvious allelopathic responses to different concentrations of corn straw water extract during the seed germination period. Among them, 100 g/L corn straw extract had the best inhibitory effect on the seed germination of the four weeds and significantly inhibited the growth of *S. viridis*, *E. crus-galli*, and *A. retroflexus*, but promoted the growth of *C. album*.

For foxtail millet field planting, a suitable amount of corn straw mulching can effectively inhibit the growth of weeds in foxtail millet fields in a short time, reduce the amount of herbicides sprayed, and reduce the environmental pollution caused by pesticides. Here, 3000 kg/ha and 6000 kg/ha corn straw mulching effectively inhibited the occurrence of weeds in the early stage of foxtail millet growth, but this effect weakened with an increase in rainfall. Corn straw mulching at 12,000 kg/ha had the strongest allelopathic inhibitory effect on the four weeds and effectively reduced the presence of *C. album* and *E. crus-galli*.

Corn straw mulching positively impacted the plant height, leaf area, and photosynthetic parameters of Zhangzagu 10, with the net photosynthetic rate and transpiration rate being the most significantly impacted. The yield of foxtail millet was positively impacted under all straw coverage; specifically, its panicle length and 1000-grain weight were improved. The growth parameters of foxtail millet were most positively impacted under corn straw coverages of 6000 and 12,000 kg/ha, respectively, with the growth of foxtail millet indicating that an appropriate amount of corn straw coverage would have an allelopathic effect on the growth of Zhangzagu 10.

Thus, corn straw mulching at 12,000 kg/ha has been identified as the most effective method for controlling malignant weeds in foxtail millet fields and can promote the growth of Zhangzagu 10. The incorporation of crop straw effectively reduces the growth of some weed species. Future studies should focus on investigating the allelopathic effects of corn straw on other weeds and millet varieties. Additionally, determining the appropriate concentrations of corn straw mulch to control weeds in the field is an important aspect that requires further exploration.

Author Contributions: Conceptualization, S.D. and C.H.; methodology, S.D. and M.L.; validation, J.D. and P.L.; formal analysis, M.L., P.L. and B.C.; data curation, Z.G. and X.S.; writing—original draft preparation, B.C. and Z.G.; writing—review and editing, S.D. and J.D.; visualization, X.S. and Y.M.; supervision, X.Y. and Y.M.; project administration, C.H.; funding acquisition, X.Y. All authors have read and agreed to the published version of the manuscript.

Funding: This research was funded by the National Key Research and Development Program sub-project (2021YFD1901103-5), China Agriculture Research System of MOF and MARA (CARS-06-14.5-A28), National Natural Science Foundation of China (32272229), Earmarked fund for Modern Agro-industry Technology Research System of Shanxi Province (2024CYJSTX04), the Special Plan for Scientific and Technological Innovation Talent Team of Shanxi Province (202204051002036).

Data Availability Statement: The data that support this study are available upon reasonable request from the corresponding author.

Acknowledgments: Thank you to all those who helped with this study and to the research projects that sponsored it.

Conflicts of Interest: The authors declare no conflict of interest.

References

1. Liang, Y.; Han, Y. Pan-genome brings opportunities to revitalize ancient crop foxtail millet. *Plant Commun.* **2023**, *5*, 100735. [CrossRef] [PubMed]
2. Wu, C.; Zhang, M.; Liang, Y.; Zhang, L.; Diao, X. Salt stress responses in foxtail millet: Physiological and molecular regulation. *Crop J.* **2023**, *11*, 1011–1021. [CrossRef]
3. Wang, L.; Fu, H.; Zhao, J.; Wang, J.; Dong, S.; Yuan, X.; Li, X.; Chen, M. Genome-wide identification and expression profiling of Glutathione S-Transferase gene family in foxtail millet (*Setaria italica* L.). *Plants* **2023**, *12*, 1138. [CrossRef] [PubMed]
4. Vetriventhan, M.; Upadhyaya, H.D.; Anandakumar, C.R.; Senthilvel, S.; Parzies, H.K.; Bharathi, A.; Varshney, R.K.; Gowda, C.L.L. Assessing genetic diversity, allelic richness and genetic relationship among races in ICRISAT foxtail millet core collection. *Plant Genet. Resour.* **2012**, *10*, 214–223. [CrossRef]
5. Upadhyaya, H.; Pundir, R.; Gowda, C.; Reddy, V.; Singh, S. Establishing a core collection of foxtail millet to enhance the utilization of germplasm of an underutilized crop. *Plant Genet. Resour.* **2009**, *7*, 177–184. [CrossRef]
6. Han, F.; Sun, M.; He, W.; Guo, S.; Feng, J.; Wang, H.; Yang, Q.; Pan, H.; Lou, Y.; Zhuge, Y. Transcriptome analysis reveals molecular mechanisms under salt stress in leaves of foxtail millet (*Setaria italica* L.). *Plants* **2022**, *11*, 1864. [CrossRef] [PubMed]

7. Cui, J.; Xia, X.; Zhao, Y.; Liu, M.; Xiao, N.; Guo, S.; Lu, Y.; Li, J.; Wei, Z.; Gao, F.; et al. Interpreting variety–location–fertilizer interactions to enhance foxtail millet productivity in northern China. *Agronomy* **2022**, *12*, 2216. [CrossRef]
8. Kalsi, R.; Bhasin, J. Nutritional exploration of foxtail millet (*Setaria italica*) in addressing food security and its utilization trends in food system. *eFood* **2023**, *4*, e111. [CrossRef]
9. Zhou, S.; Ye, F.; Xia, D.; Liu, Z.; Wu, Y.; Chen, F. Climate change impacts assessment and developing adaptation strategies for rainfed foxtail millet in northern Shanxi, China. *Agric. Water Manag.* **2023**, *290*, 108575. [CrossRef]
10. Guo, M.; Wang, Y.; Yuan, X.; Dong, S.; Wen, Y.; Song, X.; Guo, P. Responses of the antioxidant system to fluroxypyr in foxtail millet (*Setaria italica* L.) at the seedling stage. *J. Integr. Agric.* **2017**, *17*, 554–565. [CrossRef]
11. Song, X.; Wang, H.; Dong, Q.; Qiu, T.; Shi, C.; Li, X.; Dong, S.; Zhao, J.; Guo, P.; Yuan, X. Comprehensive evaluation and main identification indexes of herbicide resistance of high-quality foxtail millet (*Setaria italica* L.). *Agronomy* **2023**, *13*, 3033. [CrossRef]
12. Ullah, H.; Khan, N.; Khan, I.A. Complementing cultural weed control with plant allelopathy: Implications for improved weed management in wheat crop. *Acta Ecol. Sin.* **2023**, *43*, 27–33. [CrossRef]
13. Wang, C.; Qi, J.; Liu, Q.; Wang, Y.; Wang, H. Allelopathic potential of aqueous extracts from fleagrass (*Adenosma buchneroides* Bonati) against two crop and three weed species. *Agriculture* **2022**, *12*, 1103. [CrossRef]
14. Hussain, M.I.; Danish, S.; SánchezMoreiras, A.M.; Vicente, Ó.; Jabran, K.; Chaudhry, U.K.; Branca, F.; Reigosa, M.J. Unraveling sorghum allelopathy in agriculture: Concepts and implications. *Plants* **2021**, *10*, 1795. [CrossRef] [PubMed]
15. Bashar, H.M.K.; Juraimi, A.S.; Ahmad-Hamdani, M.S.; Uddin, M.K.; Asib, N.; Anwar, M.P.; Rahaman, F.; Haque, M.A.; Hossain, A. Evaluation of allelopathic effects of *Parthenium hysterophorus* L. methanolic extracts on some selected plants and weeds. *PLoS ONE* **2023**, *18*, e0280159. [CrossRef] [PubMed]
16. Okrushko, S.E. Allelopathic effect of couch grass (*Elymus repens* L.) on germination of common wheat seeds. *Zemdirbyste* **2022**, *109*, 323–328. [CrossRef]
17. Kong, C.; Li, Z.; Li, F.; Xia, X.; Wang, P. Chemically mediated plant–plant interactions: Allelopathy and allelobiosis. *Plants* **2024**, *13*, 626. [CrossRef]
18. Li, F.; Zhang, G.; Chen, J.; Song, Y.; Geng, Z.; Li, K.; Siddique, K. Straw mulching for enhanced water use efficiency and economic returns from soybean fields in the loess plateau China. *Sci. Rep.* **2022**, *12*, 17111. [CrossRef] [PubMed]
19. Cheruiyot, W.K.; Zhu, S.; Indoshi, S.N.; Wang, W.; Ren, A.; Cheng, Z.; Zhao, Z.; Zhang, J.; Lu, J.; Zhang, X.; et al. Shallow-incorporated straw returning further improves rainfed maize productivity, profitability and soil carbon turnover on the basis of plastic film mulching. *Agric. Water Manag.* **2023**, *289*, 108535. [CrossRef]
20. Sun, X.; Li, S.; Du, H. The influence of corn straw mulching on soil moisture, temperature and organic matter. *J. Biobased Mater. Bioenergy* **2017**, *11*, 662–665. [CrossRef]
21. Fu, X.; Wang, J.; Peng, Z.; Yang, X.; Zhang, S. Late topdressing can sustain wheat grain yield under straw mulching in the loess plateau of China. *Agron. J.* **2022**, *114*, 3487–3497. [CrossRef]
22. Kato-Noguchi, H.; Kato, M. Allelopathy and allelochemicals of *Solidago canadensis* L. and *S. altissima* L. for their naturalization. *Plants* **2022**, *11*, 3235. [CrossRef]
23. Zhang, X.; Qian, Y.; Cao, C. Effects of straw mulching on maize photosynthetic characteristics and rhizosphere soil micro-ecological environment. *Chil. J. Agric. Res.* **2015**, *75*, 481–487. [CrossRef]
24. Ullah, R.; Aslam, Z.; Maitah, M.; Zaman, Q.u.; Bashir, S.; Hassan, W.; Chen, Z. Sustainable weed control and enhancing nutrient use efficiency in crops through Brassica (*Brassica compestris* L.) allelopathy. *Sustainability* **2020**, *12*, 5763. [CrossRef]
25. Choudhary, C.S.; Behera, B.; Raza, M.B.; Mrunalini, K.; Bhoi, T.K.; Lal, M.K.; Nongmaithem, D.; Pradhan, S.; Song, B.; Das, T.K. Mechanisms of allelopathic interactions for sustainable weed management. *Rhizosphere* **2023**, *25*, 100667. [CrossRef]
26. Qin, X.; Huang, T.; Lu, C.; Dang, P.; Zhang, M.; Guan, X.; Wen, P.; Wang, T.; Chen, Y.; Siddique, K.H. Benefits and limitations of straw mulching and incorporation on maize yield, water use efficiency, and nitrogen use efficiency. *Agric. Water Manag.* **2021**, *256*, 107128. [CrossRef]
27. Cai, T.; Zhang, C.; Huang, Y.; Huang, H.; Yang, B.; Zhao, Z.; Zhang, J.; Jia, Z. Effects of different straw mulch modes on soil water storage and water use efficiency of spring maize (*Zea mays* L.) in the Loess Plateau of China. *Plant Soil Environ.* **2016**, *61*, 253–259. [CrossRef]
28. Valvi, N.A.; Tripathi, V.C.; Rane, A.; Narvankar, V.; Narkhede, S.S. Allelopathic effect of aqueous leaf extract of Tectona grandis on germination and growth of *Oryza sativa*. *Int. J. Farm Sci.* **2022**, *12*, 127–132. [CrossRef]
29. Akhtar, K.; Ain, N.U.; Wang, W.; Ren, G.; Feng, Y.; Djalovic, I.; Prasad, P.V.V.; Yang, G.; He, B.; Wen, R. Straw mulch decreased nitrogen fertilizer requirements via regulating soil moisture and temperature to improve physiology, nitrogen, and water use efficiency of wheat. *Agron. J.* **2023**, *115*, 3106–3118. [CrossRef]
30. Crawford, L.E.; Williams, M.M.; Wortman, S.E. An Early-killed rye (*Secale cereale*) cover crop has potential for weed management in edamame (*Glycine max*). *Weed Sci.* **2018**, *66*, 502–507. [CrossRef]
31. Akhtar, K.; Wang, W.; Khan, A.M.; Ren, G.; Afridi, M.Z.; Feng, Y.; Yang, G. Wheat straw mulching offset soil moisture deficient for improving physiological and growth performance of summer sown soybean. *Agric. Water Manag.* **2019**, *211*, 16–25. [CrossRef]
32. Zhu, J.; Liang, W.; Yang, S.; Wang, H.; Shi, C.; Wang, S.; Zhou, W.; Lu, Q.; Islam, F.B.; Xu, W.; et al. Safety of oilseeds rape straw mulch of different lengths to rice and its suppressive effects on weeds. *Agronomy* **2020**, *10*, 201. [CrossRef]
33. Mehra, L.K.; Cowger, C.; Weisz, R.; Ojiambo, P.S. Quantifying the effects of wheat residue on severity of Stagonospora nodorum blotch and yield in winter wheat. *Phytopathology* **2015**, *105*, 1417–1426. [CrossRef]

34. Dadkhah, A. Allelopathic potential of canola and wheat to control weeds in soybean (*Glycine max*). *Russ. Agric. Sci.* **2015**, *41*, 111–114. [CrossRef]
35. Bertholdsson, N. Allelopathy—A tool to improve the weed competitive ability of wheat with Herbicide-resistant Black-grass (*Alopecurus myosuroides* Huds.). *Agronomy* **2012**, *2*, 284–294. [CrossRef]
36. Alghamdi, S.; AlNehmi, A.; Ibrahim, O. Potential Allelopathic Effect of wheat straw aqueous extract on Bermudagrass noxious weed. *Sustainability* **2022**, *14*, 15989. [CrossRef]
37. Đorđević, T.; Sarić-Krsmanović, M.; Gajić Umiljendić, J. Phenolic compounds and allelopathic potential of fermented and unfermented wheat and corn straw extracts. *Chem. Biodivers.* **2019**, *16*, e1800420. [CrossRef]
38. Nagy, K.N.; Kardos, L.V.; Orbán, Z.; Bakacsy, L. The allelochemical potential of an invasive ornamental plant, the Indian blanket flower (*Gaillardia pulchella* Foug.). *Plant Spec. Biol.* **2023**, *39*, 102–108. [CrossRef]
39. Chung, I.M.; Park, S.K.; Ali, M.; Prabakaran, M.; Oh, Y.T.; Kim, S.H.; Siddiqui, N.A.; Ahmad, A. Flavonoid glycosides from leaves and straw of *Oryza sativa* and their effects of cytotoxicity on a macrophage cell line and allelopathic on weed germination. *Saudi Pharm. J.* **2018**, *26*, 375–387. [CrossRef] [PubMed]
40. Ullah, R.; Aslam, Z.; Attia, H.; Sultan, K.; Alamer, K.H.; Mansha, M.Z.; Althobaiti, A.T.; Al Kashgry, N.A.T.; Algethami, B.; Zaman, Q.U. Sorghum allelopathy: Aternative weed management strategy and its impact on mung bean productivity and soil rhizosphere properties. *Life* **2022**, *12*, 1359. [CrossRef] [PubMed]
41. Mészáros, E.; Bodor, A.; Szierer, Á.; Kovács, E.; Perei, K.; Tölgyesi, C.; Bátori, Z.; Feigl, G. Indirect effects of COVID-19 on the environment: How plastic contamination from disposable surgical masks affect early development of plants. *J. Hazard. Mater.* **2022**, *436*, 129255. [CrossRef]
42. Zou, X.; Niu, W.; Liu, J.; Li, Y.; Liang, B.; Guo, L.; Guan, Y. Effects of residual mulch film on the growth and fruit quality of tomato (*Lycopersicon esculentum* Mill.). *Water Air Soil Poll.* **2017**, *228*, 71. [CrossRef]
43. Sun, S.; Chen, Z.; Jiang, H.; Zhang, L. Black film mulching and plant density influencing soil water temperature conditions and maize root growth. *Vadose Zone J.* **2018**, *17*, 1–12. [CrossRef]
44. Liao, Z.; Zhang, C.; Yu, S.; Lai, Z.; Wang, H.; Zhang, F.; Li, Z.; Wu, P.; Fan, J. Ridge-furrow planting with black film mulching increases rainfed summer maize production by improving resources utilization on the Loess Plateau of China. *Agric. Water Manag.* **2023**, *289*, 108558. [CrossRef]
45. Akhtar, K.; Wang, W.; Ren, G.; Khan, A.S.; Feng, Y.; Yang, G. Changes in soil enzymes, soil properties, and maize crop productivity under wheat straw mulching in Guanzhong, China. *Soil Tillage Res.* **2018**, *182*, 94–102. [CrossRef]
46. Zhang, W.; Yang, S.; Jin, Y.; Liu, P.; Lou, S. The effects of straw mulching combined with nitrogen applications on the root distributions and nitrogen utilization efficiency of summer maize. *Sci. Rep.* **2020**, *10*, 21082. [CrossRef]
47. Yang, H.; Wu, G.; Mo, P.; Chen, S.; Wang, S.; Xiao, Y.; Ma, H.; Wen, T.; Guo, X.; Fan, G. The combined effects of maize straw mulch and no-tillage on grain yield and water and nitrogen use efficiency of dry-land winter wheat (*Triticum aestivum* L.). *Soil Tillage Res.* **2020**, *197*, 104485. [CrossRef]
48. Zhang, Z.; Li, N.; Sun, Z.; Yin, G.; Zhang, Y.; Bai, W.; Feng, L.; John, Y. Fall straw incorporation with plastic film cover increases corn yield and water use efficiency under a semi-Arid climate. *Agriculture* **2022**, *12*, 2151. [CrossRef]
49. Wen, Y.; Chen, G.; Feng, F.; Guo, Y.; Hu, F.; Chen, G.; Zhao, C.; Yu, A.; Chai, Q. Straw retention combined with plastic mulching improves compensation of intercropped maize in arid environment. *Field Crops Res.* **2017**, *204*, 42–51.
50. Hu, Y.; Ma, P.; Duan, C.; Wu, S.; Feng, H.; Zou, Y. Black plastic film combined with straw mulching delays senescence and increases summer maize yield in northwest China. *Agric. Water Manag.* **2020**, *231*, 106031. [CrossRef]
51. Behzadnejad, J.; Tahmasebi-Sarvestani, Z.; Aein, A.; Mokhtassi-Bidgoli, A. Wheat straw mulching helps improve yield in sesame (*Sesamum indicum* L.) under drought stress. *Int. J. Plant Prod.* **2020**, *14*, 389–400. [CrossRef]

plants

MDPI

Article

Primary Metabolic Response of *Aristolochia contorta* to Simulated Specialist Herbivory under Elevated CO_2 Conditions

Hyeon Jin Jeong [1,2], Bo Eun Nam [1,3], Se Jong Jeong [1,4], Gisuk Lee [5], Sang-Gyu Kim [5] and Jae Geun Kim [1,6,*]

[1] Department of Biology Education, Seoul National University, Seoul 08826, Republic of Korea; hjjeong123@korea.kr (H.J.J.)
[2] Division of Forest Biodiversity, Korea National Arboretum, Pocheon 11187, Republic of Korea
[3] Research Institute of Basic Science, Seoul National University, Seoul 08826, Republic of Korea
[4] Seoul National University Elementary School, Seoul 03087, Republic of Korea
[5] Department of Biological Sciences, Korea Advanced Institute for Science and Technology (KAIST), Daejeon 34141, Republic of Korea
[6] Center for Education Research, Seoul National University, Seoul 08826, Republic of Korea
* Correspondence: jaegkim@snu.ac.kr

Abstract: This study explores how elevated carbon dioxide (CO_2) levels affects the growth and defense mechanisms of plants. We focused on *Aristolochia contorta* Bunge (Aristolochiaceae), a wild plant that exhibits growth reduction under elevated CO_2 in the previous study. The plant has *Sericinus montela* Gray (Papilionidae) as a specialist herbivore. By analyzing primary metabolites, understanding both the growth and defense response of plants to herbivory under elevated CO_2 conditions is possible. The experiment was conducted across four groups, combining two CO_2 concentration conditions (ambient CO_2 and elevated CO_2) with two herbivory conditions (herbivory treated and untreated). Although many plants exhibit increased growth under elevated CO_2 levels, *A. contorta* exhibited reduced growth with lower height, dry weight, and total leaf area. Under herbivory, *A. contorta* triggered both localized and systemic responses. More primary metabolites exhibited significant differences due to herbivory treatment in systemic tissue than local leaves that herbivory was directly treated. Herbivory under elevated CO_2 level triggered more significant responses in primary metabolites (17 metabolites) than herbivory under ambient CO_2 conditions (five metabolites). Several defense-related metabolites exhibited higher concentrations in the roots and lower concentrations in the leaves in response to the herbivory treatment in the elevated CO_2 group. This suggests a potential intensification of defensive responses in the underground parts of the plant under elevated CO_2 levels. Our findings underscore the importance of considering both abiotic and biotic factors in understanding plant responses to environmental changes. The adaptive strategies of *A. contorta* suggest a complex response mechanism to elevated CO_2 and herbivory pressures.

Keywords: climate change; defense response; folivorous herbivory; insect herbivory; plant growth; primary metabolite; *Sericinus montela*

1. Introduction

Climate change exerts a significant influence on the ecological and physiological characteristics of plants. One of the primary factors causing climate change is the increasing concentration of carbon dioxide (CO_2). As a greenhouse gas, CO_2 traps heat in the atmosphere, which is called greenhouse effect. Greenhouse effect is essential for life as it keeps the earth warm enough to be habitable [1]. However, human activities, such as burning of fossil fuels, industrial process, and deforestation have dramatically increased the concentrations of CO_2 in the atmosphere. It results in shifts in temperature, which impact the timing and rate of plant growth [2,3]. These environmental changes disrupt metabolic processes, altering enzyme activities and biochemical pathways within plants. Furthermore, the influences of climate change extend beyond the plant itself, affecting the other organisms interacting with the plant [4]. Changes in plant physiology can affect the availability and quality of plant resources as food for herbivorous insects [5,6]. Fluctuations in the types and quantities of metabolites produced by plants under changing environmental conditions impact the feeding patterns, reproductive cycles, and overall population dynamics of herbivorous insects.

Elevated atmospheric CO_2 concentrations not only act as a driver of climate change but also directly affect plant physiology and metabolism. Elevated CO_2 concentrations impact the photosynthetic rate, water use efficiency, and nutrient concentrations in plants [7,8]. Previous studies have highlighted that elevated CO_2 levels often promote plant growth through carbon fertilization, particularly in crop plants such as wheat, tomato, and soybeans [9–11]. They showed increased photosynthetic rates, enhanced biomass production, and altered allocation patterns of carbohydrates. These often result in increased leaf area, plant height, and overall biomass accumulation [12]. However, responses to increased CO_2 levels were dissimilar among taxa [13]. In the case of *Aristolochia contorta* Bunge, belonging to Aristolochiaceae, growth reduction in elevated CO_2 conditions has been reported [14]. *Aristolochia contorta* is an herbaceous perennial vine that mostly inhabits waterfront areas [15]. Height and photosynthetic rates of the species grown under elevated CO_2 levels were significantly lower compared to those grown in ambient CO_2 levels. Further investigation is required to better understand the responses of the plant to elevated CO_2 levels.

The significant reduction in the growth of *A. contorta* in the elevated CO_2 level suggested that other metabolic processes of *A. contorta*, including its growth and defense mechanism, could also be affected by increased CO_2 concentration. *Aristolochia contorta* is known for producing aristolochic acid, a defensive secondary metabolite against herbivores [16,17]. Despite this, the *Sericinus montela* Gray (Papilionidae) specially target and feed on this plant [18]. The larvae of *S. montela* incorporate the aristolochic acid into their bodies, utilizing it as a defense mechanism against predators [19,20]. It is understood that changes in plant quality can influence the condition of specialist herbivores in response [4]. As such, variations in the metabolites of *A. contorta* caused by elevated CO_2 level could affect the growth and survival of *S. montela*. *Sericinus montela* is classified as a vulnerable species in the red data book of Korea [21]. Considering the close ecological relationship between the plant and the butterfly, it is essential to predict how the dynamics of these two species will change in response to elevated CO_2.

The response of *A. contorta* to herbivory, when grown under the elevated CO_2 conditions, has been previously studied, particularly focusing on secondary metabolites and phytohormones [14]. Although the growth of *A. contorta* significantly declined in elevated CO_2 levels, there was evidence of stronger defense response in that group when exposed to herbivory. Jasmonic acid, known for its role in activating defense responses, was significantly higher under the elevated CO_2 than ambient CO_2 group in first-year plants. Furthermore, the first-year senescent plants exhibit a 30-fold higher aristolochic acid I compared to the ambient CO_2 group.

The inhibition of *A. contorta*'s growth under elevated CO_2 conditions necessitates a thorough examination of the plant's metabolic responses in such environments, with a specific focus on primary metabolites which directly influencing growth. Since primary metabolites are involved not only in growth but also in numerous other metabolic processes, these changes can affect other metabolic activities of the plant, such as the defense response [22,23]. Resistance and tolerance are two fundamental components of induced defense mechanisms of plants against herbivory [22–25]. Resistance involves the ability of plants to deter herbivores or reduce the damage through various means, such as the production of physical barriers or chemical deterrents. This strategy makes the plant less palatable or more difficult for the herbivore to consume [26]. Tolerance refers to a plant's capacity to withstand or recover from the damage inflicted by herbivores. Instead of preventing herbivory, tolerance mechanisms allow plants to maintain or quickly regenerate their growth and reproductive output despite being consumed [27,28]. Plants simultaneously invest resources in both resistance and tolerance, thereby displaying a composite strategy of defense [29,30].

What defensive strategy does *A. contorta* adopt? As a perennial plant, *A. contorta* has the ability to store metabolites produced during the growing season to its roots for the following year. Moreover, the specific herbivory by *S. montela* is highly destructive to the above-ground parts of *A. contorta*. The larvae of *S. montela* hatch in clusters and consume significant portions of the leaves during their growth period [31]. Direct observations have shown that the leaves near the hatching sites of *S. montela* are nearly entirely eaten, particularly in small plants, leaving only the stems. Faced with such extensive consumption, *A. contorta* might lean towards a tolerance defense strategy, such as reallocating nutrients to its roots as a systemic response. On the other hand, research has shown that herbivory of *S. montela* triggers an increase in defensive secondary metabolites in the leaves of *A. contorta*, such as magnocurarine and magnoflorine [32,33], signifying a resistance defense response [34].

The investment in these defensive strategies varies depending on the environmental conditions of the plant. Since a rapid reduction in growth has been observed in elevated CO_2 environments, the investment in defense strategies may also change under this condition. Because the defense strategies of *A. contorta* under elevated CO_2 conditions have been studied in terms of secondary metabolites and phytohormone levels [14], investigating primary metabolites, which are directly involved in growth and also play a role in defense, is necessary. By analyzing the variations of primary metabolites under elevated CO_2 and herbivory conditions, we can identify how the plant utilizes or stores substances involved in growth and defense.

We aim to investigate the primary metabolic responses of *A. contorta* under the following conditions: firstly, an elevated CO_2 concentration as an abiotic factor; secondly, herbivory by *S. montela* as a biotic factor; and thirdly, the interaction of these two factors. Our focus will be on the responses of local and systemic inductions by tissue types, including local leaves directly subjected to simulated herbivory, systemic leaves not directly subjected to herbivory, and roots as another systemic tissue. This research will contribute to understanding how defense response of plants to specialist herbivore is affected by elevated CO_2 conditions.

2. Results

2.1. Growth Characteristics and Chlorophyll Content of the Plants

When comparing the growth characteristics, plants in elevated CO_2 conditions (eCO_2; EC: Elevated CO_2 and Control, EH: Elevated CO_2 and Herbivory treatment) displayed a significant reduction in size and mass compared to those in ambient CO_2 (aCO_2; AC: Ambient CO_2 and Control, AH: Ambient CO_2 and Herbivory treatment). Plants in eCO_2 were shorter, with an average height of 43.2 cm, compared to 72.8 cm in the aCO_2 group (Figure 1a). Similarly, the dry weight and leaf area was lower in the elevated group (0.14 g, 28.8 cm^2) than the ambient group (0.54 g, 108.5 cm^2; Figure 1b,e). However, the lengths of the roots and the number of leaves were similar under two CO_2 conditions (eCO_2: 17.3 cm, 30.5 leaves; aCO_2: 19.1 cm, 33.0 leaves; Figure 1c,d). Specific leaf area (SLA), measured as the ratio of leaf area to leaf dry mass, was lower in plants under elevated CO_2 (eCO_2: 36.2 $m^2 \cdot kg^{-1}$, aCO_2: 49.3 $m^2 \cdot kg^{-1}$), indicating denser but smaller leaves (Supplementary Figure S2). The chlorophyll content showed no significant difference across the four groups, although a notable interaction between CO_2 levels and herbivory treatment was observed (Supplementary Table S2). Herbivory appeared to slightly reverse the reduction in chlorophyll content associated with elevated CO_2 (23.68 SPAD in EC, 26.80 SPAD in AC, 25.53 SPAD in EH, 23.06 SPAD in AH; Figure 1f).

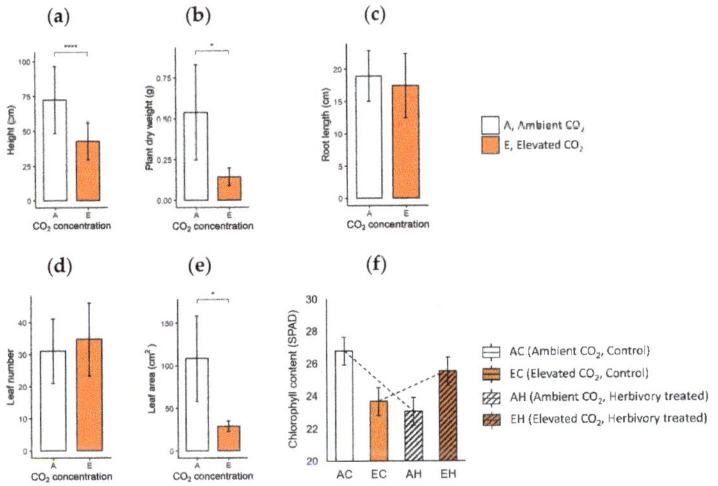

Figure 1. Growth characteristics of the plants in two CO_2 concentrations are: (**a**) height, (**b**) dry weight, (**c**) root length, (**d**) leaf number, and (**e**) leaf area of the plant in Ambient CO_2 (A) and Elevated CO_2 (E) groups; (**f**) chlorophyll contents of four experimental groups across two CO_2 levels and two herbivory treatments are: Ambient CO_2 and Control (AC), Ambient CO_2 and Herbivory treatment (AH), Elevated CO_2 and Control (EC), and Elevated CO_2 and Herbivory treatment (EH). The results of t-test are indicated on the graph in a, b, c, d, e, and the results of the two-way analysis of variance are presented through an interaction diagram on the graph f (* for p-value < 0.05; **** for p-value < 0.001).

2.2. Carbon and Nitrogen Concentrations

The two-way analysis of variance of carbon and nitrogen concentrations revealed several significant effects of CO_2 levels and the combined effect of CO_2 with herbivory. Both CO_2 levels and the combination of CO_2 with herbivory significantly influenced the carbon concentrations (weight percent; wt%) in the systemic leaves (SL; 1st to 4th leaves; Figure 2). Particularly, systemic leaves of plants subjected to ambient CO_2 with herbivory (AH) exhibited higher carbon concentrations compared to other groups. In contrast, carbon levels were lower in both the local leaves (LL; 5th to 6th leaves where herbivory was applied) and roots (R) of plants in elevated CO_2 environments. Nitrogen levels in the roots were higher under elevated CO_2 conditions. The carbon to nitrogen (C:N) ratio in the roots was lower in elevated CO_2 conditions.

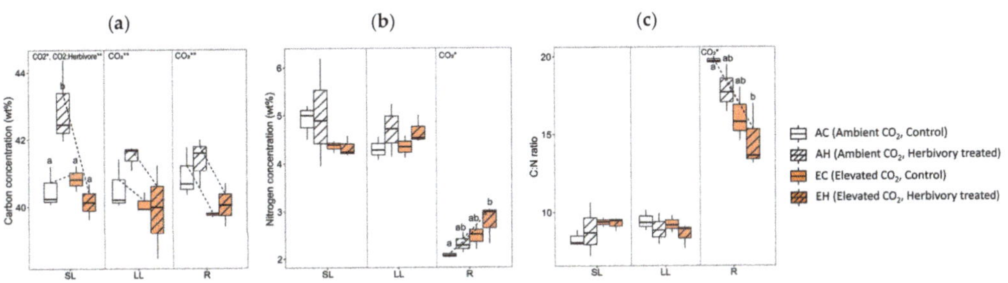

Figure 2. Carbon and nitrogen concentrations of three tissue types (local leaves that herbivory was treated, LL; systemic leaves, SL; roots, R) from four treatment groups (AC, AH, EC, EH). (**a**) carbon concentration, (**b**) nitrogen concentration, and (**c**) C:N ratio. The results of the two-way analysis of variance and Tukey's HSD were presented through an interaction diagram, letters, and text above each graph (* for p-value < 0.05; ** for p-value < 0.01).

2.3. Primary Metabolite Concentrations

To assess the metabolic changes in *A. contorta*, gas chromatography-mass spectrometry (GC-MS) method was used to measure the types and quantities of primary metabolites. Samples were taken from three plant tissues: systemic leaves (SL), local leaves (LL) and roots (R), across the four experimental groups (AC, AH, EC, EH). A total of 46 metabolites were identified and categorized into 4 groups: sugars and sugar alcohols, amino acids, organic acids, and miscellaneous. There were 15 sugars and sugar alcohols, 10 amino acids, 11 organic acids, and 10 other metabolites (Supplementary Table S1). Principal component analysis (PCA) was conducted to examine the patterns of metabolite concentrations across different tissues. This analysis revealed distinct clusters, with roots clearly separating from leaf tissues (Figure 3a). Systemic and local leaves displayed similar metabolic profiles. The first principal component (PC1) accounted for 77.9% of the total variance, and the second component (PC2) explained another 13.0%. A one-way multivariate analysis of variance (MANOVA) confirmed significant differences in metabolite profiles based on tissue types, with principal components 1–45 as dependent variables (Table 1). The PCA results for primary metabolites of each tissue, based on the treatment groups, did not exhibit clear clustering (Figure 3b,c,d).

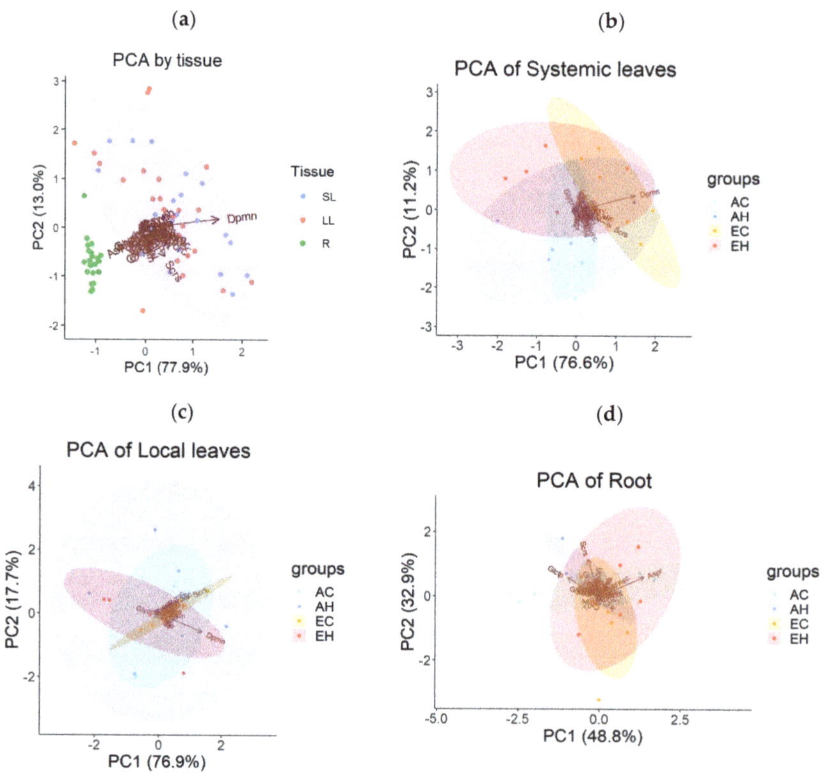

Figure 3. Results of the principal component analysis (PCA) on the concentrations of primary metabolites: (**a**) across three tissue types (local leaves treated with herbivory, LL; systemic leaves, SL; roots, R), (**b**) across four treatment groups (AC, AH, EC, EH) in SL, (**c**) in LL, and (**d**) in R. Abbreviations are used to describe the names of the primary metabolites.

Table 1. Results of the one-way multivariate analysis of variance (MANOVA) using principal components (PCs) of primary metabolite concentrations across tissue types as dependent variables. MANOVA retained seven eigenvalues.

	Df	Hotelling-Lawley	approx F	num Df	den Df	Pr (>F)
Tissue	2	10.74	99.726	14	260	<0.001
Residuals	137					

2.3.1. Effects of Elevated CO_2 on Primary Metabolites

Elevated CO_2 levels significantly altered the concentrations of several metabolites across tissue types. In comparing the AC (Ambient CO_2, Control without herbivory treatment) and EC (Elevated CO_2, Control without herbivory treatment) groups, there were significant changes in twelve metabolites: nine in SL, two in LL, and five in R (Figure 4). In terms of sugars and sugar alcohols, the concentrations of glucose and lactose were reduced in the roots under elevated CO_2 (Figure 4d,e), while myo-inositol increased in systemic leaves (Figure 4i). For amino acids, tryptophan levels rose in all tissue types under elevated CO_2 conditions (Figure 4o). Among the organic acids, linoelaidic acid level was higher in systemic leaves in EC (Figure 4s), and there was a mixed response of monopalmitin, which was higher in systemic leaves but lower in roots in EC (Figure 4t). Palmitic acid showed a similar pattern: higher in SL, but lower in R in EC (Figure 4u). Additionally, glyceryl-glycoside and triacontanol concentrations were higher in both SL and LL in elevated CO_2 (Figure 4x,ab). The concentrations of octacosanol and tocopherol were higher in SL in elevated CO_2 (Figure 4y,aa).

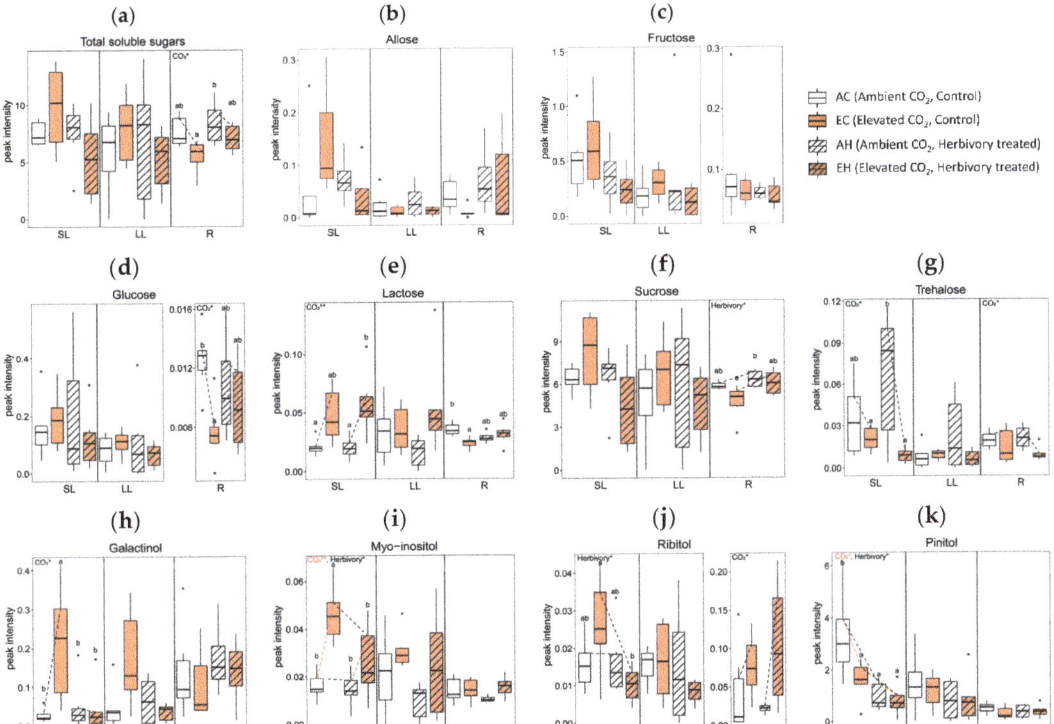

Figure 4. *Cont.*

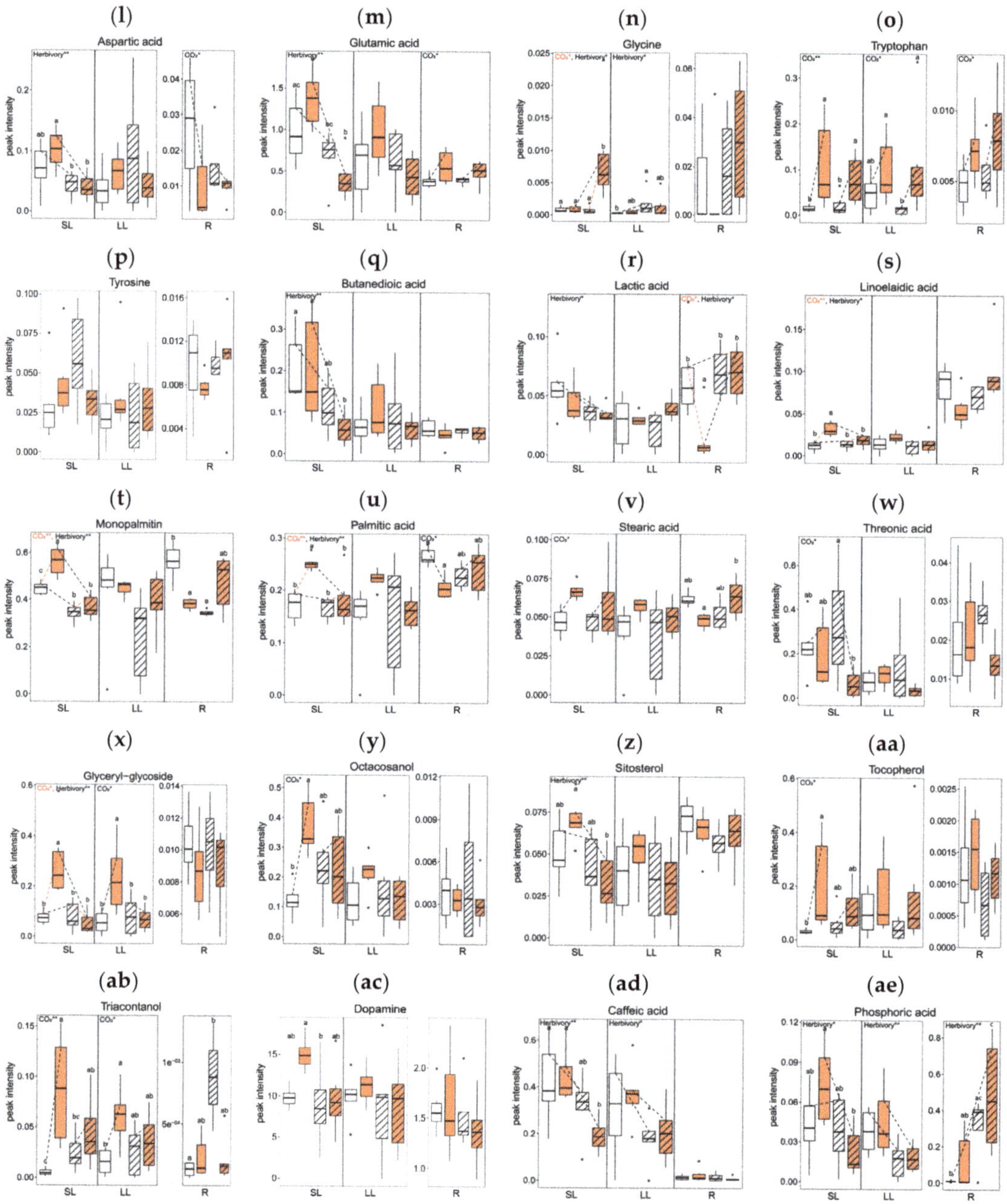

Figure 4. Concentration of primary metabolites by tissue types (local leaves that herbivory was treated, LL; systemic leaves, SL; roots, R) and four treatment groups (AC, EC, AH, EH). The results of the two-way analysis of variance and Tukey's HSD are presented through an interaction diagram, letters, and text above each plot (* for p-value < 0.05; ** for p-value < 0.01).

2.3.2. Effects of Herbivory on Primary Metabolites under Ambient CO_2 Conditions

Herbivory treatment significantly altered the concentrations of five primary metabolites. When comparing plants treated with herbivory under ambient CO_2 conditions (AH) to those without herbivory (AC), significant differences in the levels of two metabolites in SL, one in LL, and three in R were observed. For sugars and sugar alcohols, the concentration of pinitol was lower in AH than in AC in SL (Figure 4k). Although there were no significant differences between individual groups, the sucrose concentration in the roots tended to increase with herbivory treatment (Figure 4f).

In amino acids, glycine concentration was higher in AH than AC in LL (Figure 4n). Among organic acids, monopalmitin showed lower concentrations in both SL and R in AH (Figure 4t), while the concentrations of triacontanol in the roots were higher in AH than in AC (Figure 4ab). The concentration of phosphoric acid was significantly affected by herbivory treatment in all tissue types (Figure 4ae).

2.3.3. Combined Effects of Elevated CO_2 and Herbivory on Primary Metabolites

The interaction of elevated CO_2 and herbivory treatments significantly affected the primary metabolites of *A. contorta*, particularly in systemic tissues. Comparing plants treated with both elevated CO_2 and herbivory (EH) to those with elevated CO_2 but no herbivory (EC), notable differences in 17 metabolites were observed: 15 in SL and 4 in R. In the category of sugars and sugar alcohols, concentration of galactinol, myo-inositol, and ribitol were lower in the EH group compared to EC group in SL (Figure 4h,i,j). For amino acids, levels of aspartic acid and glutamic acid decreased, while glycine increased in EH compared to EC in SL (Figure 4l,m,n). Among organic acids, butanedioic acid, monopalmitin, and palmitic acid showed decreased levels in EH in SL (Figure 4q,t,u), whereas lactic acid and stearic acid were higher in EH in R (Figure 4r,v). Additionally, concentrations of linoelaidic acid, sitosterol, and caffeic acid were lower in EH in SL (Figure 4s,z,ad). Glyceryl-glycoside was lower in EH in both SL and LL (Figure 4x). The phosphoric acid concentration was lower SL but higher R in EH compared to the EC group (Figure 4ae).

Overall, the number of metabolites with significant differences was highest in SL (26), followed by R (12), and LL (3). In terms of treatment, the combined effect of elevated CO_2 and herbivory induced the most significant changes in metabolites, with a total of 19 significant differences observed. This contrasted with the lesser impact seen with herbivory alone, which showed the fewest differences (6). Additional data on primary metabolites not mentioned in the text can be found in Supplementary Figure S1, where the original peak intensities were displayed, regardless of transformation to achieving normality.

3. Discussion

3.1. *Effects of Elevated CO_2: Growth Inhibition and Changes of Stress-Related Primary Metabolites*

The effects of elevated CO_2 are estimated by comparing groups with different CO_2 concentrations: one under ambient CO_2 level (AC; Ambient CO_2, Control without herbivory treatment) and another under elevated CO_2 level (EC; Elevated CO_2, Control without herbivory treatment). Typically, plants exhibit increased photosynthesis under elevated CO_2 conditions, leading to higher carbon fixation rates and increased growth. However, *A. contorta* showed inhibited growth characteristics, contrasting with the general expectation. The height, dry weight, and total leaf area were all lower in the elevated CO_2 group than the ambient CO_2 group (Figure 1a,b,e). Furthermore, the carbon concentration within the plant was significantly lower in the EC group in LL and R compared to AC (Figure 2a). Previous study exposing *A. contorta* to the elevated CO_2 concentration noted reduced maximum carboxylation rate (Vc_{max}), indicating decreased rubisco activity [34]. Additionally, this study found that the amount of chlorophyll, known to correlate with the photosynthetic capacity, was slightly lower in the EC group than in the AC group (Figure 1f) [35,36]. This suggests that elevated CO_2 level did not enhance *A. contorta*'s photosynthetic capacity and carbon fixation, potentially acting as a stressor instead. The diminished carbon fixation and growth could result from *A. contorta* diverting energy to produce non-essential metabolites for growth in response to elevated CO_2 concentrations. Indeed, concentrations of several abiotic stress-response-related metabolites were significantly higher in the EC group than in the AC group. Myo-inositol, involved in the production of stress-related molecules [37], was significantly higher in EC than in AC in SL (Figure 4i). Octacosanol, associated with stress resistance by maintaining an outer protective layer, epicuticular waxes of leaves [38,39], was higher in EC in leaves (Figure 4y). Galactinol is a precursor for stress tolerance molecules such as raffinose family oligosaccharides (RFOs) which are involved in abiotic stress responses, including drought and cold [40]. The concentration of galactinol under elevated CO_2 conditions had not been previously studied, but was found to be higher in EC in leaves (Figure 4h). The concentration of tocopherol, which is involved in stress tolerance by deactivating the reactive oxygen species (ROS) [41], was higher in EC than in AC in SL (Figure 4aa). Triacontanol, which modulates the activation of stress tolerance molecules, exhibited a higher concentration in EC in leaves (Figure 4ab) [42].

3.2. Effects of Herbivory: Local and Systemic Responses

When plants are attacked by insect herbivores, they often trigger a response that involves the production of defense metabolites [43]. Plants initiate signaling pathways to enhance defense mechanisms, thereby activating the production of specialized metabolites in response to herbivore attacks. This study used simulated herbivory methods to induce defense responses, but limitations exist as they do not fully mimic actual herbivore bites or account for ecological interactions, requiring careful interpretation of results [44]. The local leaves that are directly attacked may produce a higher concentration of defensive compounds at the site of damage [45,46]. This localized response usually aims to confine the damage and repel the herbivore [47]. In the local leaves (LL) of *A. contorta*, glycine exhibited a significantly higher concentration in AH (Ambient CO_2, with Herbivory treatment) than in AC (Ambient CO_2, Control without herbivory treatment; Figure 4n). The glycine-rich proteins (GRPs) superfamily participates in the mechanisms of signaling and cellular stress response [48]. Additionally, allose, known to upregulate defense-related and pathogenesis-related protein genes [49], was found at slightly higher concentrations in AH than in AC (Figure 4b). Triacontanol, involved in modulating stress-related signaling pathways [42], showed slightly higher concentrations in AH groups than in AC groups in LL (Figure 4ab).

Aristolochia contorta triggered a systemic response as well as localized response throughout the plant. When comparing the concentrations of primary metabolites in the group that received herbivory treatment and the group that did not, under ambient CO_2 conditions (AH vs. AC), more significant differences in primary metabolites were observed within systemic tissues such as systemic leaves (SL; 2) and roots (R; 3) than in local leaves (LL; 1) where herbivory was directly applied. This indicates that the plant responded systemically to herbivory. The systemic response often involves signaling molecules such as jasmonic acid that travel from the damaged leaf to other parts of the plant [50,51]. In the previous study of *A. contorta*, the increase in jasmonic acid concentration was reported when herbivory of *S. montela* was treated [14]. The movement of jasmonic acid triggers a systemic defense response in other leaves, inducing the production of defense-related metabolites even in undamaged parts.

In systemic leaves (SL), the concentration of total soluble sugars, including allose, cellobiose, fructose, galactopyranose, glucose, lactose, maltose, sucrose, and trehalose, was slightly higher in AH than AC (Figure 4a). The signaling role of sugars in defense response by inducing resistance genes has been reported [52]. Trehalose, known to act as an elicitor of defense responses [53], exhibited slightly higher levels in AH than AC in SL (Figure 4g). Threonic acid, which increases in response to wounding stress [54], was slightly higher in AH than in AC in SL (Figure 4w). The concentrations of triacontanol, which is involved in regulating stress-related signaling pathways [42], was higher in AH than in AC in systemic leaves, similar to the patterns observed in local leaves (Figure 4ab). Tyrosine roles as a precursor of aristolochic acid, which is an important defensive second metabolite of *A. contorta* [17,55]. The concentration of tyrosine was slightly higher in AH compared to AC in SL (Figure 4p).

Recent studies have highlighted the critical role of roots in defense responses [56–58]. In the case of *A. contorta*, a perennial herb, examining responses of roots is essential for understanding the systemic responses. Signaling compounds triggered by herbivore attack can travel to the roots, prompting changes in root exudates and potentially influencing interactions with soil microbes or even signaling neighboring plants about potential threats [59]. In the roots (R) of *A. contorta*, several metabolites exhibited different concentrations between the AH and AC groups. Galactinol, associated with signaling during abiotic and biotic stress conditions [40], displayed slightly higher concentrations in AH than AC (Figure 4h). Elevated CO_2 lowers the concentration of total soluble sugars in R (Figure 4a). The concentrations of triacontanol and allose were higher in AH than in AC in R, exhibiting a similar pattern to that observed in SL and LL (Figure 4b,ab).

3.3. Combined Effects of Elevated CO_2 and Herbivory: Shifts in Defense Response

In our investigation of herbivory-induced responses under varying CO_2 conditions, we observed differences in responses between the ambient CO_2 (aCO_2) and the elevated CO_2 (eCO_2) groups. The concentrations of primary metabolites exhibited significantly greater differences in the eCO_2 group subjected to herbivory treatment compared to those in the aCO_2 group. A comparison of groups grown under elevated CO_2 conditions with and without herbivory treatment (EH vs. EC) revealed significant differences in 17 metabolites. Conversely, the comparison between the ambient CO_2 groups with and without herbivory treatment (AH vs. AC) revealed only five metabolites with statistically significant differences. These results underscore the significant influence of elevated CO_2 levels in restructuring plant responses to herbivory, leading to more pronounced shifts in primary metabolite concentrations under elevated CO_2 conditions when exposed to herbivory.

When herbivory treatment was applied to both ambient CO_2 and elevated CO_2 groups, we observed different trends in several primary metabolites. Comparing AH with AC, the concentration of total soluble sugars was slightly higher in AH across all tissue types (Figure 4a). Conversely, when comparing EH with EC, the concentration was lower in EH in leaves but higher in roots. The concentrations of sucrose in the leaves exhibit similar responses, showing a slight increase with herbivory treatment in ambient CO_2 groups but a decrease in elevated CO_2 groups in leaves (Figure 4f). The concentration of glucose in roots (R) exhibited an opposite pattern. It was lower in AH than in AC, but higher in EH than in EC (Figure 4d). Several signaling-related metabolites under stress conditions exhibited varied trends in their herbivory-induced responses between ambient and elevated CO_2 groups. Galactinol, a precursor of RFOs, functions as a signaling molecule in response to pathogen attacks and wounding [40]. Although there was no meaningful difference in galactinol concentration between AH and AC, the concentration was lower in EH than in EC in SL, but higher in R (Figure 4h). As mentioned, threonic acid is known to increase wounding stress [54]. It was slightly higher in all tissue types in AH than AC, but slightly lower in all tissue types in EH than EC (Figure 4w). Myo-inositol, involved in the production of stress-related molecules [37], did not exhibited a meaningful difference between the group AH and AC (Figure 4i). However, it exhibited lower concentration in leaves, particularly in SL in EH compared to EC. As indicated, the superfamily of glycine-rich proteins (GRPs) engaged in stress response mechanisms [48]. The concentration of glycine was significantly higher in AH than in AC in LL, but this difference was not observed in SL and R (Figure 4n). In EH, compared to EC, the concentration was higher in SL and R. Regarding the elicitor metabolites involved in defense response, trehalose was slightly higher in AH than in AC in leaves but slightly lower in EH than EC (Figure 4g) [53]. Allose, involved in upregulating pathogen-related genes [49], was marginally higher in AH compared to AC in SL but was slightly lower in EH compared to EC (Figure 4b). Tyrosine, a precursor of aristolochic acid [55,60], showed higher levels in AH than AC in SL but was comparable in SL and higher in R in EH than EC (Figure 4p). Dopamine, another precursor of aristolochic acid [60], showed no meaningful differences between the ambient CO_2 groups in SL but was slightly lower in EH than EC in SL (Figure 4ac).

Numerous studies have addressed the reduced photosynthetic rate of plants in response to herbivory [61,62]. This reduction is often attributed to enhanced water loss and stomatal closure at sites of leaf attack. These studies reported changes in the expression of genes involved in photosynthesis, leading to a reduced carbon fixation rate [63–65]. In our study, a decrease in chlorophyll content was observed in *A. contorta* under ambient CO_2 conditions when subjected to herbivory by *S. montela*. However, in elevated CO_2 environments, plants subjected to herbivory displayed higher chlorophyll content compared to the group without herbivory [66]. By regulating their chlorophyll content, plants optimize the use of light energy. This adaptation enables them to continuously adjust to their environment. Although the exact photosynthetic rate was not measured in this study, an increase in chlorophyll content might indicate an effort to enhance the carbon fixation rate, possibly facilitating the synthesis of defense-related substances [67]. Indeed, comparing the EH and EC groups revealed more significant differences in the concentration of primary metabolites than the comparison between AH and AC groups. This observation is consistent with findings from previous research, which have noted increases in secondary metabolites and phytohormones in *A. contorta* subjected to herbivory under elevated CO_2 conditions [34]. In response to herbivory in the elevated CO_2 environment, *A. contorta* seems to exhibit greater resistance responses by producing more stress-related metabolites than in the ambient CO_2 environment. This outcome aligns with prior research, where several crop genotypes that did not exhibit a decrease in photosynthesis rates in the face of herbivory demonstrated more resistant responses [68,69].

Several defense-related primary metabolites, including galactinol, threonic acid, myo-inositol, glycine, trehalose, and allose, exhibited higher concentrations in the roots and lower concentrations in the leaves when *A. contorta* was exposed to herbivory under elevated CO_2 conditions. This indicates a more pronounced response in the roots under elevated CO_2 compared to ambient CO_2. Since the concentrations of carbon did not increase in the roots in EH, it is difficult to attribute the rise of primary metabolites in the roots to storage purpose (Figure 2a,b). A previous study dealing with the secondary metabolites of *A. contorta* also reported a significantly higher concentration of aristolochic acid, a well-known defensive metabolite, in the roots compared to the leaves [70]. This suggests that *A. contorta* may invest more defensive responses in its roots than in its leaves. Specifically, the increased concentration of primary metabolites in the roots under elevated CO_2 conditions indicates that these defensive responses may be intensified at higher CO_2 level. To understand the specific functions and interactions of primary metabolites, and to elucidate why the roots exhibited a stronger response rather than the leaves, where herbivory of *S. montela* occurred, further investigation is needed. Nonetheless, the observed trends in metabolite concentrations offer valuable insights into the complex dynamics of plant defense responses under varying CO_2 conditions.

4. Materials and Methods

4.1. Plant Material and Growth Condition

Seeds of *A. contorta*, collected from Pyeongtaek-si and Gapyeong-gun in 2020, were stored at 4 °C, and sown in March 2021 in a soil mixture similar to their natural habitat (2:1 volumetric ratio of sand to loam) [71]. Sprouts were transplanted into pots and those over 7 cm tall were selected in July. These plants were placed in hexagonal open top chambers (OTCs) in a rooftop greenhouse with natural light exposure. The average height of the plants in each chamber were equalized. A shading net was installed on the ceiling of the greenhouse with about 50% relative light intensity (RLI), a suitable shade for *A. contorta* [72]. The variance in light intensity among the OTCs was negligible (RLI 34.5% to 44.4%, p-value 0.24). Plants were maintained in OTCs for about 45 days, watered 3 times a week for saturation, and received uniform amounts of fertilizer. Temperature and relative humidity values of each chamber, measured by sensors (HOBO Pro v2, Onset, Bourne, MA, USA), were similar (29.0 to 29.3 °C, 68.5 to 72.8%).

4.2. CO_2 and Herbivory Treatment

Plants were divided into two groups with different CO_2 levels: ambient (A, aCO_2) and elevated CO_2 (E, eCO_2; Figure 5). The concentration of elevated CO_2 was set to 540 ppm, based on the Representative Concentration Pathway (RCP) 4.5 scenario [73]. Levels of CO_2 were regulated using a system equipped with a sensor transmitter coupled with a CO_2 controller (0–2000 ppm CO_2, SH-MVG260, Soha-tech, Seoul, Republic of Korea), solenoid valve, and CO_2 gas tank (99.99%, 40L). In the aCO_2 group, a similar ventilation system was installed, as air is emitted evenly from all directions of the chamber. Three OTCs were used for each CO_2 concentration, totaling six OTCs. The actual CO_2 concentrations, measured by CO_2 data logger (CDL 210, Wöhler, Bad Wünnenberg, Germany), were 429.2 ppm in the aCO_2 group and 525.7 ppm in the eCO_2 group.

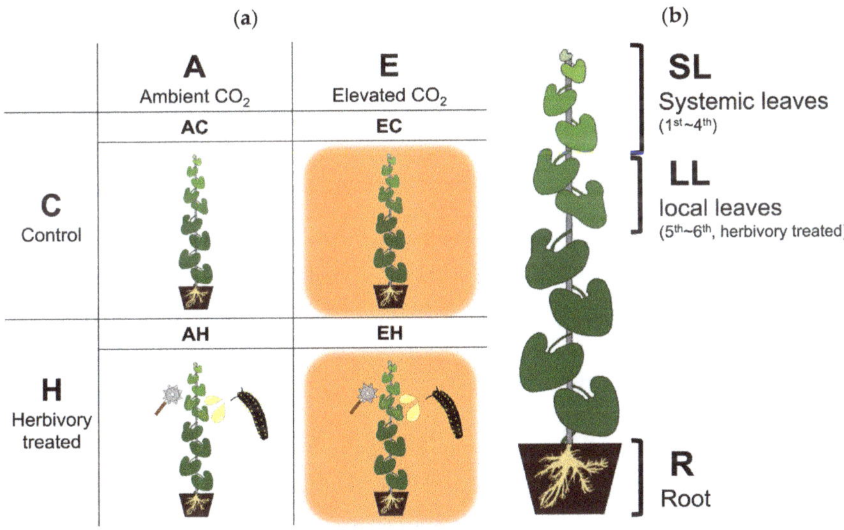

Figure 5. A schematic diagram of the experimental groups and collected tissues. (**a**) Four experimental groups, each with two CO_2 levels (Ambient and Elevated CO_2) and Herbivory treatments (Control and Herbivory-treated). (**b**) Three types of collected tissue. Systemic leaves (SL), Local leaves (LL), and Roots (R).

To induce the defense response in *A. contorta*, herbivory was mimicked using the oral secretion of specialist herbivore *S. montela* applied on the leaves of the plants. Insect materials were initially collected at the egg stage from common gardens in Pyeongtaek-si, Gyeonggi-do, Republic of Korea (37°04′06″ N, 127°00′27″ E) and raised until needed. The herbivory simulation was conducted in August at noon, aligning with the feeding time *S. montela* hatchlings [14]. For this simulation, the fifth and sixth leaves of *A. contorta* were wounded with a pattern wheel, and 20 µL of *S. montela*'s oral secretion, diluted 20 times with deionized water, was applied. This approach, involving wounding and the application of the herbivore's oral secretion, replicates effective defense induction method from previous research [74–76]. The oral secretion was obtained directly from third-instar larvae or older using pipette tips.

The experiment was structured into four groups to assess responses influenced by CO_2 and herbivory. The groups included: AC (ambient CO_2, Control), AH (ambient CO_2, Herbivory treatment), EC (elevated CO_2, Control), and EH (elevated CO_2, Herbivory treatment). Each group consisted of twelve *A. contorta* plants evenly distributed across three OTCs, with each OTC containing four plants from each treatment group. Plant samples were collected five hours after the herbivory treatment at noon to analyze metabolic changes facilitated by photosynthesis. Three parts of the plants were collected for analysis: (1) the first to fourth leaves from the top, (2) the fifth to sixth leaves, and (3) the roots. The fifth to sixth leaves, where herbivory treatment was applied, were collected as local leaves (LL), while the first to fourth leaves were collected to assess the response of systemic leaves (SL). The root (R) was collected to examine the response in the underground parts, considering the significance of the underground part of perennial herb [77].

4.3. Measuring Chlorophyll Content and Biomass

Chlorophyll content in the leaves was measured non-destructively using a chlorophyll meter (SPAD-502plus, Konica Minolta, Tokyo, Japan) before collection. It is influenced by environmental factors such as nutrient availability and environmental stresses [78]. In the group treated with herbivory, the chlorophyll content was measured five hours after the treatment. The relative content of chlorophyll is calculated from the absorbance values to obtain Soil Plant Analysis Development (SPAD) value [79].

After harvesting, growth metrics including the height of the above-ground part of the plant, the length of the roots, the number of leaves, the total leaf area, and the dry weight were measured. The total leaf area per plant was measured with LI-3000C portable leaf area meter and LI-3050C transparent belt conveyors (LI-COR, Lincoln, NE, USA). The height of the above-ground part, the length of the root, and the number of leaves were measured in 12 plants from each treatment group (a total of 48 plants). Measurements of leaf area and dry weight were conducted exclusively for three plants from each ambient and elevated CO_2 groups, to avoid the destruction which could impact primary metabolites. Specific leaf area (SLA) was calculated as the ratio of leaf area to leaf dry mass.

4.4. Carbon and Nitrogen Concentrations

Carbon and nitrogen concentration were measured by tissues (SL, LL, R) to assess the relative resource availability of the plants and to examine the impact of the stresses on the plants [80,81]. For the measurements of nitrogen and carbon content, twelve plants from each treatment group (AC, AH, EC, EH) were exclusively prepared, and four plants from each group were pooled into a single sample. The total carbon and nitrogen concentration were analyzed using an elemental analyzer (Flash EA 1112, Thermo Fisher Scientific, Waltham, MA, USA), at the National Instrumentation Center for Environmental Management (NICEM) at Seoul National University. The concentration was calculated based on the dry weight of each tissue.

4.5. Metabolic Profiling of Primary Metabolite

The concentration of primary metabolites in 12 plants from each experimental group, a total of 48 plants across 4 groups, was measured (Figure 5). Three parts of the plant were collected: Local leaves (LL), Systemic leaves (SL), and Roots (R). To prevent the change of metabolite after harvest, harvested tissues were immediately placed into the liquid nitrogen and stored at −80 °C. The Gas Chromatography-Mass Spectrometry (GC-MS) method was used to measure the concentration of primary metabolites in samples. Before analysis, the samples were lyophilized. For lyophilization, the samples were ground using pellet pestles and motors (Kimble chase, Vineland, NJ, USA) with liquid nitrogen. Ten milligrams of each sample were lyophilized. After lyophilization, the samples were prepared for GC-MS analysis. They were extracted with 1 ml of 100% methanol (HPLC grade, Sigma-Aldrich, St. Louis, MO, USA) and vortexed 1 min. After that, they were sonicated for 40 min and centrifuged with 110× *g* for 30 s. The supernatants of each sample were collected and filtered through 0.45 μm PTFE (polytetrafluoroethylene) syringe filters (Membrane solution, Plano, TX, USA). After filtration, 100 μL of each sample was transferred to amber GC vials (Agilent, Santa Clara, CA, USA) and dried with nitrogen gas for 5 min. To the dried samples, 30 μL of 20,000 μg/mL methoxyamine hydrochloride in pyridine (Sigma-Aldrich, USA) and 50 μl of BSTFA (*N,O*-bis (trimethylsilyl) trifluoroacetamide; Alfa Aesar, Ward Hill, MA, USA) containing 1% of trimethylchlorosilane (Sigma-Aldrich, USA) were added for oximation. For quantification, 10 μL of 300 μg/mL 2-chloronaphtalene in pyridine

(Sigma-Aldrich, USA) was added as an internal standard (IS). The samples were evaporated in 65 °C for 1 h, and 100 μL of each sample was transferred to another amber GC vials contained insert. Prepared samples were stored at 4 °C before GC-MS analysis.

The concentrations of primary metabolites were assessed using a Shimadzu gas chromatography system (GC-MS QP2020, Shimadzu Corp., Kyoto, Japan). This system was equipped with an MSD detector, an autosampler, a split/splitless injector, an injection module, and GCMSsolution software (https://www.shimadzu.co.kr/products/gas-chromatograph-mass-spectrometry/gc-ms-software/gcmssolution/index.html). To identify metabolites in the samples, the National Institute of Standards and Technology (NIST) mass spectral search library was employed, using a threshold of greater than 80% match quality for peak assignment. Subsequently, normalization was conducted by dividing the peak area of each compound by that of the internal standard (IS), thus enabling comparison of the relative abundance of selected metabolites across samples.

4.6. Statistical Analysis

Statistical methods were used to analyze plant height, length, dry weight, leaf count, and area. First, an F-test determined the homogeneity of variance among groups, and either Welch's t-test or an independent samples t-test was followed depending on the results. For primary metabolites, chlorophyll content, and carbon to nitrogen ratio, significant differences among the four groups (AC, EC, AH, EH) were analyzed using a two-way analysis of variance (ANOVA), followed by Tukey's honestly significant difference (HSD) test for detailed comparison. Normality of data was checked using the Shapiro–Wilk test. Concentrations of metabolites with non-normal distribution were transformed using logarithmic or exponential methods to meet normality requirements (Supplementary Table S1). Principal component analysis (PCA) was performed to evaluate the effect of tissue types and treatment on concentration of primary metabolites. Differences among tissue types were verified using a one-way multivariate analysis of variance (MANOVA), using principle components as the variables. Statistical analysis and illustration of the results were conducted using R version 4.2.1 [82] with the packages 'ggpubr' [83], 'Momocs' [84], and 'ggplot2' [85].

5. Conclusions

Our research provides insight into how *Aristolochia contorta* responds to elevated CO_2 levels and herbivory, revealing complex interactions between plant growth, defense mechanisms, and environmental changes. Contrary to the general expectation of enhanced growth, the growth of *A. contorta* was inhibited under elevated CO_2, suggesting unique responses of species to increased CO_2 levels. The presence of herbivory by *Sericinus montela* also acts as a stress factor, triggering both local and systemic defense responses in the plant.

The study showed that elevated CO_2 conditions significantly influenced the plant's metabolic response to herbivory, indicating a shift towards defense. This included increased concentrations of defense-related metabolites, especially in the roots, suggesting a potential intensification of defensive responses in the underground parts of the plant under elevated CO_2 levels.

Our findings underscore the importance of considering both abiotic and biotic factors in understanding plant responses to environmental changes. The adaptive strategies of *A. contorta*, balancing resistance and tolerance, suggest a complex response mechanism to elevated CO_2 and herbivory pressures. This research contributes to our knowledge of plant response to environmental changes, emphasizing the need for further studies to comprehend the broader implications of these interactions on ecosystem dynamics.

Supplementary Materials: The following supporting information can be downloaded at: https://www.mdpi.com/article/10.3390/plants13111456/s1, Table S1: Names and categories of all primary metabolites observed in *A. contorta*. Scale transformation to achieve normality for ANOVA is described; Table S2: Table of results for chlorophyll content from a two-way analysis of variance; Figure S1: Concentrations of primary metabolites not documented in the manuscript, categorized by tissues types (local leaves that herbivory was treated, LL; systemic leaves, SL; roots, R) and four treatment groups (AC, EC, AH, EH); Figure S2: Specific leaf area of the plants in two CO_2 concentrations.

Author Contributions: Conceptualization and sample preparation, J.G.K., H.J.J., B.E.N. and S.J.J.; primary metabolites analysis, G.L. and S.-G.K.; data analysis and original draft preparation, H.J.J.; revising and reviewing, B.E.N., S.J.J., S.-G.K. and J.G.K.; supervising and funding acquisition, J.G.K. All authors have read and agreed to the published version of the manuscript.

Funding: This research was supported by Basic Science Research Program through National Research Foundation of Korea (NRF) funded by Ministry of Education (NRF2021R1I1A2041895) and by Korea Environment Industry and Technology Institute (KEITI) through 'Wetland Ecosystem Value Evaluation and Carbon Absorption Value Promotion Technology Development Project', funded by Korea Ministry of Environment (MOE) (RS-2022-KE002025).

Data Availability Statement: The data are not publicly available due to the regulation of funding agency. The raw data supporting the conclusions of this article will be made available by the first author on request.

Acknowledgments: We thank Yu Seong Choi, So Ho Kim, Ho Yeong Yu, Ye Eun Chae, and Si-Hyun Park for sample preparation and measurement. We also thank Man Je Kim at Pyeongtaek Cultural Center for supplying larvae and eggs of *Sericinus montela*.

Conflicts of Interest: The authors declare no conflicts of interest.

References

1. Cassia, R.; Nocioni, M.; Correa-Aragunde, N.; Lamattina, L. Climate Change and the Impact of Greenhouse Gasses: CO_2 and NO, Friends and Foes of Plant Oxidative Stress. *Front. Plant Sci.* **2018**, *9*, 273. [CrossRef]
2. Körner, C. Significance of temperature in plant life. In *Plant Growth and Climate Change*; Blackwell Publishing Ltd.: Oxford, UK, 2006; pp. 48–69.
3. Ziska, L.H.; Bunce, J.A. Plant responses to rising atmospheric carbon dioxide. In *Plant Growth and Climate Change*; Blackwell Publishing Ltd.: Oxford, UK, 2006; pp. 17–47.
4. DeLucia, E.H.; Nabity, P.D.; Zavala, J.A.; Berenbaum, M.R. Climate change: Resetting plant-insect interactions. *Plant Physiol.* **2012**, *160*, 1677–1685. [CrossRef] [PubMed]
5. Bruce, T.J.A. Interplay between insects and plants: Dynamic and complex interactions that have coevolved over millions of years but act in milliseconds. *J. Exp. Bot.* **2015**, *66*, 455–465. [CrossRef] [PubMed]
6. Jamieson, M.A.; Trowbridge, A.M.; Raffa, K.F.; Lindroth, R.L. Consequences of climate warming and altered precipitation patterns for plant-insect and multitrophic interactions. *Plant Physiol.* **2012**, *160*, 1719–1727. [CrossRef] [PubMed]
7. De Graaff, M.-A.; Van Groenigen, K.-J.; Six, J.; Hungate, B.; Van Kessel, C. Interactions between plant growth and soil nutrient cycling under elevated CO_2: A meta-analysis. *Glob. Chang. Biol.* **2006**, *12*, 2077–2091. [CrossRef]
8. Eamus, D. The interaction of rising CO_2 and temperatures with water use efficiency. *Plant Cell Environ.* **1991**, *14*, 843–852. [CrossRef]
9. Hellmers, H.; Sionit, N.; Strain, B.R. Effects of different concentrations of atmospheric CO_2 on growth and yield components of wheat. *J. Agric. Sci.* **1981**, *97*, 335–339. [CrossRef]
10. Nilsen, S.; Hovland, K.; Dons, C.; Sletten, S.P. Effect of CO_2 enrichment on photosynthesis, growth and yield of tomato. *Sci. Hortic.* **1983**, *20*, 1–14. [CrossRef]
11. Rogers, H.H.; Cure, J.D.; Thomas, J.F.; Smith, J.M. Influence of elevated CO_2 on growth of Soybean plants. *Crop Sci.* **1984**, *24*, 361–366. [CrossRef]
12. Toreti, A.; Deryng, D.; Tubiello, F.N.; Müller, C.; Kimball, B.A.; Moser, G.; Boote, K.; Asseng, S.; Pugh, T.A.M.; Vanuytrecht, E.; et al. Narrowing uncertainties in the effects of elevated CO_2 on crops. *Nat. Food* **2020**, *1*, 775–782. [CrossRef]
13. Wang, D.; Heckathorn, S.A.; Wang, X.; Philpott, S.M. A meta-analysis of plant physiological and growth responses to temperature and elevated CO_2. *Oecologia* **2012**, *169*, 1–13. [CrossRef] [PubMed]
14. Park, H.J.; Nam, B.E.; Lee, G.; Kim, S.-G.; Joo, Y.; Kim, J.G. Ontogeny-dependent effects of elevated CO_2 and watering frequency on interaction between *Aristolochia contorta* and its herbivores. *Sci. Total Environ.* **2022**, *838*, 156065. [CrossRef] [PubMed]
15. Nakonechnaya, O.V.; Kholina, A.B.; Koren, O.G.; Zhuravlev, Y.N. Genetic diversity of a rare species *Aristolochia contorta* Bunge (Aristolochiaceae) in Primorsky Krai. *Russ. J. Genet.* **2012**, *48*, 152–162. [CrossRef]
16. Rücker, G.; Chung, B. Aristolochiasäuren aus *Aristolochia manshuriensis*. *Planta Med.* **1975**, *27*, 68–71. [CrossRef]
17. Tada, A.; Sase, K.; Ohmura, I.; Shoji, J.; Tanaka, O. Studies on the Aristolochiaceous plants. I. The isolation of aristolochic acid from *Aristolochia griffithii* Hook f. et Thoms. *Syoyakugaku Zasshi* **1969**, *23*, 99–101.
18. Li, X.; Luo, Y.; Yang, H.; Yang, Q.; Settele, J.; Schweiger, O. On the ecology and conservation of *Sericinus montelus* (Lepidoptera: Papilionidae)–Its threats in Xiaolongshan forests area (China). *PLoS ONE* **2016**, *11*, e0150833. [CrossRef] [PubMed]
19. Nishida, R. Sequestration of plant secondary compounds by butterflies and moths. *Chemoecology* **1994**, *5*, 127–138. [CrossRef]

20. Nishida, R.; Fukami, H. Ecological adaptation of an Aristolochiaceae-feeding swallowtail butterfly, *Atrophaneura alcinous*, to aristolochic acids. *J. Chem. Ecol.* **1989**, *15*, 2549–2563. [CrossRef] [PubMed]
21. National Institute of Biological Resources. *Red Data Book of Endangered Insects in Korea I*; National Institute of Biological Resources: Incheon, Republic of Korea, 2012; Available online: https://archive.nationalredlist.org/files/2016/04/7.-Red-Data-Book-of-Endangered-Insects-in-Korea-I-Korean.pdf (accessed on 20 May 2024). (In Korean)
22. Wang, M.; Gu, Z.; Wang, R.; Guo, J.; Ling, N.; Firbank, L.G.; Guo, S. Plant primary metabolism regulated by nitrogen contributes to plant–pathogen interactions. *Plant Cell Physiol.* **2019**, *60*, 329–342. [CrossRef]
23. Salam, U.; Ullah, S.; Tang, Z.-H.; Elateeq, A.A.; Khan, Y.; Khan, J.; Khan, A.; Ali, S. Plant metabolomics: An overview of the role of primary and secondary metabolites against different environmental stress factors. *Life* **2023**, *13*, 706. [CrossRef]
24. Watts, S.M.; Dodson, C.D.; Reichman, O.J. The roots of defense: Plant resistance and tolerance to belowground herbivory. *PLoS ONE* **2011**, *6*, e18463. [CrossRef] [PubMed]
25. Mitchell, C.; Brennan, R.M.; Graham, J.; Karley, A.J. Plant defense against herbivorous pests: Exploiting resistance and tolerance traits for sustainable crop protection. *Front. Plant Sci.* **2016**, *7*, 1132. [CrossRef] [PubMed]
26. Janzen, D.H. When is it coevolution. *Evolution* **1980**, *34*, 611–612. [CrossRef] [PubMed]
27. Stowe, K.A.; Marquis, R.J.; Hochwender, C.G.; Simms, E.L. The evolutionary ecology of tolerance to consumer damage. *Annu. Rev. Ecol. Evol. Syst.* **2000**, *31*, 565–595. [CrossRef]
28. Strauss, S.Y.; Agrawal, A.A. The ecology and evolution of plant tolerance to herbivory. *Trends Ecol. Evol.* **1999**, *14*, 179–185. [CrossRef] [PubMed]
29. Núñez-Farfán, J.; Fornoni, J.; Valverde, P.L. The evolution of resistance and tolerance to herbivores. *Annu. Rev. Ecol. Evol. Syst.* **2007**, *38*, 541–566. [CrossRef]
30. War, A.R.; Taggar, G.K.; Hussain, B.; Taggar, M.S.; Nair, R.M.; Sharma, H.C. Plant defence against herbivory and insect adaptations. *AoB Plants* **2018**, *10*, ply037. [CrossRef]
31. Park, S.-H.; Kim, J.H.; Kim, J.G. Effects of human activities on *Sericinus montela* and its host plant *Aristolochia contorta*. *Sci. Rep.* **2023**, *13*, 8289. [CrossRef]
32. El Tahir, K.E.H. Pharmacological Actions of Magnoflorine and Aristolochic Acid-1 Isolated from the Seeds of *Aristolochia bracteata*. *Int. J. Pharm.* **1991**, *29*, 101–110. [CrossRef]
33. McKenzie, A.; Price, J. The alkaloids of *Gyrocarpus americanus* Jacq. *Aust. J. Chem.* **1953**, *6*, 180–185. [CrossRef]
34. Park, H.J. Effects of Environmental Factors on Host Plant and Its Specialist Herbivore, *Aristolochia contorta* and *Sericinus montela*. Ph.D. Thesis, Seoul National University Graduate School, Seoul, Republic of Korea, 2021.
35. Dewar, R.C. The correlation between plant growth and intercepted radiation: An interpretation in terms of optimal plant nitrogen content. *Ann. Bot.* **1996**, *78*, 125–136. [CrossRef]
36. Smith, N.G.; Keenan, T.F.; Colin Prentice, I.; Wang, H.; Wright, I.J.; Niinemets, Ü.; Crous, K.Y.; Domingues, T.F.; Guerrieri, R.; Yoko Ishida, F.; et al. Global photosynthetic capacity is optimized to the environment. *Ecol. Lett.* **2019**, *22*, 506–517. [CrossRef] [PubMed]
37. Valluru, R.; Van den Ende, W. Myo-inositol and beyond–Emerging networks under stress. *Plant Sci.* **2011**, *181*, 387–400. [CrossRef] [PubMed]
38. Pollard, A.; Chibnall, A.C.; Piper, S.H. The isolation of n-octacosanol from wheat wax. *Biochem. J.* **1933**, *27*, 1889–1893. [CrossRef] [PubMed]
39. Simões, R.; Rodrigues, A.; Ferreira-Dias, S.; Miranda, I.; Pereira, H. Chemical composition of cuticular waxes and pigments and morphology of leaves of *Quercus suber* trees of different provenance. *Plants* **2020**, *9*, 1165. [CrossRef] [PubMed]
40. Sengupta, S.; Mukherjee, S.; Basak, P.; Majumder, A.L. Significance of galactinol and raffinose family oligosaccharide synthesis in plants. *Front. Plant Sci.* **2015**, *6*, 656. [CrossRef] [PubMed]
41. Munné-Bosch, S. The role of alpha-tocopherol in plant stress tolerance. *J. Plant Physiol.* **2005**, *162*, 743–748. [CrossRef]
42. Islam, S.; Mohammad, F. Triacontanol as a dynamic growth regulator for plants under diverse environmental conditions. *Physiol. Mol. Biol. Plants* **2020**, *26*, 871–883. [CrossRef] [PubMed]
43. Kliebenstein, D.J. Plant defense compounds: Systems approaches to metabolic analysis. *Annu. Rev. Phytopathol.* **2012**, *50*, 155–173. [CrossRef]
44. Baldwin, I.T. Herbivory simulations in ecological research. *Trends Eco. Evol.* **1990**, *5*, 91–93. [CrossRef]
45. Green, T.R.; Ryan, C.A. Wound-induced proteinase inhibitor in plant leaves: A possible defense mechanism against insects. *Science* **1972**, *175*, 776–777. [CrossRef] [PubMed]
46. Karban, R.; Baldwin, I.T. *Induced Responses to Herbivory*; University of Chicago Press: Chicago, IL, USA, 2007.
47. Jaremo, J.; Tuomi, J.; Nilsson, P. Adaptive status of localized and systemic defense responses in plants. In *The Ecology and Evolution of Inducible Defenses*; Princeton University Press: Princeton, NJ, USA, 1999; pp. 33–44.
48. Ashraf, M.; Foolad, M.R. Roles of glycine betaine and proline in improving plant abiotic stress resistance. *Environ. Exp. Bot.* **2007**, *59*, 206–216. [CrossRef]
49. Kano, A.; Gomi, K.; Yamasaki-Kokudo, Y.; Satoh, M.; Fukumoto, T.; Ohtani, K.; Tajima, S.; Izumori, K.; Tanaka, K.; Ishida, Y.; et al. A rare sugar, d-allose, confers resistance to rice bacterial blight with upregulation of defense-related genes in *Oryza sativa*. *Phytopathology* **2010**, *100*, 85–90. [CrossRef] [PubMed]

50. Myers, R.J., Jr.; Fichman, Y.; Zandalinas, S.I.; Mittler, R. Jasmonic acid and salicylic acid modulate systemic reactive oxygen species signaling during stress responses. *Plant Physiol.* **2022**, *191*, 862–873. [CrossRef] [PubMed]
51. Wasternack, C.; Stenzel, I.; Hause, B.; Hause, G.; Kutter, C.; Maucher, H.; Neumerkel, J.; Feussner, I.; Miersch, O. The wound response in tomato–Role of jasmonic acid. *J. Plant Physiol.* **2006**, *163*, 297–306. [CrossRef] [PubMed]
52. Ahn, J.H.; Lee, J.S. Sugar acts as a regulatory signal on the wound-inducible expression of *SbHRGP3::GUS* in transgenic plants. *Plant Cell Rep.* **2003**, *22*, 286–293. [CrossRef] [PubMed]
53. Fernandez, O.; Béthencourt, L.; Quero, A.; Sangwan, R.S.; Clément, C. Trehalose and plant stress responses: Friend or foe? *Trends Plant Sci.* **2010**, *15*, 409–417. [CrossRef]
54. Park, S.-H.; Scheffler, J.; Scheffler, B.; Cantrell, C.L.; Pauli, C.S. Chemical defense responses of upland cotton, *Gossypium hirsutum* L. to physical wounding. *Plant Direct* **2019**, *3*, e00141. [CrossRef]
55. Sharma, V.; Jain, S.; Bhakuni, D.S.; Kapil, R.S. Biosynthesis of aristolochic acid. *J. Chem. Soc. Perkin Trans.* **1982**, *1*, 1153–1155. [CrossRef]
56. Erb, M.; Lenk, C.; Degenhardt, J.; Turlings, T.C.J. The underestimated role of roots in defense against leaf attackers. *Trends Plant Sci.* **2009**, *14*, 653–659. [CrossRef]
57. Bezemer, T.M.; van Dam, N.M. Linking aboveground and belowground interactions via induced plant defenses. *Trends Ecol. Evol.* **2005**, *20*, 617–624. [CrossRef]
58. Kaplan, I.; Halitschke, R.; Kessler, A.; Rehill, B.J.; Sardanelli, S.; Denno, R.F. Physiological integration of roots and shoots in plant defense strategies links above- and belowground herbivory. *Ecol. Lett.* **2008**, *11*, 841–851. [CrossRef] [PubMed]
59. Balmer, D.; Mauch-Mani, B. More beneath the surface? Root versus shoot antifungal plant defenses. *Front. Plant Sci.* **2013**, *4*, 54921. [CrossRef]
60. Comer, F.; Tiwari, H.P.; Spenser, I.D. Biosynthesis of aristolochic acid. *Can. J. Chem.* **1969**, *47*, 481–487. [CrossRef]
61. Aldea, M.; Hamilton, J.G.; Resti, J.P.; Zangerl, A.R.; Berenbaum, M.R.; DeLucia, E.H. Indirect effects of insect herbivory on leaf gas exchange in soybean. *Plant Cell Environ.* **2005**, *28*, 402–411. [CrossRef]
62. Bueno, A.D.F.; Bueno, R.C.O.D.F.; Nabity, P.D.; Higley, L.G.; Fernandes, O.A. Photosynthetic response of soybean to twospotted spider mite (Acari: Tetranychydae) injury. *Braz. Arch. Biol. Technol.* **2009**, *52*, 825–834. [CrossRef]
63. Hermsmeier, D.; Schittko, U.; Baldwin, I.T. Molecular interactions between the specialist herbivore *Manduca sexta* (Lepidoptera, Sphingidae) and its natural host *Nicotiana attenuata*. I. Large-scale changes in the accumulation of growth- and defense-related plant mRNAs1. *Plant Physiol.* **2001**, *125*, 683–700. [CrossRef] [PubMed]
64. Lawrence, S.D.; Novak, N.G.; Ju, C.J.T.; Cooke, J.E.K. Potato, *Solanum Tuberosum*, defense against colorado potato beetle, *Leptinotarsa Decemlineata* (Say): Microarray gene expression profiling of potato by Colorado potato beetle regurgitant treatment of wounded leaves. *J. Chem. Ecol.* **2008**, *34*, 1013–1025. [CrossRef]
65. Ralph, S.G.; Yueh, H.; Friedmann, M.; Aeschliman, D.; Zeznik, J.A.; Nelson, C.C.; Butterfield, Y.S.N.; Kirkpatrick, R.; Liu, J.; Jones, S.J.M.; et al. Conifer defence against insects: Microarray gene expression profiling of Sitka spruce (*Picea sitchensis*) induced by mechanical wounding or feeding by spruce budworms (*Choristoneura occidentalis*) or white pine weevils (*Pissodes strobi*) reveals large-scale changes of the host transcriptome. *Plant Cell Environ.* **2006**, *29*, 1545–1570. [CrossRef]
66. Li, Y.; He, N.; Hou, J.; Xu, L.; Liu, C.; Zhang, J.; Wang, Q.; Zhang, X.; Wu, X. Factors influencing leaf chlorophyll content in natural forests at the biome scale. *Front. Ecol. Evol.* **2018**, *6*, 64. [CrossRef]
67. Kerchev, P.I.; Fenton, B.; Foyer, C.H.; Hancock, R.D. Plant responses to insect herbivory: Interactions between photosynthesis, reactive oxygen species and hormonal signalling pathways. *Plant Cell Environ.* **2012**, *35*, 441–453. [CrossRef] [PubMed]
68. Gutsche, A.R.; Heng-Moss, T.M.; Higley, L.G.; Sarath, G.; Mornhinweg, D.W. Physiological responses of resistant and susceptible barley, *Hordeum vulgare* to the Russian wheat aphid, *Diurpahis noxia* (Mordvilko). *Arthropod-Plant Interact.* **2009**, *3*, 233–240. [CrossRef]
69. Haile, F.J.; Higley, L.G.; Ni, X.; Quisenberry, S.S. Physiological and growth tolerance in wheat to Russian wheat aphid (Homoptera: Aphididae) injury. *Environ. Entomol.* **1999**, *28*, 787–794. [CrossRef]
70. Jeong, S.J.; Nam, B.E.; Jeong, H.J.; Jang, J.Y.; Joo, Y.; Kim, J.G. Age-dependent resistance of a perennial herb, *Aristolochia contorta* against specialist and generalist leaf-chewing herbivores. *Front. Plant Sci.* **2023**, *14*, 1145363. [CrossRef] [PubMed]
71. Park, H.J.; Park, H.; Son, G.Y.; Nam, B.E.; Kim, J.G. Biotic and abiotic effects on the growth and reproduction of *Aristolochia contorta*. *J. Wet. Res.* **2020**, *22*, 113–120.
72. Park, S.-H.; Nam, B.E.; Kim, J.G. Shade and physical support are necessary for conserving the *Aristolochia contorta* population. *Ecol. Eng.* **2019**, *135*, 108–115. [CrossRef]
73. Meinshausen, M.; Smith, S.J.; Calvin, K.; Daniel, J.S.; Kainuma, M.L.T.; Lamarque, J.F.; Matsumoto, K.; Montzka, S.A.; Raper, S.C.B.; Riahi, K.; et al. The RCP greenhouse gas concentrations and their extensions from 1765 to 2300. *Clim. Chang.* **2011**, *109*, 213. [CrossRef]
74. Halitschke, R.; Schittko, U.; Pohnert, G.; Boland, W.; Baldwin, I.T. Molecular interactions between the specialist herbivore *Manduca sexta* (Lepidoptera, Sphingidae) and its natural host *Nicotiana attenuata*. III. Fatty acid-amino acid conjugates in herbivore oral secretions are necessary and sufficient for herbivore-specific plant responses. *Plant Physiol.* **2001**, *125*, 711–717. [CrossRef]
75. McGale, E.; Diezel, C.; Schuman, M.C.; Baldwin, I.T. Cry1Ac production is costly for native plants attacked by non-Cry1Ac-targeted herbivores in the field. *New Phytol.* **2018**, *219*, 714–727. [CrossRef]

76. Ohnmeiss, T.E.; Baldwin, I.T. The allometry of nitrogen to growth and an inducible defense under nitrogen-limited growth. *Ecology* **1994**, *75*, 995–1002. [CrossRef]
77. Munné-Bosch, S. Perennial roots to immortality. *Plant Physiol.* **2014**, *166*, 720–725. [CrossRef] [PubMed]
78. Palta, J.P. Leaf chlorophyll content. *Remote Sens. Rev.* **1990**, *5*, 207–213. [CrossRef]
79. Süß, A.; Danner, M.; Obster, C.; Locherer, M.; Hank, T.; Richter, K. *Measuring Leaf Chlorophyll Content with the Konica Minolta SPAD-502Plus–Theory, Measurement, Problems, Interpretation*; EnMAP Field Guides Technical Report; GFZ Data Services: Potsdam, Germany, 2015.
80. Royer, M.; Larbat, R.; Le Bot, J.; Adamowicz, S.; Robin, C. Is the C:N ratio a reliable indicator of C allocation to primary and defence-related metabolisms in tomato? *Phytochemistry* **2013**, *88*, 25–33. [CrossRef] [PubMed]
81. Sardans, J.; Rivas-Ubach, A.; Peñuelas, J. The C:N:P stoichiometry of organisms and ecosystems in a changing world: A review and perspectives. *Perspect. Plant Ecol. Evol. Syst.* **2012**, *14*, 33–47. [CrossRef]
82. R Core Team. *R: A Language and Environment for Statistical Computing*; R Foundation for Statistical Computing: Vienna, Austria, 2022.
83. Kassambara, A. ggpubr: 'ggplot2' Based Publication Ready Plots. 2023. Available online: https://rpkgs.datanovia.com/ggpubr/ (accessed on 20 May 2024).
84. Vincent Bonhomme, S.P.; Gaucherel, C.; Claude, J. Momocs: Outline analysis using R. *J. Stat. Softw.* **2014**, *56*, 1–24.
85. Wickham, H. *ggplot2: Elegant Graphics for Data Analysis*; Springer: New York, NY, USA, 2016. Available online: https://ggplot2.tidyverse.org (accessed on 20 May 2024).

plants

Article

OsRCI-1-Mediated GLVs Enhance Rice Resistance to Brown Planthoppers

Kaiming Mao [1,†], Chengzhe Li [1,†], Huacai Zhai [1,†], Yuying Wang [1], Yonggen Lou [2], Wenhua Xue [1,*] and Guoxin Zhou [1,*]

1 Key Lab for Biology of Crop Pathogens and Insect Pests and Their Ecological Regulation of Zhejiang Province, The Key Laboratory for Quality Improvement of Agricultural Products of Zhejiang Province, College of Advanced Agricultural Sciences, Zhejiang A & F University, Hangzhou 311300, China; mkm759288843@163.com (K.M.); chengzheli@163.com (C.L.); huacaizhai@163.com (H.Z.); wyy3112666500@163.com (Y.W.)

2 State Key Laboratory of Rice Biology, Ministry of Agriculture Key Lab of Molecular Biology of Crop Pathogens and Insects, Institute of Insect Sciences, Zhejiang University, Hangzhou 310058, China; yglou@zju.edu.cn

* Correspondence: whxue@zafu.edu.cn (W.X.); gxzhou@zafu.edu.cn (G.Z.); Tel./Fax: +86-571-63743089 (G.Z.)

† These authors contributed equally to this work.

Abstract: Green leaf volatiles (GLVs) play pivotal roles in plant anti-herbivore defense. This study investigated whether the rice 13-lipoxygense gene *OsRCI-1* is involved in GLV production and plant defense in rice. The overexpression of *OsRCI-1* (*oeRCI* lines) in rice resulted in increased wound-induced levels of two prominent GLVs, *cis*-3-hexen-1-ol and *cis*-3-hexenal. In a previous study, we found that the overexpression of *OsRCI-1* reduced the colonization by the rice brown planthopper (BPH, *Nilaparvata lugens*) but increased the attractiveness to the egg parasitoid *Anagrus nilaparvatae* compared to wild-type (WT) plants. This study found that when *cis*-3-hexen-1-ol, but not *cis*-3-hexenal, was added to WT plants, it could change the BPH's colonization preference, i.e., more BPHs preferred to colonize the *oeRCI* lines. The exogenous application of *cis*-3-hexen-1-ol or *cis*-3-hexenal to BPH-infested WT plants could weaken or overturn the preference of *A. nilaparvatae* for *oeRCI* lines. However, field experiments revealed that only *cis*-3-hexenal was attractive to the parasitoid and increased the parasitism rates of BPH eggs. These results indicate that *OsRCI-1* is involved in rice GLV production and therefore modulates both direct and indirect defense in rice.

Keywords: rice; green leaf volatiles (GLVs); *cis*-3-hexen-1-ol; *cis*-3-hexenal; *Nilaparvata lugens*; *Anagrus nilaparvatae*

1. Introduction

Lipoxygenase (LOX), the pivotal enzyme within the fatty acid metabolism pathway, can be divided into *9-LOXs* and *13-LOXs* based on the position of oxygenation in the substrate carbon chain, participating in various aspects of plant growth, development, and responses to both biotic and abiotic stresses [1]. Two branch pathways of the LOX pathway, the allene oxide synthase pathway (AOS) and the hydroperoxide lyase pathway (HPL), have been well studied in induced plant defense against herbivores. Green leaf volatiles (GLVs) including C6 alcohols, aldehydes, and esters, acting through the HPL pathway, play important roles in plant resistance against herbivores, as well as the well-known signal molecules, jasmonates (JAs), produced from the AOS pathway [2].

GLVs, known for their distinct and strong aromas, reminiscent of fresh grass and green leaves, are mostly produced during the early stage of plant development and are rapidly released when cells are mechanically damaged or exposed to herbivores for feeding or oviposition [3]. Research shows that GLVs play a crucial role in both direct and indirect plant defense against insect pests [4]. In particular, emissions of the GLV compound *cis*-3-hexen-1-ol from plants such as rice (*Oryza sativa* L.; Gramineae: Oryza), tomato (*Solanum lycopersicum* L.; Solanaceae: *Solanum*), and tea (*Camellia sinensis* (L.) Kuntze; Theaceae: Camellia) have been shown to decrease pest numbers by attracting their natural enemies [5–7]. Specifically, studies focusing on key genes of the oxylipin pathway, such as rice *OsPLDα4*, *α5*, and *OsHPL3*, have shown that *cis*-3-hexen-1-ol, released by rice plants, acts as a repellent against the rice brown planthopper (BPH, *Nilaparvata lugens* Stål; Hemiptera: Delphacidae), while attracting its natural enemy [5,8]. In contrast, the release of *cis*-3-hexenal from rice plants was found to reduce their resistance to the white-backed planthopper (WBPH, *Sogatella furcifera* Horváth; Hemiptera: Delphacidae) [9].

In rice, 14 LOX genes have been identified, of which seven genes have been cloned and characterized, including *OsRLL/OsHI-LOX, LOX-2, LOX-3, Osr9-LOX1, OsRCI-1*, and *OsLOX1* [10–15]. The resistance of rice against the rice brown planthopper was reduced in the antisense-suppressed lines of *OsLOX1*, while it was enhanced in the overexpression lines, and *OsLOX1* was related to the synthesis of JA and *cis*-3-hexen-1-ol [16]. Zhou et al. (2009) cloned the *OsHI-LOX* gene from the rice *13-LOXs*, and the antisense-suppressed mutant rice exhibited enhanced resistance to piercing-sucking brown planthoppers but reduced resistance to chewing insects like the striped stem borer (*Chilo suppressalis* Walker; Lepidoptera: Pyralidae) and rice leaf folder (*Cnaphalocrocis medinalis* Guenée; Lepidoptera: Pyralidae), but they found that *OsHI-LOX* was only involved in the synthesis of JA, rather than the production of GLVs [13]. *OsRCI-1* has been identified as a member of the *13-LOX* family [15]. *OsRCI-1* is induced by compounds such as benzo-(1,2,3)-thiadiazole-7-carbothioic S-methyl ester (BTH) and benzo-(1,2,3)-thiadiazole-7-carbothioic S-methyl ester (INA). The overexpression of *OsRCI-1* in rice transiently resulted in the upregulation of the pathogenesis-related protein PR1 gene [17]. *OsRCI-1* is also reported to be involved in BPH-induced JA bursts and plays a positive role in rice defense against BPH [18].

The biosynthesis of GLVs is orchestrated through the oxygenation of acyl groups by pathway-specific lipoxygenases; subsequent cleavage reactions facilitated by hydroperoxide lyases contribute to the formation of these characteristic volatile compounds [19]. In this study, we used overexpression *OsRCI-1* rice (*oeRCI* lines) [18] to investigate whether *OsRCI-1* was involved in rice GLV production and whether these GLVs played a role in the regulation of BPH colonization and the behavior of its egg parasitoid wasp, *Anagrus nilaparvatae* Pang et Wang (Hymenoptera: Mymaridae). We hypothesized that the overexpression of *OsRCI-1* in rice would enhance the production of GLVs, therefore playing a role in regulating BPH colonization and the behavior of its egg parasitoid *A. nilaparvatae*.

2. Results

2.1. Overexpressing OsRCI-1 Enhances Production of Mechanical Damage-Induced GLVs in Rice

To investigate the impact of *OsRCI-1* on the production of rice GLVs, the levels of GLVs from *OsRCI-1* overexpression rice (*oeRCI* lines) and wild-type rice plants were estimated after mechanical damage treatment. After the mechanical damage treatment, the content of three GLVs showed varying degrees of increase in the overexpression lines (Figure 1A–C). It is interesting that, in comparison to WT rice plants, the content of *cis*-3-hexen-1-ol significantly rose in the R1, R6, and R8 lines, with increases of 3.98, 2.36, and 6.03 times, respectively (Figure 1A). Another GLV, *cis*-3-hexenal, was also significantly increased in these three *oeRCI* lines, with an increase of 2.41–4.37 times (Figure 1B). However, the content of *trans*-2-hexenal was only significantly increased in the R6 line and not in the R1 and R8 lines (Figure 1C). These results indicate that the overexpression of *OsRCI-1* may be involved in rice GLV production.

2.2. Effects of cis-3-hexen-1-ol and cis-3-hexenal on Feeding and Oviposition Preferences of BPH

Here, we evaluated BPH selection in WT rice plants supplemented with *cis*-3-hexen-1-ol or *cis*-3-hexenal, compared with *oeRCI* lines, to assess the activity of GLVs on BPH's feeding and egg laying preference (Figures 2 and 3). When 125 nmol of *cis*-3-hexen-1-ol was applied to the leaf sheaths of WT plants, BPH still showed a preference to feed and lay eggs on WT plants over *oeRCI* lines (R1 or R6) (Figure 2A,C). The preference of BPH was not evident in the comparison between *cis*-3-hexen-1-ol-treated WT plants and the R8 line (Figure 2E). In addition, when 250 nmol of *cis*-3-hexen-1-ol was applied to WT plants, a reversal in BPH's feeding and oviposition preference was observed; BPH escaped feeding and oviposition on the R1 or R6 lines (Figure 2B,D). However, the application of either 125 or 250 nmol of another GLV, *cis*-3-hexenal, on the leaf sheaths of WT rice plants did not affect BPH, preferring *cis*-3-hexenal-treated WT plants ($p < 0.05$, Figure 3A–F) as well as untreated WT plants. These findings suggest that *cis*-3-hexen-1-ol but not *cis*-3-hexenal showed a repellent effect on the female adult BPH.

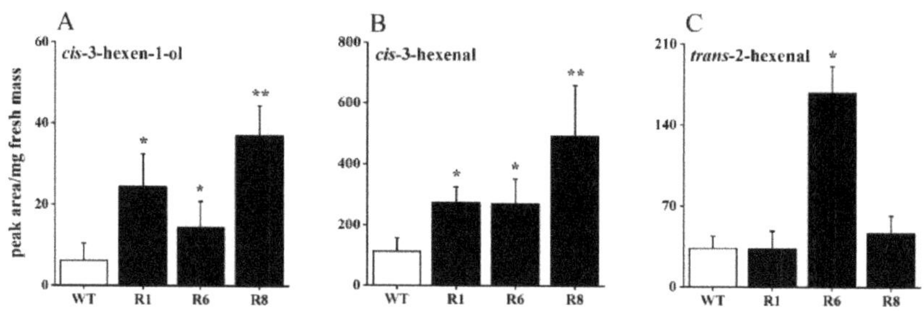

Figure 1. The content of *cis*-3-hexen-1-ol (**A**), *cis*-3-hexenal (**B**), and *trans*-2-hexenal (**C**) in wild-type (WT) and *oeRCI* lines (R1, R6, and R8) after mechanical damage treatment. Asterisks indicate significant differences between pairs of samples within time points (*, $p < 0.05$; **, $p < 0.01$; Student's *t*-test).

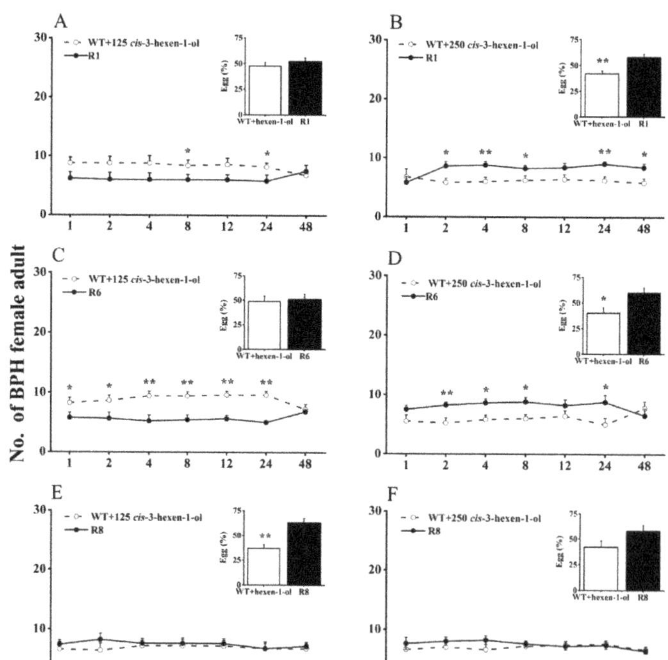

Figure 2. Effects of 125 (**A**,**C**,**E**) or 250 nmol (**B**,**D**,**F**) *cis*-3-hexen-1-ol on feeding and oviposition preference of BPH female adults (+SE, $n = 5$) in WT and *OsRCI-1* overexpression lines. Asterisks denote significant differences between pairs of groups within each time point (*, $p < 0.05$; **, $p < 0.01$; Student's *t*-test).

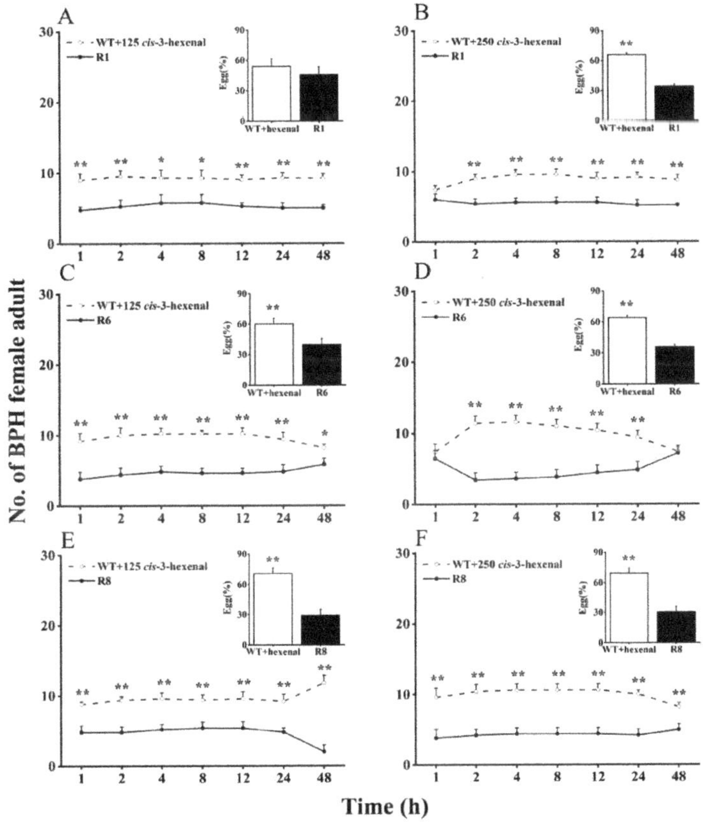

Figure 3. Effects of 125 (**A**,**C**,**E**) or 250 nmol (**B**,**D**,**F**) *cis*-3-hexenal on feeding and oviposition preference of BPH female adults (+SE, n = 5) in WT and *OsRCI-1* overexpression lines. Asterisks denote significant differences between pairs of members within time points (*, $p < 0.05$; **, $p < 0.01$; Student's t-test).

2.3. GLVs Modify Rice's Attractiveness to Parasitoids after BPH Infestation

To explore the potential impact of GLVs in rice on the parasitic preference of the parasitoid *A. nilaparvatae* (BPH egg parasitoid), we initially introduced the parasitoids into the *OsRCI-1* overexpression *oeRCI* lines and WT plants infested with BPH in a laboratory setting. It is interesting that the parasitism rates of BPH eggs on *OsRCI-1*-overexpressing plants were significantly higher than those on WT plants ($p < 0.05$), with the parasitism rates of BPH eggs on the R1, R6, and R8 lines being 3.35, 2.52, and 2.95 times higher than those on WT plants, respectively (Figure 4A). Furthermore, whether *cis*-3-hexen-1-ol or *cis*-3-hexen-al could increase the parasitism rates of BPH eggs on WT plants was assessed in the field. The application of *cis*-3-hexen-1-ol (5–500 nmol) on WT plants showed no significant difference in the parasitism rates of BPH eggs between treated and untreated WT plants (Figure 4B); however, the application of *cis*-3-hexenal at concentrations ranging from 5 to 500 nmol significantly elevated the parasitism rates of BPH eggs on WT plants in the field ($p < 0.05$; Figure 4C). The application of 250 nmol *cis*-3-hexen-1-ol to WT plants had no significant effect on the BPH egg parasitism rates in comparison to *oeRCI* lines in indoor experiments ($p > 0.05$; Figure 4D); unexpectedly, applying 250 nmol *cis*-3-hexenal to WT plants could deactivate or reverse *A. nilaparvatae*'s preference, parasitizing BPH eggs on *oeRCI* lines (Figure 4E), and the parasitism rates of BPH eggs on *cis*-3-hexenal-treated WT plants were apparently higher than those on the R6 and R8 rice plants ($p < 0.05$; Figure 4E).

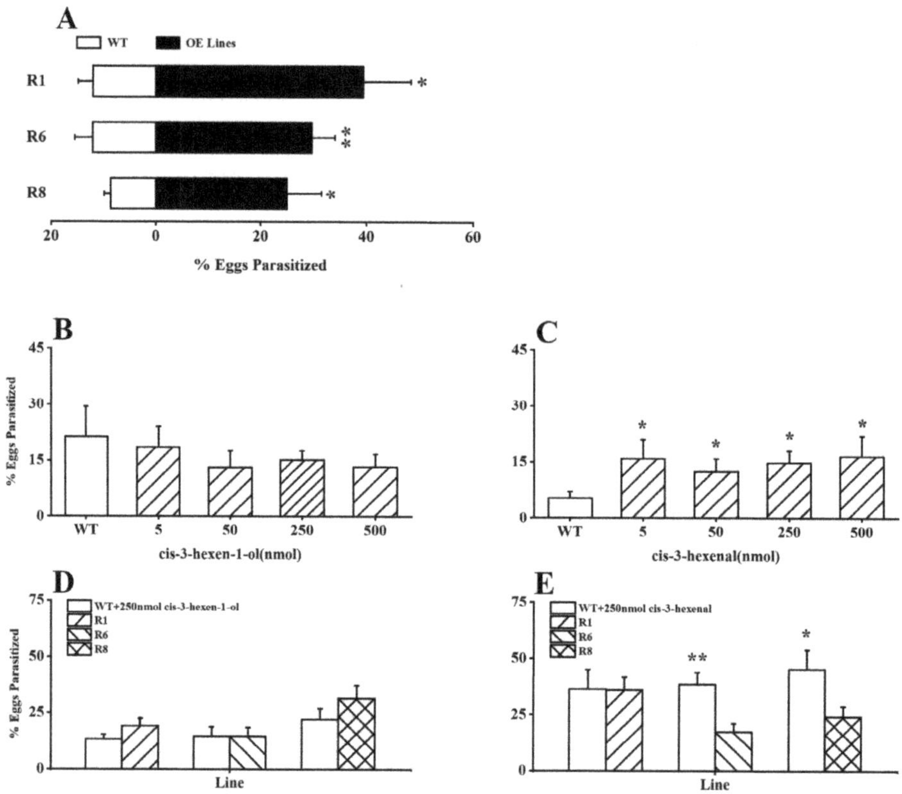

Figure 4. (**A**) The percentage (+SE, $n = 7$) of BPH eggs parasitized by *A. nilaparvatae* on R1, R6, R8, and WT plants. (**B**,**C**) The mean (+SE, $n = 7$) parasitism rates (%) of BPH eggs on WT plants treated with 5–500 nmol *cis*-3-hexen-1-ol or *cis*-3-hexenal in the field. (**D**,**E**) The mean (+SE, $n = 7$) parasitism rates (%) of BPH eggs in indoor experiments, with 250 nmol *cis*-3-hexen-1-ol or *cis*-3-hexenal applied to WT plants versus *oeRCI* lines. Asterisks indicate significant differences in R lines compared with WT plants (*, $p < 0.05$; **, $p < 0.01$; Student's *t*-test).

3. Discussion

The expression of the *OsRCI-1* gene in rice can be rapidly and consistently induced by *Chilo suppressalis* and BPH infestation. *C. suppressalis* feeding on *OsRCI-1*-silenced lines increased in body weight by 1.58–2.15 times compared to feeding on WT plants [20]. The overexpression of *OsRCI-1* not only increased the levels of BPH-induced JA and JA-Ile but also decreased the feeding and oviposition preferences of BPH female adults, ultimately reducing BPH's performance [18]. In this study, we found that the overexpression of *OsRCI-1* increased the levels of GLVs, including *cis*-3-hexen-1-ol and *cis*-3-hexenal. The increased GLVs, especially *cis*-3-hexenal and *cis*-3-hexen-1-ol (but not *cis*-3-hexenal alone), significantly regulated the BPH's feeding and oviposition performance. Additionally, we observed that the wasp *A. nilaparvatae* preferred to parasitize BPH eggs on *OsRCI-1* overexpression rice plants. This preference of the wasp could be partly attributed to the increased levels of GLVs produced by the *OsRCI-1* overexpression plants. In conclusion, the rice lipoxygenase *OsRCI-1* is involved in GLV production of the HPL branch pathway, as well as the JA generation of the AOS branch pathway, in the rice jsamonate pathway, and the GLVs play an important role in protecting rice from invasive pests.

The different GLVs showed disparate biological functions in the interactions between plants, herbivores, and the enemies of herbivores. In the *OsRCI-1* overexpression lines, the levels of *cis*-3-hexen-1-ol and *cis*-3-hexenal but not *trans*-2-hexenal significantly increased, but it was found that *cis*-3-hexen-1-ol and *cis*-3-hexenal had different effects on the behavior of BPH and *A. nilaparvatae* (Figures 2–4). The supplementation of *cis*-3-hexen-1-ol in WT rice plants could apparently alter the feeding and oviposition preferences of BPH adults (Figure 2), working as a repellent, and its repellent activity appeared to be dependent on its concentration (Figure 2). However, when adding *cis*-3-hexen-1-ol to WT rice plants in the field, no difference was found in the parasitism rates of BPH eggs on control rice and *cis*-3-hexen-1-ol-treated WT plants (Figure 4). Other studies showed that *cis*-3-hexen-1-ol is involved in regulating direct and indirect defense pathways in tea plants [21] and can activate defense responses against insect pests in plants like maize [22–24]. In contrast, *cis*-3-hexenal emitted from *OsRCI-1*-overexpressed rice lines did not change the feeding and oviposition preference of BPH adults (Figure 3), but significantly increased the parasitism rate of BPH eggs (Figure 4). Therefore, the rice lipoxygenase gene *OsRCI-1* could repel BPH by releasing *cis*-3-hexen-1-ol, functioning as a "push" strategy, and attract *A. nilaparvatae* by releasing *cis*-3-hexenal, therefore increasing the parasitism rate of BPH eggs, functioning as a "pull" strategy. Similar phenomena have been reported; Lamy et al. [25] found that the volatile dimethyl disulfide exhibited a "push" strategy by reducing the eggs laid on broccoli by the cabbage root fly (*Delia radicum* Linnaeus; Diptera: Anthomyiidae). Meanwhile, *cis*-3-hexenyl acetate exhibited a "pull" strategy by attracting the natural enemy of *D. radicum*.

The lipoxygenase pathway genes can regulate the emission of GLVs; therefore, they play an important role in the direct and indirect defense responses of rice. For example, compared to wild-type (WT) plants, mutants of antisense-suppressed phospholipase D (PLD) genes (*OsPLDa4* and *a5*) exhibited reduced levels of trypsin inhibitors and GLVs, specifically *cis*-3-hexenal and *cis*-3-hexen-1-ol [5]. This resulted in the increased growth and reproduction of BPH and *C. suppressalis*, along with decreased attraction to the natural enemy of *C. suppressalis*. Applying these two GLVs externally restored the resistance of the mutants to BPH and *C. suppressalis* [5]. The loss-of-function or antisense inhibition mutants of the rice hydrogen peroxide lyase gene (*OsHPL3*) exhibited reduced levels of *cis*-3-hexen-1-ol and *cis*-3-hexenal, resulting in decreased resistance to BPH but increased resistance against *WBPH*, and also increased attraction to the parasitoid *A. nilaparvatae* [8,26]. In this study, we found that the overexpression of the rice lipoxygenase gene *OsRCI-1* increased the levels of *cis*-3-hexen-1-ol and *cis*-3-hexenal (Figure 1), therefore reducing the feeding and oviposition preferences of BPH but enhancing the attraction of *A. nilaparvatae* (Figure 4E). These findings suggest that elevated levels of green leaf volatiles (*cis*-3-hexen-1-ol and *cis*-3-hexenal) contribute to enhancing the direct and indirect resistance of rice against BPH.

In summary, our study provides compelling evidence for the involvement of *OsRCI-1* in regulating the release of GLVs including *cis*-3-hexenal and *cis*-3-hexen-1-ol in rice; *OsRCI-1* is another LOX gene that contributes to both JA and GLV synthesis in rice as *OsLOX1* [16]. *OsRCI-1* overexpression in rice lines could significantly reduce BPH's feeding and oviposition preferences and increase the attractiveness to egg parasitoid *A. nilaparvatae* [18]. Another new insight is the increase in *cis*-3-hexenal in rice plants, which functions as a repellent to BPH, and *cis*-3-hexen-1-ol in rice functions as an attractant towards the herbivore's natural enemy. These results highlight the role of GLVs mediated by *OsRCI-1* in regulating both the direct and indirect defenses of rice against BPH. The rice lipoxygenase gene *OsRCI-1* and its overexpression in rice have potential applications in pest-resistant rice breeding.

4. Materials and Methods

4.1. Rice Materials

The rice genotype Xiushui 11 (WT) and three *OsRCI-1*-overexpressing transgenic *oeRCI* lines (R1, R6, and R8) [18] were used in this study. The rice seeds were first germinated in a growth chamber and then transplanted into 25-liter plastic pots filled with rice culture solution. Both seed germination and plant growth were monitored under controlled conditions (27 ± 2 °C, relative humidity 70–80%, and light–dark = 16:8). Rice seedlings at 35–40 days old were used for experiments.

4.2. Insect Materials

BPH populations were collected from rice fields in Lin'an, Zhejiang, China, and maintained on TN1 rice seedlings for at least ten generations in a controlled climate room at 27 °C ± 2 °C, 70–80% relative humidity, and a 12 h light phase, as reported by Liao et al. [18]. TN1 rice plants with BPH eggs were then placed in a rice field at Zhejiang A & F University to attract the parasitoid *A. nilaparvatae*. In the growth chamber, BPH female adults lay eggs on rice seedlings and then we offered these seedlings to the parasitoid.

4.3. Reagents and Instruments

The reagents used in this study were *cis*-3-hexen-1-ol, *cis*-3-hexenal, *trans*-2-hexenal, and lanolin, which were purchased from Shanghai Aladdin Biochemical Technology Co., Ltd., Shanghai, China. The instruments used were the RXZ-380B intelligent growth chamber (Ningbo Jiangnan Instrument Factory, Ningbo, Zhejiang, China) and the Motic binocular dissecting microscope (Meike Audi Industrial Group Co., Ltd., Xiamen, Fujian, China).

4.4. GLV Analysis

GLVs (*cis*-3-hexenal, *cis*-3-hexen-1-ol, and *trans*-2-hexenal) were analyzed with a portable gas analyzer (zNose 4200, Electronic Sensor Technology, Newbury Park, CA, USA) using the method described by Zhou et al. [6]. Rice leaves (0.02 g) from the three overexpression lines R1, R6, and R8 and their wild-type XS11 were cut into small pieces (about 2 mm long) and immediately placed in a 50 mL glass chamber (Jiangyin Elite Experimental Instrument Co., Ltd., Jiangyin, Jiangsu, China), respectively; after 5 min, GLVs were sampled for 40 s and analyzed by the zNose 4200. At least eight replicates were carried out.

4.5. BPH Bioassays

To estimate the impact of GLVs, specifically *cis*-3-hexen-1-ol or *cis*-3-hexenal, on the colonization and oviposition preference of BPH between the WT and oeRCI rice lines, pots with two plants were used. One plant was an *oeRCI* line treated with 10 μL lanolin and the other was a WT rice plant treated with 10 μL lanolin containing 125 or 250 nmol GLVs (*cis*-3-hexen-1-ol or *cis*-3-hexenal). The pots were confined within a glass cylinder (diameter, 4 cm; height, 8 cm; with 24 small holes (diameter 0.8 mm) evenly distributed on the cylinder wall) sealed with a sponge, into which 15 gravid BPH females were introduced. The number of BPHs on each plant was recorded at 1, 2, 4, 8, 12, 24, and 48 h. After 48 h, the number of BPH eggs on each plant was counted under a dissecting microscope. The experiment was repeated at least five times for each pair of plants.

4.6. Field Experiment

To evaluate the role of GLVs (*cis*-3-hexen-1-ol or *cis*-3-hexenal) in rice–parasitoid interactions in the field, WT plants (one plant per pot) were individually treated with 0, 5, 50, 250, or 500 nmol *cis*-3-hexen-1-ol or *cis*-3-hexenal in 10 μL lanolin and were infested with 8 BPH females for 24 h; then, the BPHs were removed as described by Liao et al. [19]. Five potted plants (each of 0, 5, 50, 250, or 500 nmol *cis*-3-hexen-1-ol- or *cis*-3-hexenal-treated WT plants) were placed and arranged within a 40 cm circular radius as located in a rice field. The location was 1 m from the edge of the rice field, and the distance between each location was kept at 2 m. Two days later, the plants were transferred to a controlled climate room at 27 ± 2 °C with a 14 h light phase and 70% relative humidity for five days. Then, the total number of BPH eggs and eggs parasitized by the wasps (these eggs became red) were counted. Each treatment was repeated at least seven times.

4.7. Wasp Bioassays in Controlled Climate Chamber

The WT and *oeRCI* rice lines were infested with 12 BPH females and oviposited for 12 h, respectively, and then the BPHs were removed. Pots with two plants (one *oeRCI* line with BPH eggs vs. a WT plant with BPH eggs) were kept 1 cm apart and were confined within glass cylinders (diameter 4 cm, height 8 cm, with 24 small holes), into which six newly emerged *A. nilaparvatae* wasps (3♀ and 3♂) were introduced, as described by Liao et al. [19]. After 24 h, the parasitoid wasps were removed, and the rice plants were kept in a controlled climate room at 27 ± 2 °C with a 14 h light phase and 70% relative humidity for five days. Then, the total number of BPH eggs and eggs parasitized by the wasps were counted. Each treatment was repeated at least seven times.

Author Contributions: Conceptualization, K.M., G.Z. and W.X.; methodology, H.Z.; software, K.M. and C.L.; validation, C.L. and Y.W.; formal analysis, K.M. and C.L.; investigation, K.M.; resources, G.Z. and W.X.; data curation, K.M.; writing—original draft preparation, G.Z., Y.L., K.M., W.X. and H.Z.; writing—review and editing, G.Z., Y.L., K.M., W.X. and C.L.; visualization, K.M. and W.X.; supervision, G.Z., Y.L. and W.X.; project administration, G.Z.; funding acquisition, G.Z., Y.L. and W.X. All authors have read and agreed to the published version of the manuscript.

Funding: This research was funded by the National Natural Science Foundation of China (32072508, 31572020), the "Pioneer" and "Leading Goose" R&D Program of Zhejiang Province (2022C02020), the "San Nong Jiu Fang" Program of Zhejiang Provincial Department of Agriculture and Rural Affairs (2022SNJF011), and the Natural Science Foundation of Zhejiang Province (LQ23C140006; LQ23C140005). This project was also supported by the Student Scientific Research Training Program of Zhejiang A&F University (2023KX008).

Data Availability Statement: Data are contained within the article.

Acknowledgments: The authors thank the anonymous reviewers for their constructive comments on a draft of the paper.

Conflicts of Interest: The authors declare no conflicts of interest.

References

1. Feussner, I.; Wasternack, C. The lipoxygenase pathway. *Annu. Rev. Plant Biol.* **2002**, *53*, 275–297. [CrossRef] [PubMed]
2. Greco, M.; Chiappetta, A.; Bruno, L.; Bitonti, M.B. In *Posidonia oceanica* cadmium induces changes in DNA methylation and chromatin patterning. *J. Exp. Bot.* **2012**, *63*, 695–709. [CrossRef]
3. Selli, S.; Kelebek, H.; Ayseli, M.T.; Tokbas, H. Characterization of the most aroma-active compounds in cherry tomato by application of the aroma extract dilution analysis. *Food Chem.* **2014**, *165*, 540–546. [CrossRef] [PubMed]
4. Ameye, M.; Allmann, S.; Verwaeren, J.; Smagghe, G.; Haesaert, G.; Schuurink, R.C.; Audenaert, K. Green leaf volatile production by plants: A meta-analysis. *New Phytol.* **2018**, *220*, 666–683. [CrossRef] [PubMed]
5. Qi, J.; Zhou, G.; Yang, L. The chloroplast-localized phospholipases D α4 and α5 regulate herbivore-induced direct and indirect defense in rice. *Plant Physiol.* **2011**, *157*, 1987–1999. [CrossRef] [PubMed]
6. Xin, Z.; Zhang, L.; Zhang, Z.; Chen, Z.; Sun, X. A tea hydroperoxide lyase gene, *CsiHPL1*, regulates tomato defense response against *Prodenia litura* (Fabricius) and *Alternaria alternata* f. sp. *Lycopersici* by modulating green leaf volatiles (GLVs) release and jasmonic acid (JA) gene expression. *Plant Mol. Biol. Rep.* **2013**, *32*, 62–69.
7. Yang, F.; Zhang, Q.; Yao, Q.; Chen, G.; Tong, H.; Zhang, J.; Li, C.; Su, Q.; Zhang, Y. Direct and indirect plant defenses induced by (Z)-3-hexenol in tomato against whitefly attack. *J. Pest Sci.* **2020**, *93*, 1243–1254. [CrossRef]
8. Tong, X.; Qi, J.; Zhu, X.; Mao, B.; Zeng, L.; Wang, B.; Li, Q.; Zhou, G.; Xu, X.; Lou, Y.; et al. The rice hydroperoxide lyase *OsHPL3* functions in defense responses by modulating the oxylipin pathway. *Plant J.* **2012**, *71*, 763–775. [CrossRef] [PubMed]
9. Wang, B.; Zhou, G.; Xin, Z.; Ji, R.; Lou, Y. (Z)-3-hexenal, one of the green leaf volatiles, increases susceptibility of rice to the white-backed planthopper *Sogatella furcifera*. *Plant Mol. Biol.* **2014**, *33*, 377–387. [CrossRef]
10. Peng, Y.L.; Shirano, Y.; Ohta, H.; Hibino, T.; Tanaka, K.; Shibata, D. A novel lipoxygenase from rice: Primary structure and specific expression upon incompatible infection with rice blast fungus. *J. Biol. Chem.* **1994**, *269*, 3755–3761. [CrossRef]
11. Ohta, H.; Shirano, Y.; Tanaka, K.; Morita, Y.; Shibata, D. cDNA cloning of rice lipoxygenase L-2 and characterization using an active enzyme expressed from the cDNA in *Escherichia coli*. *Eur. J. Biochem.* **1992**, *206*, 331–336. [CrossRef]
12. Suzuki, Y.; Miura, K.; Shigemune, A.; Sasahara, H.; Ohta, H.; Uehara, Y.; Ishikawa, T.; Hamada, S.; Shirasawa, K. Marker-assisted breeding of a LOX-3-null rice line with improved storability and resistance to preharvest sprouting. *Theor. Appl. Genet.* **2015**, *128*, 1421–1430. [CrossRef]
13. Zhou, G.; Qi, J.; Ren, N.; Cheng, J.; Erb, M.; Mao, B.; Lou, Y. Silencing *OsHI-LOX* makes rice more susceptible to chewing herbivores, but enhances resistance to a phloem feeder. *Plant J.* **2009**, *60*, 638–648. [CrossRef]

14. Mizuno, K.; Iida, T.; Takano, A.; Yokoyama, M.; Fujimura, T. A new 9-lipoxygenase cDNA from developing rice seeds. *Plant Cell Physiol.* **2003**, *44*, 1168–1175. [CrossRef] [PubMed]
15. Schaffrath, U.; Zabbai, F.; Dudler, R. Characterization of RCI-1, a chloroplastic rice lipoxygenase whose synthesis is induced by chemical plant resistance activators. *Eur. J. Biochem.* **2000**, *267*, 5935–5942. [CrossRef] [PubMed]
16. Wang, R.; Shen, W.; Liu, L.; Jiang, L.; Liu, Y.; Su, N.; Wan, J. A novel lipoxygenase gene from developing rice seeds confers dual position specificity and responds to wounding and insect attack. *Plant Mol. Biol. Rep.* **2008**, *66*, 401–414. [CrossRef]
17. Zabbai, F.; Jarosch, B.; Schaffrath, U. Over-expression of chloroplastic lipoxygenase RCI1 causes PR1 transcript accumulation in transiently transformed rice. *Physiol. Mol. Plant Pathol.* **2004**, *64*, 37–43. [CrossRef]
18. Liao, Z.; Wang, L.; Li, C.; Cao, M.; Wang, J.; Yao, Z.; Zhou, S.; Zhou, G.; Zhang, D.; Lou, Y. The lipoxygenase gene *OsRCI-1* is involved in the biosynthesis of herbivore-induced JAs and regulates plant defense and growth in rice. *Plant Cell Environ.* **2022**, *45*, 2827–2840. [CrossRef]
19. Matsui, K. Green leaf volatiles: Hydroperoxide lyase pathway of oxylipin metabolism. *Curr. Opin. Plant Biol.* **2006**, *9*, 274–280. [CrossRef]
20. Yao, Z.; Cao, M.; Wang, X.; Miao, J.; Shen, L.; Deng, J.; Wu, H.; Xu, Z.; Lou, Y.; Zhou, G. Rice lipoxygenase gene *OsRCI*-1 mediated pathway positively regulates resistance to rice striped stem borer. *J. Environ. Entomol.* **2014**, *36*, 507–515.
21. Liao, Y.; Tan, H.; Jian, G.; Zhou, X.; Huo, L.; Jia, Y.; Zeng, L.; Yang, Z. Herbivore-induced (Z)-3-hexen-1-ol is an airborne signal that promotes direct and indirect defenses in tea (*Camellia sinensis*) under light. *J. Agric. Food Chem.* **2021**, 3, 43. [CrossRef] [PubMed]
22. Engelberth, J.; Alborn, H.T.; Schmelz, E.A.; Tumlinson, J.H. Airborne signals prime plants against insect herbivore attack. *Proc. Natl. Acad. Sci. USA* **2004**, *101*, 1781–1785. [CrossRef] [PubMed]
23. Engelberth, J.; Contreras, C.F.; Dalvi, C.; Li, T.; Engelberth, M.; Heil, M. Early transcriptome analyses of Z-3-hexenol-treated Zea mays revealed distinct transcriptional networks and anti-herbivore defense potential of green leaf volatiles. *PLoS ONE* **2013**, *8*, e77465. [CrossRef] [PubMed]
24. Farag, M.A.; Fokar, M.; Abd, H.; Zhang, H.; Allen, R.D.; Pare, P.W. (Z)-3-Hexenol induces defense genes and downstream metabolites in maize. *Planta* **2005**, *220*, 900–909. [CrossRef] [PubMed]
25. Lamy, F.; Dugravot, S.; Cortesero, A.M.; Chaminade, V.; Faloya, V.; Poinsot, D. One more step toward a push-pull strategy combining both a trap crop and plant volatile organic compounds against the cabbage root fly *Delia radicum*. *Environ. Sci. Pollut. Res. Int.* **2018**, *25*, 29868–29879. [CrossRef]
26. Wei, J.; Wang, L.; Zhu, J.; Zhang, S.; Nandi, O.I.; Kang, L. Plants attract parasitic wasps to defend themselves against insect pests by releasing hexenol. *PLoS ONE* **2007**, *2*, e852. [CrossRef]

plants

MDPI

Article

Identification of Attractants from Three Host Plants and How to Improve Attractiveness of Plant Volatiles for *Monochamus saltuarius*

Yifan Dong †, Dongping Chen †, Siye Zhou, Zhengyi Mao and Jianting Fan *

School of Forestry and Biotechnology, National Joint Local Engineering Laboratory for High-Efficient Preparation of Biopesticide, Zhejiang A & F University, Lin'an 311300, China; dyf@stu.zafu.edu.cn (Y.D.); klopq@foxmail.com (D.C.); z15857071498@163.com (S.Z.); 15382323773@163.com (Z.M.)

* Correspondence: fanjt@zafu.edu.cn

† These authors contributed equally to this work.

Abstract: As a new vector insect of pine wood nematodes in China, the *Monochamus saltuarius* (Coleoptera: Cerambycidae) vectors pine wilt nematodes into healthy pine trees through feeding and oviposition, resulting in huge economic losses to forestry. A promising control strategy is to develop safe and efficient attractants. This study aims to screen for the key active volatiles of *Pinus koraiensis* (Pinales: Pinaceae), *Pinus tabuliformis* (Pinales: Pinaceae), and *Picea asperata* (Pinales: Pinaceae) that can attract *M. saltuarius*, and to study the synergistic attraction of the main attractant plant volatiles with ethanol and insect aggregation pheromones. The preference of *M. saltuarius* for three hosts is *P. koraiensis* > *P. tabuliformis* > *Picea asperata*. We detected 18 organic volatiles from three host plants. Through EAG assays and indoor Y-tube behavioral experiments, 3-carene, (-)-camphor, β-pinene, α-phellandrene, terpinolene, α-pinene, D-limonene, and myrcene were screened to have attractive effects on *M. saltuarius*. We found that 3-carene, β-pinene, and α-pinene are the most attractive kairomones in field experiments, which may play a crucial role in the host localization of *M. saltuarius*. Ethanol has a synergistic effect on the attractant activity of 3-carene and β-pinene, and the synergistic effect on β-pinene is the best. The mixture of ethanol, 2-undecyloxy-1-ethanol, and ipsdienol can significantly enhance the attraction effect of β-pinene on *M. saltuarius*. These new findings provide a theoretical basis for the development of attractants for adult *M. saltuarius* and contribute to the green control of *M. saltuarius*.

Keywords: *Monochamus saltuarius*; attractants; plant volatiles; aggregation pheromone

1. Introduction

Monochamus saltuarius (Coleoptera: Cerambycidae) is an important forest pest in Eurasia, feeding on important forest trees, such as *Pinus koraiensis* (Pinales: Pinaceae), *Pinus tabuliformis* (Pinales: Pinaceae), and *Picea asperata* (Pinales: Pinaceae) [1]. It affects tree growth and lumber quality and further reduces the economic value of wood. Recently, *M. saltuarius* has been identified as a vector for pine wood nematodes (PWNs) *Bursaphelenchus xylophilus* (Tylenchida: aphelenchidae) in Japan, Korea, and China, introducing PWNs into host plants through feeding and reproduction [2]. *B. xylophilus* is a plant-parasitic nematode native to North America that later invaded Europe and Asia and widely distributed in East Asia, causing deadly pine wood nematode disease and increasing the risk of tree death [3,4]. The spread of PWNs between host trees is mainly mediated by vector insects, and the *Monochamus* spp. are the main vectors [5–9]. In Northeast China, *M. saltuarius* plays a role in the transmission of pine wood nematodes and causes damage to important economic trees [8]. It is of great significance to develop attractants that can trap *M. saltuarius* efficiently for forest protection and management.

At present, the common strategies of insect pest control are chemical control, biological control, and breeding-resistant varieties [10–13]. Chemical control is the main method used to control pests, but there are also many problems with using chemical pesticide, such as polluting the environment, affecting non-target organisms, producing drug resistance, and so on [14–17]. It has been shown that insects determine the host location by detection of precise volatile ratios [18,19], and subtle changes in the volatile ratios of host plants can alter the behavior of insects and alter their perception and orientation [20]. Developing efficient attractants to trap *M. saltuarius* is a feasible and eco-friendly method for its controlling.

Plant volatiles can serve as semiochemicals to attract or repel insects. The volatiles of different host plants are significantly different, and are specific signal substances between plants in the same ecological environment, and are also important semiochemicals that directly affect the feeding and laying choice of adult insects. In short, plant volatiles play a very important role in the specific host and mate location for insects [21–23]. For example, the specific combination of α-pinene, camphene, and β-pinene was reported to be attractive to female *Conogethes punctiferalis* (Lepidoptera: Pyralidae) [24]. The compound 1-Octen-3-ol, a volatile from mango, induced the oviposition of female *Bactrocera dorsalis* Hendel (Diptera: Tephritidae) [25]. (Z)-3-Hexenyl-acetate in maize volatiles was reported to attract female *Spodoptera frugiperda* (Lepidoptera: Noctuidae) and stimulate oviposition [26]. 3-carene was found to be attractive for *Dendroctonus* and *Hylurgops* [27]. In addition, ethanol combined with host plant volatiles enhances the attraction of *Monochamus* species [28,29]. There are two types of volatile attractant pheromones in longhorn beetles (Coleoptera: Cerambycidae), aggregation pheromones and sex pheromones [30], which play an important role in pest control. In *Monochamus titillator* (Coleoptera: Cerambycidae) adults, males release aggregation pheromones and females release non-volatile contact pheromones [31], which enable male and female adults to recognize each other in sexual communication. The male adults of *Monochamus alternatus* (Coleoptera: Cerambycidae), *M. titillator*, and *M. saltuarius* lure female adults by releasing aggregation pheromones [32–34]. Moreover, pheromones of the same species in different subfamilies or genera have similar pheromone components [35]. The pheromones released by different species with the same host can also serve as mutualistic attractants for feeding or oviposition localization [33]. It is meaningful to find the key volatiles from the host plant and study the synergistic effect of plant volatiles and other pheromones for the development of *M. saltuarius* attractants.

Pinus koraiensis, *P. tabuliformis*, and *Picea asperata* are the main host plants of *M. saltuarius*. This article mainly analyzes the volatiles of these three host plants and selects host plant volatile organic compounds (VOCs) that have a seductive effect on *M. saltuarius*. At the same time, combining insect aggregation pheromones, the composite formulas of pheromones and VOCs were studied to screen attractants that can effectively attract adults of *M. saltuarius*. This plays an important role in green controlling the population of *M. saltuarius* and preventing the further spread of pine wilt disease.

2. Results

2.1. Feeding Preferences of M. saltuarius

The feeding areas of three host plants were as follows: *P. koraiensis* (959.00 ± 238.73 mm^2/6 head/3d) > *P. tabuliformis* (452.25 ± 50.79 mm^2/6 head/3d) > *Picea asperata* (269.25 ± 72.59 mm^2/6 head/3d). There were significant differences in the feeding amount of the three plants, and the adults of *M. saltuarius* ate more *P. koraiensis* than *P. tabuliformis* and *Picea asperata* ($p < 0.05$) (Figure 1).

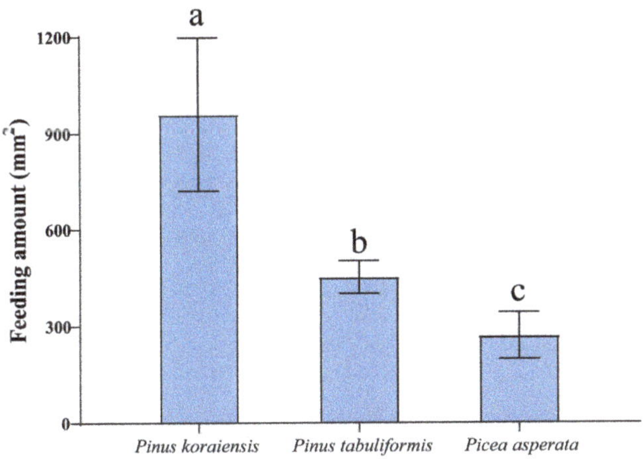

Figure 1. Feeding preference assay of *Monochamus saltuarius* to three host plants: *Pinus koraiensis*, *Pinus tabuliformis*, and *Picea asperata*. Different lowercase letters indicate significant differences at $p < 0.05$.

2.2. Volatiles Identification of 3 Host Plants

There are mainly 18 volatile organic compounds in the volatiles of the three host plants of *P. koraiensis*, *P. tabuliformis*, and *P. asperata*, which can be divided into five categories: terpenes, aldehydes, ketones, alkanes, and esters (Table 1).

Table 1. Volatiles identification of *Pinus koraiensis*, *Pinus tabuliformis*, and *Picea asperata*.

No.	Compounds	Category	No. CAS	Relative Content (%)		
				Pinus koraiensis	*Pinus tabuliformis*	*Picea asperata*
1	Nonane	Alkane	111-84-2	0.34 ± 0.05	0	0
2	α-Pinene	Terpenes	7785-26-4	11.41 ± 0.61 b	21.36 ± 5.61 a	9.49 ± 3.30 b
3	4-Carene	Terpenes	29050-33-7	1.36 ± 0.81 a	1.24 ± 0.21 a	0
4	Camphene	Terpenes	79-92-5	0.54 ± 0.04	0	0
5	Benzaldehyde	Aldehyde	100-52-7	0	0	2.02 ± 1.47
6	β-Pinene	Terpenes	18172-67-3	7.20 ± 0.65 b	34.62 ± 6.71 a	12.92 ± 3.49 b
7	β-Myrcene	Terpenes	123-35-3	22.59 ± 2.67 a	0	11.57 ± 2.89 b
8	α-Phellandrene	Terpenes	99-83-2	3.01 ± 0.53	0	0
10	β-Thujene	Terpenes	28634-89-1	0.24 ± 0.06	0	0
11	γ-Terpinene	Terpenes	99-85-4	4.68 ± 2.95 a	0.97 ± 0.39 a	0
12	β-Terpinene	Terpenes	99-84-3	0	7.57 ± 3.25	0
13	3-Carene	Terpenes	13466-78-9	22.75 ± 3.74 a	0	20.76 ± 6.18 b
14	D-Limonene	Terpenes	7705-14-8	28.00 ± 3.85 b	25.43 ± 6.46 b	40.73 ± 8.14 a
15	Terpinolene	Terpenes	586-62-9	0.54 ± 0.11 a	0.23 ± 0.09 a	0
16	Undecane	Alkane	1120-21-4	0	4.32 ± 0.14	0
17	(-)-Camphor	Ketones	464-48-2	0.77 ± 0.25 b	0	3.93 ± 1.60 a
18	Bornyl acetate	Ester	76-49-3	0.1 ± 0.06	0	0

Different lowercase letters indicate significant differences among different host plants ($p < 0.05$ by Duncan's multiple analysis).

The volatile compounds of *P. koraiensis* contain four groups of compounds: terpenes (accounting for 98.83%), ketones (accounting for 0.74%), alkanes (accounting for 0.33%), and esters (accounting for 0.10%). Among the volatiles of *P. koraiensis*, terpenes make up the highest proportion, with D-limonene having the highest relative content of 28.00 ± 3.85%, followed by 3-carene, with a relative content of 22.75 ± 3.74%. The volatile compounds of *P. tabuliformis* contain two types of volatile compounds: terpenes (95.49%) and alkanes (4.51%), among which, β-pinene accounts for the largest proportion, at 34.62 ± 6.71%.

There are three types of compounds in plant volatiles of *P. asperata*: terpenes (94.13%), aldehydes (1.99%), and ketones (3.87%). Among them, D-limonene makes up the highest relative content at 40.73 ± 8.14%, followed by 3-carene at 20.76 ± 6.18% (Table 1).

Limonene, α-pinene, and β-pinene were identified in three host plants, and their relative contents were high. The content of D-limonene in *P. asperata* was significantly higher than that of *P. tabuliformis* and *P. koraiensis* ($p < 0.05$). The contents of α-pinene and β-pinene in *P. tabuliformis* were significantly higher than those in *P. koraiensis* and *P. asperata* ($p < 0.05$). 3-carene, β-myrcene, and (-)-camphor were only present in *P. tabuliformis* and *P. asperata*, and the content of the three volatiles among the two plants was significantly different ($p < 0.05$). This may have contributed to the feeding preference of *M. saltuarius* (Table 1).

2.3. Electrophysiological Responses of *M. saltuarius* Adults to Volatiles in 3 Host Plants

From the analyzed plant volatiles, 11 compounds with a high relative content or those that have been reported to have attractant activity were selected for EAG experiments. As shown in Figure 2, the adults of *M. saltuarius* exhibited different degrees of EAG reaction to the 11 compounds, and the EAG relative response values of *M. saltuarius* to these volatiles at 100 μg/μL were greater than 10 μg/μL. At concentrations of 100 μg/μL, α-pinene, β-pinene, 3-carene, (-)-camphor, terpinolene, D-limonene, α-phellandrene, and myrcene, the relative antennal responses were 150% greater in both female and male insects (Figure 2).

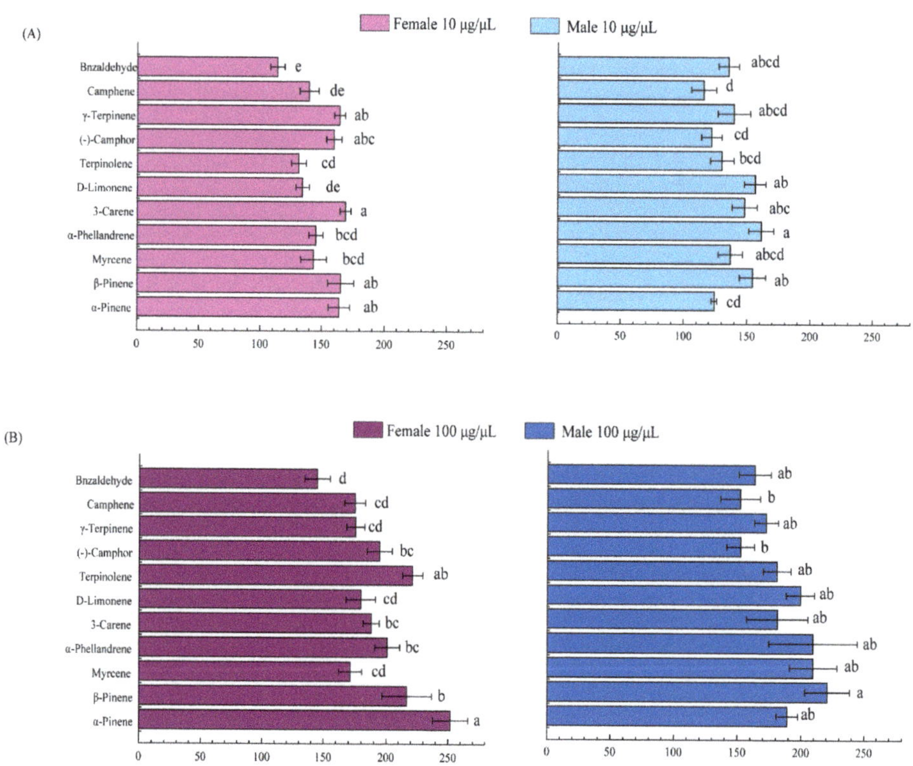

Figure 2. Electroantennography (EAG) relative response values of female and male M. saltuarius adults to the 11 active volatiles. (**A**) The 10 μg/μL active volatiles. (**B**) The 100 μg/μL active volatiles. Different lowercase letters on the bars are significantly different among EAG relative response values at different volatiles ($p < 0.05$ by Duncan's multiple analysis).

2.4. Olfaction Selection Preference on Antennal-Active Host Volatiles by Y-Tube Assay

To evaluate the effect of host plant volatiles on the directional behavior of M. saltuarius, eight host plant volatiles that can induce strong EAG responses were used for indoor Y-tube experiments. As shown in Figure 3, at the concentration of 10 μg/μL and 100 μg/μL, female insects can be significantly attracted by D-limonene ($p < 0.05$), but when the concentration of D-limonene is 100 μg/μL, female insects can be significantly attracted ($p < 0.01$) and male insects can be significantly repelled ($p < 0.01$). The 3-carene, β-pinene, and α-pinene at concentrations of 10 μg/μL and 100 μg/μL were highly attractive to both males and females ($p < 0.01$). α-Phellandrene at concentrations of 10 μg/μL and 100 μg/μL can significantly attract female adults ($p < 0.01$), but has no significant attractant or repellent effect on male adults ($p > 0.05$). When the concentration of myrcene was 10 μg/μL, it had no significant attractant or repellent effect on female and male *M. saltuarius* ($p > 0.05$), but when the concentration of myrcene was 100 μg/μL, it had a significant attractant effect on male ($p < 0.05$) and female ($p < 0.01$) insects. Camphor and terpinolene at concentrations of 10 μg/μL and 100 μg/μL had no significant attractant or repellent effects on female or adult insects ($p > 0.05$) (Figure 3). In conclusion, 3-carene, β-pinene, and α-pinene had the best trapping effect on M. saltuarius. At the same time, males and females showed different selectivity to D-limonene.

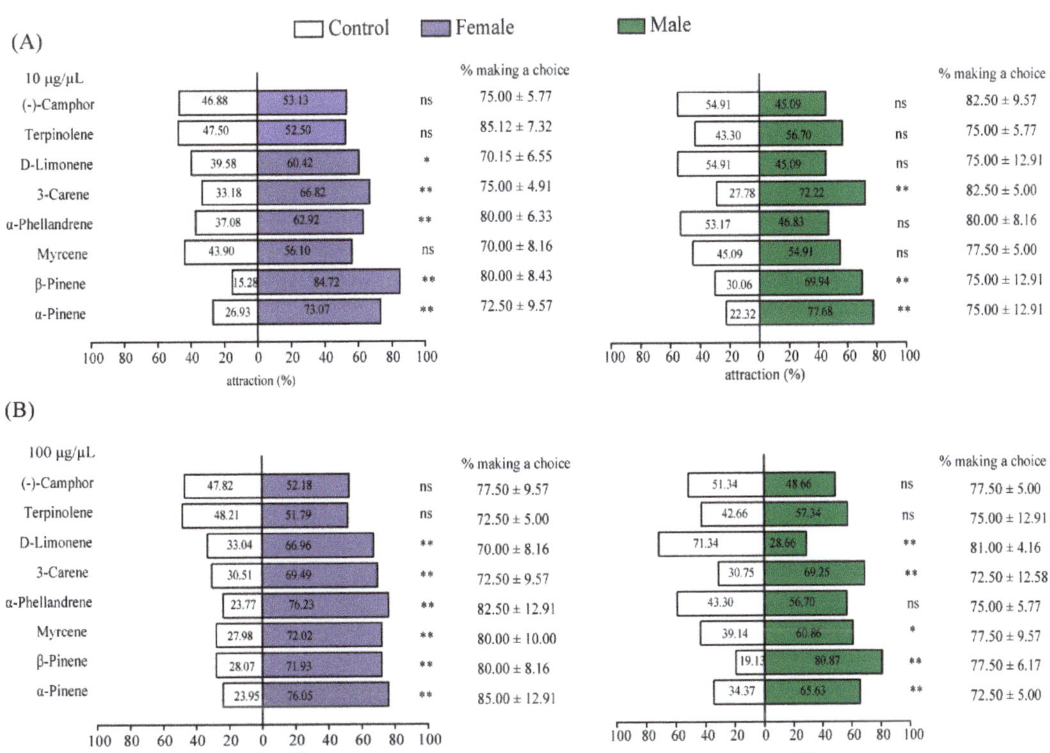

Figure 3. Olfactory behavioral responses of female and male Monochamus saltuarius adults to 8 antennal-active host volatiles at 10 μg/μL (**A**) and 100 μg/μL (**B**). Asterisks (** and *) in the figure indicate significant differences at the $p < 0.01$ and $p < 0.05$ levels, respectively, by the Chi-squared test. 'ns' indicates no significant difference.

2.5. *Forest Trapping Effects of 8 Active Volatiles to M. saltuarius Adults*

We tested the trapping effect of eight antennal active volatiles on *M. saltuarius* adults in the forest. It was found that different volatiles had different trapping effects on adults of *M. saltuarius*. As presented in Figure 4, during 12 days of continuous observation, camphor, terpinolene, and D-limonene had a poor trapping effect on *M. saltuarius* which was not significantly different from the control ($p > 0.05$). α-Phellandrene, myrcene, 3-carene, β-pinene, and α-pinene had a significant trapping effect on *M. saltuarius* that was significantly higher than that of the control ($p < 0.05$) at 6 d, 9 d, and 12 d (Figure 4). Moreover, the trapping effects of 3-carene, β-pinene, and α-pinene were significantly greater than α-phellandrene and myrcene at 9 d and 12 d (Figure 4).

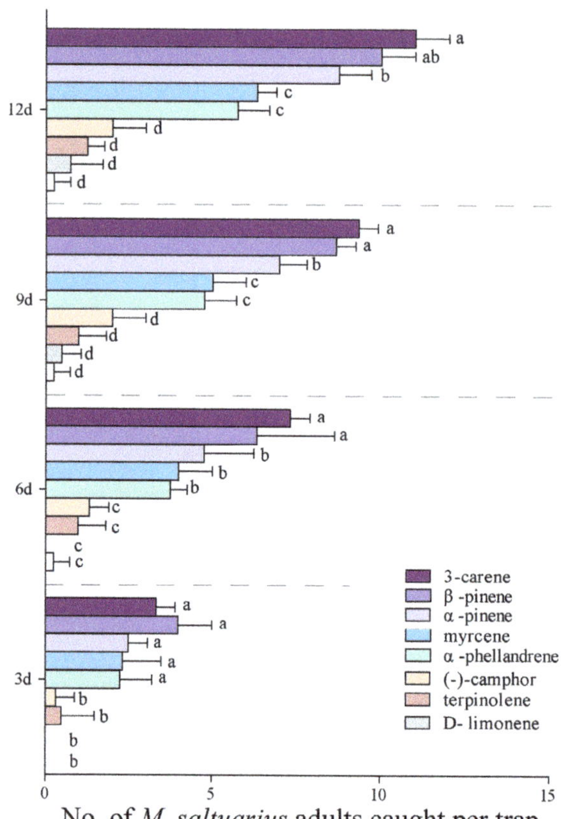

Figure 4. Effects of forest trapping of 8 antennally-active volatiles on *M. saltuarius* adults at 3 d, 6 d, 9 d, and 12 d. Different letters on the bars indicate significant differences among the active volatiles at the same conditions ($p < 0.05$, Duncan's multiple analysis).

2.6. *Field Trapping Effects of Blends on M. saltuarius Adults*

2.6.1. Responses of *M. saltuarius* Adults to Blends of Single Volatile and Ethanol

We studied the synergistic effects of ethanol on 3-carene, β-pinene, and α-pinene, respectively. It was found that ethanol had a good synergistic effect. When the lures were used for 9 d and 12 d in the forest, ethanol had a significant synergistic effect on 3-carene, β-pinene, and α-pinene, and the synergistic effect on β-pinene was the best (Figure 5). The price of β-pinene is cheaper than 3-carene, and it is more suitable for the development of pheromone products. Therefore, the synergistic effect of 2-undecyloxy-1-ethanol, ipsenol, and ipsdienol on β-pinene was further studied.

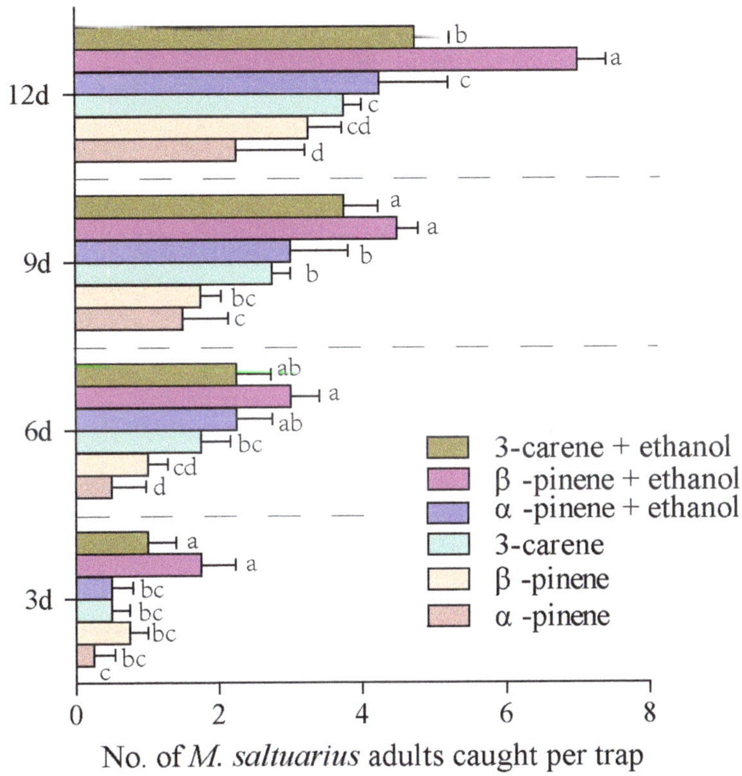

Figure 5. Mean catches of *Monochamus saltuarius* adults in traps baited from 3 d to 12 d, with 180 mL α-pinene, 180 mL β-pinene, 180 mL 3-carene, 135 mL α-pinene + 45 mL ethanol, 135 mL β-pinene + 45 mL ethanol, 135 mL 3-carene + 45 mL ethanol. Different lowercase letters on the bars mean significantly differences in the means by Duncan's multiple analysis ($p < 0.05$).

2.6.2. Responses of *M. saltuarius* Adults to Blends of β-Pinene, Ethanol and 2-Undecyloxy-1-ethanol

Based on the synergistic effect of ethanol, we compared the synergistic effects of β-pinene and ethanol combined with different doses of 2-undecyloxy-1-ethanol on the attraction of *M. saltuarius* adults. It was found that β-pinene, ethanol, and 2-undecyloxy-1-ethanol had a synergistic effect. According to the trapping effect on day 6 and day 12, the number of trapped adults changed with the increase in the concentration of 2-undecyloxy-1-ethanol. The trapping efficacy was as follows: 0 μL < 20 μL < 50 μL < 200 μL < 100 μL 2-undecyloxy-1-ethanol (Figure 6). The trapping effect was best when 100 μL of 2-undecyloxy-1-ethanol was mixed with β-pinene and ethanol, and the mean number of adults captured was respectively, 12.6 ± 1.8 adults/6 d/trap and 19.6 ± 1.0 adults/12 d/trap, respectively (Figure 6).

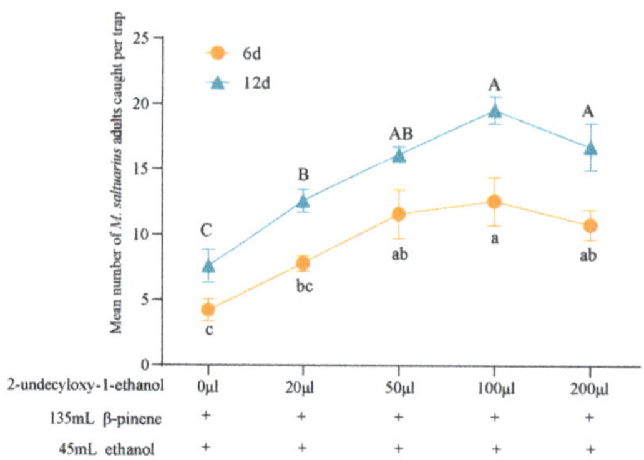

Figure 6. Mean catches of *M. saltuarius* adults in traps baited with 135 mL β-pinene, 45 mL ethanol, and different volumes of 2-undecyloxy-1-ethanol (0 μL, 20 μL, 50 μL, 100 μL, 200 μL). The data on the Y-axis are mean ± standard error ($p < 0.05$, Duncan's multiple analysis).

2.6.3. Responses of *M. saltuarius* Adults to Blends of β-Pinene, Ethanol, 2-Undecyloxy-1-ethanol, and Pheromones of Bark Beetles

Ipsenol and ipsdienol are aggregation pheromones of bark beetles that have been reported to synergistically enhance attraction to some *Monochamus* species with plant volatiles and pheromones. Therefore, based on the above results, we compared the synergistic effects of ipsenol and ipsdienol on the mixture of 135 mL β-pinene, 45 mL ethanol, and 100 μL 2-undecyloxy-1-ethanol. It was found that, after the addition of 20 μL or 40 μL ipsdienol, the lure effect of the mixture on M. saltuarius was as follows: 40 μL ipsdienol > 40 μL ipsdienol+ 40 μL ipsenol > 20 μL ipsdienol > 20 μL ipsenol > 40 μL ipsenol (Figure 7). The addition of 40 μL ipsenol did not indicate a significant synergistic effect with the mixture of 135 mL β-pinene + 45 mL ethanol + 100 μL 2-undecyloxy-1-ethanol at 6 d and 12 d, while 20 μL ipsdienol and 40 μL ipsdienol showed a significant synergistic effect (Figure 7). Thus, ipsdienol and plant-derived compounds β-pinene, ethanol, and monochamol have a good synergistic effect, which is suitable for the development of efficient trapping agents for *M. saltuarius*.

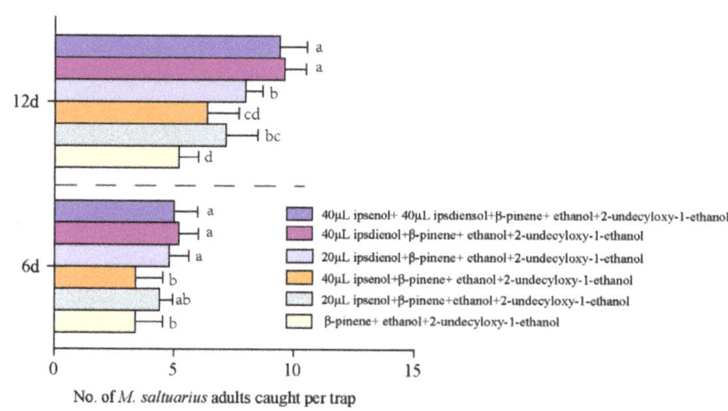

Figure 7. Mean number of *M. saltuarius* adults caught in traps baited with (1) 20 μL ipsdienol+ 135 mL β-pinene + 45 mL ethanol+ 100 μL 2-undecyloxy-1-ethanol, (2) 40 μL ipsdienol+ 135 mL β-pinene + 45 mL ethanol+ 100 μL 2-undecyloxy-1-ethanol, (3) 20 μL ipsenol+ 135 mL β-pinene + 45 mL ethanol+ 100 μL 2-undecyloxy-1-ethanol, (4) 40 μL ipsenol+ 135 mL β-pinene + 45 mL ethanol+ 100 μL 2-undecyloxy-1-ethanol, (5) 40 μL ipsdienol +40 μL ipsenol + 135 mL β-pinene + 45 mL ethanol+ 100 μL 2-undecyloxy-1-ethanol, (6) 135 mL β-pinene + 45 mL ethanol+ 100 μL 2-undecyloxy-1-ethanol. Different letters on the bars indicate significantly differences by Duncan's multiple analysis ($p < 0.05$).

3. Discussion

The selection and location of host plants are crucial for the survival and reproduction of herbivorous insects, which can distinguish nutritional and toxic compounds through their olfactory and gustatory systems [36–38]. Adults can detect and distinguish volatile substances from specific plants through VOCs [39–42], thus helping them search for host plants for oviposition or feeding.

This study collected and identified 18 VOCs from the following three host plants: *P. koraiensis*, *P. tabuliformis*, and *P. asperata*. These compounds mainly include terpenes, aldehydes, ketones, alkanes, and esters, with terpenes accounting for the largest proportion. α-Pinene β-pinene, and D-limonene were found in all three plants. Among terpenes, 3-carene and myrcene are present in *P. koraiensis* and *P. asperata*, while phellandrene and camphene are only identified in *P. koraiensis*. α-Pinene, β-pinene, 3-carene, myrcene, phellandrene, camphene, camphor, terpinolene, benzaldehyde, and γ-terpinene are relatively high in plants and some of them have been reported to attract other herbivorous insects [43,44]. D-Limonene is used as a repellent in some herbivorous insects. So, it is speculated that these 11 compounds may be potential attractants or repellent components for *M. saltuarius*. And, *M. saltuarius* revealed different degrees of EAG response to the 11 compounds, and the relative values of EAG responses to the 11 VOCs increase with increasing concentration. The EAG reaction cannot determine the selection or avoidance of compounds by *M. saltuarius*. From the analyzed compounds, 8 substances with strong EAG reaction values in *M. saltuarius* were selected for Y-tube behavior experiments. These results indicate that at two concentrations (10 μg/μL, 100 μg/μL), α- pinene, 3-carene, and β-pinene have significant attractant activity for both male and female adults of *M. saltuarius*, which may be the main plant-derived volatiles affecting host location.

The volatiles of the host plant of *M. saltuarius* are complex, and the composition and proportion of host volatiles may be influenced by various factors, such as plant age and physiological conditions, season, and biological and abiotic damage, all of which may affect the attractiveness to *M. saltuarius* [45,46]. Specific compounds from non-host plants may also be beneficial for host recognition [46–50]. This study indicates that, in forest trapping, 3-carene, β-pinene, and α-pinene have a strong attractive effect on *M. saltuarius*, with 3-carene having the highest number of individuals caught in traps.

Ethanol is a volatile indicator of stress and dying trees. When trees are stressed, the volatilization of ethanol from trees increases significantly. But, ethanol is attractive to some insects [51–54]. And, it has been reported that ethanol can work synergistically with α-pinene to increase the attraction effect of some beetles [49]. This study showed that β-pinene combined with ethanol had the best trapping effect on *M. saltuarius*. And, compared to using β-pinene, 3-carene, or α-pinene alone, the three volatiles combined with ethanol as a lure, have a higher number of individuals caught in traps. There is a synergistic effect between ethanol and the three compounds that enhances the attraction of *M. saltuarius*.

Compared with host volatiles, the aggregation pheromone of *M. saltuarius* is considered to have better attractant activity due to a higher sensitivity and selectivity [30,35]. The trapping effect of *A. glabripennis* on aggregation pheromones combined with plant volatiles was greater than that of aggregation pheromones alone [55]. Plant volatile compounds increase the attraction of sex pheromones to *Holotrichia parallela* (Coleoptera: Scarabaeidae), and a mixture of (E)-2-hexenyl acetate and sex pheromones results in significantly higher catches than sex pheromones alone [56]. It is reported that the combined use of different types of pheromones can effectively improve the attraction effect of *Xylotrechus pyrrhoderus* (Coleoptera: Cerambycidae), and when the sex pheromone (3-hydroxy-2-hexanone and 3-hydroxy-2-octanone) and the host pheromone (methyl benzoate) are mixed, its attraction effect is better [57]. The results showed that the aggregation pheromone 2-undecyloxy-1-ethanol had a synergistic effect on the combination of β-pinene and ethanol in the forest trapping. The trapping effect of the mixture with 2-undecyloxy-1-ethanol was significantly higher than that of the treatment without 2-undecyloxy-1-ethanol.

Different herbivorous insects may parasitize or lay eggs on the same host plant. To identify the same species on the host plant, more specific olfactory recognition pheromones are needed. As evolution progress, different insects feeding on the same host plant may also recognize each other's pheromones. The suitable habitat of *M. saltuarius*, *Monochamus galloprovincialis* (Coleoptera: Cerambycidae), and bark beetles overlaps. The pheromones of bark beetles, ipsenol and ipsdienol, were also identified as synergists of the aggregation pheromone for *M. galloprovincialis* adults [58]. This study found that ipsdienol can also synergistically increase the trapping effect by mixing with aggregation pheromones of *M. saltuarius* and host plant volatiles.

This article describes a study on the feeding preferences of three hosts, the components of host volatiles, and the responses of *M. saltuarius* to host volatiles. The synergistic effects of host volatiles, ethanol, 2-undecyloxy-1-ethanol, and the pheromones of bark beetles were also evaluated. However, there are still some issues that need further exploration. The volatile compounds of host plants in the forest, as background odors, are usually complex in composition, and further exploration is needed to investigate the effects of the interactions between the components of host volatiles on the attraction of *M. saltuarius*. The different ratios of volatile organic compounds from different host plants on the attraction effect on *M. saltuarius* require further research. For other species with the same host, pheromones released by them can also serve as mutualists, which might be recognized by *M. saltuarius* for feeding or oviposition localization. Further research is needed to investigate the kairomonal attraction of pheromone components from other species to *M. saltuarius* or their interactions with identified host plant volatiles.

Developing attractants for longhorn beetles is a promising control strategy, but there are still some potential limitations and challenges. At present, most of the attractants developed can only trap sexually matured longhorn beetles, which have already spread pine nematodes through feeding and caused certain harm to trees. Therefore. it is urgent to develop attractants to trap longhorn beetles during their maturation feeding period. In addition, attractants were generally used to catch longhorn beetles to help control the pest population, but the control effects were limited. However, the attractants and chemical pesticides can be combined to improve the utilization efficiency of pesticides and reduce their use. In short, the development of attractants is promising, and we should make efforts to improve the effectiveness of attractants and overcome the challenges.

4. Materials and Methods

4.1. Insects and Plants

The insects of *M. saltuarius* used in the experiment were obtained from a farm in Qingyuan Manchu Autonomous County, Fushun City, Liaoning Province, China. They were lured and caught by traps, kept in insect cages, and fed with branches of *P. koraiensis*, and fresh branches were replaced every 2 days. Five-year-old healthy *P. koraiensis*, *P. tabuliformis*, and *P. asperata* were selected as test plants.

4.2. Chemicals

The CAS number, purity, and sources of chemicals involved in the research are listed in Table 2.

Table 2. Information of compounds in the study.

No.	Compounds	CAS	Purity	Sources
1	α-Pinene	7785-26-4	98%	J&K SCIENTIFIC
2	β-Pinene	18172-67-3	98%	J&K SCIENTIFIC
3	Myrcene	123-35-3	90%	J&K SCIENTIFIC
4	3-Carene	13466-78-9	90%	TCI AMERICA
5	α-Phellandrene	99-83-2	98%	J&K SCIENTIFIC
6	D-Limonene	5989-27-5	95%	J&K SCIENTIFIC
7	(-)-Camphor	464-48-2	96%	MACKLIN

Table 2. *Cont.*

No.	Compounds	CAS	Purity	Sources
8	Terpinolene	586-62-9	95%	J&K SCIENTIFIC
9	Camphene	79-92-5	78%	TCI AMERICA
10	γ-Terpinene	99-85-4	95%	J&K SCIENTIFIC
11	Benzaldehyde	100-52-7	98%	J&K SCIENTIFIC
12	Paraffin liquid	8012-95-1	99%	HUSHI

4.3. Host Plant Preference Assays of *M. saltuarius*

Different plants were used to study the feeding tendency of *M. saltuarius*. Host materials were obtained by cutting 30 cm long fresh branches from the tops of three types of plants: *P. koraiensis*, *P. tabuliformis*, and *P. asperata*. The cut branches were placed in insect cages (1 m × 1 m × 0.5 m), with a lighting setting of 16 L: 8 D. In each cage, 6 adults of *M. saltuarius*, 3 females and 3 males, who had been hungry for 6 h and were still active were used for host plant preference assays. We fed the adults of *M. saltuarius* for 3 d, replaced fresh branches daily, and calculated the feeding area of the branches. Each experiment was repeated 8 times. The feeding marks on the branches were replicated using sulfuric acid paper, and the feeding area of the branches was calculated using coordinate paper.

4.4. Volatile Analysis of Three Host Plants

The plant volatiles from 3 healthy host plants in healthy states were collected using the methods of Fan et al. [59]. The side branches of tested plants were sealed in oven bags (Polypropylene, 482 × 569 mm, Reynolds, Patrick County, VA, USA) for 6 h. We inserted two polytetrachloroethylene pipes were inserted into the bag opening and tied tightly with iron wire. The bag opening was wrapped and sealed with parafilm sealing film. One served as the exhaust rubber pipe and the other served as the intake rubber pipe. One end of the intake rubber tube was connected to an activated carbon sampling tube and the activated carbon tube was also connected to the air inlet of the atmospheric sampler through a polytetrachloroethylene tube. The outlet rubber tube was connected to a TENAX tube (which contains 200 mg of adsorbent), and the other end of the TENAX tube was also connected to the air inlet of the atmospheric sampler through a rubber tube, forming a closed loop that enabled air to flow out sequentially from the inlet of the atmospheric sampler, activated carbon tube, sampling bag, TENAX tube, and the outlet of the atmospheric sampler. The TENAX tube was activated at 300 °C for 3 h in advance as an adsorption tube. The atmospheric sampler had a flow rate of 0.5 L/min and, after 6 h of collection, the adsorption tube was removed and marked. The two ends were sealed tightly with parafilm sealing film and stored in a −80 °C freezer for testing. Volatile substances were analyzed using GC-MS-QP 2010 gas chromatography–mass spectrometry (Agilent, Santa Clara, CA, USA). Each experiment was repeated 3 times.

4.5. Electroantennography

For the electroantennography (EAG) assays, an EAG detector (Syntech-IDAC 4, Kirchzarten, Germany) was used to record the olfactory response of *M. saltuarius* to volatile compounds. We fixed the head was fixed and the complete antennae of the *M. saltuarius* was cut. The antenna was connected to the electrode using Spectra 360 conductive adhesive (Parker Laboratories, Fairfield, NJ, USA), with the base of the antenna connected to the reference electrode and the tip of the tail of the antenna connected to the recording electrode. The different chemicals were dissolved in paraffin oil at concentrations of 10 μg/μL and 100 μg/μL, respectively, and paraffin oil was used as control. An aliquot of 20 μL of solvent was applied to a 5 mm × 20 mm filter paper strip and the filter paper strip was placed completely inside the Pasteur pipette. The Pasteur pipette was connected to the airflow control device and the tip was inserted into the side hole of the mixing tube. A continuous airflow flow rate, set at 120 mL/min, stimulated the airflow flow rate to 20 mL/min, and the stimulation source was about 1.5 cm away from the antenna. The duration of each stimulation was 0.2 s, with a 1 min interval between the two stimuli to avoid antenna fatigue or adaptation. Paraffin oil was applied at the beginning and end of the trial (one antenna) to monitor antenna status. The EAG response values were normalized relative to the average response to paraffin oil. There were 8 replicates for each male and female adult.

4.6. Olfactory Behavioral Responses to Synthetic Active Volatiles

A total of 8 chemicals were selected from the volatiles of the main host plants of *M. saltuarius* for Y-tube olfactometer experiments. These compounds were as follows: (-)-camphor, terpinolene, D-limonene, 3-carene, α-phellandrene, myrcene, α-pinene, and β-pinene. The standards of each compound were prepared with n-hexane as solvent at concentrations of 10 μg/μL and 100 μg/μL, respectively, and n-hexane was used as the control. Specifications of Y-tube olfactometer were as follows: the inner diameter and length of main arm were, respectively, 9 cm and 40 cm, and the inner diameter and length of two lateral arms with a 60° angle at the Y-junction were 6 cm and 25 cm, respectively. Air was pushed through activated charcoal filter and a bottle with distilled water by a vacuum pressure pump (Beijing Municipal Institute of Labour Protection, Beijing, China). The air flow rate was about 500 mL/min. The adults of *M. saltuarius* were placed in a Y-shaped glass olfactometer to observe behaviors. One adult was taken each time for behavior selection measurement and observed for 10 min. For those who entered the arm with volatiles or the control arm for more than 1/3 of the distance and stayed for more than 15 s, they were recorded as *M. saltuarius* made a choice; otherwise, they were noted as unresponsive. After testing 5 insects, arm positions were switched to eliminate the influence of the experimental environment on the olfactory behavior of the spruce spotted beetle. A total of 40 males and 40 females were tested for each concentration of volatile matter. After each experiment, the experimental apparatus was cleaned with ethanol and dried.

4.7. Field Experiments

The experiments were conducted in Qingyuan Manchu Autonomous County, Fushun City, Liaoning Province, China (41°59′ N, 124°31′ E). Traps were hung in the sample plot, and the compounds were released through sustained-release bottles.

4.7.1. Trapping Experiments of Single Compounds in Forests

(-)-Camphor, terpinolene, D-limonene, 3-carene, α-phellandrene, myrcene, α-pinene, and β-pinene were separately placed inside slow-release bottle at a volume of 180 mL. The slow-release bottles were used as lure core to put in traps. The empty slow-release bottles were used as a control. The 9 traps with different individual substance were put in a block with traps 20 m apart. A total of 5 blocks were set up for 5 repeated experiments. The interval between different blocks was 50 m. The survey was conducted every 3 days to count the adults of *M. saltuarius* caught by each lure core from 21 July to 1 August 2022, and then the adults in the trap collection cup were cleaned up. The lure core containing a single volatile was replaced by another trap during the same block every 3 days in order to reduce the impact of geographical location on the test results.

4.7.2. The Synergies of Ethanol on the Trapping Effect of *M. saltuarius* in the Forest

Catches of *M. saltuarius* were compared in traps baited with 180 mL α-pinene, 180 mL β-pinene, 180 mL 3-carene, 135 mL α-pinene + 45 mL ethanol, 135 mL β-pinene + 45 mL ethanol, and 135 mL 3-carene + 45 mL ethanol. There were 5 replicates for each treatment. These assays were conducted from 14 to 23 May 2023, and the number of *M. saltuarius* was counted every 3 days.

4.7.3. Trapping Assays on the Synergistic Effect of Aggregation Pheromones of *M. saltuarius*

The mixture combinations with aggregation pheromones for attracting *M. saltuarius* are shown in Table 3. These formulations were used to compare the synergistic effect of different concentrations of aggregation pheromones of *M. saltuarius*. The experiments were conducted from 29 May to 13 June 2023, and the number of captured adults of *M. saltuarius* was recorded every 6 days. Five replicates were carried out for each treatment.

Table 3. Formulation of attracting *Monochamus saltuarius* with aggregation pheromone in forest.

Serial Number	Formulations
1	135 mL β-pinene + 45 mL ethanol
2	135 mL β-pinene + 45 mL ethanol + 20 μL 2-undecyloxy-1-ethanol
3	135 mL β-pinene + 45 mL ethanol + 50 μL 2-undecyloxy-1-ethanol
4	135 mL β-pinene + 45 mL ethanol + 100 μL 2-undecyloxy-1-ethanol
5	135 mL β-pinene + 45 mL ethanol + 200 μL 2-undecyloxy-1-ethanol

4.7.4. Field Attraction Tests to Explore Synergies of Bark Beetle Pheromones—Ipsdienol and Ipsenol

To explore the potential synergies of bark beetle pheromones, the formulations used to attract *M. saltuarius* in the forest were made as shown in Table 4. The survey was conducted from 12–24 July 2023, and the number of *M. saltuarius* caught by each trap was recorded every 6 days. Each treatment was repeated 5 times.

Table 4. Formulation of pheromones of bark beetles used to attract *Monochamus saltuarius* in the forest.

Serial Number	Formulations
1	135 mL α-pinene + 45 mL ethanol + 2-undecyloxy-1-ethanol
2	135 mL β-pinene + 45 mL ethanol + 2-undecyloxy-1-ethanol
3	135 mL β-pinene + 45 mL ethanol + 100 μL 2-undecyloxy-1-ethanol + 20 μL ipsdienol
4	135 mL β-pinene + 45 mL ethanol + 100 μL 2-undecyloxy-1-ethanol + 20 μL ipsenol
5	135 mL β-pinene + 45 mL ethanol + 100 μL 2-undecyloxy-1-ethanol + 40 μL ipsdienol
6	135 mL β-pinene + 45 mL ethanol + 100 μL 2-undecyloxy-1-ethanol + 40 μL ipsenol
7	135 mL β-pinene + 45 mL ethanol + 100 μL 2-undecyloxy-1-ethanol + 40 μL ipsenol + 40 μL ipsdienol

4.8. Data Analysis

All data were analyzed by SPSS 20.0 software, and the graphing was made by GraphPad Prism 9.5. The data from feeding experiments and forest experiments was conducted using one-way ANOVA and Duncan's multiple comparisons. In the EAG experiment, a one-way ANOVA with Duncan's multiple comparisons was also used to compare the relative EAG response values of male and female adults of *M. saltuarius* to different concentrations of different compounds. The indoor behavior data of Y-tube experiments were analyzed by using Chi-squared (χ^2) goodness-of-fit tests.

5. Conclusions

In summary, we have identified plant volatiles that has attractive effects on *M. saltuarius*, identified key plant source volatiles for *M. saltuarius* host location, and further explored multi-component formulations using insect pheromones to optimize attractant formulations and improve trapping efficiency. We found that α-pinene, 3-carene, and β-pinene have significant attractant activity in adults of *M. saltuarius*, which may play an important role in the host location. β-Pinene, ethanol, 2-undecyloxy-1-ethanol, and ipsdienol had the best synergistic effect on luring *M. saltuarius*. This study provides a theoretical basis for controlling the population of *M. saltuarius* and preventing pine wood nematodes.

Author Contributions: Conceptualization, J.F.; methodology, Y.D. and D.C.; software, Y.D. and D.C.; formal analysis, Y.D., S.Z. and Z.M.; investigation, Y.D., S.Z. and Z.M.; resources, J.F.; data curation, Y.D. and D.C.; writing—original draft preparation, Y.D. and D.C.; writing—review and editing, Y.D., D.C. and J.F.; supervision, J.F.; project administration, J.F. All authors have read and agreed to the published version of the manuscript.

Funding: This research was funded by Central Finance Forestry Science and Technology Demonstration and Promotion Project ([2022]TS01-4).

Data Availability Statement: All data are incorporated into the article.

Acknowledgments: We would like to thank the staff who helped us to collect and investigate *Monochamus saltuarius*.

Conflicts of Interest: The authors declare no conflicts of interest.

References

1. Ge, S.X.; Li, J.X.; Jiang, Z.H.; Zong, S.X.; Ren, L.L. Cradle for the newborn *Monochamus saltuarius*: Microbial associates to ward off entomopathogens and disarm plant defense. *Insect Sci.* **2023**, *30*, 1165–1182. [CrossRef] [PubMed]
2. Wang, J.; Zhang, S.; Zheng, Y. Feeding Preferences and Responses of *Monochamus saltuarius* to Volatile Components of Host Pine Trees. *Insects* **2022**, *13*, 888. [CrossRef] [PubMed]
3. Kim, J.; Jung, Y.; Lee, S. Diel rhythmicity of field responses to synthetic pheromone lures in the pine sawyer *Monochamus saltuarius*. *Insects* **2021**, *12*, 441. [CrossRef] [PubMed]
4. Li, J.; Fu, N.; Ren, L.; Luo, Y. Identification and Validation of Reference Genes for Gene Expression Analysis in *Monochamus saltuarius* Under *Bursaphelenchus xylophilus* Treatment. *Front. Physiol.* **2022**, *13*, 882792. [CrossRef] [PubMed]
5. Hou, Z.; Shi, F.; Ge, S.; Tao, J.; Ren, L.; Wu, H.; Zong, S. Comparative transcriptome analysis of the newly discovered insect vector of the pine wood nematode in China, revealing putative genes related to host plant adaptation. *BMC Genom.* **2021**, *22*, 189. [CrossRef]
6. Kishi, Y. *Pine Wood Nematode and the Japanese Pine Sawyer*; Thomas Company Limited: Tokyo, Japan, 1995.
7. Kwon, T.S.; Lim, J.H.; Sim, S.J.; Kwon, Y.D.; Son, S.K.; Lee, K.Y.; Kim, Y.T.; Park, J.W.; Shin, C.H.; Ryu, S.B. Distribution patterns of *Monochamus alternatus* and *M. saltuarius* (Coleoptera: Cerambycidae) in Korea. *J. Korean Soc. For. Sci.* **2006**, *95*, 543–550.
8. Li, M.; Li, H.; Sheng, R.C.; Sun, H.; Sun, S.H.; Chen, F.M. The first record of *Monochamus saltuarius* (Coleoptera; Cerambycidae) as vector of *Bursaphelenchus xylophilus* and its new potential hosts in China. *Insects* **2020**, *11*, 636. [CrossRef] [PubMed]
9. Linit, M. Nemtaode-vector relationships in the pine wilt disease system. *J. Nematol.* **1988**, *20*, 227.
10. Buszewski, B.; Bukowska, M.; Ligor, M.; Staneczko-Baranowska, I. A holistic study of neonicotinoids neuroactive insecticides properties, applications, occurrence, and analysis. *Environ. Sci. Pollut. Res.* **2019**, *26*, 34723–34740. [CrossRef]
11. Moon, J.H.; Won, S.J.; Maung, C.E.H.; Choi, J.H.; Choi, S.I.; Ajuna, H.B.; Ahn, Y.S. *Bacillus velezensis* CE 100 inhibits root rot diseases (*Phytophthora* spp.) and promotes growth of Japanese cypress (*Chamaecyparis obtusa* Endlicher) seedlings. *Microorganisms* **2021**, *9*, 821. [CrossRef]
12. Bordat, A.; Marchand, G.; Langlade, N.B.; Pouilly, N.; Muños, S.; Dechamp-Guillaume, G.; Vincourt, P.; Bret-Mestries, E. Different genetic architectures underlie crop responses to the same pathogen: The {*Helianthus annuus* Phoma macdonaldii*} interaction case for black stem disease and premature ripening. *BMC Plant Biol.* **2017**, *17*, 167. [CrossRef] [PubMed]
13. Chen, J.; Li, J.; Ma, M.; Li, B.; Zhou, Y.; Pan, Y.; Fan, Y.; Yi, B.; Tu, J. Improvement of resistance to clubroot disease in the Ogura CMS restorer line R2163 of *Brassica napus*. *Plants* **2022**, *11*, 2413. [CrossRef] [PubMed]
14. Huang, K.; Jiang, Q.; Liu, L.; Zhang, S.; Liu, C.; Chen, H.; Ding, W.; Zhang, Y. Exploring the key microbial changes in the rhizosphere that affect the occurrence of tobacco root-knot nematodes. *AMB Express* **2020**, *10*, 72. [CrossRef] [PubMed]
15. Li, M.; Yang, T.; Kandul, N.P.; Bui, M.; Gamez, S.; Raban, R.; Bennett, J.; Sánchez, C.H.M.; Lanzaro, G.C.; Schmidt, H. Development of a confinable gene drive system in the human disease vector *Aedes aegypti*. *eLife* **2020**, *9*, e51701. [CrossRef] [PubMed]
16. Pilz-Junior, H.L.; de Lemos, A.B.; de Almeida, K.N.; Corção, G.; Schrekker, H.S.; Silva, C.E.; da Silva, O.S. Microbiota potentialized larvicidal action of imidazolium salts against *Aedes aegypti* (Diptera: Culicidae). *Sci. Rep.* **2019**, *9*, 16164. [CrossRef] [PubMed]
17. Wang, Q.; Liu, Y.; Hu, J.; Zhang, Y.; Xie, K.; Wang, B.; Tuyen, L.Q.; Song, Z.; Wu, H.; Liu, Y.; et al. Detection of Quantitative Trait Loci (QTLs) for Resistances to Small Brown Planthopper and Rice Stripe Virus in Rice Using Recombinant Inbred Lines. *Int. J. Mol. Sci.* **2013**, *14*, 8406–8421. [CrossRef] [PubMed]
18. Bruce, T.J.; Wadhams, L.J.; Woodcock, C.M. Insect host location: A volatile situation. *Trends Plant Sci.* **2005**, *10*, 269–274. [CrossRef] [PubMed]
19. Kamala Jayanthi, P.D.; Kempraj, V.; Aurade, R.M.; Venkataramanappa, R.K.; Nandagopal, B.; Verghese, A.; Bruce, T.J. Specific volatile compounds from mango elicit oviposition in gravid *Bactrocera dorsalis* females. *J. Chem. Ecol.* **2014**, *40*, 259–266. [CrossRef] [PubMed]
20. Beck, J.J.; Light, D.M.; Gee, W.S. Electroantennographic bioassay as a screening tool for host plant volatiles. *J. Vis. Exp.* **2012**, *63*, e3931. [CrossRef]
21. Pérez-Aparicio, A.; Torres-Vila, L.M.; Gemeno, C. EAG responses of adult *Lobesia botrana* males and females collected from *Vitis vinifera* and *Daphne gnidium* to larval host-plant volatiles and sex pheromone. *Insects* **2019**, *10*, 281. [CrossRef]
22. Fettköther, R.; Reddy, G.V.; Noldt, U.; Dettner, K. Effect of host and larval frass volatiles on behavioural response of the old house borer, *Hylotrupes bajulus* (L.) (Coleoptera: Cerambycidae), in a wind tunnel bioassay. *Chemoecology* **2000**, *10*, 1–10. [CrossRef]
23. Das, A.; Lee, S.H.; Hyun, T.K.; Kim, S.W.; Kim, J.Y. Plant volatiles as method of communication. *Plant Biotechnol. Rep.* **2013**, *7*, 9–26. [CrossRef]

24. Du, Y.; Zhang, J.; Yan, Z.; Ma, Y.; Yang, M.; Zhang, M.; Zhang, Z.; Qin, L.; Cao, Q. Host preference and performance of the yellow peach moth (*Conogethes punctiferalis*) on chestnut cultivars. *PLoS ONE* **2016**, *11*, e0157609. [CrossRef] [PubMed]
25. Xu, L.; Jiang, H.B.; Yu, J.L.; Pan, D.; Tao, Y.; Lei, Q.; Chen, Y.; Liu, Z.; Wang, J.J. Two odorant receptors regulate 1-octen-3-ol induced oviposition behavior in the oriental fruit fly. *Commun. Biol.* **2023**, *6*, 176. [CrossRef] [PubMed]
26. Wang, J.; Wei, J.; Yi, T.; Li, Y.-Y.; Xu, T.; Chen, L.; Xu, H. A green leaf volatile, (Z)-3-hexenyl-acetate, mediates differential oviposition by *Spodoptera frugiperda* on maize and rice. *BMC Biol.* **2023**, *21*, 140. [CrossRef] [PubMed]
27. Kelsey, R.G.; Westlind, D.J. Attraction of red turpentine beetle and other SS colytinae to ethanol, 3-carene or ethanol+ 3-carene in an oregon pine forest. *Agric. For. Entomol.* **2018**, *20*, 272–278. [CrossRef]
28. Ikeda, T.; Oda, K. The occurrence of attractiveness for *Monochamus alternatus* Hope (Coleoptera: Cerambycidae) in nematode-infected pine trees. *J. Jpn. For. Soc.* **1980**, *62*, 432–434.
29. Ikeda, T.; Enda, N.; Yamane, A.; Oda, K.; Toyoda, T. Attractants for the Japanese pine sawyer, *Monochamus alternatus* hope (Coleoptera: Cerambycidae). *Appl. Entomol. Zool.* **1980**, *15*, 358–361. [CrossRef]
30. Millar, J.; Hanks, L. Chemical Ecology of Cerambycid Beetles. In *Cerambycidae of the World: Biology and Pest Management*; CRC Press/Taylor & Francis: Boca Raton, FL, USA, 2016.
31. Löbl, I.; Löbl, D. *Catalogue of Palaearctic Coleoptera. Volume 1: Archostemata-Myxophaga-Adephaga*; Brill Academic Publishers: Leiden, The Netherlands, 2018.
32. Teale, S.A.; Wickham, J.D.; Zhang, F.; Su, J.; Chen, Y.; Xiao, W.; Hanks, L.M.; Millar, J.G. A male-produced aggregation pheromone of *Monochamus alternatus* (Coleoptera: Cerambycidae), a major vector of pine wood nematode. *J. Econ. Entomol.* **2011**, *104*, 1592–1598. [CrossRef]
33. Lee, H.R.; Lee, S.C.; Lee, D.H.; Choi, W.S.; Jung, C.S.; Jeon, J.H.; Kim, J.E.; Park, I.K. Identification of the aggregation-sex pheromone produced by male *Monochamus saltuarius*, a major insect vector of the pine wood nematode. *J. Chem. Ecol.* **2017**, *43*, 670–678. [CrossRef]
34. Allison, J.D.; McKenney, J.L.; Millar, J.G.; Mcelfresh, J.S.; Mitchell, R.F.; Hanks, L.M. Response of the woodborers *Monochamus carolinensis* and *Monochamus titillator* (Coleoptera: Cerambycidae) to known cerambycid pheromones in the presence and absence of the host plant volatile α-pinene. *Environ. Entomol.* **2012**, *41*, 1587–1596. [CrossRef] [PubMed]
35. Hanks, L.M.; Millar, J.G. Sex and aggregation-sex pheromones of cerambycid beetles: Basic science and practical applications. *J. Chem. Ecol.* **2016**, *42*, 631–654. [CrossRef] [PubMed]
36. Song, C.; Ma, L.; Zhao, J.; Xue, Z.; Yan, X.; Hao, C. Electrophysiological and behavioral responses of *Plutella xylostella* (Lepidoptera: Plutellidae) to volatiles from a non-host plant, Geranium, *Pelargonium* × *hortorum* (Geraniaceae). *J. Agric. Food Chem.* **2022**, *70*, 5982–5992. [CrossRef] [PubMed]
37. Zhao, Y.Q.; Zhao, J.R.; Mao, L.J.; Shi, Z.H. Effects of the volatiles from different tomato varieties on host selection behavior of B-biotype *Bemisia tabaci*. *J. Appl. Ecol.* **2012**, *23*, 2509–2514.
38. Yang, X.; Fan, F.; An, L.; Li, M.; Wei, G. Oviposition preferences in *Grapholita molesta*: The relative importance of visual and olfactory cues. *Entomol. Exp. Et Appl.* **2019**, *167*, 722–728. [CrossRef]
39. Piesik, D.; Wenda-Piesik, A. *Sitophilus granarius* responses to blends of five groups of cereal kernels and one group of plant volatiles. *J. Stored Prod. Res.* **2015**, *62*, 36–39. [CrossRef]
40. Fürstenau, B.; Rosell, G.; Guerrero, A.; Quero, C. Electrophysiological and behavioral responses of the black-banded oak borer, *Coroebus florentinus*, to conspecific and host-plant volatiles. *J. Chem. Ecol.* **2012**, *38*, 378–388. [CrossRef] [PubMed]
41. Cloonan, K.R.; Abraham, J.; Angeli, S.; Syed, Z.; Rodriguez-Saona, C. Advances in the chemical ecology of the spotted wing drosophila (*Drosophila suzukii*) and its applications. *J. Chem. Ecol.* **2018**, *44*, 922–939. [CrossRef]
42. Wenda-Piesik, A.; Piesik, D.; Nowak, A.; Wawrzyniak, M. *Tribolium confusum* responses to blends of cereal kernels and plant volatiles. *J. Appl. Entomol.* **2016**, *140*, 558–563. [CrossRef]
43. Chénier, J.; Philogene, B. Field responses of certain forest Coleoptera to conifer monoterpenes and ethanol. *J. Chem. Ecol.* **1989**, *15*, 1729–1745. [CrossRef]
44. Nehme, M.; Keena, M.; Zhang, A.; Baker, T.; Xu, Z.; Hoover, K. Evaluating the use of male-produced pheromone components and plant volatiles in two trap designs to monitor *Anoplophora glabripennis*. *Environ. Entomol.* **2010**, *39*, 169–176. [CrossRef] [PubMed]
45. Li, J. Comparative analysis of volatile compounds from different host plants of *Anoplophora glabripennis* (Motsch.). *J. Beijing For. Univ.* **2002**, *24*, 165–169.
46. Zhu, N.; Zhang, D.Y.; Wu, L.P.; Hu, Q.; Fan, J.T. Attractiveness of aggregation pheromones and host plant volatiles to *Anoplophora glabripennis* and *A. chinensis* (Coleoptera: Cerambycidae). *Acta Entomol. Sin.* **2017**, *60*, 421–430.
47. Fan, L.Q.; Yan, S.C.; Sun, Z.H.; Meng, Z.J. EAG and behavioral responses of Asian longhorn beetle *Anoplophora glabripennis* (Coleoptera: Cerambycidae) to plant volatiles. *Chin. J. Ecol.* **2013**, *32*, 142–148.
48. Li, S.; Gao, W.; Cheng, X.; Zhou, Y.; Cui, W. Electroantennogram response of *Anoplophora glabripennis* (Motsch.) to Acer negundo volatiles. *For. Pest Dis.* **2016**, *35*, 9–14.
49. Nehme, M.; Keena, M.; Zhang, A.; Baker, T.; Hoover, K. Attraction of *Anoplophora glabripennis* to male-produced pheromone and plant volatiles. *Environ. Entomol.* **2009**, *38*, 1745–1755. [CrossRef] [PubMed]
50. Yu, H.; Wang, Z.; Qin, H.; Wang, D.; Shi, J. Effects of lures to trap *Anoplophora glabripennis* (Motschulsky)(Coleoptera: Cerambycidae) in the coastal protection forest in Zhejiang, China. *J. Environ. Entomol.* **2017**, *39*, 694–700.

51. Cavaletto, G.; Faccoli, M.; Ranger, C.M.; Rassati, D. Ambrosia beetle response to ethanol concentration and host tree species. *J. Appl. Entomol.* **2021**, *145*, 800–809. [CrossRef]
52. Ranger, C.M.; Reding, M.E.; Persad, A.B.; Herms, D.A. Ability of stress-related volatiles to attract and induce attacks by *Xylosandrus germanus* and other ambrosia beetles. *Agric. For. Entomol.* **2010**, *12*, 177–185. [CrossRef]
53. Ranger, C.M.; Reding, M.E.; Gandhi, K.J.; Oliver, J.B.; Schultz, P.B.; Cañas, L.; Herms, D.A. Species dependent influence of (−)-α-pinene on attraction of ambrosia beetles (Coleoptera: Curculionidae: Scolytinae) to ethanol-baited traps in nursery agroecosystems. *J. Econ. Entomol.* **2011**, *104*, 574–579. [CrossRef]
54. Kendra, P.E.; Montgomery, W.S.; Narvaez, T.I.; Carrillo, D. Comparison of trap designs for detection of *Euwallacea* nr. fornicatus and other Scolytinae (Coleoptera: Curculionidae) that vector fungal pathogens of avocado trees in Florida. *J. Econ. Entomol.* **2020**, *113*, 980–987. [PubMed]
55. Lyu, F.; Hai, X.; Wang, Z. A review of the host plant location and recognition mechanisms of Asian longhorn beetle. *Insects* **2023**, *14*, 292. [CrossRef] [PubMed]
56. Ju, Q.; Guo, X.Q.; Li, X.; Jiang, X.J.; Jiang, X.G.; Ni, W.I.; Qu, M.J. Plant volatiles increase sex pheromone attraction of *Holotrichia parallela* (Coleoptera: Scarabaeoidea). *J. Chem. Ecol.* **2017**, *43*, 236–242. [CrossRef] [PubMed]
57. Koczor, S.; Tóth, M. Field attraction of *Eurydema ornata* (Hemiptera: Pentatomidae) to allyl isothiocyanate. *Sci. Rep.* **2023**, *13*, 11063. [CrossRef] [PubMed]
58. Pajares, J.; Ibeas, F.; Diez, J.; Gallego, D. Attractive responses by *Monochamus galloprovincialis* (Col., Cerambycidae) to host and bark beetle semiochemicals. *J. Appl. Entomol.* **2004**, *128*, 633–638. [CrossRef]
59. Fan, J.T.; Zheng, K.W.; Dong, Y.F.; Gu, Y.T.; Wickham, J.D. Electrophysiological and Behavioral Responses of *Batocera horsfieldi* Hope to Volatiles from Pistacia chinensis Bunge. *Insects* **2023**, *14*, 911. [CrossRef]

plants

MDPI

Article

First Clarification of the Mechanism of Action of the Apple Glycosyltransferase MdUGT91AJ2 Involved in the Detoxification Metabolism of the Triketone Herbicide Sulcotrione

Aijuan Zhao [1,†], Xiao Teng [2,†], Yingxin Ma [1], Lijun Mu [1], Shibo Han [1], Shumin Wang [1], Kang Lei [1], Lusha Ji [1,*] and Pan Li [1,*]

[1] State Key Laboratory for Macromolecule Drugs and Large-Scale Manufacturing, School of Pharmaceutical Sciences, Liaocheng University, Liaocheng 252059, China; zaj15105403245@163.com (A.Z.); m13021553065@163.com (Y.M.); mlj15215354741@163.com (L.M.); hshibo1999@163.com (S.H.); 13686366329@163.com (S.W.); leikang@lcu.edu.cn (K.L.)

[2] Rizhao Academy of Agricultural Science, Rizhao 276500, China; teng-xiao-rzaas@hotmail.com

* Correspondence: jilu2060@163.com (L.J.); lipan202301@163.com (P.L.)

† These authors contributed equally to this work.

Abstract: Sulcotrione is a member of triketone herbicides, a class of HPPD (4-hydroxyphenylpyruvate dioxygenase) inhibitors with broad-spectrum herbicidal activity. Modifications of glycosylation mediated by glycosyltransferases (GT) are involved in plant detoxification. In this study, we analyzed chip data published online and found that eight glycosyltransferases from group A of the apple glycosyltransferase family 1 may be involved in the metabolic mechanism of detoxification of triketone herbicides. To verify this prediction, we induced apple seedlings with six types of triketone herbicides, and then detected the expression levels of eight glycosyltransferase genes through real-time PCR. We found that triketone herbicides induced up-regulation of eight glycosyltransferase genes to varying degrees, with *MdUGT91AJ2* being the most significantly up-regulated by sulcotrione-induced glycosyltransferase gene expression. Then, through in vitro enzymatic reactions and HPLC identification of glycoside substrates, it was found that the glycosyltransferase MdUGT91AJ2 had the highest specific enzyme activity against the triketone herbicide sulcotrione. Furthermore, the in vivo mechanism of the glycosyltransferase MdUGT91AJ2 in the detoxification metabolism of sulcotrione was further validated by overexpressing the strain in the plant. HPLC analysis showed that the content of sulcotrione glycosides in the overexpressing strain of *MdUGT91AJ2* was significantly higher than that in the wild type. This result indicated that the apple glycosyltransferase MdUGT91AJ2 can still glycosylate and modify sulfotrione in plants, and participate in its detoxification metabolism. In summary, this study identified for the first time a novel apple glycosyltransferase MdUGT91AJ2 and elucidated its mechanism of action in the detoxification and metabolism of the triketone herbicide sulfotriene.

Keywords: glycosyltransferase; MdUGT91AJ2; triketone herbicides; sulcotrione; detoxification metabolism

1. Introduction

Triketone herbicides are another class of HPPD (p-hydroxyphenylpyruvate dioxygenase) inhibitors developed by Jellicam after pyrazole herbicides. Their action is characterized by broad-spectrum herbicidal activity and can be used both pre-emergence and postemergence. Currently, several varieties have been successfully developed, including sulcotrione, furazosulfuron, bicyclosulfuron, nitrosulfuron, cycloheximide, and flupyradifuron [1]. The target of the action of the triketone herbicides is the HPPD enzyme, which exists in a wide range of organisms, and is the key enzyme in the synthesis of plastoquinone and α-tocopherol in the plant (see Figure 1). HPPD inhibitors inhibit the synthesis of HPPA (4-hydroxyphenylpyruvic acid) into HGA (homogentisate), which leads to the blockage of the synthesis of plastoquinone, and then impedes the catalysis of the phytoene desaturase (PDS), affecting the biosynthesis of carotenoids. The decrease in carotenoid synthesis leads to a lack of protective pigments in chlorophyll, resulting in weeds yellowing or death due to bleaching symptoms [2,3]. Sulcotrione is a foliar herbicide that can also be absorbed through the root system causing the weeds to discolor quickly and die slowly after application. However, the mode of action is still not fully understood and it is likely that chlorophyll synthesis is directly affected, acting finally on carotenoid synthesis [4]. Pyrazosulfuron-methyl inhibits the synthesis of pigments essential for plant growth, which is absorbed and rapidly transmitted through roots, stems, young shoots, and leaves, ultimately affecting the biosynthesis of carotenoids, and the weed is subjected to whitening of the foliage followed by necrosis of the meristematic tissue [5]. Dicyclosulfocarbazone, which has a bicyclic and phenylthioether structure, is absorbed by weed roots and stems, and then transported to the whole plant, where it affects the synthesis of plastoquinone through inhibition of the HPPD, and then the action of plastoquinone on octahydroxy phytoene desaturase (PDS), which ultimately affects the biosynthesis of carotenoids and bleaching of the foliage [6]. Cyclosulfamuron is effective at controlling major broadleaf grasses and some grassy weeds and is readily conducted in the xylem and phloem of plants; it is both touched and persistent [7]. Cyclosulfamuron is mainly used in maize (*Zea mays* L.) fields and blocks isoprenoid quinone biosynthesis in the plant, causing green loss, discoloration, tissue necrosis, and eventual death within two weeks [8]. Flumioxazin can cause symptoms of bleaching and eventual death of plant meristematic tissues by inhibiting carotenoid biosynthesis [9].

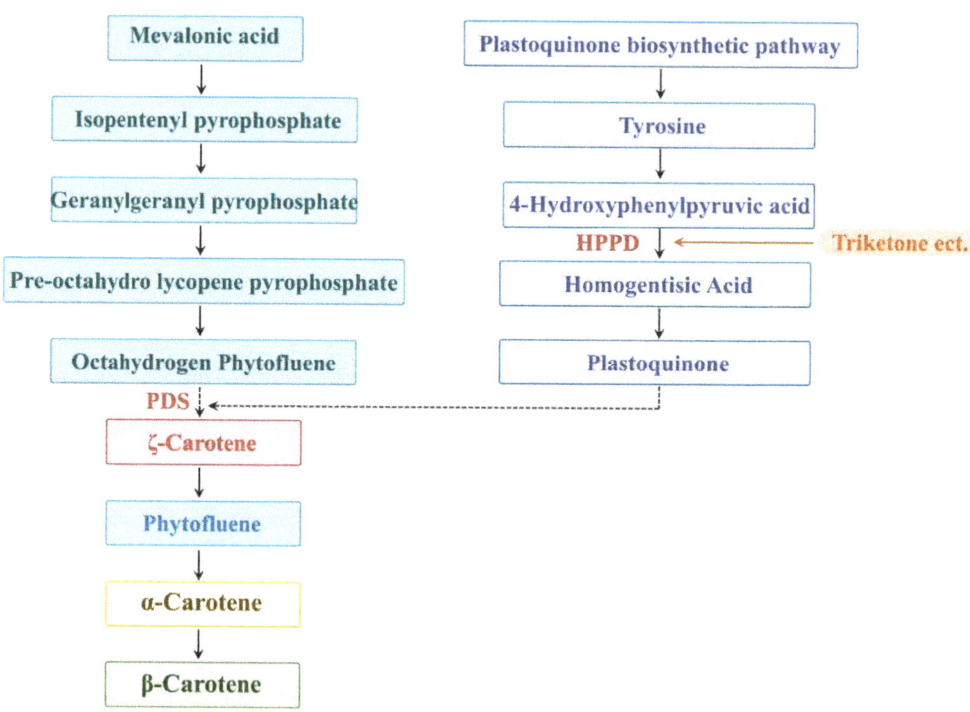

Figure 1. Mechanism of herbicidal action of triketones.

Herbicides of the triketone group are widely used in agricultural fields such as maize (*Zea mays* L.) to control grass weeds and broadleaf weeds [10]. The herbicide is weakly acidic, its aqueous solution is highly stable, not easy to photolyze and volatilize, and it is widely used in large quantities, which can be harmful to the ecological environment of the soil when it is used inappropriately or repeated in large quantities over a long period of time [11]. Despite the low toxicity of sulcotrione itself, it has been shown that the metabolites of sulcotrione in the environment are higher than the parent itself, and studies on onion root meristematic tissues under light conditions found that the cytotoxicity and genotoxicity of sulcotrione photolysis products were significantly higher than those of the sulcotrione parent [12]. In addition to the product CMBA (2-chloro-4-meta phenylbenzoic acid), sulfoxaflor photolysis also generates a class of cyclisation product CP (cyclization product) [13]. Halle et al., compared the toxicity of sulcotrione, CMBA, and CP to microbial bacteria (*Vibrio fischeri*), algae (*Pseudokirchneriella subcapitata*), and protists (*Tetrahymena pyriformis*), respectively. The results indicated that sulcotrione produces higher toxicity to algae and greater CP toxicity to bacteria (*Vibrio fischeri*) and protists (*Tetrahymena pyriformis*) [14]. The toxicity of the photolytic mixture of sulcotrione is higher than that of the parent and CMBA, therefore, the role of its cyclized metabolites in the environmental behavior and risk assessment of sulcotrione deserves continued attention [15]. Pesticides enter the soil system and their degradation involves a complex series of chemical processes that can be biotransformed or abiotransformed [16]. Bioremediation, also known as biomanagement, refers to the use of the life metabolic activities of organisms to absorb, degrade, and transform pollutants in the environment so that the polluted environment can be improved [17]. It can be divided into microbial remediation, phytoremediation, and animal remediation according to the type of organisms [18]. Phytoremediation mainly involves remediation of pollutant residues in the environment through plant uptake, plant extraction, plant volatilization, and root secretions in association with inter-root microorganisms [19]. Qu et al. have shown that in order to cope with the harmful effects of exogenous toxin herbicides, the plant body has developed a series of detoxification mechanisms to convert toxic exogenous toxins into nontoxic substances during the long-term evolutionary process, which in turn reduces the damage to plant cells [20]. The detoxification mechanism consists of two phases: the first phase is mainly an activation reaction, i.e., hydrolysis or oxidation [21]; and the second phase is mainly a conjugation reaction, which usually involves linking the metabolite or heterologue activated in the first phase directly to an endogenous hydrophilic molecule, e.g., glucose, malonate, or glutathione, by means of covalent bonding [22].

Plant UDP-glycosyltransferases are an important class of cytochrome P450 enzyme coenzymes involved in a variety of biological processes such as plant growth and development, secondary metabolism, and environmental adaptation. Numerous studies have shown that glycosyltransferases can participate in the detoxification reactions of organisms by transferring enzyme glucose or other sugar groups [23–25]. UGTs play an important role by facilitating the binding of harmful substances and limiting the effects of harmful substances on normal physiological and metabolic processes [23]. Blomstedt et al. succeeded in identifying a sorghum (*Sorghum bicolor* (L.) *Moench*) glycosyltransferase gene, *UGT85B1*, and found that the mutation in this gene resulted in reduced plant vigor, dwarf plants, poor roots development, low fertility, and no production of the toxic substance dhurrin. Further investigation of the mechanism revealed that the reason why this mutant did not show toxicity was sorghum did not have a feedback mechanism to inhibit the initial steps of dhurrin biosynthesis when lacking the glucosyltransferase activity required to complete dhurrin synthesis [24]. Su et al. proposed a TCP (trichlorophenol) phytoremediation system based on sugar conjugation by overexpressing poplar (*Populus euphratica Oliv*) UDP-glucosyltransferase. It was found that among the many glycosyltransferases, PtUGT72B1 had the highest binding activity to TCP, and its overexpression of *Arabidopsis thaliana* lines showed significantly enhanced tolerance to 2,4,5-TCP and 2,4,6-TCP [25]. Although the involvement of plant glycosyltransferases in the detoxification process has been reported in some cases, the involvement of apple (*Malus* × *domestica*) glycosyltransferases in the metabolic mechanism of detoxification of triketone herbicides has hardly been studied. Therefore, in this study, a new apple (*Malus* × *domestica*) glycosyltransferase,

MdUGT91AJ2, was identified for the first time, and its mechanism of action in the detoxification metabolism of the triketone herbicide sulcotrione was clarified, which will provide scientific data for the reduction of the potential risks of pesticides to ecosystems, crops, animals, and humans.

2. Results

2.1. Induction of mRNA Expression Level of Apple Glycosyltransferase Gene by Triketone Herbicides

We analyzed the microarray data published online (https://www.ncbi.nlm.nih.gov/, accessed on 10 December 2023) and found that eight glycosyltransferases in group A of apple (*Malus* × *domestica*) glycosyltransferase family 1, *MdUGT91AJ1* (*MD02G1083600*), *MdUGT91AJ2* (*MD02G1084000*), *MdUGT91AJ3* (*MD15G1211400*), *MdUGT91AJ4* (*MD10G1101200*), *MdUGT91AJ5* (*MD02G1083100*), *MdUGT91AJ6* (*MD02G1083300*), *MdUGT91AJ7* (*MD02G1083800*), and *MdUGT91C7* (*MD08G1200700*) may be involved in the mechanism of triketone herbicide detoxification metabolism. To verify the prediction, we exogenously sprayed apple (*Malus* × *domestica*) seedlings with six triketone herbicides, sulcotrione, tefuryltrione, benzobicyclon, mesotrione, tembotrione, and bicyclopyrone, extracted the RNAs, and reverse-transcribed them into cDNAs, and then detected the expression levels of eight glycosyltransferase genes via real-time PCR; the results are shown in Figure 2. Triketone herbicides induced the up-regulation of the expression of eight glycosyltransferase genes to varying degrees (Figure 2A, Table S4), with sulcotrione inducing the most significant up-regulation of the glycosyltransferase gene, *MdUGT91AJ2* (* $p < 0.05$, ** $p < 0.01$) (Figure 2B).

(A)

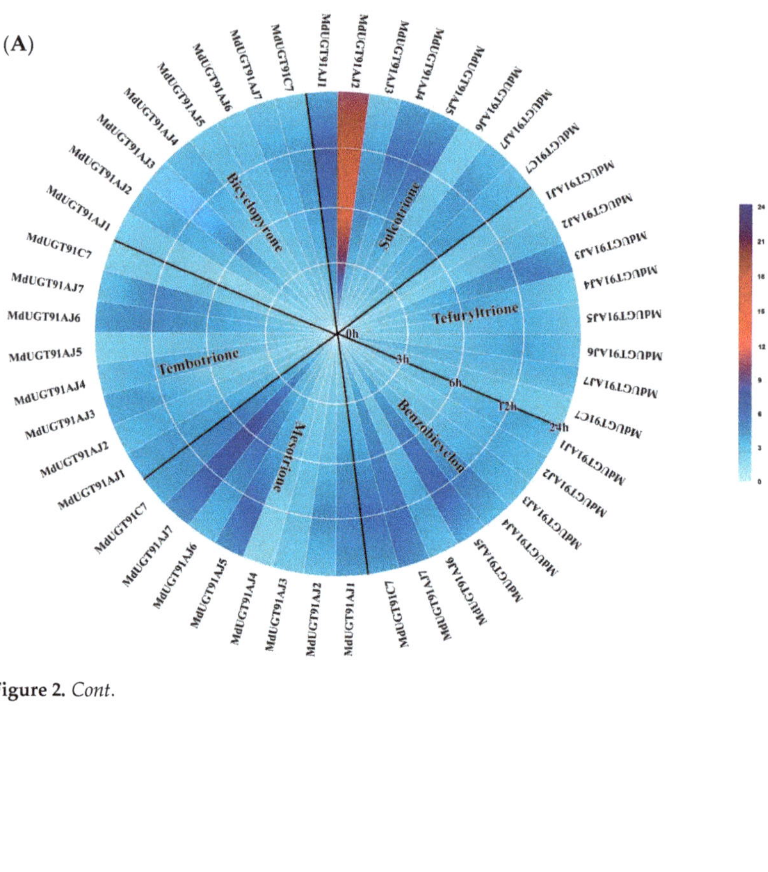

Figure 2. *Cont.*

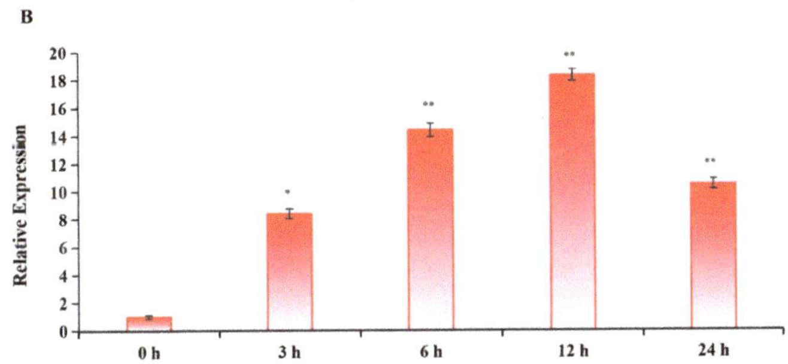

Figure 2. Real-time PCR detection of six triketone herbicides, as sulcotrione, tefuryltrione, benzobicyclon, mesotrione, tembotrione, and bicyclopyrone after induction for 0, 3, 6, 12 and 24 h. In apple (*Malus* × *domestica*) glycosyltransferase family 1, 8 glycosyltransferase genes in group A of *MdUGT91AJ1 (MD02G1083600)*, *MdUGT91AJ2 (MD02G1084000)*, *MdUGT91AJ3 (MD15G1211400)*, *MdUGT91AJ4 (MD10G1101200)*, *MdUGT91AJ5 (MD02G1083100)*, *MdUGT91AJ6 (MD02G1083300)*, *MdUGT91AJ7 (MD02G1083800)*, *MdUGT91C7 (MD08G1200700)* mRNA expression levels. (**A**) Real-time PCR to detect the mRNA expression levels of eight glycosyltransferase genes in group A of apple (*Malus* × *domestica*) glycosyltransferase family 1 induced by each of the six triketone herbicides; (**B**) real-time PCR to detect the mRNA expression levels of the apple (*Malus* × *domestica*) glycosyltransferase gene, *MdUGT91AJ2*, induced by the triketone herbicide sulcotrione. Note: * $p < 0.05$, ** $p < 0.01$, $n = 3$.

2.2. In Vitro Enzyme Activity Assay of Apple Glycosyltransferase Glycosylated Modified Triketone Herbicides

In the above study, we found that triketone herbicides induced different degrees of mRNA levels of eight glycosyltransferase genes in group A of apple (*Malus* × *domestica*) glycosyltransferase family 1, which are enzymes in plants that can transfer sugar groups to different small molecule compounds. For this reason, we obtained eight glycosyltransferase proteins successively through prokaryotic expression vector construction and in vitro enzyme-protein purification system, performed in vitro enzymatic reactions against six triketone herbicides, and identified glycosidic substrates by HPLC, and found that glycosyltransferases were selective and preferential to triketone herbicides (Table 1). The glycosyltransferase MdUGT91AJ2 had the highest specific enzyme activity for the triketone herbicide sulcotrione compared to other glycosylation modifications, i.e., MdUGT91AJ2 was the most active for sulcotrione glycosylation modification (Figure 3A,B).

Table 1. Specific enzyme activity determination of eight apple glycosyltransferases against triketones herbicides.

Name	Substrates	Specific Activity (nkat/mg Protein)
MdUGT91AJ1	Sulcotrione	1.24 ± 0.19
	Tefuryltrione	ND
	Benzobicyclon	ND
	Mesotrione	ND
	Tembotrione	ND
	Bicyclopyrone	ND

Table 1. *Cont.*

Name	Substrates	Specific Activity (nkat/mg Protein)
MdUGT91AJ2	Sulcotrione	3.56 ± 0.25
	Tefuryltrione	ND
	Benzobicyclon	ND
	Mesotrione	ND
	Tembotrione	ND
	Bicyclopyrone	ND
MdUGT91AJ3	Sulcotrione	ND
	Tefuryltrione	0.36 ± 0.11
	Benzobicyclon	ND
	Mesotrione	ND
	Tembotrione	ND
	Bicyclopyrone	ND
MdUGT91AJ4	Sulcotrione	0.76 ± 0.20
	Tefuryltrione	ND
	Benzobicyclon	ND
	Mesotrione	ND
	Tembotrione	ND
	Bicyclopyrone	ND
MdUGT91AJ5	Sulcotrione	ND
	Tefuryltrione	ND
	Benzobicyclon	ND
	Mesotrione	ND
	Tembotrione	ND
	Bicyclopyrone	ND
MdUGT91AJ6	Sulcotrione	ND
	Tefuryltrione	ND
	Benzobicyclon	ND
	Mesotrione	0.31 ± 0.04
	Tembotrione	ND
	Bicyclopyrone	ND
MdUGT91AJ7	Sulcotrione	ND
	Tefuryltrione	ND
	Benzobicyclon	ND
	Mesotrione	0.45 ± 0.08
	Tembotrione	ND
	Bicyclopyrone	ND

Table 1. *Cont.*

Name	Substrates	Specific Activity (nkat/mg Protein)
MdUGT91C7	Sulcotrione	ND
	Tefuryltrione	ND
	Benzobicyclon	ND
	Mesotrione	0.39 ± 0.07
	Tembotrione	ND
	Bicyclopyrone	ND

Note: ND, none detected.

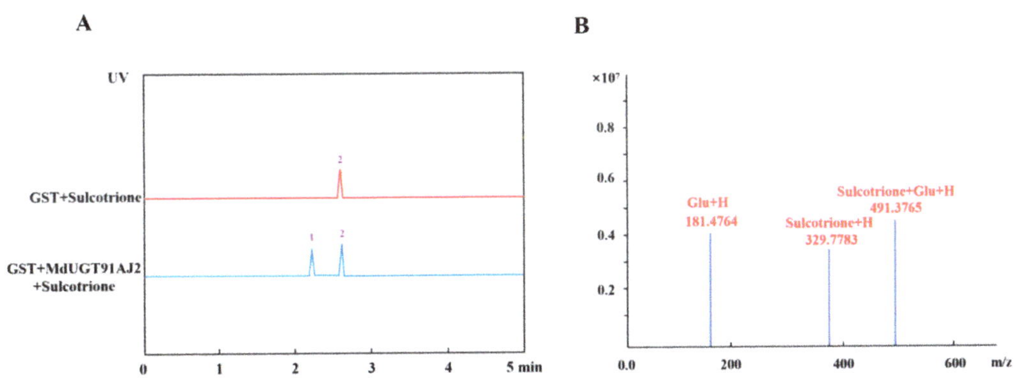

Figure 3. HPLC detection of MdUGT91AJ2 glycosylation modification substrate sulcotrione. (**A**) HPLC detection of MdUGT91AJ2 glycosylation modification product sulcotrione glucoside; (**B**) mass spectrometry detection of MdUGT91AJ2 glycosylation modification product sulcotrione glucoside molecular weight. Note: 1: sulcotrione; 2: sulcotrione glucoside.

2.3. Analysis of the Expression Pattern of MdUGT91AJ2 in Different Tissue Sites

In order to further understand the tissue expression pattern of *MdUGT91AJ2*, we detected the gene expression level of *MdUGT91AJ2* in different parts of apple (*Malus* × *domestica*) via real-time PCR and the results are shown in Figure 4. The expression level of *MdUGT91AJ2* varied considerably in different parts of the apple (*Malus* × *domestica*), with the highest expression level in the root system, followed by that in mature leaves, while in mature seeds the expression level was the lowest. Combined with the above results that MdUGT91AJ2 was able to glycosylate and modify sulcotrione herbicide, we speculated that MdUGT91AJ2 may play a detoxification metabolism role in the root system and mature leaves.

2.4. Glycosylation of Apple Glycosyltransferase MdUG91AJ2 In Vivo in Plant to Modify Sulforaphane Enzyme Activity Assays

In the above study, we demonstrated that MdUGT91AJ2 can glycosylate sulcotrione in vitro and its activity was much higher than that of the other seven glycosyltransferases; MdUGT91AJ2 was also able to glycosylate sulcotrione in plants. To this end, we overexpressed *MdUGT91AJ2* into apples (*Malus* × *domestica*) and obtained transgenic overexpression lines *MdUGT91AJ2-OE11* and *MdUGT91AJ2-OE23* after screening (Figure 5A). After exogenous spraying of sulcotrione, HPLC determination of sulcotrione glycoside content in each strain. The result revealed a significant increase in sulcotrione glycoside content in the overexpression lines (Figure 5B). This study suggested that the apple (*Malus* × *domestica*) glycosyltransferase MdUG91AJ2 can still glycosylate modified sulcotrione in plants to participate in detoxification metabolism.

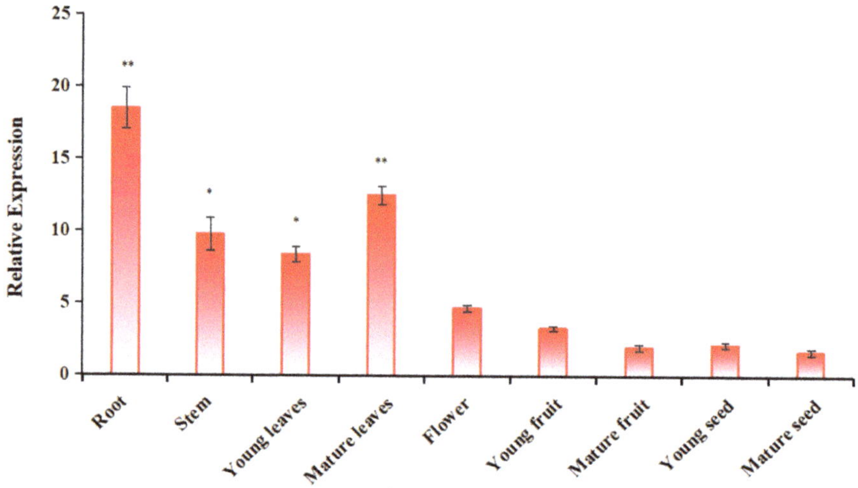

Figure 4. Real-time PCR to detect the expression of different parts of apple glycosyltransferase gene MdUG91AJ2. Note: compared with the mature seed group, * $p < 0.05$, ** $p < 0.01$, $n = 3$.

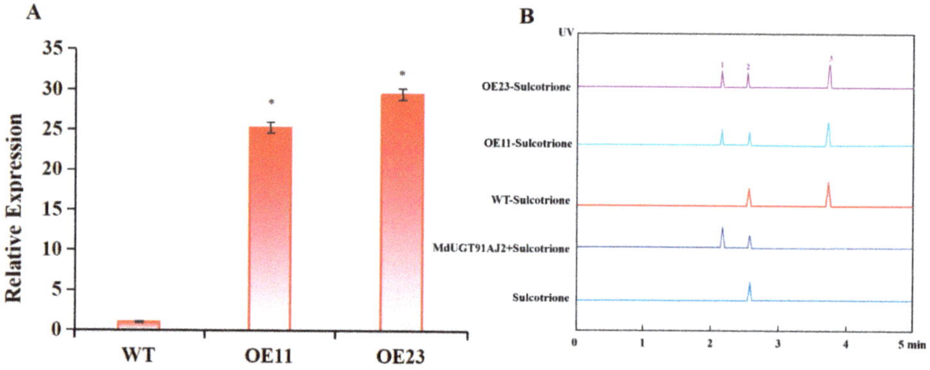

Figure 5. *MdUGT91AJ2* overexpression lines obtained and sulcotrione glycoside content assayed. (**A**) Detection of *MdUGT91AJ2* gene expression level in overexpression lines *OE11* and *OE23*; (**B**) determination of sulcotrione and sulcotrione glycoside content in *MdUGT91AJ2* overexpression lines *OE11* and *OE23*, 1: sulcotrione glycoside, 2: sulcotrione, 3: internal reference. Note: * $p < 0.05$, $n = 3$.

3. Discussion

With the gradual development of agriculture, herbicides have become one of the most common weed control tools in the world. Triketones, derived chemically from a natural phytotoxin (leptospermone), are a good example of allelochemicals as lead molecules for the development of new herbicides. Triketone herbicides are the latest generation of herbicides developed to inhibit a novel key enzyme involved in carotenoid biosynthesis, HPPD (4-hydroxyphenylpyruvate dioxygenase herbicide) [26]. HPPD belongs to α-ketate dependent, nonsemi-containing, and Fe^{2+} dependent enzymes that are important enzymes in tyrosine degradation. They can catalyze the formation of the aromatic precursor HGA (homogentisic acid) of plasmic quinones and vitamin E. The reaction involves decarboxylation, substituent migration, and aromatic oxidation in a single catalytic cycle [27]. Its inhibition leads to the consumption of plant plastid quinones and vitamin E, leading to bleaching syndrome, which is crucial for the discovery of new bleaching herbicides [28].

The decrease in carotenoid levels in weeds leads to rapid degradation of chlorophyll and visible bleaching of sensitive plants, which is due to uncontrolled free radical production and chlorophyll oxidation altering cell structure [29]. Corn growers often use β-triketone herbicides that are used to prevent the growth of weeds in the field. This has brought a lot of convenience to farmers. However, the application of herbicides can pollute the atmosphere and be intercepted and absorbed by weeds and crops. Plants containing pesticide residues may reach the soil during the crop cycle or after harvest, causing pollution of soil, surface water, and groundwater, posing a further threat to the ecological environment [30].

The pollution of toxic substances in the environment has become a concern for agricultural countries. Photolysis is an important way for pesticides to dissipate on crops, so people achieve ideal results by increasing the use of pesticides [31]. Sulcotrione is a trione herbicide used to control dicotyledonous weeds in maize cultivation. These herbicides are very potent for the selective pre- and in some cases postemergence control of a wide range of broadleaf and grass weeds in maize (*Zea mays* L.) and rice (*Oryza sativa*). It can rapidly photolyze on plant leaves and produce two main light products: oxoanthane-1,9-dione-3,4-dihydro-6-methylsulfonyl (XDD) and 2-chloro-4-meta phenylbenzoic acid (CMBA) [32]. However, an increase in the concentration of XDD and CMBA can significantly inhibit the mitosis of plant cells, leading to a continuous increase in cells with chromosomal abnormalities [13]. At the same time, it can also damage the normal growth of crops, so it is necessary to study an antidote to herbicides.

Previous studies have found that glycosylation modifications led by glycosyltransferases in plants are involved in detoxification [33–35]. Huang et al. found that the glycosyltransferase UGT91C1 can significantly reduce the toxicity of the triketone herbicide sulcotrione, enhance the resistance of arabidopsis to sulcotrione, and thus reduce its own damage and chlorophyll destruction [33]. Li et al. found that the glycosyltransferase Group I family has a significant detoxification effect on (acetyl CoA carboxylase) inhibitor herbicides, with the glycosyltransferase MdUGT83K2 showing the most significant detoxification effect on the ACCase herbicide QPP-7 [34]. Zhang et al. found that the glycosyltransferase Group D family has a significant detoxification effect on (ALS) inhibitor herbicides. Among them, the glycosyltransferase MdUGT73CG22 has the strongest specificity and most significant detoxification effect on the ALS herbicide nicosulfuron [35]. In this study, we demonstrated that the glycosyltransferase MdUGT91AJ2 in apples (*Malus* × *domestica*) can glycosylate the herbicide sulcotrione in vitro and in vivo, and this enzyme had substrate specificity for carotene. After using the overexpressed strains, we also found that the glycosyltransferase MdUGT91AJ2 enhanced plant resistance to such herbicides. In summary, this work not only revealed the role of a new specific glycosyltransferase in apples (*Malus* × *domestica*) in detoxifying the triketone herbicide sulcotrione, but also opened a new door for addressing the use of herbicides in apple (*Malus* × *domestica*) cultivation and development.

4. Materials and Methods

4.1. Plant Material

Roots, stems, leaves, flowers, fruits, and seeds of a 10-year-old "*Gala*" apple (*Malus* × *domestica*) tree (*Malus* × *domestica Borkh.*) planted in the School of Life Sciences of Shandong Agricultural University were selected to investigate the expression characteristics of MdUGT91AJ2 in different tissues of apple (*Malus* × *domestica*). After sampling, they were quickly frozen in liquid nitrogen and stored for use at −80 °C. The apple (*Malus* × *domestica*) seedling material used in this study was "*Gala*" apple (*Malus* × *domestica*) seedlings, which were bred and preserved in our laboratory. The apple (*Malus* × *domestica*) seedlings were cultured in MS (Murashige and Skoog) medium and the overexpression of MdUGT91AJ2 transgenic seedlings were cultured in MS medium containing 25 mg/L Kanamycin. After 2 weeks, the seedlings were transplanted to a mixed (substrate: vermiculite: perlite 4:1:1) nutrient substrate in an incubator at 24 °C with 16 h of light/8 h of darkness. The planting conditions of plant materials referred to reference [36], but there were some changes. After three weeks, the seedlings were used for experiments.

4.2. Exogenous Treatment of Apple Seedlings with Triketone Herbicides

Apple (*Malus* × *domestica*) seedlings, which had been normally cultured for 5 weeks, were randomly divided into six groups and sprayed with 10 μmol/L sulcotrione, tefuryltrione, benzobicyclon, mesotrione, tembotrione, and bicyclopyrone (Table 2) for different times of treatment (0, 3, 6, 12, 24 h) and immediately liquid nitrogen snap-frozen and stored at −80 °C for subsequent RNA extraction.

Table 2. Specific information on triketone herbicides.

Chemical Compound	Structure Name	Structural Formula	Manufacturer	CAS
Sulcotrione	2-(2-Chloro-4-(methylsulfonyl)benzoyl)-1,3-cyclohexanedione	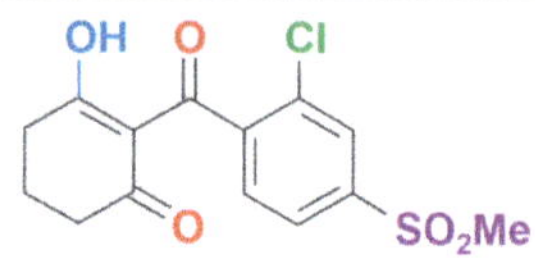	JLK Pharmaceutical Technology Co., Ltd (Beijing, China)	99105-77-8
Tefuryltrione	2-{2-chloro-4-(methylsulfonyl)-3-[(tetrahydrofuran-2-ylmethoxy)methyl]benzoyl}cyclohexane-1,3-dione	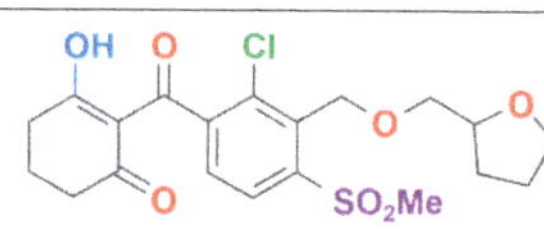	Bayer CropScience Co., Ltd (Hangzhou, China)	473278-76-1
Benzobicyclon	3-[2-chloro-4-(methylsulfonyl)benzoyl]-4-(phenylthio)bicyclo[3.2.1]oct-3-en-2-one	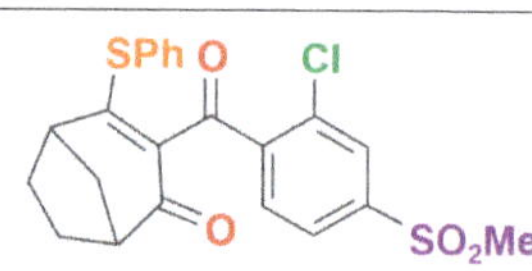	One Plus One Biotechnology Co., Ltd (Wuhan, China)	156963-66-5
Mesotrione	2-(4-Methylsulfonyl-2-nitrobenzoyl)-1,3-cyclohexanedione	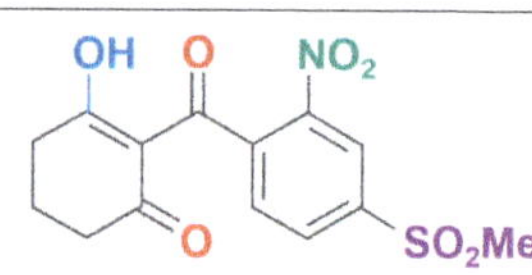	Xianda Agrochemical Co., Ltd (Binzhou, China)	104206-82-8
Tembotrione	2-{2-chloro-4-mesyl-3-[(2,2,2-trifluoroethoxy)methyl]benzoyl}cyclohexane-1,3-dione	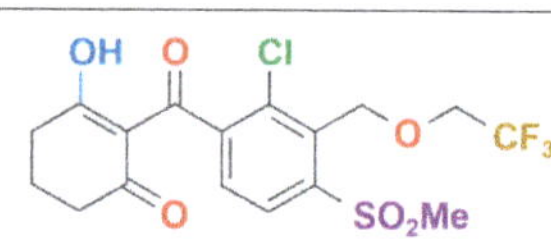	Bayer CropScience Co., Ltd (Hangzhou, China)	335104-84-2
Bicyclopyrone	4-Hydroxy-3-{2-[(2-methoxyethoxy)methyl]-6-(trifluoromethyl)-3-pyridinylcarbonyl}bicyclo[3.2.1]oct-3-en-2-one	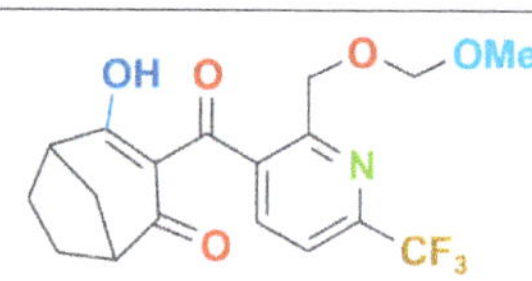	Huisheng Agricultural Technology Co., Ltd (Beijing, China)	352010-68-5

4.3. Real-Time PCR

RNA extraction: plant material was subjected to RNA extraction using an RNA extraction kit (DP437, Tiangen Biochemical Science and Technology Co. Ltd, Beijing, China).

cDNA reverse transcription: cDNA first strand synthesis was performed using PrimeScript^{TM} RT reagent Kit with a gDNA Eraser kit (DRR047A, Bao Bioengineering Co., Ltd, Dalian, China).

Real-time PCR: The UltraSYBR Mixture kit (CW0659S, Kangwei Century Technology Co., Ltd., Beijing, China) was used, and the fluorescence quantitative PCR instrument model was the BIO-RAD IQ5. The primer sequences are shown in Table S1. A total of 20 μL of the PCR reaction system was composed of: 2× Ultra SYBR Mixture 10.0 μL, Forward primer (10 μmol·L^{-1}) 1.0 μL, Reverse primer (10 μmol·L^{-1}) 1.0 μL, cDNA 1.0 μL, and ddH$_2$O 7.0 μL. The reaction procedure: predenaturation at 94 °C for 10 min; denaturation at 94 °C for 15 s, annealing at 56 °C for 15 s, 40 cycles, and then the data were analyzed via $2^{-\Delta\Delta}$CT algorithm.

4.4. Expression Vector Construction

Download the coding sequences of the eight glycosyltransferase genes(*MdUGT91AJ1 (MD02G1083600)*; *MdUGT91AJ2 (MD02G1084000)*; *MdUGT91AJ3 (MD15G1211400)*; *MdUGT91AJ4 (MD10G1101200)*; *MdUGT91AJ5 (MD02G1083100)*; *MdUGT91AJ6 (MD02G1083300)*; *MdUGT91AJ7 (MD02G1083800)*, *MdUGT91C7 (MD08G1200700))* from the apple genome database GDDH13 v1.1 (Table S2). Full-length primers for MdUGT91AJ2 gene were designed (Table S1). The coding region fragments of MdUGT91AJ2 gene were cloned using 'Gala' cDNA as a template. PCR amplification was performed using the high-fidelity 2×PCR Solution PrimeSTAR^{TM} HS Premix (DR040A, Bao Bioengineering Co., Ltd., Dalian, China). The PCR products were electrophoresed, recovered, and ligated into T vector (pEASYBlunt Zero, Beijing Total Gold), and the positive clones were selected after PCR verification of the bacterial liquid and sent to Shanghai Sangong Biotechnology Co. Ltd. for sequencing. One hundred percent of the sequencing was correctly performed and the products of the digestion were ligated into the same endonuclease-digested pGEX-2T [37] and PBI121 vectors [38], respectively. After verification via bacteriological PCR and double enzymatic digestion, the positive recombinant plasmids MdUGT91AJ1-pGEX-2T, MdUGT91AJ2-pGEX-2T, MdUGT91AJ3-pGEX-2T, MdUGT91AJ4-pGEX-2T, MdUGT91AJ5-pGEX-2T, MdUGT91AJ6-pGEX-2T, MdUGT91AJ7-pGEX-2T, MdUGT91C7-pGEX-2T, and MdUGT91AJ2-pBI121 were transformed into E. coli BL-21 and R. rhizogenes strain K599 for subsequent protein expression and transgenic strain screening.

4.5. Glycosyltransferase Protein Purification and In Vitro Enzyme Activation Reaction

The *E. coli BL21* strain containing positive recombinant plasmid was inoculated into 2 mL LB medium containing Amp$^+$ antibiotic, and incubated at 37 °C with shaking overnight. It was then diluted into 100 mL triangular flasks according to the ratio of 1:20, and shaking cultivation continued for about 2 h. When the OD_{600} value was about 0.6, the expression was induced by adding the final concentration of 1 mmol·L^{-1} IPTG, and the induction temperature was 20 °C; the induction incubation time was 16 h. Centrifugation was performed at 4200 rpm·min^{-1} for 10 min, 10 mL of the bacterial bodies was collected, and 1 mL of 1×PBS buffer was used for the incubation. A total of 1 mL of 1×PBS buffer was used to resuspend the organisms, and the organisms were treated via repeated freeze-thawing method, centrifuged at 12,000 rpm·min^{-1} for 15 min, and then all the supernatant was taken. The recombinant proteins were purified using a His-tag Protein Purification Kit Tagged Protein Purification Kit (BYX8228C, Beiyuanxin Biotechnology Co., Ltd., Changzhou, China), according to the instructions. The amino acid sequences of the eight glycosyltransferases in this study are shown in Table S3.

In vitro enzymatic reaction: 100 μL in vitro enzymatic reaction system containing 10 μL 0.5 M Tris-HCl (pH 8.0), 5 μL 50 mM $MgSO_4$, 5 μL 200 mM KCl, 2.5 μL 0.1 M UDP-glucose, 1 μL 10% β-mercaptoethanol, 1 μL 100 mM sulcotrione, 2 μL MdUGT91AJ2 enzyme protein, and 73.5 μL ddH_2O [34].

4.6. Determination of Triketone Herbicide Glycosides via HPLC

Glycoside extraction: 5-week-old apple (*Malus × domestica*) seedlings treated with 10 μM sulcotrione, tefuryltrione, benzobicyclon, mesotrione, tembotrione, and bicyclopyrone for 12 h were milled to powder with liquid nitrogen. They were transferred to 10 mL of 80% methanol in a triangular vial and 2 μL of endosulfan (Piclorom) was added. The samples were placed in a shaker for 6 h at 60 r·min^{-1} and shaken slowly. Then, they were transferred to a rotary evaporator for vacuum evaporation and the residual powder on the inner wall of the centrifuge tube was dissolved with 1 mL of methanol solution and centrifuged at 15,000 rpm for 20 min; the glycosides were detected via HPLC [39].

HPLC conditions: ACQUITY UPLC BEH Amide column (100 mm × 2.1 mm × 1.7 μm), column temperature 40 °C, temperature in the sample chamber 10 °C, injection volume 5.0 μL, mobile phase A was 0.1% ammonia-acetonitrile, and mobile phase B was 1 mmol/L ammonium acetate (containing 0.1% ammonia) at a flow rate of 0.3 mL·min^{-1}. The gradient elution procedure was 0~1.0 min, 100% A; 1.0~2.5 min, 100% ~60% A; 2.0~4.0 min, 60% ~100% A; and 4.0~5.0 min, 100% A. The elution was carried out at a flow rate of 0.3 mL·min^{-1}.

Mass spectrometry conditions: ESI source, positive ion mode, multiple reaction detection scanning (MRM), capillary voltage 3.5 kV, cone-well voltage 35 V, ion source temperature 100 °C, desolvent gas temperature 300 °C, cone-well backblast nitrogen flow rate of 50 L·h^{-1}, desolvent gas flow rate of 600 L·h^{-1}, photomultiplier voltage of 650 V, and collision chamber vacuum 0.076 Pa. The conditions were the same as those of HPLC. The elution method was the same, and the surface-induced decomposition intensity was 30 eV.

4.7. Transgenic Apple Seedling Screening

The apple genetic transformation experiment used "*Gala*" apple (*Malus × domestica*) seedlings as materials, and obtained the line "GL-3" group-cultured seedlings with strong regeneration ability and high transformation efficiency. The leaves of "GL-3" were transformed with Agrobacterium tumefaciens carrying the sequence of MdUGT91AJ2, and the transgenic apple (*Malus × domestica*) lines *OE11* and *OE23* were obtained. The specific operation was referred to in reference [40,41], which included the steps of preparation of bacterial solution, infiltration, bacterial washing, screening, and identification. After obtaining the positive strains, they were propagated, rooted, and transplanted to soil, and then grown to a certain stage for subsequent tests.

4.8. Data Statistics and Analyses

The experimental data are expressed as "mean ± standard error." The data were processed and statistically analyzed using Excel 2019 and SPSS 21.0, and graphs were produced; the difference was tested for significance using the Least Significant Difference (LSD) method (* $p < 0.05$, **$p < 0.01$).

Supplementary Materials: The following supporting information can be downloaded at: https://www.mdpi.com/article/10.3390/plants13131796/s1, Table S1: primer sequences; Table S2: coding sequences of 8 glycosyltransferase genes; Table S3: amino acid sequences of 8 glycosyltransferase genes; Table S4: expression of eight glycosyltransferase genes induced by triketone herbicides.

Author Contributions: All authors contributed to the study's conception and design. Conceptualization, A.Z. and X.T.; methodology, P.L. and K.L.; software, X.T. and L.M.; validation, Y.M.; formal analysis, S.W. and S.H.; investigation, P.L. and K.L.; resources, K.L. and L.J.; data curation, A.Z. and S.W.; writing—original draft preparation, A.Z.; writing—review and editing, X.T. and S.H.; visualization, Y.M.; supervision, L.M.; project administration, A.Z. and L.J.; funding acquisition, P.L. and L.J. All authors have read and agreed to the published version of the manuscript.

Funding: This research was funded by the Shandong Provincial Natural Science Foundation of China, grant number No. ZR202102180037, Guangyue Young Scholar Innovation Team of Liaocheng University, grant number No. LCUGYTD2022-04.

Data Availability Statement: Data will be made available on request.

Acknowledgments: Thank you very much for the plant materials provided by the College of Life Sciences of Shandong Agricultural University.

Conflicts of Interest: The authors declare no conflicts of interest.

References

1. Wang, H.; Liu, W.; Jin, T.; Peng, X.; Zhang, L.; Wang, J. Bipyrazone: A new HPPD-inhibiting herbicide in wheat. *Sci. Rep.* **2020**, *10*, 5521. [CrossRef] [PubMed]

2. Lu, H.; Yu, Q.; Han, H.; Owen, M.J.; Powles, S.B. Evolution of resistance to HPPD-inhibiting herbicides in a wild radish population via enhanced herbicide metabolism. *Pest. Manag. Sci.* **2020**, *76*, 1929–1937. [CrossRef] [PubMed]
3. Zhang, F.; Qiao, Z.; Yao, C.; Sun, S.; Liu, W.; Wang, J. Effects of the novel HPPD-inhibitor herbicide QYM201 on enzyme activity and microorganisms, and its degradation in soil. *Ecotoxicology* **2021**, *30*, 80–90. [CrossRef] [PubMed]
4. Baćmaga, M.; Wyszkowska, J.; Borowik, A.; Kucharski, J. Bacteria, Fungi, and Enzymes in Soil Treated with Sulcotrione and Terbuthylazine. *Int. J. Mol. Sci.* **2023**, *24*, 14469. [CrossRef] [PubMed]
5. Ma, K.; Zhang, W.; Zhang, L.; He, X.; Fan, Y.; Alam, S.; Yuan, X. Pyrazosulfuron-Methyl on the Photosynthetic Characteristics and Antioxidant Systems of Foxtail Millet. *Front. Plant Sci.* **2021**, *12*, 696169. [CrossRef] [PubMed]
6. Brabham, C.; Norsworthy, J.K.; González-Torralva, F. Presence of the HPPD Inhibitor Sensitive 1 Gene and ALSS653N Mutation in Weedy Oryza sativa Sensitive to Benzobicyclon. *Plants* **2020**, *9*, 1576. [CrossRef] [PubMed]
7. Dai, S.; Georgelis, N.; Bedair, M.; Hong, Y.J.; Qi, Q.; Larue, C.T.; Sitoula, B.; Huang, W.; Krebel, B.; Shepard, M.; et al. Ectopic expression of a rice triketone dioxygenase gene confers mesotrione tolerance in soybean. *Pest. Manag. Sci.* **2022**, *78*, 2816–2827. [CrossRef] [PubMed]
8. Mužinić, V.; Katić, A.; Kašuba, V.; Micek, V.; Milić, M.; Želježić, D. Assessment of transplacental and lactational genotoxicity of tembotrione in Wistar rats at different developmental stages by alkaline comet assay. *Toxicology* **2021**, *463*, 152983. [CrossRef] [PubMed]
9. Kumar, V.; Liu, R.; Boyer, G.; Stahlman, P.W. Confirmation of 2,4-D resistance and identification of multiple resistance in a Kansas Palmer amaranth (Amaranthus palmeri) population. *Pest Manag. Sci.* **2019**, *75*, 925–2933. [CrossRef]
10. Šojić Merkulov, D.V.; Despotović, V.N.; Banić, N.D.; Armaković, S.J.; Finčur, N.L.; Lazarević, M.J.; Četojević-Simin, D.D.; Orčić, D.Z.; Radoičić, M.B.; Šaponjić, Z.V.; et al. Photocatalytic decomposition of selected biologically active compounds in environmental waters using TiO_2/polyaniline nanocomposites: Kinetics, toxicity and intermediates assessment. *Environ. Pollut.* **2018**, *239*, 457–465. [CrossRef]
11. Chen, P.; Shi, M.; Liu, X.; Wang, X.; Fang, M.; Guo, Z.; Wu, X.; Wang, Y. Comparison of the binding interactions of 4-hydroxyphenylpyruvate dioxygenase inhibitor herbicides with humic acid: Insights from multispectroscopic techniques, DFT and 2D-COS-FTIR. *Ecotoxicol. Environ. Saf.* **2022**, *239*, 113699. [CrossRef] [PubMed]
12. Goujon, E.; Maruel, S.; Richard, C.; Goupil, P.; Ledoigt, G. Transformation of the Herbicide Sulcotrione into a Root Growth Enhancer Compound by Sequential Photolysis and Hydrolysis. *J. Agric. Food Chem.* **2016**, *64*, 563–569. [CrossRef]
13. Goujon, E.; Richard, C.; Goupil, P.; Ledoigt, G. Cytotoxicity on Allium cepa of the two main sulcotrione photoproducts, xanthene-1,9-dione-3,4-dihydro-6-methylsulphonyl and 2-chloro-4-mesylbenzoic acid. *Pestic. Biochem. Physiol.* **2015**, *124*, 37–42. [CrossRef]
14. ter Halle, A.; Lavieille, D.; Richard, C. The effect of mixing two herbicides mesotrione and nicosulfuron on their photochemical reactivity on cuticular wax film. *Chemosphere* **2010**, *79*, 482–487. [CrossRef]
15. Wiszniowski, J.; Halle, A.T.; Richard, C.; Bonnemoy, F.; Bohatier, J. Toxicity of sulcotrione photoproducts mixture towards vibrio fischeri in the aquatic environment. *Arch. Environ. Prot.* **2011**, *37*, 15–23.
16. Rani, R.; Juwarkar, A. Adsorption of Phorate, an Organophosphorus Pesticide, on Vertisol. *Arch. Environ. Contam. Toxicol.* **2010**, *58*, 927–934. [CrossRef]
17. Tang, K.H.D.; Kristanti, R.A. Bioremediation of perfluorochemicals: Current state and the way forward. *Bioprocess. Biosyst. Eng.* **2022**, *45*, 1093–1109. [CrossRef] [PubMed]
18. Morillo, E.; Madrid, F.; Lara-Moreno, A.; Villaverde, J. Soil bioremediation by cyclodextrins. A review. *Int. J. Pharm.* **2020**, *591*, 119943. [CrossRef]
19. Borthakur, D.; Rani, M.; Das, K.; Shah, M.P.; Sharma, B.K.; Kumar, A. Bioremediation: An alternative approach for detoxification of polymers from the contaminated environment. *Lett. Appl. Microbiol.* **2022**, *75*, 744–758. [CrossRef]
20. Qu, L.; Xu, J.; Dai, Z.; Elyamine, A.M.; Huang, W.; Han, D.; Dang, B.; Xu, Z.; Jia, W. Selenium in soil-plant system: Transport, detoxification and bioremediation. *J. Hazard. Mater.* **2023**, *452*, 131272. [CrossRef]
21. Bhat, S.A.; Singh, S.; Singh, J.; Kumar, S.; Bhawana; Vig, A.P. Bioremediation and detoxification of industrial wastes by earthworms: Vermicompost as powerful crop nutrient in sustainable agriculture. *Bioresour. Technol.* **2018**, *252*, 172–179. [CrossRef] [PubMed]
22. Yaashikaa, P.R.; Kumar, P.S.; Jeevanantham, S.; Saravanan, R. A review on bioremediation approach for heavy metal detoxification and accumulation in plants. *Environ. Pollut.* **2022**, *301*, 119035. [CrossRef]
23. Ferreira, P.; Fernandes, P.A.; Ramos, M.J. The Catalytic Mechanism of the Retaining Glycosyltransferase Mannosylglycerate Synthase. *Chemistry* **2021**, *27*, 13998–14006. [CrossRef] [PubMed]
24. Blomstedt, C.K.; O'Donnell, N.H.; Bjarnholt, N.; Neale, A.D.; Hamill, J.D.; Møller, B.L.; Gleadow, R.M. Metabolic consequences of knocking out UGT85B1, the gene encoding the glucosyltransferase required for synthesis of dhurrin in Sorghum bicolor (L. Moench). *Plant Cell Physiol.* **2016**, *57*, 373–386. [CrossRef] [PubMed]
25. Su, Z.H.; Xu, Z.S.; Peng, R.H.; Tian, Y.S.; Zhao, W.; Han, H.J.; Yao, Q.H.; Wu, A.Z. Phytoremediation of trichlorophenol by Phase II metabolism in transgenic Arabidopsis overexpressing a Populus glucosyltransferase. *Environ. Sci. Technol.* **2012**, *46*, 4016–4024. [CrossRef] [PubMed]
26. Dumas, E.; Giraudo, M.; Goujon, E.; Halma, M.; Knhili, E.; Stauffert, M.; Batisson, I.; Besse-Hoggan, P.; Bohatier, J.; Bouchard, P.; et al. Fate and ecotoxicological impact of new generation herbicides from the triketone family: An overview to assess the environmental risks. *J. Hazard. Mater.* **2017**, *325*, 136–156. [CrossRef] [PubMed]

27. Moran, G.R. 4-Hydroxyphenylpyruvate dioxygenase. *Arch. Biochem. Biophys.* **2005**, *433*, 117–128. [CrossRef]
28. Matringe, M.; Sailland, A.; Pelissier, B.; Rolland, A.; Zink, O. p-Hydroxyphenylpyruvate dioxygenase inhibitor-resistant plants. *Pest Manag. Sci.* **2005**, *61*, 269–276. [CrossRef] [PubMed]
29. Jiang, Z.B.; Gao, S.; Hu, W.; Sheng, B.R.; Shi, J.; Ye, F.; Fu, Y. Design, synthesis and biological activity of novel triketone herbicides containing natural product fragments. *Pestic. Biochem. Physiol.* **2023**, *194*, 105493. [CrossRef]
30. He, H.; Liu, Y.; You, S.; Liu, J.; Xiao, H.; Tu, Z. A Review on Recent Treatment Technology for Herbicide Atrazine in Contaminated Environment. *Int. J. Environ. Res. Public Health* **2019**, *16*, 5129. [CrossRef]
31. Trivella, A.; Richard, C. New insights into pesticide photoprotection. *Environ. Sci. Pollut. Res. Int.* **2014**, *21*, 4828–4836. [CrossRef] [PubMed]
32. Goujon, E.; Sta, C.; Trivella, A.; Goupil, P.; Richard, C.; Ledoigt, G. Genotoxicity of sulcotrione pesticide and photoproducts on Allium cepa root meristem. *Pestic. Biochem. Physiol.* **2014**, *113*, 47–54. [CrossRef] [PubMed]
33. Huang, X.X.; Zhao, S.M.; Zhang, Y.Y.; Li, Y.J.; Shen, H.N.; Li, X.; Hou, B.K. A novel UDP-glycosyltransferase 91C1 confers specific herbicide resistance through detoxification reaction in Arabidopsis. *Plant Physiol. Biochem.* **2021**, *159*, 226–233. [CrossRef] [PubMed]
34. Li, P.; Zhao, A.; Li, R.; Han, S.; Li, N.; Ji, L.; Lei, K. Identifying a Detoxifying Uridine Diphosphate Glucosyltransferase (UGT), MdUGT83K2, Which Can Glycosylate the Aryloxyphenoxypropionate Herbicide. *Agronomy* **2023**, *13*, 306. [CrossRef]
35. Zhang, Y.; Zhao, A.; Mu, L.; Teng, X.; Ma, Y.; Li, R.; Lei, K.; Ji, L.; Wang, X.; Li, P. First Clarification of the Involvement of Glycosyltransferase MdUGT73CG22 in the Detoxification Metabolism of Nicosulfuron in Apple. *Plants* **2024**, *13*, 1171. [CrossRef] [PubMed]
36. Yang, Y.Y.; Zheng, P.F.; Ren, Y.R.; Yao, Y.X.; You, C.X.; Wang, X.F.; Hao, Y.J. Apple MdSAT1 encodes a bHLHm1 transcription factor involved in salinity and drought responses. *Planta* **2021**, *253*, 46. [CrossRef] [PubMed]
37. Forcella, M.; Manzoni, M.; Benaglia, G.; Bonanomi, M.; Giacopuzzi, E.; Papini, N.; Bresciani, R.; Fusi, P.; Borsani, G.; Monti, E. Characterization of three sialidases from Danio rerio. *Biochimie* **2021**, *187*, 57–66. [CrossRef] [PubMed]
38. Beihaghi, M.; Marashi, H.; Bagheri, A.; Sankian, M. Transient expression of CCL21as recombinant protein in tomato. *Biotechnol. Rep.* **2017**, *17*, 10–15. [CrossRef] [PubMed]
39. Pastor, V.; Vicent, C.; Cerezo, M.; Mauch-Mani, B.; Dean, J.; Flors, V. Detection, characterization and quantification of salicylic acid conjugates in plant extracts by ESI tandem mass spectrometric techniques. *Plant Physiol. Biochem.* **2012**, *53*, 19–26. [CrossRef] [PubMed]
40. Attaran, E.; Zeier, T.E.; Griebel, T.; Zeier, J. Methyl salicylate production and jasmonate signaling are not essential for systemic acquired resistance in Arabidopsis. *Plant Cell* **2009**, *21*, 954–971. [CrossRef]
41. Clough, S.J.; Bent, A.F. Floral dip: A simplified method for Agrobacterium-mediated transformation of Arabidopsis thaliana. *Plant J.* **1998**, *16*, 735–743. [CrossRef] [PubMed]

plants

MDPI

Article

The Dynamics of Allelochemicals and Phytotoxicity in *Eisenia fetida* during the Decomposition of *Eucalyptus grandis* Litter

Danju Zhang [1,*,†], Chaoyu Lv [1,†], Shaojun Fan [1], Yumei Huang [2], Na Kang [3], Shun Gao [1] and Lianghua Chen [1]

[1] Sichuan Provincial Key Laboratory of Ecological Forestry Engineering, College of Forestry, Sichuan Agricultural University, Wenjiang, Chengdu 611130, China; lvchaoyu@stu.sicau.edu.cn (C.L.); fanshaojun@stu.sicau.edu.cn (S.F.); shungao@sicau.edu.cn (S.G.); chenlh@sicau.edu.cn (L.C.)
[2] College of Landscape Architecture, Sichuan Agricultural University, Wenjiang, Chengdu 611130, China; hyumei@sicau.edu.cn
[3] Shijiazhuang Zoo, Luquan, Shijiazhuang 050200, China; kn8029@163.com
* Correspondence: zdj_8080573@sohu.com; Fax: +86-(0)28-86290957
† These authors contributed equally to this work.

Abstract: Allelopathy is an underlying and controversial mechanism for detrimental environmental effects in the management of *Eucalyptus* plantations. However, little attention has been paid to the dynamics of allelochemicals and phytotoxicity in soil fauna during litter decomposition. To explore the relationship between the dynamics of phytotoxicity and allelochemicals, a decomposition experiment was conducted using 4-year-old and 8-year-old *Eucalyptus grandis* litter (0, 10, 20, 30, and 45 days). The acute toxicity of *Eisenia fetida* was assessed, and a chemical analysis of the eucalyptus leaves was performed. Biochemical markers, including total protein, acetylcholinesterase (AChE) activity, and oxidative stress levels (SOD and MDA) were measured. A comet assay was used to determine DNA damage in *E. fetida* cells. The results showed that after 20–30 days of decomposition, *E. grandis* litter exhibited stronger phytotoxic effects on *E. fetida* in terms of growth and biochemical levels. After 20 days of decomposition, the weight and total protein content of *E. fetida* first decreased and then increased over time. SOD activity increased after 20 days but decreased after 30 days of decomposition before increasing again. MDA content increased after 20 days, then decreased or was stable. AChE activity was inhibited after 30 days of decomposition and then increased or stabilized with further decomposition. Soluble allelochemicals, such as betaine, chlorogenic acid, and isoquercitrin, significantly decreased or disappeared during the initial decomposition stage, but pipecolic acid significantly increased, along with newly emerging phenolic fractions that were present. More allelochemicals were released from 8-year-old litter than from 4-year-old *E. grandis* litter, resulting in consistently more severe phytotoxic responses and DNA damage in *E. fetida*. Scientific management measures, such as the appropriate removal of leaf litter in the early stages of decomposition, might help support greater biodiversity in *E. grandis* plantations.

Keywords: ecotoxicology; litter decomposition; *Eucalyptus*; *Eisenia fetida*; DNA damage

1. Introduction

Eucalyptus species have been introduced in many countries because of their rapid growth, broad adaptability, and high productivity for use in wood, paper, and charcoal industries [1,2]. Over 5.46 million ha of *Eucalyptus* plantations have been established in southern China [3]. However, the extensive area of *Eucalyptus* plantations and frequent harvesting during a short rotation period has led to various environmental issues, such as soil degradation, reduced biodiversity, decreased productivity, and allelopathy [1,4,5]. Allelopathy is considered to be one of the key mechanisms contributing to the decreased biodiversity associated with the development of *Eucalyptus* plantations [6–8].

Allelopathy is the effect of chemicals released by plants or microbes on the growth of other organisms [9]. Allelopathic compounds can be actively released by plants into the environment through volatilization, leaching, root exudation, or passive production during litter or residue decomposition [10–12]. Litter decomposition is one of the most important allelochemical release pathways after root exudation. Allelochemicals have undergone several abiotic and biotic processes in the soil, such as sorption and polymerization by soil organic matter and clay minerals and chemical transformation and degradation, maintaining sufficient concentrations to affect plants and other biota [13,14]. The allelopathic effects of plant litter on germination and seedling growth have been studied extensively [15–17]. However, limited information is available on the toxic effects of allelochemicals released by plants on the soil biota. Allelochemicals entering soils can be directly toxic to soil faunal communities or indirectly toxic by changing their food resource patterns and feeding habitats by affecting the soil microbial community [12,18,19]. Allelochemicals can also exert a positive influence on the soil biota community since they can serve as a carbon for the soil microbial community. Macrofauna, such as earthworms, occupy a large amount of soil biomass and play a vital role in the turnover of soil organic matter, nutrient cycling, and plant growth [20,21]. Earthworms can facilitate litter fragmentation and aerate soils by burrowing, increasing water infiltration, transporting soil organic matter, and reducing soil erosion by bioturbation [22]. Earthworms are ideal bioindicators of soil quality because of their sensitivity to the environment and contaminants [23,24].

Leaf litter decomposition is a major ecosystem process that determines nutrient cycling, primary productivity, and soil fertility [25]. Several studies have focused on the effects of litter quality, diversity, and environment on decomposition rate and nutrient release [26–28]. Different litter components have different mechanisms of litter decomposition [27,28]. Most previous studies have focused on the effects of complex phenolic compounds, such as polyphenols and tannins, on their retarding effects at the late stage of litter decomposition [29–31]. However, allelochemicals mainly consist of simple phenolic compounds such as phenolic acids and flavones, which can be produced through litter decomposition [32]. Although the allelopathic effects of plant litter have been extensively studied, limited attention has been paid to the dynamics of allelochemicals and their kinetic phytotoxicities during decomposition. Existing studies indicate that most of the allelochemicals are released rapidly and are degraded during the initial stage of field decomposition [33–36]. The resistant and newly emerging fractions of allelochemicals produced by litter decomposition undergo several physical and biochemical processes that can either increase or decrease phytotoxicity on other biota [34]. Previous studies have indicated that the decaying litter of *Eucalyptus* inhibits the germination and growth of target plants at the early stages of decomposition; however, phytotoxicity declines as decomposition proceeds [36]. However, little information is available on the allelochemical dynamics and phytotoxicity of soil fauna during the decomposition of *Eucalyptus* litter.

Eucalyptus has been widely studied for its high allelochemicals content, including water-soluble phenolics and VOCs released from decomposing litter, such as benzoic acid, hydroxybenzoic acid, vanillic acid, and terpenoids [7,37]. Many previous studies have evaluated the allelopathic effects of *Eucalyptus* litter extracts on the germination and seedling growth of target plant species [16,17]. Few studies have been conducted to evaluate the phytotoxicity of *Eucalyptus* litter to soil fauna. In addition, most previous studies have been based on laboratory, and few studies have been conducted to survey the phytotoxicity of *Eucalyptus* litter during decomposition under field conditions. *Eucalyptus grandis* is one of the main introduced, fast-growing tree species used for afforestation in Southwest China and is usually managed with a short rotation period (5–7 years) [1]. However, large areas of *E. grandis* plantations have also resulted in adverse ecological effects, such as reduced biodiversity, soil degradation, and decreased productivity [4,10,38,39]. Our previous results at the study sites showed that plant and soil biodiversity in *E. grandis* remained stable or decreased over 4–5 years, followed by a significant increase with the age of the plantation. Consistently, higher levels of VOCs and phenolic allelochemicals were detected in younger *E. grandis* litter (approximately four years old). In addition, a critical increase in allelochemicals in *E. grandis* litter was observed at the eight-year-old mark [1,7].

In this study, four- and eight-year-old *E. grandis* plantations were selected to represent the critical time point for the increase in allelochemicals. *Eisenia fetida*, a common soil fauna species used in toxicological and standard tests (OECD), was selected as the target organism, as it has been found in *E. grandis* plantations at our study sites [1,40]. The purpose of this study was to address the dynamics of phytotoxicity during the decomposition of *E. grandis* litter by measuring the growth and biochemical responses (malondialdehyde [MDA], superoxide dismutase [SOD], acetylcholinesterase [AChE], and DNA damage) of *E. fetida*. In addition, the dynamics of the identified allelochemicals were evaluated during the decomposition of the eucalyptus litter. We hypothesized that (1) allelochemicals are rapidly released and degraded in the initial stages of field decomposition, influencing the phytotoxicity dynamics of litter extracts on *E. fetida*, and (2) more allelochemicals are released from 4-year-old *E. grandis* litter than from 8-year-old litter.

2. Materials and Methods

2.1. Study Site

This study was conducted at the Experimental Station of Sichuan Agricultural University, located in the western Sichuan Province (102°57′–103°04′ E, 29°55′–29°59′ N, 570–592 m altitude). The climate is subtropical, with an average annual rainfall and temperature of 1397 mm and 17.5 °C, respectively [1]. The soil is a ferralsol with an old alluvial yellow loam and a granular structure [41]. Large *E. grandis* plantations were established at various growth stages. We selected four- and eight-year-old *E. grandis* plantations; both of the two sites were larger than 5 hectares. The study sites were planted with crops before afforestation with *E. grandis*. Cropping system, intensity, and management were typical and similar in the study region. Weed control treatments were not applied after afforestation. Three sites in each age group were selected for sampling (Table 1).

Table 1. Soil characteristics of two differently aged *E. grandis* plantations.

Stand	Soil Bulk Density	Soil pH	Soil Water Content	Soil Organic Matter	Soil Total N	Soil Available *p*
4a	1.33 ± 0.08	3.99 ± 0.03 *	29.38 ± 0.01 ***	15.90 ± 0.88	0.27 ± 0.03	2.54 ± 0.24
8a	1.22 ± 0.12	3.90 ± 0.03	28.99 ± 0.02	24.56 ± 0.58 ***	0.91 ± 0.06 ***	6.89 ± 0.26 ***

* indicates significant differences between soil properties depending on forest ages. *, $p < 0.05$; ***, $p < 0.001$; mean ± standard error (SE).

2.2. Biodiversity Investigation

Vegetation surveys were conducted within each plot. Three 5 × 5 m quadrats containing one 1 × 1 m sub-plot were established to quantify the woody and herbaceous plants and record the species name, number of plants, and coverage of shrubs and herbaceous plants, respectively. Three 2 × 2 m plots were randomly selected, and soil macrofauna in the litter layer and 0–5 cm and 5–10 cm soil layers were hand-sorted within a sampling area of 0.25 m^2 (50 × 50 cm). Three sampling points within each plot were selected, litter samples (10 × 10 cm) were collected, and soil samples from 0 to 5 cm and 5 to 10 cm were collected using a cylindrical sampler (5 cm in diameter, 10 cm in height). Tullgren funnels (24 h, 38 °C) were used to extract microarthropods. All the fauna specimens extracted were preserved in 75% ethanol. Individuals were counted and identified at the family level. Rhizosphere soil samples within 2 mm of the root surface of *E. grandis* were collected and stored at −80 °C for soil microbial high-throughput sequencing [40].

2.3. Leaf Litter and Soil Sampling

Intact and newly fallen leaves, recognizable by their color, were collected from two mature *E. grandis* plantations using litter traps (1 m × 1 m, 1 m above the ground) in June 2021. The *E. grandis* fallen leaves were air-dried and used in laboratory and field experiments. Fifty litter bags made of plastic with a 1.5 mm mesh, allowing access to soil microbes and fauna and containing 20 g of newly fallen air-dried litter, were placed on the soil surface in the plantations. The litter bags were recovered after 0, 10, 20, 30, and 45 days of *E. grandis* litter decomposition and air-dried.

Air-dried leaf litter was cut into small pieces (<5 mm). Twenty grams of the leaf litter was shaken for 24 h in 160 mL of distilled water at a mass: volume ratio of 1:8 at room temperature. The mixture was then centrifuged (Beckman Coulter, Inc., Brea, CA, USA), filtered, and diluted to concentrations of 10 mg mL^{-1}, 25 mg mL^{-1}, 75 mg mL^{-1}, and 125 mg mL^{-1} based on the original solution. Sterilized distilled water was used as control. Natural soil samples from surface horizons (0–15 cm) were collected from abandoned croplands near *E. grandis* plantations. The soil samples were thoroughly mixed and sieved through a 2 mm mesh to measure soil physicochemical properties.

2.4. Earthworms

E. fetida was cultivated according to OECD guidelines [42]. *E. fetida* was identified according to the pictorial keys to the soil animals of China. Earthworms were reared in a container with a substrate composed of peat moss, manure, and $CaCO_3$ (the pH value was adjusted to 6–7). The earthworms were fed twice per month. Adult worms with a clitellum and similar weights (300–400 mg) were selected. The earthworms were removed from the culture, rinsed with water, placed on dampened filter paper, kept in the dark at 18 °C for depuration, and acclimated to the test soil before treatment.

2.5. Identification of Secondary Metabolites

Ten grams of air-dried pure *E. grandis* litter samples at the initial stage and two months after decomposition from each plantation were subjected to extraction with 95 mL of 80% methanol (Thermo Fisher Scientific, Waltham, MA, USA). Before ultrasonic extraction (Zhengzhou Shengyuan Instrument Co., Ltd., Zhengzhou, China) (30 kHz for 40 min), the solutions stored in 250 mL brown flasks were shaken (Changzhou Tianrui Instrument Co., Ltd., Suzhou, China) for 24 h for a more complete extraction. The suspension was then filtered through a 0.45 μm micro-porous membrane and stored at −70 °C until subsequent ultra-performance liquid chromatography–tandem mass spectrometry (UPLC-MS/MS) analysis [32].

Metabolites of all thawed litter samples were extracted using methanol at 4 °C. Extraction solution (180 μL) from each sample and 20 μL of 100 ppm internal standards (naringenin and baicalein) were added to a 2 mL centrifuge tube. Subsequently, 20 μL of each sample was mixed with a quality control (QC) sample, and the remaining samples were subjected to UPLC-MS detection. Liquid chromatography was performed using a Thermo Ultimate 3000 system (Thermo Fisher Scientific) equipped with an AXQUTTY UPLC® HSS T3 column (150 × 2.1 mm, 1.8 μm; Waters, Milford, MA, USA). Mass spectrometry was performed using a Thermo Q Exactive Focus mass spectrometer (Thermo Fisher Scientific). High-energy collision-dissociation cells were used for secondary dissociation to acquire MS/MS information, and unnecessary information was removed by dynamic exclusion.

After converting the raw data into XCMS format with Proteowizard software (V3.0.8789), the identification, filtration, and alignment of peaks were processed using the XCMS package in R (V3.3.2). To compare different magnitudes within the data, we obtained a dimensionless unit, that is, the normalized intensity, by normalizing the peak areas. All data were processed for QC to obtain reliable and high-quality metabolomic data. The peaks were matched using Metlin (http://metlin.scripps.edu, accessed on 20 August 2021), MoNA (http://mona.fiehnlab.ucdavis.edu, accessed on 20 August 2021), and a proprietary database (BioNovoGene; Suzhou, China). The OPLS-DA model was used to determine the differential metabolites (DMs) between groups based on thresholds of variable importance in projection (VIP) $\geq$ 1 and $p < 0.05$ [1,32].

2.6. Acute Toxicity

Brown glass bottles (200 mL) filled with 200 g of air-dried and sieved abandoned cropland soils were prepared. The litter extracts (20 mL) were mixed into the soil samples and stirred using glass sticks, with a range of concentrations (CK and C1-C4) as follows: 0.00, 1.00, 2.50, 7.50, and 12.50 g kg^{-1} sets for the acute toxicity tests. In total, the toxicity test involved five litter decomposition times × two forest ages × five test concentrations × three replicates = 150 bottles. Ten healthy *E. fetida* individuals were randomly placed into each bottle. These bottles were sealed with sterile, breathable sealing films. The bottles were then placed in a cabinet (Zhengzhou Shengyuan Instrument Co., Ltd.) maintained at 20 °C, with 1000 illumination and a 12/12 h photoperiod. Water was added to the samples, and the soil water content was adjusted to 30% of the water-holding capacity. Soil samples dried at 105 °C for 24 h were used to determine the soil water content. The water-holding capacity was determined as described in the ISO guidelines [43]. Water was added to ensure constant substrate water content by weighing the containers.

The mortality of the earthworms was determined after 7 and 14 days of exposure. Earthworms were sorted by hand and considered dead if they did not respond to a gentle mechanical stimulation. Living earthworms were rinsed with distilled water, dried on a filter paper, and weighed.

2.7. Physiological Properties

Two live *E. fetida* from each replicate of each treatment were randomly selected for the measurement of physiological properties. Normal saline (0.9%) (determination of total protein, MDA content, and SOD activity) or extracting solution (determination of AChE activity) was added, as well as the remains of the earthworms (to 1:9 weight-to-volume ratio). The mixture was then homogenized under ice-cold conditions and centrifuged at 8000 rpm for 15 min at 4 °C [44]. The supernatant was collected and stored to evaluate enzymatic activity and protein content. Physiological properties were determined using test kits following the manufacturer's instructions. Test kits for total protein content, superoxide dismutase (T-SOD) activity, and MDA content were purchased from Nanjing Jiancheng Bioengineering Institute (Nanjing, China). Test kits for AChE) activity were purchased from Beijing Solarbio Science & Technology Co., Ltd. (Beijing, China).

Total protein content was determined using the bicinchoninic acid (BCA) method [45]. Under alkaline conditions, the protein reduces Cu^{2+} to Cu^{+}, forming purple complexes with BCA. Supernatants and various reagents were pipetted into 2 mL centrifuge tubes in an orderly fashion, swirled well, and allowed to stand for 5 min. Subsequently, 200 µL was pipetted from the tubes into wells of a 96-well transparent plate. The absorbance of the microplate was measured at 562 nm (AU_{562nm}) using a microplate reader [46].

SOD activity was determined using the xanthine oxidase–hydroxylamine method; O_2^{-} was produced by the xanthine and xanthine oxidase reaction, which oxidized hydroxylamine to form nitrite and presented purple red under the action of a chromogenic agent. Color was read at 550 nm using a microplate reader [47]. MDA reacts with thiobarbituric acid at 95 °C for 80 min to form a red product, and the color was read using a spectrophotometer at 532 nm [48]. AChE catalyzes the hydrolysis of Ach to produce choline, which reacts with 5,5′-dithiobis-2-nitrobenzoic acid (DTNB) to produce TNB, and the absorbance was then measured at 412 nm using a microplate reader [49].

2.8. Comet Assay

Based on the following results, the effects of litter extracts on the growth and physiological properties of earthworms were stronger after 30 days of litter decomposition. Earthworms were cultivated in brown bottles treated with *E. grandis* litter extracts for 30 days of decomposition. After 14 days, earthworm coelomocytes were extracted. Three *E. fetida* individuals were selected from each bottle, rinsed with distilled water, and soaked in chilled extrusion medium (5% ethanol, 95% saline, 2.5 mg mL^{-1} Na_2 EDTA, and 10 mg mL^{-1} guaiacol glyceryl ether, pH 7.3) for 3 min [50]. The obtained coelomocyte cells were released, washed three times in phosphate-buffered saline, and centrifuged at 4 °C for 5 min. The final cell concentration was adjusted to approximately 10^6 cells mL^{-1} with phosphate-buffered saline.

The cell suspension was mixed with melted comet agarose (37 °C) at 1:10 (volume ratio) in an EP tube. The mixture (20 µL) was added to the wells of a 6-Well Comet Slide, covered with a coverslip, and incubated at 4 °C for 15 min under yellow light to mitigate UV-induced DNA damage. Alkaline electrophoresis was conducted in a buffer of 300 mA NaOH and 1mM Na_2EDTA (pH > 13) for 30 min at 25 V and 300 mA, using a horizontal electrophoresis apparatus. After washing twice with distilled water for 5 min each, the slides were dehydrated in ethanol (10 min) and dried. Then, 40 µL of dyeing solution was added to each agarose spot and stained in a humidified dark box for 10 min at room temperature. A fluorescence microscope (BX41 Olympus, Tokyo, Japan) was used for observation and capture of comet images. The captured images were analyzed using Comet A 1.0 software (Cell Biolabs, San Diego, CA, USA). The obtained data of tail DNA (%), tail length, tail moment, and OTM data were utilized for the assessment of DNA damage in earthworms. OTM refers to the tail moment (distance between the center of mass of the tail and that of the head) multiplied by the percentage of DNA within the tail. It is also the most sensitive parameter reflecting the quality and quantity of DNA damage [14].

2.9. Statistical Analysis

The survival rate of *E. fetida* was calculated as follows:

$$SR = (S_0 - D_t)/S_0 \times 100\%$$

where S_0 is the initial number of earthworms, and D_t is the number of dead earthworms after t days.

The rate of weight inhibition of *E. fetida* was calculated as follows:

$$IR = (W_0 - W_t)/W_0 \times 100\%$$

where W_0 is the initial weight of earthworms and W_t is the weight of earthworms after t days [44].

The diversity indices were calculated as follows:

$$\text{Shannon–Wiener diversity index } (H) = -\sum_{i=1}^{n} P_i \ln P_i$$

$$\text{Pielou evenness index } (E) = H/\ln S$$

$$\text{Berger–Parker index } (D_{B\text{-}P}) = N_{\max} - N$$

$$\text{Simpson index } (D) = 1 - \sum P_i^2$$

where N_i is the number of individuals in the i-th group, N is the number of individuals in all groups in the community, N_{max} is the number of individuals of the species with the highest abundance in the sample unit, $P_i = N_i/N$, and S is the number of groups.

All statistical analyses were performed using the SPSS software (version 25.0, SPSS Inc., Chicago, IL, USA), PAST 4.02, and R 4.3.1. Graphics were plotted using Origin 2021 and R version 4.3.1. The exposure time was taken as a within-subjects variable, and the decomposition time, age, and dose were taken as between-subjects factors to conduct repeated measurement analysis of variance and interaction analysis for all indicators. One-way analysis of variance (ANOVA) and Tukey's test were used to analyze the significance of the differences in growth, physiological properties, and DNA damage depending on litter decomposition time and concentration. The data normality and homogeneity of the variations were tested using Shapiro–Wilk and Levene's test, respectively. Data sets failing the tests were Ln transformed or analyzed using non-parametric tests before further analysis to help satisfy the requirement of normality and variance homogeneity. An independent sample *t*-test was used to analyze the significance of the differences in the above-mentioned properties of earthworms depending on forest age and exposure time. Analysis of similarity based on Bray–Curtis distances was used to determine differences in the composition of metabolic compounds in litter in the two aged *E. grandis* plantation stands during litter decomposition. Non-metric multidimensional scaling (NMDS) ordination was used to plot differences. RDA analysis was used to determine the correlations between the main differential potential allelochemicals in litter and understory environmental factors in the two aged plantations, such as pH, water content, organic matter, and N and P contents, and biodiversity was determined before and after two months of litter decomposition. RDA analysis was also used to determine the correlations between the growth and physiological properties of *E. fetida* and the main differential potential allelochemicals in litter.

3. Results

3.1. Soil Properties and Plant and Soil Biodiversity

The soil bulk density did not differ significantly between the two mature plantations. Soil water content and soil pH were higher in the 4-year-old stands than in the 8-year-old stands. The soil organic matter, total N, and available P contents were higher in the 8-year-old stands than in the 4-year-old stands (Table 1). Regarding woody species, the species number, density, and Shannon–Wiener index were higher in 8-year-old stands than in 4-year-old stands, whereas the Berger–Parker indices showed the opposite trend. For macrofauna and microfauna, the Shannon–Wiener index was higher in 8-year-old stands than in 4-year-old stands; however, the Berger–Parker indices of macrofauna exhibited the opposite trend (Table 2).

Table 2. Plant, soil microbial, and soil fauna diversity in the two different aged *E. grandis* plantations.

		Species Number	Density (Ind m^{-2})	Shannon–Wiener	Pielou	Berger–Parker
Plant's diversity	Shrub (4a)	3.00 ± 0.87	0.51 ± 0.18	0.75 ± 0.14	0.76 ± 0.08	0.68 ± 0.03 ***
	Shrub (8a)	6.50 ± 0.43 **	1.11 ± 0.18 *	1.65 ± 0.05 ***	0.82 ± 0.02	0.35 ± 0.00
	Herbaceous (4a)	7.50 ± 2.18	48.00 ± 18.99	1.57 ± 0.21	0.73 ± 0.06	0.40 ± 0.03
	Herbaceous (8a)	7.75 ± 1.30	54.25 ± 4.63	1.64 ± 0.20	0.79 ± 0.07	0.39 ± 0.03
Soil faunal diversity	Macrofauna (4a)	6.42 ± 0.14	54.00 ± 11.27	1.58 ± 0.09	0.80 ± 0.05	0.40 ± 0.05 *
	Macrofauna (8a)	7.50 ± 1.30	62.33 ± 19.73	1.79 ± 0.09 *	0.86 ± 0.07	0.31 ± 0.02
	Microfauna (4a)	9.08 ± 0.95	8441.67 ± 1203.21	1.66 ± 0.02	0.64 ± 0.07	0.40 ± 0.02
	Microfauna (8a)	10.67 ± 0.14	12141.67 ± 2546.61	1.79 ± 0.07 *	0.59 ± 0.05	0.37 ± 0.04
		OTUs	Shannon–Wiener	Simpson	ACE	Chao1
Soil microbial diversity	(4a)	1623.08 ± 20.12	5.94 ± 0.07	0.99 ± 0.00	2474.72 ± 35.69	2312.38 ± 16.72
	(8a)	1537.58 ± 91.82	5.72 ± 0.17	0.98 ± 0.01	2398.02 ± 127.48	2208.14 ± 145.15

* indicates significant differences between soil properties depending on forest ages. *, $p < 0.05$; **, $p < 0.01$; ***, $p < 0.001$; mean ± standard error (SE).

3.2. Secondary Metabolites during *E. grandis* Litter Decomposition

A total of 328 and 381 primary and secondary metabolites, respectively, were identified in the litter of the two mature *E. grandis* plantations before and after two months of litter decomposition. The identified metabolites mainly included low-molecular-weight primary compounds, such as carbohydrates, amino acids, peptides, nucleotides and purines, cofactors, and vitamins. In addition, a wide variety of secondary metabolites, such as phenolic acids, flavones, terpenoids, alkaloids, and non-protein amino acids, were observed in *E. grandis* litter (Figure 1). The composition of metabolites in the litter varied between the two *E. grandis* plantations during litter decomposition (Figure 2).

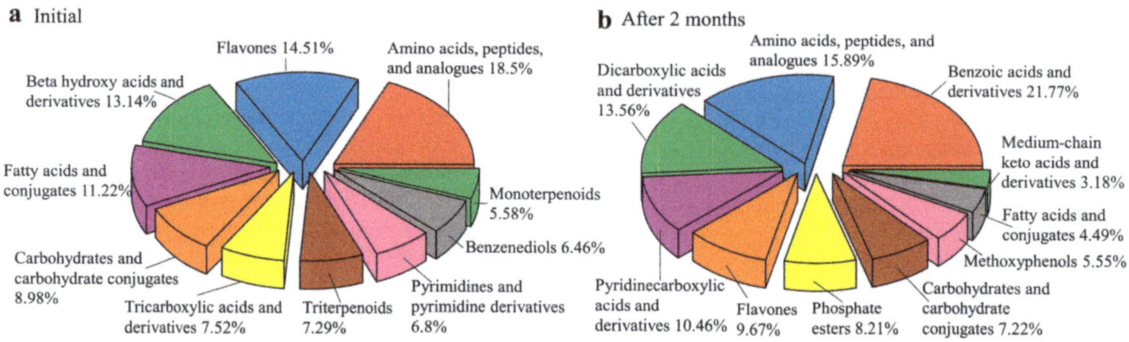

Figure 1. Top 10 categories of metabolites identified in litters of 4- and 8-year-old *E. grandis* at the initial decomposition stage (**a**) and two months after decomposition (**b**).

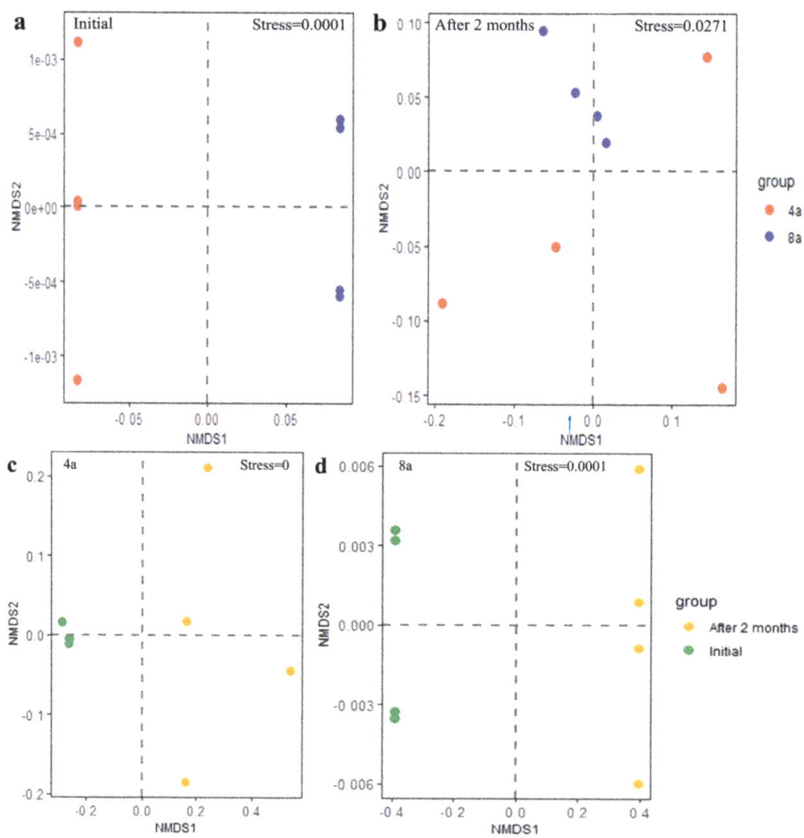

Figure 2. Non-metric multidimensional scale ordination for metabolites of 4- and 8-year-old *E. grandis* at the initial stage of leaf litter decomposition (**a**) and two months after decomposition (**b**); non-metric multidimensional scale ordination for metabolites at the initial stage of leaf litter decomposition and two months after decomposition at 4-year-old (c) and 8-year-old *E. grandis* (**d**).

A total of 154 and 120 DMs were identified in the litter of the two mature *E. grandis* plantations during the litter decomposition, respectively. A total of 301 and 310 DMs were detected before and after two months of decomposition in the 4- and 8-year-old stands, respectively. Notably, the levels of amino acids and peptides were higher in the litter of 4-year-old *E. grandis* than in 8-year-old *E. grandis*, a pattern also observed for lipids after two months of decomposition. Conversely, other primary metabolites were more abundant in the litter of 8-year-old *E. grandis* than in the 4-year-old *E. grandis*. Among the secondary metabolites, higher levels of terpenoids and amines were observed in the initial litter of the 4-year-old than in the 8-year-old *E. grandis*. Other potential allelochemicals, such as phenolic acids, flavonoids, alkaloids, and non-protein amino acids, were more abundant in 8-year-old stands than in 4-year-old stands. A similar trend was observed for amines in the litter after two months of decomposition. Notably, no significant variation in terpenoid levels was observed between the two mature *E. grandis* after two months of decomposition. Except for lipids in the 4-year-old *E. grandis* and amines in the 8-year-old *E. grandis*, there were no significant differences during decomposition; nucleotides, alkaloids, and non-protein amino acids exhibited a significant increase after two months of litter decomposition, whereas the other primary and secondary metabolites decreased significantly after litter decomposition for the same months (Figure 3). SIMPER analysis revealed that betaine, pipecolic acid, phenol, gamma-aminobutyric acid, and the other 13 potential compounds contributed to the variation in allelochemical composition during *E. grandis* litter decomposition and between the two mature plantations (Table 3).

Table 3. Results of analysis of similarity (ANOSIM) for differential potential allelochemicals of *E. grandis* between forest ages and decomposition times and analysis of similarity percentage (SIMPER).

Effects	R and Significance	Main Contributive Compounds for Variation
		Two-way ANOSIM
Forest age	(1) ***	Betaine, pipecolic acid, phenol, gamma-aminobutyric acid, catechin, chlorogenic acid, eucalyptol, isoquercitrin, kaempferol, pulegone, sinapic acid, sinapoyl aldehyde, and sphinganine
Decomposition time	(1) ***	Gamma-aminobutyric acid, phenol, pipecolic acid, betaine, catechin, chlorogenic acid, eucalyptol, isoquercitrin, kaempferol, sinapoyl aldehyde, pulegone, sinapic acid, sphinganine, and phytosphingosine
		One-way ANOSIM of forest age effects
Initial	(1) *	Betaine, methyleugenol, catechin, isoquercitrin, pulegone, sinapic acid, Sphinganine, 4-hydroxycinnamic acid, eucalyptol, Taraxerol, chlorogenic acid, sinapoyl aldehyde, and pipecolic acid
After 2 months	(1) *	Betaine, pipecolic acid, phenol, gamma-aminobutyric acid, and saccharopine
		One-way ANOSIM of decomposition time effects
4 years	(1) *	Betaine, 2-aminophenol, Taraxerol, gamma-aminobutyric acid, phenol, pipecolic acid, chlorogenic acid, eucalyptol, catechin, isoquercitrin, and pulegone
8 years	(1) *	Betaine, 2-aminophenol, phenol, gamma-aminobutyric acid, Taraxerol, catechin, pipecolic acid, chlorogenic acid, isoquercitrin, and eucalyptol

Forest ages: 4 and 8 years old. Data in parentheses are the R value. Significant effects are indicated in superscript: *, $p < 0.05$; ***, $p < 0.001$. The rightmost column is the contributive compounds (SIMPER analysis, contributive rate ≧ 2%) to the variation in differential potential allelochemicals composition among study sites or decomposition times.

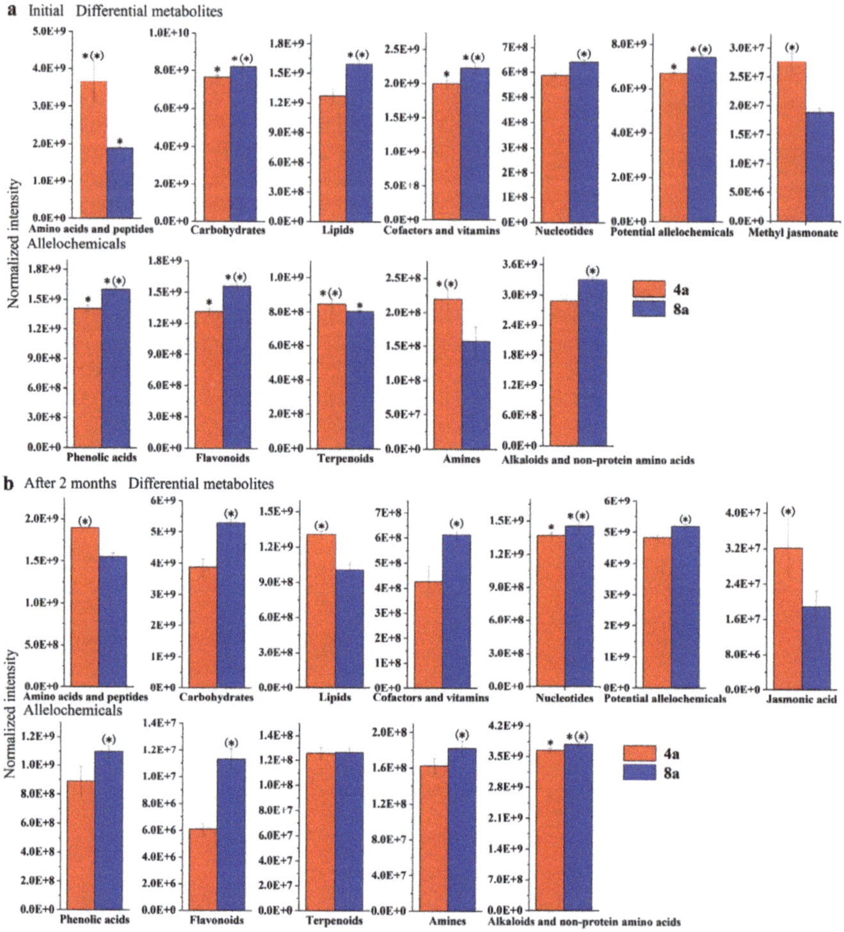

Figure 3. Normalization intensity of differential metabolites and potential allelochemicals in 4- and 8-year-old *E. grandis* at the initial stage of leaf litter decomposition (**a**) and two months after decomposition (**b**). * indicates significant differences between the two decomposition times; (*) indicates significant differences between the two aged *E. grandis* stands; $p < 0.05$. Error bars indicate the standard errors.

The levels of some key differentially contributing metabolites in *E. grandis* litter during decomposition, such as betaine, chlorogenic acid, isoquercitrin, and kaempferol, generally decreased during decomposition. However, the pipecolic acid content increased after two months of decomposition. Catechin, eucalyptol, pulegone, sinapic acid, sinapoyl aldehyde, and sphinganine disappeared as the decomposition progressed, whereas phenol, phytosphingosine, and gamma-aminobutyric acid appeared as the decomposition process (Figure 4).

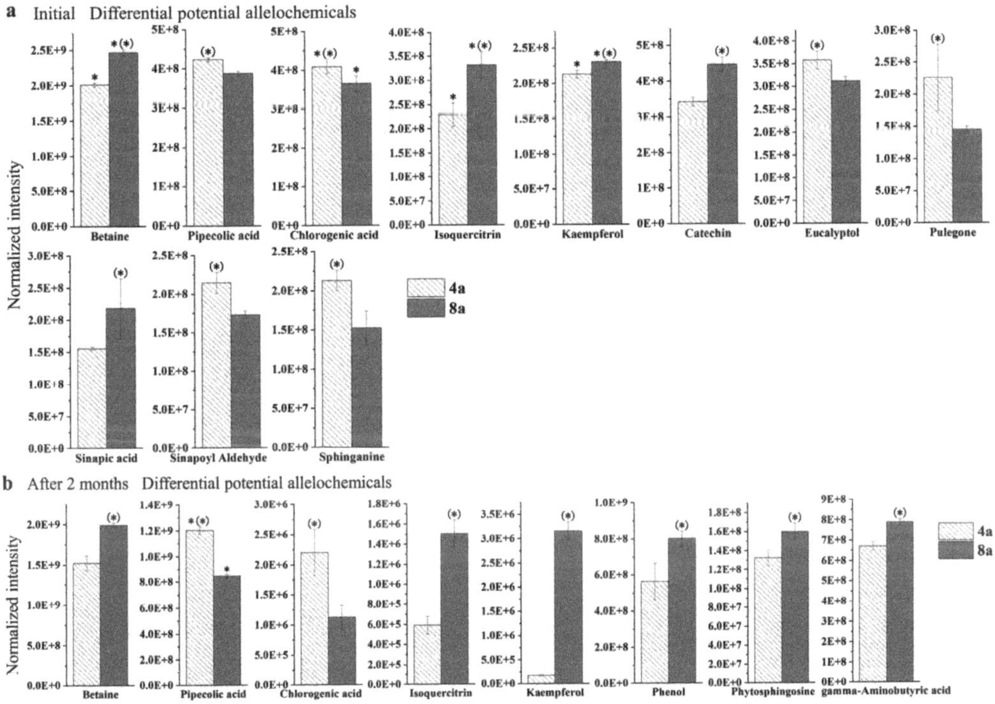

Figure 4. Normalization intensity of differential potential allelochemicals with contributive rate ≧ 2% in two-way ANOSIM and SIMPER analysis (Table 3) of 4- and 8-year-old *E. grandis* at the initial stage of leaf litter decomposition (**a**) and two months after decomposition (**b**). * indicates significant differences between the two decomposition times; (*) indicates significant differences between the two aged *E. grandis* stands; $p < 0.05$. Error bars indicate the standard errors.

3.3. Effects of Leaf Litter Extract on Earthworms

3.3.1. Survival Rate and Growth of *E. fetida*

Repeated ANOVA revealed that decomposition, exposure time, and dose significantly affected the survival and weight inhibition rates of *E. fetida* (Table 4). After 14 days of exposure to the 4-year-old litter treatment, the survival rate of *E. fetida* at C2 significantly increased after 10 days of decomposition and then remained stable over time (Table S1). The weight inhibition rate of *E. fetida* at C3 significantly increased after 20 days of decomposition and subsequently decreased over time. The weight inhibition rate of *E. fetida* decreased with increasing doses of litter extract under several treatments (Table S2).

Table 4. Repeated measures ANOVA of the effects of forest age, decomposition time, exposure time, and dose on growth and physiological characteristics of *E. fetida*.

	Survival Rate		Weight Inhibition Rate		Total Protein Content		SOD Activity		MDA Content		AChE Activity	
	F	*p*	*F*	*p*	*F*	*p*	*F*	*p*	*F*	*p*	*F*	*p*
Decomposition time	17.476	0.000 ***	5.498	0.000 ***	41.199	0.000 ***	69.155	0.000 ***	29.708	0.000 ***	43.625	0.000 ***
Age	0.129	0.720	0.088	0.767	5.130	0.026 *	10.673	0.001 ***	2.596	0.110	2.815	0.097
Dose	0.587	0.673	4.680	0.002 **	3.034	0.021 *	7.769	0.000 ***	2.708	0.034 *	6.349	0.000 ***
Exposure time	134.20	0.000 ***	420.00	0.000 ***	22.767	0.000 ***	2.395	0.125	94.713	0.000 ***	8.834	0.004 **
Age × Dose	1.863	0.123	2.372	0.057	0.816	0.518	2.345	0.060	0.532	0.712	1.100	0.361
Age × Decomposition time	0.662	0.620	2.455	0.051	0.905	0.464	0.865	0.488	1.065	0.378	3.667	0.008 **

* indicates significant differences. *, $p < 0.05$; **, $p < 0.01$; ***, $p < 0.001$.

3.3.2. Physiological Properties of *E. fetida*

The content of total protein, MDA, and activities of SOD and AChE changed significantly with decomposition time, forest age, dose, and exposure time. Significant interactive effects on AChE activity were observed (Table 4). Under 4-year-old litter treatments, after seven days at C1 and C3 and 14 days of exposure at C1, C3, and C4 doses, the total protein content of *E. fetida* decreased significantly after 10–20 days of decomposition but remained stable or increased after 30 days and then decreased with time. Under 8-year-old litter treatments and seven days of exposure at C1 and C2, the total protein content of *E. fetida* decreased after 10 days of decomposition and then remained stable over time. After 14 days of exposure to C4, the protein content was initially stable but decreased after 45 days of decomposition (Figure S1).

After 7 or 14 days of exposure at C1 and C4 under the two mature litter treatments, the SOD activity of *E. fetida* fluctuated. It increased after 10–20 days of decomposition, decreased after 30 days, and then increased again after 45 days of decomposition. However, after seven days of exposure at C2 and C4, the SOD activity of *E. fetida* increased after 10 days of decomposition and then remained stable over time (Figure S2). After 14 days of exposure, the MDA content of *E. fetida* stimulated under 4-year-old litter treatments at C1, C2, and C4 increased after 20 days but decreased after 30 days of decomposition and then increased with time. For the treatments of the litter of 8-year-old trees, the MDA content at C3 after seven days of exposure was stimulated after ten days of decomposition and then stabilized over time. After 14 days of exposure at C1 and C2, the MDA content of *E. fetida* fluctuated and was stimulated after 20 or 45 days of decomposition (Figure S3).

After seven days of exposure, the AChE activity of *E. fetida* under 4-year-old litter treatments at C1 and C2 decreased after ten days of decomposition and then stabilized over time. After 14 days of exposure at C1 and C4, AChE activity decreased after 20–30 days, then stabilized with decomposition time. Under the treatment of 8-year-old litter at C2 after seven days of exposure, AChE activity decreased after 10 days and increased after 20 days of decomposition. However, after 14 days of exposure to C1, AChE activity was initially stable and then decreased after 30 or 45 days of decomposition (Figure S4).

The effects of dose and forest age on *E. fetida* were observed in several treatments. After 20 days of decomposition, the weight inhibitory rate of *E. fetida* increased with increasing doses of litter extracts of 4-year-old *E. grandis* and reached a maximum at C3. SOD activity of earthworms treated with 8-year-old litter increased at C1 and then stabilized with increasing doses. The total protein content of *E. fetida* was higher in the litter extracts of 4-year-old trees than in 8-year-old trees. However, SOD activity was higher in the litter of 8-year-old *E. grandis* than in 4-year-old *E. grandis* litter.

3.4. *Effect of* E. grandis *Leaf Litter Extract on DNA Damage in* E. fetida

Based on the above results, the toxic effects of *E. grandis* on *E. fetida* were stronger after 20–30 days of litter decomposition. We selected *E. grandis* litter after 30 days of decomposition to evaluate DNA damage in *E. fetida* using a comet assay. After 14 days of exposure, the litter extracts of 8-year-old *E. grandis* caused DNA damage to *E. fetida*, and no significant DNA damage to *E. fetida* was observed after treatment with litter extracts of 4-year-old trees. Compared with the control, the tail DNA content, tail moment, and OTM decreased significantly at C1 and then increased with increasing doses. At the highest concentration (C4), tail DNA content, tail length, and OTM were significantly higher in the litter of 8-year-old *E. grandis* than in the 4-year-old *E. grandis* litter (Table 5).

Table 5. DNA damage effects of *E. grandis* litter extracts from different aged stands evaluated via a comet assay on coelomocytes from *E. fetida*.

Dose	Tail DNA (%)		Tail Moment		Olive Tail Moment		Tail Length	
	4a	8a	4a	8a	4a	8a	4a	8a
CK	1.51 ± 0.81	1.51 ± 0.81 (ab)	0.19 ± 0.12	0.19 ± 0.12 (ab)	0.46 ± 0.21	0.46 ± 0.21 (ab)	7.46 ± 2.31	7.46 ± 2.31
C1	1.63 ± 0.48	1.17 ± 0.29 (b)	0.34 ± 0.2	0.12 ± 0.03 (b)	0.50 ± 0.17	0.29 ± 0.07 (b)	6.28 ± 0.69	5.02 ± 0.49
C2	2.16 ± 0.87	1.52 ± 0.13 (ab)	0.52 ± 0.32	0.52 ± 0.18 (ab)	0.75 ± 0.33	0.56 ± 0.04 (ab)	9.04 ± 4.48	7.98 ± 1.44
C3	1.50 ± 0.54	2.84 ± 1.32 (ab)	0.24 ± 0.1	0.79 ± 0.56 (a)	0.46 ± 0.21	0.97 ± 0.55 (a)	6.85 ± 2.02	11.75 ± 6.19
C4	1.03 ± 0.24	3.20 ± 0.74 (a)(*)	0.12 ± 0.07	0.82 ± 0.58 (a)	0.27 ± 0.08	1.04 ± 0.27 (a)(*)	5.33 ± 1.34	11.70 ± 3.28 (*)

Lowercase letters in parentheses indicate significant differences among various concentrations; (*) indicates significant differences between two aged *E. grandis* stands; $p < 0.05$; mean ± standard error (SE).

3.5. Correlation Analysis

Environmental changes between the two aged *E. grandis* affected the potential allelochemicals released by the litter during litter decomposition. The RDA showed that the first ordinate axes represented changes in most environmental factors, except for the diversity of herb species. At the initial decomposition stage, the second ordinate axes were closely correlated with the diversity of herb species. The contents of alkaloids, non-protein amino acids, and flavonoids in the initial litter were positively correlated with soil available P, total N, organic matter, shrub, and soil faunal diversity in the two-aged *E. grandis* plantations. The contents of terpenoids and phenolic acids were positively correlated, but that of amines was negatively correlated with changes in herb diversity. After two months of litter decomposition, the content of alkaloids and non-protein amino acids was positively correlated with changes in soil organic matter, available P, total N, soil faunal diversity, and shrub diversity. The phenolic acid content in *E. grandis* litter was positively correlated with changes in herb diversity. The amine, terpenoid, and flavonoid contents were not significantly correlated with the environmental factors (Figure 5, Table S3).

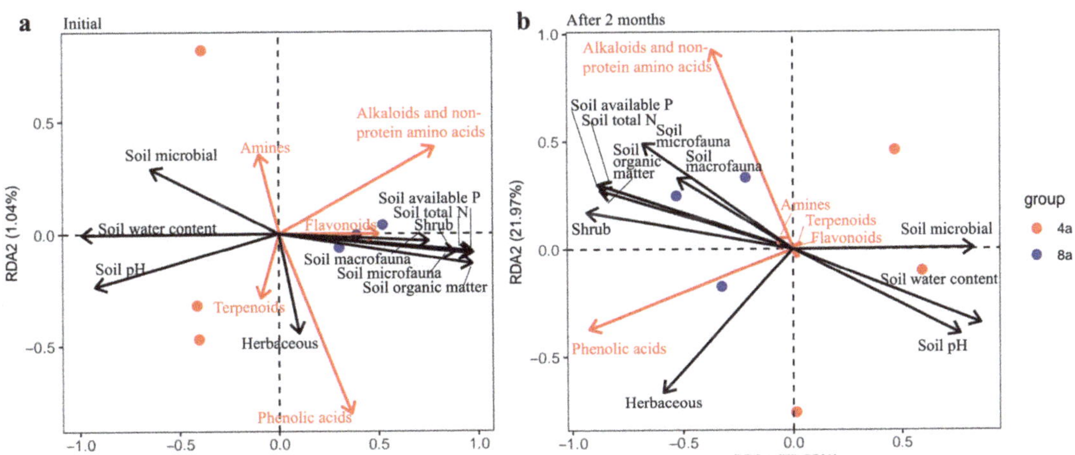

Figure 5. Redundancy analysis (RDA) diagram illustrating the relationship between soil physicochemical traits, soil biodiversity, and differential potential allelochemicals in 4- and 8-year-old *E. grandis* at the initial stage of leaf litter decomposition (**a**) and two months after decomposition (**b**).

RDA also indicated that the responses of growth and physiological properties of *E. fetida* were correlated with the release of potential allelochemicals after two months of decomposition. For the 4-year-old stands, the first ordinate axis represents the changes in potential allelochemicals, whereas the second ordinate axis represents the changes in amines during litter decomposition in 8-year-old stands. Under the treatments of 4-year-old litter, the responses of SOD, AChE, and MDA in *E. fetida* were closely related to the changes in amines, phenolic acids, flavonoids, and terpenoids during litter decomposition. For the litter treatments of 8-year-old *E. grandis*, the changes in amines were responsible for the changes in the total protein and weight inhibitory rate of *E. fetida* (Figure 6, Table S4).

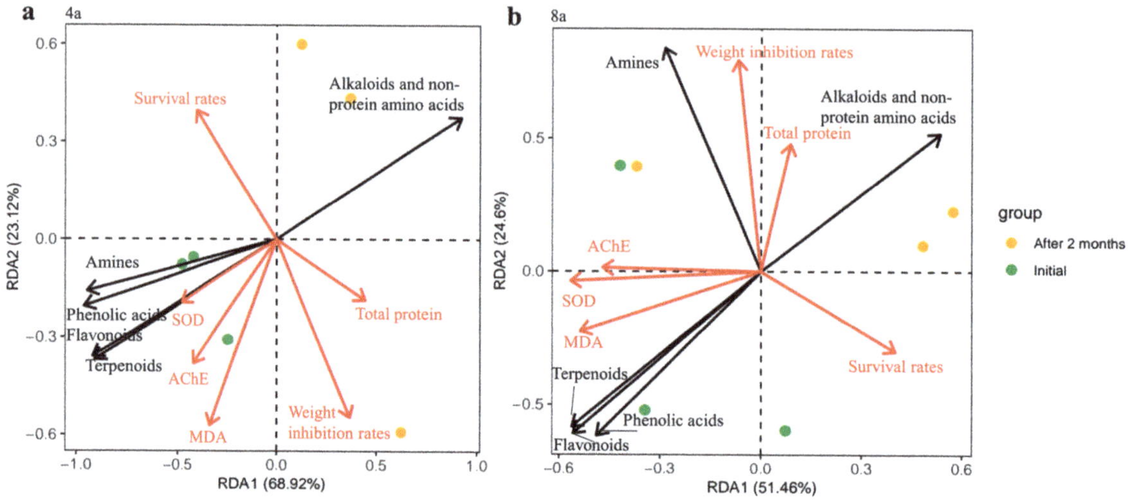

Figure 6. Redundancy analysis (RDA) diagram illustrating the relationship between differential potential allelochemicals and growth and physiological properties of *E. fetida* in the initial stage of leaf litter decomposition and two months after decomposition at 4-year-old (**a**) and 8-year-old *E. grandis* (**b**). The growth and physiological properties of *E. fetida* were standardized by Z-Score, in which the growth and physiological properties of *E. fetida* were 0 and 30 days of decomposition and 14 days of exposure to C4 concentration, respectively.

4. Discussion

Earthworms are epidermal respiratory soil macrofauna that can be easily absorbed into the body through respiration or can cause toxic effects through direct skin contact [51]. In this study, under most treatments, the survival rate of earthworms did not change significantly. The results indicated that allelochemicals in *E. grandis* litter were not maintained at sufficient concentrations, and earthworms activated the surface defense system in response to stimuli to avoid death [52]. However, inhaled and absorbed toxic chemicals inhibited this increase in earthworm weight. The weight inhibition rate of *E. fetida* increased significantly when treated with litter extracts after 20 days of decomposition. Earthworms reduce their food intake while avoiding the intake of toxic substances that are involved in the inhibitory effects on the weight of *E. fetida* [53].

Earthworms also adapt to environmental stress by adjusting their physiological activities. In this study, the total protein content and activities of SOD and AChE decreased, but the MDA content of *E. fetida* increased after 20 or 30 days of decomposition. The results indicated that the phytotoxicity of the litter extracts of *E. grandis* on *E. fetida* was severe at the initial stage of approximately 20–30 days of decomposition, followed by weaker toxicity as decomposition progressed. These results were consistent with previous studies, indicating that the most inhibitory effects occurred in the early stage of decomposition and then weakened in phytotoxicity with decomposition [11,33]. The initial decomposition phase of litter consists of comminution and release of the soluble fraction, followed by the degradation of cellulose and hemicellulose and retarded decomposition caused by inhibitory compounds such as tannins and lignin at the late stage of decomposition [26,54,55]. Several studies have reported a rapid loss of soluble phenolics from the *Eucalyptus* litter within the first week or month [33,36]. The rapid degradation of phenolics might be attributed to chemical processes after leaf fall, such as higher concentrations of oxidizing enzymes, particularly phenolases [34]. Subsequently, the rapid loss of phenolics from the *Eucalyptus* litter could be ascribed to leaching and sugar and microbial degradation [34,36,56]. A delay has been demonstrated between nutrient release from decomposing organic matter and root colonization of decomposing litter, which could be related to the occurrence of phytotoxic "windows" during the early stages of decomposition, which inhibits the development of the soil microbial community [11,57]. Despite a rapid decrease in phenolic compounds within a few days of leaf fall, Puig et al. (2018) observed that lettuce germination was completely inhibited at every sampling time [36], which demonstrated that *E. globulus* litter showed a continuous release of phenolic and volatile compounds during the 30 days of decomposition. As the decomposition period was prolonged, the phytotoxin levels declined when allelochemical inactivation was not balanced by new production. Previous studies have shown that there is a more resistant phenolic fraction after the initial rapid disappearance of phenolics and volatile compounds [36,58]. The relationship between allelochemicals and biological features, such as soil fauna, should refer to the resistant fraction that might be leached into the soil [11,34]. These soluble phenolics and volatiles entering the soil might be directly or indirectly toxic to soil fauna by changing the soil microbial community.

A total of 328 and 381 primary and secondary metabolites, respectively, were identified in the litter of the two mature *E. grandis*, and their presence, quantities, and composition varied significantly, depending on individual compounds as well as the decomposition process. Except for lipids in 4-year-old litter and amine in the litter of 8-year-old trees, which were stable, and the levels of nucleotides, alkaloids, and non-protein amino acids, which increased significantly, the levels of other primary and secondary metabolites reduced significantly after litter decomposition for two months. The secondary metabolites identified, such as catechin, eucalyptol, sinapic acid, and epicatechin, were partly in agreement with the allelochemicals previously identified in *Eucalyptus* litter [32,36,58]. In contrast, a few compounds, such as betaine, aminobutyric acid, and isoquercitrin, were not detected in previously identified *E. grandis* litter. This can be explained by the fact that allelochemicals are subjected to various biotic and abiotic processes, reduce their concentrations, and induce chemical transformation into other compounds [59,60]. The quantities of some of the main differentially contributing metabolites in *E. grandis* litter, such as betaine, chlorogenic acid, isoquercitrin, and kaempferol, generally decrease during decomposition. However, the quantity of pipecolic acid increased after two months of decomposition. Catechin, eucalyptol, pulegone, sinapic acid, sinapoyl aldehyde, and sphinganine disappeared as the decomposition proceeded, whereas phenol, phytosphingosine, and gamma-aminobutyric acid appeared as the decomposition proceeded. Although metabolites from *E. grandis* litter have not been consistently identified at each specific time point of decomposition, we also observed that the dynamics of allelochemical release were closely aligned with the phytotoxicity of *E. grandis* litter throughout the decomposition process. The main allelochemicals, including amines, phenolic acids, flavonoids, and terpenoids, explain the variation in the phytotoxicity of *E. fetida* during decomposition. Despite the insolubility of volatile organic compounds, such as terpenoids, their lower solubility and retention in soil particles have also been reported, which partly explains the phytotoxicity of eucalyptus litter on *E. fetida* [14,36].

Biochemical responses of earthworms have been regarded as a warning system for environmental stress [61]. Proteins are important components of an organism and regulate various physio-metabolic processes [62]. In this study, the total protein content of *E. fetida* decreased significantly after 20 days of litter decomposition. After the initial fragmentation of the litter, the rapid release of potential allelochemicals promoted the secretion of protein enzymes from the cell wall and intestinal tract and altered membrane permeability, consuming total protein from *E. fetida* [63,64]. Subsequent release of the soluble C fraction and nutrients gradually reduced the stimulation of earthworms, provided food resources, and increased total protein [20]. SOD is an important enzyme in the antioxidant system that prevents lipid peroxidation [65]. The SOD activity of *E. fetida* was initially increased but decreased after 30 days of decomposition and then increased. The increase in SOD activity implies that earthworms suffered from oxidative stress, as reflected by the production of superoxide anion radicals to reduce the cellular stress response [66,67]. The MDA content was also stimulated after 20 days of decomposition and then stabilized or decreased with decomposition time. MDA is the final product of lipid peroxidation, caused by an increase in oxygen free radicals, reflecting the degree of lipid peroxidation and damage to the organism [68,69]. The AChE is a specific filament amino acid hydrolase that promotes neuronal development and regeneration [70]. AChE activity of *E. fetida* was initially stable but decreased after 30 days of decomposition and then stabilized or increased with decomposition. Inhibition of AChE activity and behavioral alterations stimulate cholinergic receptors, ultimately leading to uncoordinated movements and neuromuscular paralysis [71,72]. In this study, a lack of avoidance response was observed after treatment with litter after 30 days of decomposition. Therefore, leaf litter of *E. grandis* after decomposition for 20–30 days had a stronger effect on the physiological properties of *E. fetida*.

In addition, the dynamics of allelochemicals and phytotoxicity of *E. grandis* litter on *E. fetida* varied considerably depending on plantation forest age. In this study, the total protein content of *E. fetida* was higher in the litter of 4-year-old trees than in 8-year-old trees, whereas the SOD activity of earthworms showed the opposite trend. Natural antioxidant defenses can be overwhelmed when enzymes cannot scavenge free radicals from pollutants, causing serious cellular DNA damage [73,74]. Comet assays have been used to detect DNA damage in *E. fetida* exposed to *E. grandis* litter extract. Olive tail moment (OTM) is the most sensitive comet parameter, which reveals the severity and quantity of DNA damage [75,76]. In this study, the increment of tail DNA, comet tail length, and OTM indicated that litter extracts of 8-year-old *E. grandis* caused more DNA damage to *E. fetida* than 4-year-old trees. DNA damage in earthworms leads to compromised immunity, rendering *E. fetida* vulnerable [74]. In addition, the litter extracts of *E. grandis* showed more severe toxic effects on *E. fetida* with increasing doses [77]; however, DNA damage decreased at the lowest dose. Lower concentrations of allelochemicals can weaken phytotoxicity and enhance the growth and survival of the earthworm.

Previous studies at the study sites indicated that 4-year-old and 8-year-old trees were the critical turning points for the increment of potential allelochemicals in litter across a range of *E. grandis* plantations. In the present study, except for the higher amount of amino acids and peptides, terpenoids, and amines in the initial litter and lipids after 2-month decomposition, the other primary and secondary metabolites were in greater abundance in the litter of 8-year-old *E. grandis* than in 4-year-old *E. grandis* litter. In general, mature leaves produce more secondary metabolic substances than young leaves, whereas young leaves have more abundant amino and organic acids [78]. Given the higher growth rate of *E. grandis* at four years, intra- and inter-specific competition for soil resources was severe, and *E. grandis* would invest more resources in chemical defense mechanisms [7,39]. However, the rapid growth rate of the tree created a relatively open canopy before the rotation period (approximately four years), which facilitated the development of soil microbial communities and, to some extent, might promote the degradation of the accumulated production of allelochemicals. Previous studies at the study sites also observed a slight difference in the diversity of the soil microbial community between the two-aged stands; however, they were higher in 4-year-old stands than in 8-year-old stands [79]. External environmental factors could also explain the between-age differences in potential allelochemicals [12,80]. Soluble compounds may have been efficiently leached out of young *E. grandis* litter under the relatively open canopy of 4-year-old stands. The closed canopy and developed understory plant structure and diversity in 8-year-old *E. grandis* stands increased the interception of rainfall, and phenolics were less efficiently leached out and consequently persisted longer in the mature leaf litter. As the decomposition proceeded, more abundant terpenoids were released quickly in the young leaf litter but retained for longer periods in mature *E. grandis* litter [14,81]. In addition, pH can affect the solubility of allelochemicals with increasing toxicity under acidic conditions in 8-year-old stands [11,82]. All of these factors explain the severe phytotoxicity of litter extracts of 8-year-old *E. grandis* compared to 4-year-old *E. grandis*.

5. Conclusions

In conclusion, this study demonstrated the occurrence of litter phytotoxicity in *E. fetida* with clear, dynamic patterns during the initial decomposition stage. The litter of *E. grandis*, after decomposition for 20–30 days, exerted stronger phytotoxicity on the growth and biochemical characteristics of *E. fetida*, and the toxicity was weakened by decomposition. *E. fetida* individuals exhibited significant inhibition of weight and the activities of AChE under litter treatment after 20–30 days of decomposition, whereas the activities and MDA content were stimulated at this stage. The phytotoxicity dynamics of decaying litter on *E. fetida* might be involved in the resistance of phenolic and volatile fractions after the initial fast disappearance of allelochemicals. Compared with the litter of 4-year-old trees, *E. fetida* exhibited a more severe phytotoxicity response to protein content, stimulation of SOD activity, and DNA damage in the litter extracts of 8-year-old *E. grandis*. Consistently, an increased abundance of potential allelochemicals was observed in the litter of 8-year-old *E. grandis*. However, the expression of allelopathy in 8-year-old litter does not necessarily need to outcompete 4-year-old litter under natural conditions because of the developed stand structure, environmental heterogeneity, and abundant precipitation in this region. Overall, our results suggest that the use of scientific management measures, such as the appropriate removal of leaf litter in the early stages of decomposition, may help support greater biodiversity in *E. grandis* plantations.

Supplementary Materials: The following supporting information can be downloaded at https://www.mdpi.com/article/10.3390/plants13172415/s1. Table S1: Effects of litter extracts from two aged *E. grandis* stands during decomposition on the survival rate of *E. fetida* after 7 and 14 days of exposure; Table S2: Effects of litter extracts from two aged E. grandis stands during decomposition on the weight inhibition rate of *E. fetida* after 7 and 14 days of exposure; Table S3: Correlations between the first two axes and environmental factors in RDA and the determination coefficient (r^2) and significance test (Pr) of the correlation between environmental factors and differential potential allelochemicals using the envfit function (R package vegan); Table S4: Correlations between the first two axes and differential potential allelochemicals in RDA, and the determination coefficient (r^2) and significance test (Pr) of the correlation between differential potential allelochemicals and growth and physiological properties of *E. fetida* using the envfit function (R package vegan); Figure S1: Effects of litter extracts from two mature E. grandis stands during decomposition on the total protein content of *E. fetida* after 7 and 14 days of exposure; Figure S2: Effects of litter extracts from two mature *E. grandis* stands during decomposition on SOD activity of *E. fetida* after 7 and 14 days of exposure; Figure S3: Effects of litter extracts from two mature *E. grandis* stands during decomposition on MDA content of *E. fetida* after 7 and 14 days of exposure; Figure S4: Effects of litter extracts from two mature *E. grandis* stands during decomposition on AChE activity of *E. fetida* after 7 and 14 days of exposure.

Author Contributions: D.Z.: Conceptualization, data curation, formal analysis, investigation, methodology, validation, writing—review and editing, funding acquisition, project administration, and resources; C.L.: investigation, data curation, formal analysis, methodology, software, visualization, and writing—original draft; S.F.: investigation, data curation, formal analysis, and methodology; Y.H.: investigation, methodology, and supervision; N.K.: investigation, methodology, and resources; S.G.: investigation and methodology; L.C.: investigation and methodology. All authors have read and agreed to the published version of the manuscript.

Funding: This work was funded by the National Natural Science Foundation of China (32171775, 31770671).

Data Availability Statement: Data will be made available upon request.

Conflicts of Interest: The authors declare that they have no known competing financial interests or personal relationships that could have appeared to influence the work reported in this paper.

References

1. Zhang, D.J.; Li, J.J.; Huang, Y.M.; Gao, S.; Zhang, J. Root-soil facilitation in mixed *Eucalyptus grandis* plantations including nitrogen-fixing species. *For. Ecol. Manag.* **2022**, *516*, 120215. [CrossRef]
2. Massuque, J.; Roque Lima, M.D.; Müller Da Silva, P.H.; de Paula Protásio, T.; Trugilho, P.F. Potential of charcoal from non-commercial *Corymbia* and *Eucalyptus* wood for use in the steel industry. *Renew. Energy* **2023**, *211*, 179–187. [CrossRef]
3. National Forestry and Grassland Administration. *China Forest Resource Report*; National Forestry and Grassland Administration: Beijing, China, 2020.
4. Forrester, D.I.; Bauhus, J.; Cowie, A.L.; Vanclay, J.K. Mixed-species plantations of *Eucalyptus* with nitrogen-fixing trees: A review. *For. Ecol. Manag.* **2006**, *233*, 211–230. [CrossRef]
5. Rodríguez-Loinaz, G.; Amezaga, I.; Onaindia, M. Use of native species to improve carbon sequestration and contribute towards solving the environmental problems of the timberlands in Biscay, northern Spain. *J. Environ. Manag.* **2013**, *120*, 18–26. [CrossRef]
6. Ahmed, R.; Hoque, A.T.M.R.; Hossain, M.K. Allelopathic effects of leaf litters of *Eucalyptus camaldulensis* on some forest and agricultural crops. *J. For. Res.* **2008**, *19*, 19–24. [CrossRef]
7. Zhang, D.J.; Zhang, J.; Yang, W.Q.; Wu, F.Z. Potential allelopathic effect of *Eucalyptus grandis* across a range of plantation ages. *Ecol. Res.* **2010**, *25*, 13–23. [CrossRef]
8. Chu, C.J.; Mortimer, P.E.; Wang, H.C.; Wang, Y.F.; Liu, X.B.; Yu, S.X. Allelopathic effects of *Eucalyptus* on native and introduced tree species. *For. Ecol. Manag.* **2014**, *323*, 79–84. [CrossRef]
9. Rice, E.L. *Allelopathy*, 2nd ed.; Academic Press: Orlando, FL, USA, 1984.
10. Inderjit; Callaway, R.M. Experimental designs for the study of allelopathy. *Plant Soil* **2003**, *256*, 1–11. [CrossRef]
11. Bonanomi, G.; Sicurezza, M.G.; Caporaso, S.; Esposito, A.; Mazzoleni, S. Phytotoxicity dynamics of decaying plant materials. *New Phytol.* **2006**, *169*, 571–578. [CrossRef]
12. Chomel, M.; Fernandez, C.; Bousquet-Mélou, A.; Gers, C.; Monnier, Y.; Santonja, M.; Gauquelin, T.; Gros, R.; Lecareux, C.; Baldy, V. Secondary metabolites of *Pinus halepensis* alter decomposer organisms and litter decomposition during afforestation of abandoned agricultural zones. *J. Ecol.* **2014**, *102*, 411–424. [CrossRef]
13. Kruidhof, H.M.; Gallandt, E.R.; Haramoto, E.R.; Bastiaans, L. Selective weed suppression by cover crop residues: Effects of seed mass and timing of species' sensitivity. *Weed Res.* **2011**, *51*, 177–186. [CrossRef]
14. Tang, Z.Q.; Zhang, J.; Yu, J.L.; Wang, C.Z.; Zhang, D.J. Allelopathic effects of volatile organic compounds from *Eucalyptus grandis* rhizosphere soil on *Eisenia fetida* assessed using avoidance bioassays, enzyme activity, and comet assays. *Chemosphere* **2017**, *173*, 307–317. [CrossRef]
15. Molina, A.; Reigosa, M.J.; Carballeira, A. Release of allelochemical agents from litter, throughfall, and topsoil in plantations of *Eucalyptus globulus* Labill in Spain. *J. Chem. Ecol.* **1991**, *17*, 147–160. [CrossRef] [PubMed]
16. Fang, B.Z.; Yu, S.X.; Wang, Y.F.; Qiu, X.; Cai, C.X.; Liu, S.P. Allelopathic effects of *Eucalyptus urophylla* on ten tree species in south China. *Agrofor. Syst.* **2009**, *76*, 401–408. [CrossRef]
17. Mohamed, E.I.; Gad, E.K. *Eucalyptus citriodora* leaf extract as a source of allelochemicals for weed control in pea fields compared with some chemical herbicides. *J. Plant Prot. Res.* **2019**, *59*, 392–399. [CrossRef]
18. Inderjit; Wardle, D.A.; Karban, R.; Callaway, R.M. The ecosystem and evolutionary contexts of allelopathy. *Trends Ecol. Evol.* **2011**, *26*, 655–662. [CrossRef]
19. Boeno, D.; Silva, R.F.; Almeida, H.S.; Rodrigues, A.C.; Vanzan, M.; Andreazza, R. Influence of eucalyptus development under soil fauna. *Braz. J. Biol.* **2020**, *80*, 345–353. [CrossRef]
20. Rodriguez-Campos, J.; Dendooven, L.; Alvarez-Bernal, D.; Contreras-Ramos, S.M. Potential of earthworms to accelerate removal of organic contaminants from soil: A review. *Appl. Soil Ecol.* **2014**, *79*, 10–25. [CrossRef]
21. Zhang, C.; Dai, J.; Chen, X.F.; Li, H.H.; Lavelle, P. Effects of a native earthworm species (*Amynthas morrisi*) and *Eisenia fetida* on metal fractions in a multi-metal polluted soil from South China. *Acta Oecologica* **2020**, *102*, 103503. [CrossRef]
22. Xie, X.C.; Qian, Y.; Wu, Y.X.; Yin, J.; Zhai, J.P. Effects of decabromodiphenyl ether (BDE-209) on the avoidance response, survival, growth and reproduction of earthworms (*Eisenia fetida*). *Ecotoxicol. Environ. Saf.* **2013**, *90*, 21–27. [CrossRef]
23. Markert, B.; Breure, A.M.; Zechmeister, H.G. *Bioindicators and Biomonitors: Principles, Concepts and Applications*; Elsevier Science: New York, NY, USA, 2003.
24. Gomez-Eyles, J.L.; Svendsen, C.; Lister, L.; Martin, H.; Hodson, M.E.; Spurgeon, D.J. Measuring and modelling mixture toxicity of imidacloprid and thiacloprid on *Caenorhabditis elegans* and *Eisenia fetida*. *Ecotoxicol. Environ. Saf.* **2009**, *72*, 71–79. [CrossRef] [PubMed]
25. Makkonen, M.; Berg, M.P.; van Logtestijn, R.S.P.; van Hal, J.R.; Aerts, R. Do physical plant litter traits explain non-additivity in litter mixtures? A test of the improved microenvironmental conditions theory. *Oikos* **2013**, *122*, 987–997. [CrossRef]
26. Gartner, T.B.; Cardon, Z.G. Decomposition Dynamics in Mixed-Species Leaf Litter. *Oikos* **2004**, *104*, 230–246. [CrossRef]
27. Hättenschwiler, S.; Tiunov, A.V.; Scheu, S. Biodiversity and Litter Decomposition in Terrestrial Ecosystems. *Annu. Rev. Ecol. Evol. Syst.* **2005**, *36*, 191–218. [CrossRef]
28. Wang, L.F.; Zhou, Y.; Chen, Y.M.; Xu, Z.F.; Zhang, J.; Liu, Y. Home-field advantage and ability alter labile and recalcitrant litter carbon decomposition in an alpine forest ecotone. *Plant Soil* **2023**, *485*, 213–225. [CrossRef]
29. Hättenschwiler, S.; Vitousek, P.M. The role of polyphenols in terrestrial ecosystem nutrient cycling. *Trends Ecol. Evol.* **2000**, *15*, 238–243. [CrossRef]

30. Coq, S.; Souquet, J.; Meudec, E.; Cheynier, V.; Hättenschwiler, S. Interspecific variation in leaf litter tannins drives decomposition in a tropical rain forest of French Guiana. *Ecology* **2010**, *91*, 2080–2091. [CrossRef]
31. Zhang, Y.; Li, X.; Zhang, D.J.; Qin, Y.; Zhou, Y.; Song, S.M.; Zhang, J. Characteristics of fungal community structure during the decomposition of mixed foliage litter from *Pinus massoniana* and broadleaved tree species in southwestern China. *J. Plant Ecol.* **2020**, *13*, 574–588. [CrossRef]
32. Li, J.J.; Huang, Y.M.; Chen, L.H.; Gao, S.; Zhang, J.; Zhang, D.J. Understory plant diversity and phenolic allelochemicals across a range of *Eucalyptus grandis* plantation ages. *J. For. Res.* **2023**, *34*, 1577–1590. [CrossRef]
33. An, M.; Pratley, J.E.; Haig, T. Phytotoxicity of Vulpia Residues: IV. Dynamics of Allelochemicals During Decomposition of Vulpia Residues and Their Corresponding Phytotoxicity. *J. Chem. Ecol.* **2001**, *27*, 395–409. [CrossRef]
34. Bernhard-Reversat, F.; Main, G.; Holl, K.; Loumeto, J.; Ngao, J. Fast disappearance of the water-soluble phenolic fraction in eucalypt leaf litter during laboratory and field experiments. *Appl. Soil Ecol.* **2003**, *23*, 273–278. [CrossRef]
35. Blum, U. *Plant-Plant Allelopathic Interactions II: Laboratory Bioassays for Water-Soluble Compounds with an Emphasis on Phenolic Acids*; Springer Science and Business Media: Cham, Switzerland, 2014.
36. Puig, C.G.; Gonçalves, R.F.; Valentão, P.; Andrade, P.B.; Reigosa, M.J.; Pedrol, N. The Consistency between Phytotoxic Effects and the Dynamics of Allelochemicals Release from *Eucalyptus globulus* Leaves Used as Bioherbicide Green Manure. *J. Chem. Ecol.* **2018**, *44*, 658–670. [CrossRef]
37. Liu, S.; Qin, F.C.; Yu, S.X. *Eucalyptus urophylla* root-associated fungi can counteract the negative influence of phenolic acid allelochemicals. *Appl. Soil Ecol.* **2018**, *127*, 1–7. [CrossRef]
38. Zhang, D.J.; Zhang, J.; Yang, W.Q.; Wu, F.Z. Effects of afforestation with *Eucalyptus grandis* on soil physicochemical and microbiological properties. *Soil Res.* **2012**, *50*, 167–176. [CrossRef]
39. Zhang, D.J.; Zhang, J.; Yang, W.Q.; Wu, F.Z.; Huang, Y.M. Plant and soil seed bank diversity across a range of ages of *Eucalyptus grandis* plantations afforested on arable lands. *Plant Soil* **2014**, *376*, 307–325. [CrossRef]
40. Zhang, D.J.; Zhang, J.; Yang, W.Q.; Wu, F.Z.; Huang, Y.M.; Zhang, Z.W.; Wang, X.; Wang, X.Q.; Zhu, L. Plant's and soil organism's diversity across a range of *Eucalyptus grandis* plantation ages. *Acta Ecol. Sin.* **2013**, *33*, 3947–3962. (In Chinese) [CrossRef]
41. Research Group of Chinese Soil Taxonomy. *Keys to Chinese Soil Taxonomy*, 3rd ed.; Press of University of Science and Technology of China: Hefei, China, 2001.
42. OECD. *Guideline for Testing of Chemicals. No.222. Earthworm Reproduction Test (Eisenia fetida/andrei)*; Organization for Economic Cooperation and Development: Paris, France, 2004.
43. *ISO/DIS 17512-1.2*; Soil Quality: Avoidance Test for Testing the Quality of Soils and Effects of Chemicals on Behavior—Part 1: Test with Earthworms (*Eisenia fetida* and *Eisenia andrei*). International Organization for Standardization: Geneva, Switzerland, 2008.
44. Yu, J.L.; Zhang, J.; Tang, Z.Q.; Wang, C.Z.; Qian, L.Z.; Zhang, D.J. Variations of growth and physiological indicators of *Eisenia foetida* in soils across a chronosequence of *Eucalyptus grandis* plantations. *Chin. J. Ecol.* **2017**, *36*, 124–131. (In Chinese) [CrossRef]
45. Smith, P.K.; Krohn, R.I.; Hermanson, G.T.; Mallia, A.K.; Gartner, F.H.; Provenzano, M.D.; Fujimoto, E.K.; Goeke, N.M.; Olson, B.J.; Klenk, D.C. Measurement of protein using bicinchoninic acid. *Anal. Biochem.* **1985**, *150*, 76–85. [CrossRef]
46. Cortés-Ríos, J.; Zárate, A.M.; Figueroa, J.D.; Medina, J.; Fuentes-Lemus, E.; Rodríguez-Fernández, M.; Aliaga, M.; López-Alarcón, C. Protein quantification by bicinchoninic acid (BCA) assay follows complex kinetics and can be performed at short incubation times. *Anal. Biochem.* **2020**, *608*, 113904. [CrossRef]
47. Ōyanagui, Y. Reevaluation of assay methods and establishment of kit for superoxide dismutase activity. *Anal. Biochem.* **1984**, *142*, 290–296. [CrossRef]
48. Kosugi, H.; Kikugawa, K. Thiobarbituric acid reaction of aldehydes and oxidized lipids in glacial acetic acid. *Lipids* **1985**, *20*, 915–921. [CrossRef]
49. Gorun, V.; Proinov, I.; Băltescu, V.; Balaban, G.; Bârzu, O. Modified Ellman procedure for assay of cholinesterases in crude enzymatic preparations. *Anal. Biochem.* **1978**, *86*, 324–326. [CrossRef]
50. Eyambe, G.S.; Goven, A.J.; Fitzpatrick, L.C.; Venables, B.J.; Cooper, E.L. A non-invasive technique for sequential collection of earthworm (*Lumbricus terrestris*) leukocytes during subchronic immunotoxicity studies. *Lab. Anim.* **1991**, *25*, 61–67. [CrossRef] [PubMed]
51. Ribeiro, S.; Sousa, J.P.; Nogueira, A.J.A.; Soares, A.M.V.M. Effect of Endosulfan and Parathion on Energy Reserves and Physiological Parameters of the Terrestrial Isopod *Porcellio dilatatus*. *Ecotoxicol. Environ. Saf.* **2001**, *49*, 131–138. [CrossRef] [PubMed]
52. Verdú, I.; Trigo, D.; Martínez-Guitarte, J.L.; Novo, M. Bisphenol A in artificial soil: Effects on growth, reproduction and immunity in earthworms. *Chemosphere* **2018**, *190*, 287–295. [CrossRef]
53. Olvera-Velona, A.; Capowiez, Y.; Mascle, O.; Ortiz-Hernandez, L.; Benoit, P. Assessment of the toxicity of ethyl-parathion to earthworms (*Aporrectodea caliginosa*) using behavioural, physiological and biochemical markers. *Appl. Soil Ecol.* **2008**, *40*, 476–483. [CrossRef]
54. Lummer, D.; Scheu, S.; Butenschoen, O. Connecting litter quality, microbial community and nitrogen transfer mechanisms in decomposing litter mixtures. *Oikos* **2012**, *121*, 1649–1655. [CrossRef]
55. Wu, F.Z.; Peng, C.H.; Yang, W.Q.; Zhang, J.; Han, Y.; Mao, T. Admixture of alder (*Alnus formosana*) litter can improve the decomposition of eucalyptus (*Eucalyptus grandis*) litter. *Soil Biol. Biochem.* **2014**, *73*, 115–121. [CrossRef]
56. Maity, S.K.; Joy, V.C. Impact of antinutritional chemical compounds of leaf litter on detritivore soil arthropod fauna. *J. Ecobiol.* **1999**, *11*, 193–202.

57. Hodge, A.; Stewart, J.; Robinson, D.; Griffiths, B.S.; Fitter, A.H. Root proliferation, soil fauna and plant nitrogen capture from nutrient-rich patches in soil. *New Phytol.* **1998**, *139*, 479–494. [CrossRef]
58. Reigosa, M.J.; Pazos-Malvido, E. Phytotoxic Effects of 21 Plant Secondary Metabolites on *Arabidopsis thaliana* Germination and Root Growth. *J. Chem. Ecol.* **2007**, *33*, 1456–1466. [CrossRef]
59. Otte, B.A.; Rice, C.P.; Davis, B.W.; Schomberg, H.H.; Mirsky, S.B.; Tully, K.L. Phenolic acids released to soil during cereal rye cover crop decomposition. *Chemoecology* **2020**, *30*, 25–34. [CrossRef]
60. Bonanomi, G.; Zotti, M.; Idbella, M.; Mazzoleni, S.; Abd-ElGawad, A.M. Microbiota modulation of allelopathy depends on litter chemistry: Mitigation or exacerbation? *Sci. Total Environ.* **2021**, *776*, 145942. [CrossRef] [PubMed]
61. Gu, H.T.; Yuan, Y.D.; Cai, M.; Wang, D.S.; Lv, W.G. Toxicity of isoprocarb to earthworms (*Eisenia fetida*): Oxidative stress, neurotoxicity, biochemical responses and detoxification mechanisms. *Environ. Pollut.* **2021**, *290*, 118038. [CrossRef]
62. Chen, L.Y.; Bai, J.; Wan, J.; Song, Y.; Xiang, G.H.; Duan, R.Y.; Zheng, Y. Endocrine system, cell growth and death, and energy metabolism induced by Sb(III) exposure in earthworm (*Pheretima guillemi*) revealed by transcriptome and metabolome analysis. *Environ. Pollut.* **2024**, *356*, 124357. [CrossRef] [PubMed]
63. Zhao, H.B.; Shi, S.Y.; Zhao, H.; Guo, J.; Yang, Z.; Gao, H.S.; Lu, F.P. Proteomic analysis of the earthworm *Eisenia fetida* exposed to oxytetracycline in soil. *RSC Adv.* **2019**, *9*, 41628–41638. [CrossRef]
64. Rong, H.; Wang, C.R.; Liu, H.T.; Zhang, M.; Yuan, Y.T.; Pu, Y.J.; Huang, J.; Yu, J.Y. Biochemical Toxicity and Potential Detoxification Mechanisms in Earthworms *Eisenia fetida* Exposed to Sulfamethazine and Copper. *Bull. Environ. Contam. Toxicol.* **2020**, *105*, 255–260. [CrossRef] [PubMed]
65. Yao, X.F.; Liang, C.L.; Lv, H.J.; Liu, W.R.; Wang, Q.; Ding, J.; Li, X.X.; Wang, J. Expanding the insight of ecological risk on the novel chiral pesticide mefentrifluconazole: Mechanism of enantioselective toxicity to earthworms (*Eisenia fetida*). *J. Hazard. Mater.* **2024**, *466*, 133585. [CrossRef]
66. Shao, Y.T.; Wang, J.H.; Wang, J.; Du, Z.K.; Li, B.; Zhu, L.S.; Juhasz, A.; Liu, X.Y.; Xu, Y.Q.; Li, W.X. Oxidative stress and genotoxic effects in earthworms induced by five imidazolium bromide ionic liquids with different alkyl chains. *Chemosphere* **2019**, *227*, 570–579. [CrossRef]
67. Song, P.P.; Gao, J.P.; Li, X.X.; Zhang, C.; Zhu, L.S.; Wang, J.H.; Wang, J. Phthalate induced oxidative stress and DNA damage in earthworms (*Eisenia fetida*). *Environ. Int.* **2019**, *129*, 10–17. [CrossRef]
68. Shi, Z.M.; Yan, J.H.; Ren, X.N.; Wen, M.; Zhao, Y.H.; Wang, C.Y. Effects of biochar and thermally treated biochar on *Eisenia fetida* survival, growth, lysosomal membrane stability and oxidative stress. *Sci. Total Environ.* **2021**, *770*, 144778. [CrossRef] [PubMed]
69. Bao, X.; Wang, Z.J.; Liu, L.; Wang, D.W.; Gu, Y.T.; Chen, L.; Chen, X.J.; Meng, Z.Y. The combined effects of azoxystrobin and different aged polyethylene microplastics on earthworms (*Eisenia fetida*): A systematic evaluation based on oxidative damage and intestinal function. *Sci. Total Environ.* **2024**, *923*, 171494. [CrossRef] [PubMed]
70. Gautam, K.; Dwivedi, S.; Verma, R.; Vamadevan, B.; Patnaik, S.; Anbumani, S. Combined effects of polyethylene microplastics and carbendazim on *Eisenia fetida*: A comprehensive ecotoxicological study. *Environ. Pollut.* **2024**, *348*, 123854. [CrossRef]
71. He, F.L.; Liu, R.T. Mechanistic insights into phenanthrene-triggered oxidative stress-associated neurotoxicity, genotoxicity, and behavioral disturbances toward the brandling worm (*Eisenia fetida*) brain: The need for an ecotoxicological evaluation. *J. Hazard. Mater.* **2023**, *450*, 131072. [CrossRef] [PubMed]
72. Ali, H.M.; Abdel-Aty, B.; El-Sayed, W.; Mariy, F.M.; Hegazy, G.M.; Mohamed, R.A.; Zoghly, H.M. Imidacloprid effects on acetylcholinesterase and nicotinic acetylcholine receptor in *Apis mellifera*. Experimental and molecular modeling approaches. *Chemosphere* **2024**, *356*, 141899. [CrossRef]
73. Zhao, S.L.; Wang, Y.L.; Duo, L. Biochemical toxicity, lysosomal membrane stability and DNA damage induced by graphene oxide in earthworms. *Environ. Pollut.* **2021**, *269*, 116225. [CrossRef]
74. Li, X.X.; Jiang, N.; Zhang, J.; Yao, X.F.; Liu, W.R.; Wang, Q.; Ding, J.; Hu, Z.R.; Zhu, L.S.; Wang, J.H.; et al. Soil health hazards of di(2-ethylhexyl) phthalate: New perspectives on earthworms from different ecological niches DNA damage, gut microbial disruption and soil enzyme changes. *J. Hazard. Mater.* **2024**, *467*, 133700. [CrossRef] [PubMed]
75. Espinosa-Reyes, G.; Costilla-Salazar, R.; Pérez-Vázquez, F.J.; González-Mille, D.J.; Flores-Ramírez, R.; Del Carmen Cuevas-Díaz, M.; Medellin-Garibay, S.E.; Ilizaliturri-Hernández, C.A. DNA damage in earthworms by exposure of Persistent Organic Pollutants in low basin of Coatzacoalcos River, Mexico. *Sci. Total Environ.* **2019**, *651*, 1236–1242. [CrossRef]
76. Yan, X.J.; Wang, J.H.; Zhu, L.S.; Wang, J.; Li, S.Y.; Kim, Y.M. Oxidative stress, growth inhibition, and DNA damage in earthworms induced by the combined pollution of typical neonicotinoid insecticides and heavy metals. *Sci. Total Environ.* **2021**, *754*, 141873. [CrossRef]
77. Xue, Y.N.; Li, Z.G.; Liu, C.; Liu, D.M.; Wang, J.H.; Liu, C.; Xia, X.M. Effect of different exposure times and doses of cyantraniliprole on oxidative stress and genotoxicity in earthworms (*Eisenia fetida*). *Chemosphere* **2023**, *319*, 138023. [CrossRef]
78. Lin, W.D.; Li, J.M. Comparative Analysis of the Main Metabolites in Tender and Matured Leaves of *Cyclocarya Paliurus* Based on Metabolomic Data. *J. Taizhou Univ.* **2021**, *43*, 55–62. (In Chinese) [CrossRef]
79. Wang, C.Z.; Zhang, D.J.; Yu, J.L.; Tang, Z.Q.; Zhang, J. Soil Microbial Community Diversity and Composition across a Range of *Eucalyptus grandis* Plantations of Different Ages. *Int. J. Agric. Biol.* **2019**, *21*, 527–537. [CrossRef]
80. Kuiters, A.T.; Sarink, H.M. Leaching of phenolic compounds from leaf and needle litter of several deciduous and coniferous trees. *Soil Biol. Biochem.* **1986**, *18*, 475–480. [CrossRef]

81. Ormeño, E.; Fernandez, C.; Mévy, J. Plant coexistence alters terpene emission and content of Mediterranean species. *Phytochemistry* **2007**, *68*, 840–852. [CrossRef] [PubMed]
82. Armstrong, J.; Armstrong, W. *Phragmites* die-back: Toxic effects of propionic, butyric and caproic acids in relation to pH. *New Phytol.* **1999**, *142*, 201–217. [CrossRef]

plants

MDPI

Article

A Nucleotide-Binding Domain Leucine-Rich Repeat Gene Regulates Plant Growth and Defense Against Chewing Herbivores

Chen Qiu [1], Xiaochen Jin [1], Yumiao Zhao [1,2], Peng Kuai [1,*] and Yonggen Lou [1,2,*]

1 State Key Laboratory of Rice Biology & Ministry of Agriculture Key Laboratory of Agricultural Entomology, Key Laboratory of Biology of Crop Pathogens and Insects of Zhejiang Province, Institute of Insect Sciences, Zhejiang University, Hangzhou 310058, China; cqiu2019@163.com (C.Q.); 12116098@zju.edu.cn (X.J.); zhaoyumiaoi@163.com (Y.Z.)

2 Hainan Institute, Zhejiang University, Sanya 572025, China

* Correspondence: kpchen7493@163.com (P.K.); yglou@zju.edu.cn (Y.L.); Tel.: +86-137-3551-5845 (P.K.); +86-571-88982622 (Y.L.)

Abstract: Plant nucleotide-binding leucine-rich repeat immune receptor genes (NLRs) play an important role in plant defenses against pathogens, pathogenic nematodes, and piercing–sucking herbivores. However, little is known about their functions in plant defenses against chewing herbivores. Here, we identified a plasma membrane-localized coiled-coil-type NLR protein, OsPik-2-like, whose transcript levels were induced by the infestation of rice leaf folder (LF, *Cnaphalocrocis medinalis*) larvae, and by treatment with mechanical wounding. Knocking out *OsPik-2-like* in rice increased the LF-induced levels of jasmonic acid (JA) and jasmonoyl–isoleucine (JA-Ile), the activity of trypsin protease inhibitors (TrypPIs), and the basal levels of some flavonoids, which in turn decreased the performance of LF larvae. Moreover, knocking out *OsPik-2-like* reduced plant growth. These findings demonstrate that OsPik-2-like regulates the symbiosis between rice and LF by balancing plant growth and defense.

Keywords: rice; *OsPik-2-like*; *Cnaphalocrocis medinalis*; jasmonic acid; jasmonoyl–isoleucine; trypsin protease inhibitors

1. Introduction

Plants are constantly being challenged by a variety of stresses, including herbivores. As a consequence, plants have evolved complicated strategies to cope with herbivore infestation. When attacked by herbivores, plants employ two tiers of immune receptors—plasma membrane-localized pattern recognition receptors (PRRs) and intracellular nucleotide-binding domain leucine-rich repeat receptors (NLRs)—to initiate specific defense responses [1–3]. PRRs mainly recognize plant-derived damage-associated molecular patterns (DAMPs) or herbivore-associated molecular patterns (HAMPs, also called herbivore-associated elicitors) to activate pattern-triggered immunity (PTI) [2,4], whereas NLRs mainly perceive herbivore-secreted effectors to initiate effector-triggered immunity (ETI) [5,6]. After recognizing patterns/effectors, plants activate a series of defense responses, including calcium influx, the production of reactive oxygen species (ROS), the activation of mitogen-activated protein kinases (MAPKs), and signaling pathways mediated by phytohormones (such as jasmonic acid (JA), salicylic acid (SA), ethylene (ET), and abscisic acid (ABA)), and the accumulation of defense compounds. Acting together, these responses enhance the direct and indirect resistance of plants to herbivores [2,7].

Plant NLRs are usually composed of three conserved domains: a variable N-terminal domain; a central NB-ARC (nucleotide-binding adaptor shared by Apaf1, certain resistance genes, and a CED4) domain; and a C-terminal LRR (leucine-rich repeat) domain [8].

Based on their N-terminal domains, plant NLRs are divided into three classes, including Toll/interleukin-1 receptor/R protein (TIR)-type NLRs (TNLs); coiled-coil (CC)-type NLRs (CNLs); and resistance to powdery mildew 8 (RPW8)-type NLRs (RNLs) [8]. Numerous studies have revealed that NLRs usually regulate plant disease resistance by forming homomultimers or heteromultimers [8,9]. For example, *PigmR* (*Pigm Resistant*) and *PigmS* (*Pigm Susceptible*) are functional gene pairs identified at the rice *Pigm* locus for rice blast resistance. PigmR forms homologous dimers conferring broad-spectrum disease resistance to rice, but severely decreasing rice yields in the process; however, PigmS binds to PigmR to form heterodimers, thereby inhibiting PigmR-mediated broad-spectrum disease resistance and offsetting the negative effect of PigmR on rice yields [10]. In *Arabidopsis* and tobacco plants, ZAR1 (HopZ-activated resistance 1) interacts with multiple members of the receptor-like cytoplasmic kinase (RLCK) subfamily to form immune receptor complexes; these sense the pathogen-derived effectors and trigger immune responses [11–14]. For instance, the ZAR1 resistosome in *Arabidopsis*—a heterodimer composed of ZAR1, the resistance-related kinase 1 (RKS1), and the PBL2^{UMP} (the acetylated form of PBL2 by the effector protein AvrAC of *Xanthomonas campestris* pv. *campestris*)—directly enters the lipid bilayer of the plasma membrane of cells, where it acts as a Ca^{2+} channel and promotes Ca^{2+} influx, thereby triggering ROS production and cell death [14].

Recently, NLRs have also been reported to play an important role in plant herbivore resistance [3]. For example, several NLR genes involved in plant resistance to aphids have been identified and cloned, such as *Vat* in *Cucumis melo*, *Ra* in *Lactuca sativa*, and *AIN* (*Acyrthosiphon-induced necrosis*) in *Medicago truncatula* [15–18]. Also, the resistance gene *Mi1.2* in *Lycopersicon peruvianum* confers broad-spectrum resistance to root-knot nematodes, whitefly (*Bemisia tabaci*), and psyllid (*Bactericerca cockerelli*) on plants [19,20]. When *Pieris brassicae* egg deposition-induced hypersensitive response (HR)-like cell death was genetically mapped in the black mustard *Brassica nigra*, a cluster of TNL genes was identified as a potential source of R genes; these genes may regulate the ability of plants to kill herbivore eggs [21]. In rice, two CNL genes, *Brown planthopper resistance 9* (*Bph9*) and *Bph14*, are well known to confer BPH resistance on rice: *Bph9* regulates rice BPH resistance by affecting JA, SA, and ET-mediated signaling pathways, and *Bph14* positively mediates the resistance of rice to BPH by promoting SA and ROS-mediated defenses [22–25]. Further studies proved that the CC and NB domains, as well as the full length of Bph14, interact with two WRKY proteins (OsWRKY46 and OsWRKY72); this interaction stabilizes the two WRKY proteins and enhances their transactivation activity, thereby increasing the expression of callose-biosynthesis genes and the deposition of calloses in the phloem of rice leaf sheaths [26]. Moreover, Bph14 interacts with the BPH-derived effector BISP (Bph14-interacting salivary protein) to activate ETI, which markedly increased rice resistance to BPH [27]. However, our understanding of the role of NLRs in plant defense against herbivores and their underlying mechanisms is limited.

Rice (*Oryza sativa* L.), one of the primary food crops in the world, is often seriously damaged in the field by multiple herbivores, including the rice leaf folder (LF, *Cnaphalocrocis medinalis*) [28]. LF larvae fold rice leaves longitudinally and feed on the green mesophyll tissues of these folded leaves, actions that reduce both the photosynthetic productivity and grain yield of rice plants [29]. It has been reported that the infestation of LF larvae activates defense-related signaling pathways mediated by JA, SA, ET, and H_2O_2 in rice; these signaling pathways jointly modulate the expression of defense genes and the accumulation of defense compounds, such as trypsin protease inhibitors (TrypPIs) and phenolamines [30,31]. Although a few NLRs have been reported to be involved in the resistance of rice to piercing–sucking herbivores, as stated above, whether NLRs participate in rice defense against chewing herbivores, such as LF, remains largely unknown.

In this study, we isolated a rice CNL gene *OsPik-2-like* (XM_015756755.2), which was induced by LF larval infestation [31], and investigated its function in interactions between rice and LF. By combining molecular tools, chemical analysis, and bioassays, we found that *OsPik-2-like* negatively regulates the biosynthesis of LF-induced JA and JA-Ile, the activity of induced TrypPIs, and the resistance of rice to LF, suggesting that NLRs participate in rice–LF interactions.

2. Results

2.1. Characterization of OsPik-2-like

Using transcriptome data [31], we determined that a putative CNL gene was up-regulated by herbivore infestation. The full-length cDNA sequence obtained from a cDNA library of rice variety XS11 using the reverse-transcription polymerase chain reaction (RT-PCR) was identical to the sequence of a rice disease resistance gene *OsPik-2-like* (annotated in the Genbank database, https://www.ncbi.nlm.nih.gov/genbank/, under the ID: XM_015756755.2; accessed 30 August 2023). The gene includes an open-reading frame of 2,892 bp that encodes a protein of 963 amino acids with a predicted molecular weight of 109.03 kDa (Figure S1). The protein harbored three conserved domains: a CC domain, a NB-ARC domain, and a LRR domain (Figure 1a). OsPik-2-like shared high similarity with CNL proteins in other plants, including ObPik-2-like (88.30% identity) in *Oryza brachyantha*, BdRPM1 (72.28% identity) in *Brachypodium distachyon*, and PaPIK6-NP-like (71.06% identity) in *Phragmites australis* (Figure 1b). These results showed that OsPik-2-like belongs to the CNL protein family.

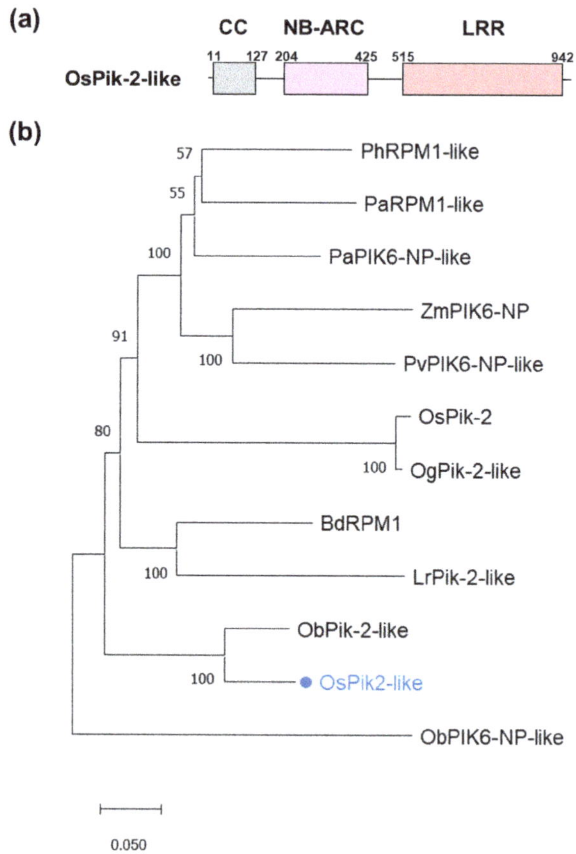

Figure 1. Structure and phylogenetic relationship of OsPik-2-like. (**a**) Schematic structure of OsPik-2-like. (**b**) Sequence alignment of OsPik-2-like and its homologs. Species abbreviations are included before

the protein names: Bd, *Brachypodium distachyon*; Lr, *Lolium rigidum*; Ob, *Oryza brachyantha*; Og, *Oryza glaberrima*; Os, *Oryza sative*; Pa, *Phragmites australis*; Ph, *Panicum hallii*; Pv, *Panicum virgatum*; and Zm, *Zea mays*. Plant species and accession numbers from the NCBI database are as follows: BdRPM1, XP_003576018.1; LrPik-2-like, XP_047085044.1; ObPIK6-NP-like, XP_040383595.1; ObPik-2-like, XP_006660848.1; OgPik-2-like, XP_052139461.1; OsPik-2, XP_015619167.2; OsPik-2-like, XP_015612241.1; PaPIK6-NP-like, XP_062198511.1; PaRPM1-like, XP_062193254.1, PhRPM1 like, XP_025803740.1; PvPIK6-NP-like, XP_039795950.1; ZmPIK6-NP, XP_008652943.2. The blue dot indicates OsPik-2-like. The scale bar represents 0.05 amino acid substitutions per site in the primary structure.

To elucidate the subcellular localization of OsPik-2-like, we generated the construct *35S::OsPik-2-like-EGFP* (the enhanced green fluorescent protein was fused to OsPik-2-like), which was driven by a *CaMV 35S* promoter. Then, the construct was transiently expressed in *Nicotiana benthamiana* leaves using *Agrobacterium tumefaciens*-mediated transformation. As shown in Figure 2, compared with the wide distribution of the control EGFP, the OsPik-2-like-EGFP mainly localized to plasma membranes of tobacco leaf cells, indicating that OsPik-2-like was a plasma membrane (PM)-localized CNL protein.

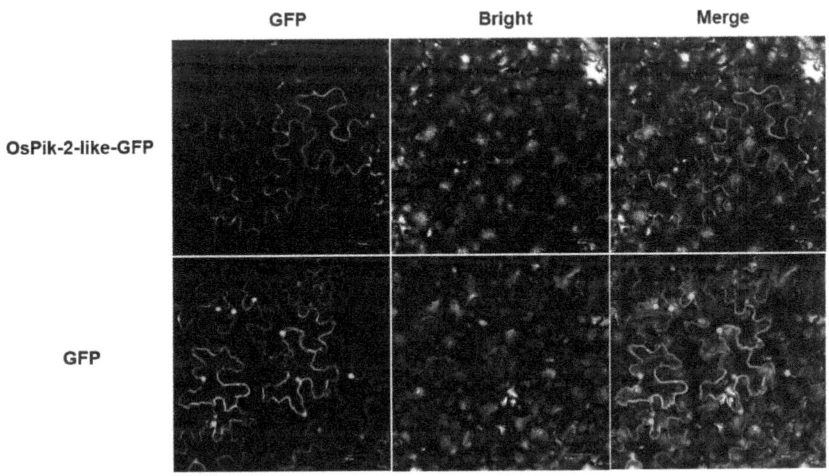

Figure 2. Subcellular localization of OsPik-2-like. Subcellular localization of the fused OsPik-2-like-GFP or GFP in *N. benthamiana* leaf cells. OsPik-2-like-GFP, green fluorescent protein (GFP) fluorescence from OsPik-2-like-GFP; Bright, bright field; Merged, the merged image of OsPik-2-like-GFP or GFP and Bright. Bar = 50 μm.

Quantitative real-time PCR (qRT-PCR) revealed that the constitutive transcript levels of *OsPik-2-like* in rice leaves were low. However, when rice plants (leaf blades) were infested by LF larvae (Figure 3a,b), the levels of *OsPik-2-like* transcripts in leaf blades were rapidly (starting 0.5 h after infestation) and lastingly up-regulated; the levels, which peaked at 72 h, were about 70-fold higher than the levels in control plants (Figure 3b; Table S1). Mechanical wounding also quickly (0.5 h after treatment) induced the expression of *OsPik-2-like* (levels, which peaked at 0.5 h, were about 10-fold higher than levels in control plants); however, the levels of *OsPik-2-like* transcripts rapidly declined to those of controls 3 h after treatment (Figure 3c; Table S1). MeJA treatment only slightly induced the expression of *OsPik-2-like* at 12 h after treatment (Figure 3d; Table S1). These results suggest that *OsPik-2-like* might be involved in LF-induced rice defense responses.

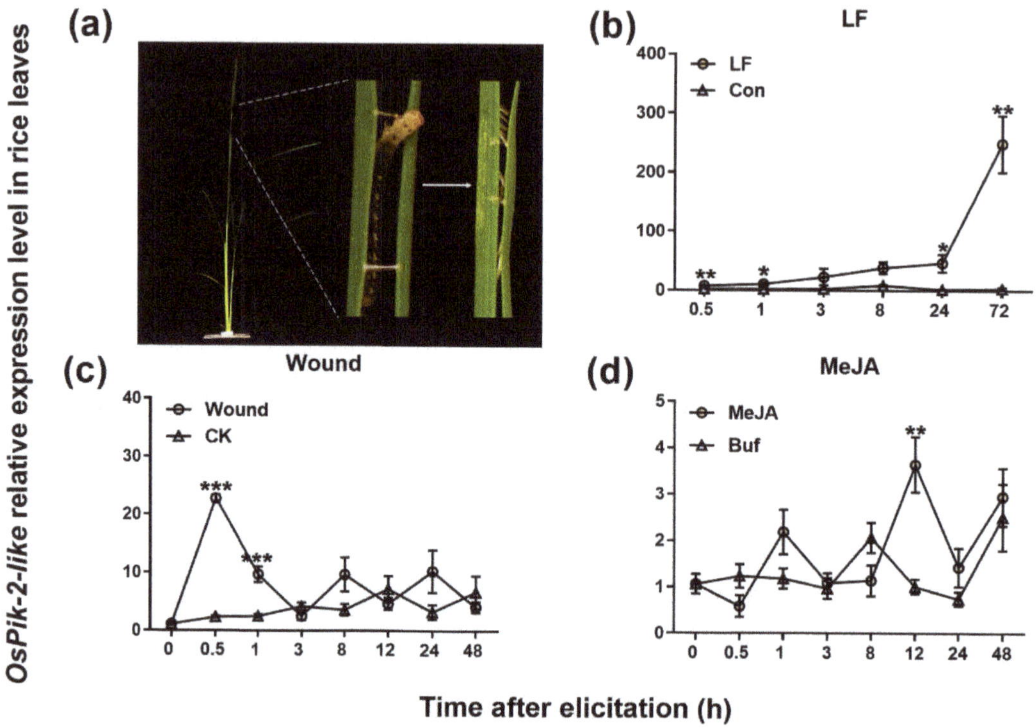

Figure 3. Relative expression levels of *OsPik-2-like* in rice leaves after different treatments. (**a**) Phenotypes of rice plants that had been infested by one 3rd-instar LF larva. (**b–d**) Mean transcript levels (±SE, n = 3~6) of *OsPik-2-like* in rice leaves that were infested with one 3rd LF larva (LF) (**b**), or punctured by a fabric pattern wheel rolling (Wound) (**c**), or treated with methyl jasmonate (MeJA) by root absorption (**d**). Con, non-treated plants; Buf: plants treated with the same concentration of ethanol in the nutrient solution as MeJA treatment. Asterisks represent significant differences between the treatments and controls at each time point (* $p < 0.05$, ** $p < 0.01$, and *** $p < 0.001$, Student's *t*-tests or *t* tests with Welch's correction).

2.2. Knocking out OsPik-2-like in Rice

To explore the function of *OsPik-2-like* in interactions between rice and LF, we obtained two T-DNA-free homozygous rice lines with knocked-out *OsPik-2-like* in the variety XS11 using a CRISPR/Cas9-mediated gene-editing system (ko-*pik2l*: *kopik2l*-1 and *kopik2l*-16). As shown in Figure 4a, compared with the target sequence of *OsPik-2-like* gene in WT plants, the line *kopik2l*-1 displayed a single "A" insertion and the line *kopik2l*-16 showed a 16-bp deletion (Figure 4a). DNA sequencing analysis revealed no mutations of off-target sites (such mutations had been predicted by the CRISPR-P server, http://cbi.hzau.edu.cn/crispr/, accessed 25 June 2023), in ko-*pik2l* lines (Figure S2), suggesting that the CRISPR/Cas9-based *OsPik-2-like*-knockout was specific.

We investigated the effect of knocking out *OsPik-2-like* on plant growth. Knocking out *OsPik-2-like* reduced shoot height (by 11.75% in *kopik2l*-1 and 1.84% in *kopik2l*-16, compared with WT plants), root length (by 18.18% and 9.91%), chlorophyll content in leaves (by 7.15% and 3.47%), stem strength (by 15.34% and 10.7%), shoot fresh mass (by 42.24% and 31.21%) and shoot dry mass (by 42.17% and 25.14%), and root fresh mass (by 42.47% and 22.99%) and root dry mass (by 45.28% and 25.01%) of 30-day-old plants (Figure 4b–i; Table S2). These data suggest that *OsPik-2-like* positively regulated rice growth.

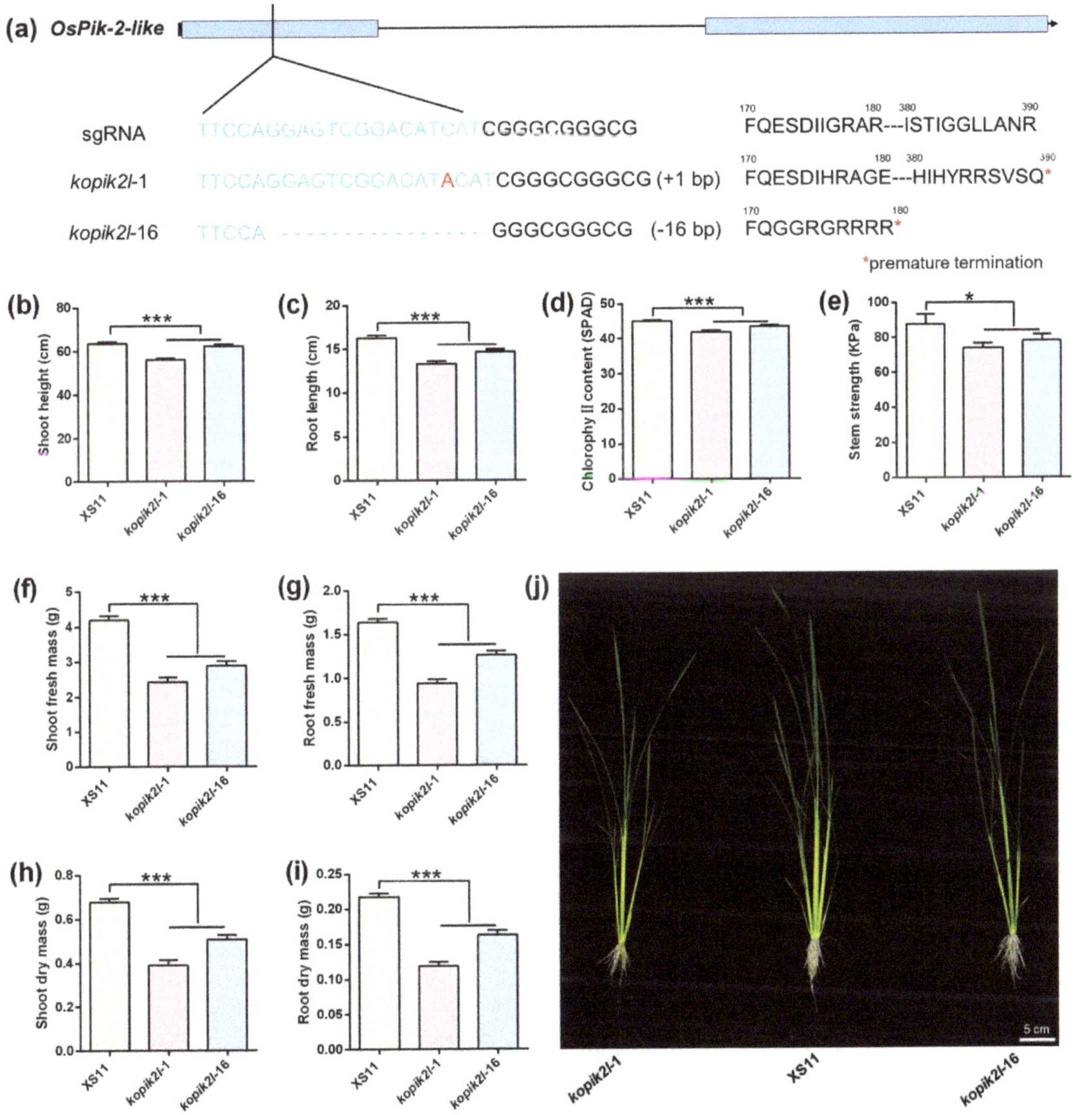

Figure 4. *OsPik-2-like* gene-knockout rice lines and their growth phenotypes. (**a**) Sanger sequencing analysis of mutation patterns in the target site of *OsPik-2-like* gene in the wild-type (WT) and *OsPik-2-like* knockout rice lines. Blue boxes represent exons. The sgRNA sequence that specifically targets *OsPik-2-like* is indicated. The mutations, an "A" insertion in line *kopik2l*-1 and a 16 bp deletion in *kopik2l*-16, lead to premature translational termination of OsPik-2-like. (**b**–**i**) Mean shoot height (**b**), root length (**c**), chlorophyll content (**d**), stem strength (**e**), shoot fresh weight (**f**), root fresh weight (**g**), shoot dry weight (**h**), and root dry weight (**i**) (+SE, n = 20) of WT plants and ko-*pik2l* lines at 30 days old in the phytotron. Asterisks indicate significant differences in ko-*pik2l* lines compared with WT plants evaluated by Bayesian analysis of variance (* $p < 0.05$ and *** $p < 0.001$). (**j**) The phenotype of 30-day-old WT and ko-*pik2l* lines in the phytotron.

2.3. Knocking out OsPik-2-like-Enhances LF-Induced JA and JA-Ile Levels

The JA-mediated signaling pathway plays a central role in rice defenses against chewing herbivores [32]. Hence, we determined the levels of JA and JA-Ile in WT plants, as well as ko-*pik2l* lines before and after LF infestation. Consistent with previous studies [31,33], LF infestation significantly increased the contents of JA and JA-Ile in rice leaf blades in WT plants (Figure 5; Table S3). Knocking out *OsPik-2-like* did not influence the constitutive levels of JA and JA-Ile in rice leaves. However, it did enhance the induced levels of JA and JA-Ile, especially JA-Ile, in rice leaves 8 h after LF infestation: the LF-induced JA-Ile levels in the two ko-*pik2l* lines *kopik2l*-1 and *kopik2l*-16 were about 2- and 1.5-fold higher than those in WT plants (Figure 5; Table S3).

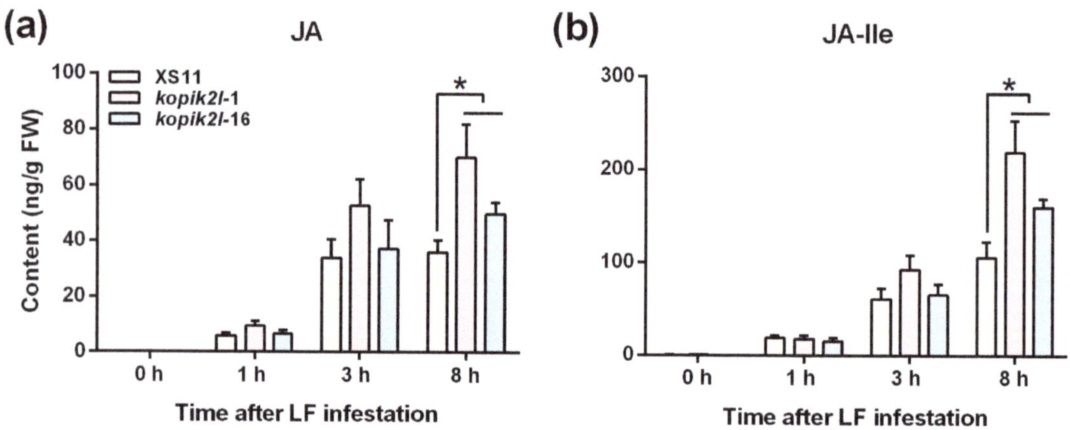

Figure 5. Knocking out *OsPik-2-like* enhanced LF-induced JAs levels in rice leaves. Mean levels (+SE, n = 4~6) of JA (**a**) and JA-Ile (**b**) in WT plants and ko-*pik2l* lines that were individually infested by a 3rd-instar LF larva on the first fully expanded leaf at the indicated time points. Asterisks indicate significant differences in ko-*pik2l* lines compared with WT plants evaluated by Bayesian analysis of variance (* $p < 0.05$).

We also found that LF larval infestation (at 8 h) enhanced the content of 3-indoleacetic acid (IAA) in rice leaf blades. However, no difference was observed in IAA content between WT and ko-*pik2l* lines (Figure S3a; Tables S3 and S4). These results indicate that the JA signaling pathway might be involved in *OsPik2-like*-mediated rice defenses against LF.

2.4. Knocking out OsPik-2-like Enhances the Level of Defense Compounds and Rice Resistance to LF

Plants employ multiple defense compounds to limit the population development of herbivores, such as flavonoids, phenolamides, and TrypPIs [32]. To evaluate whether *OsPik-2-like* regulates the production of defense compounds in rice plants, we examined the content of flavonoids and TrypPIs in WT plants and in ko-*pik2l* lines before and after LF infestation. Knocking out *OsPik-2-like* in rice constitutively increased the contents of three flavonoids—prunin, carlinoside, and isovitexin (Figure 6a–c; Table S5)—but decreased the contents of two flavonoids—astragalin and luteolin 7-*O*-glucoside (Figure S3b,c; Table S5). After LF larval infestation, knocking out *OsPik-2-like* also decreased the contents of five flavonoids: astragalin (by 49.83% in *kopik2l*-1 and 59.17% in *kopik2l*-16 compared to WT plants), luteolin 7-*O*-glucoside (by 49.89% and 59.76%), luteolin (by 65.71% and 72.89%), sakuranetin (by 100% and 100%), and isoquercitrin (by 32.61% and 28.04%) (Figure S3b–f; Table S5). Consistent with the results reported previously [31], the basal level of TrypPIs in rice leaves of WT plants was extremely low. However, when plants were infested by LF larvae, the TrypPI activity in leaves was markedly enhanced (Figure 6d,e; Table S5).

Knocking out *OsPik-2-like* did not influence the basal activity of TrypPIs in plant leaf blades, but it did significantly enhance the LF-induced activities of TrypPIs in leaf blades (Figure 6d,e; Table S5).

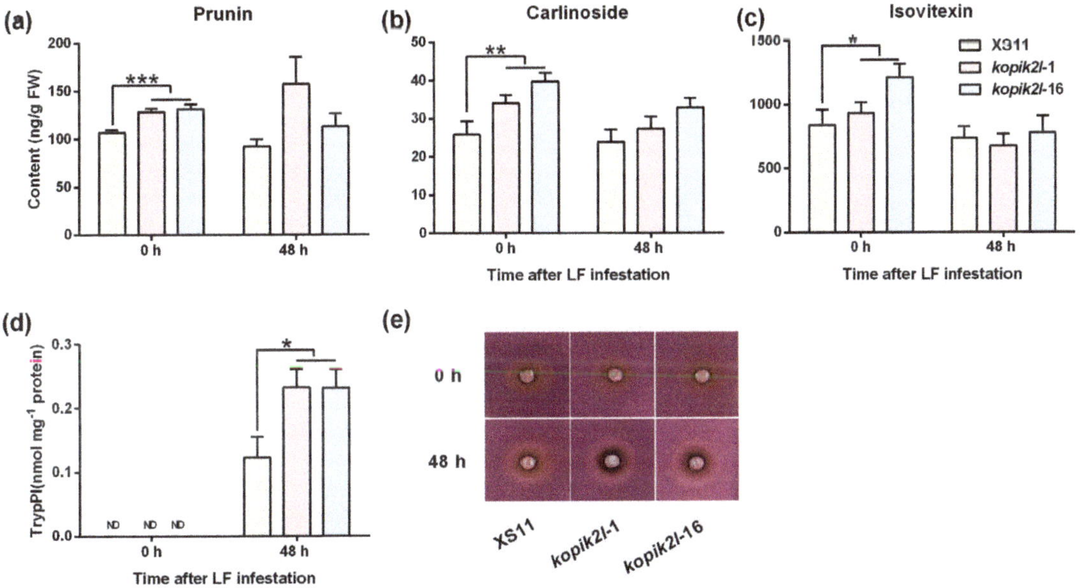

Figure 6. Flavonoid levels and TrypPI activity in the leaves of WT plants and ko-*pik2l* lines. Mean concentrations (+SE, n = 5~6) of prunin (**a**), carlinoside (**b**), isovitexin (**c**), and mean TrypPI activity (+SE, n = 7) (**d**) in WT plants and ko-*pik2l* lines that were individually infested by a 3rd-instar LF larva on the first fully expanded leaf for 0 h or 48 h. Asterisks indicate significant differences in ko-*pik2l* lines compared with WT plants evaluated by Bayesian analysis of variance (* $p < 0.05$, ** $p < 0.01$ and *** $p < 0.001$). (**e**) Representative pictures of TrypPI after different treatments measured by radial diffusion assay.

Given that knocking out *OsPik-2-like* in rice influenced the JA signaling pathway and the levels of defense compounds in plants, we asked whether knocking out *OsPik-2-like* affected rice defense against LF. As expected, LF larvae fed on ko-*pik2l* plants gained significantly less mass compared with those fed on WT plants: by day 11, the mass of LF larvae fed on the *kopik2l*-1 and *kopik2l*-16 plants decreased by about 18% and 20%, respectively, compared with the mass of larvae fed on WT plants (Figure 7; Table S6). These results indicate that *OsPik-2-like* negatively regulates rice defense against LF larvae.

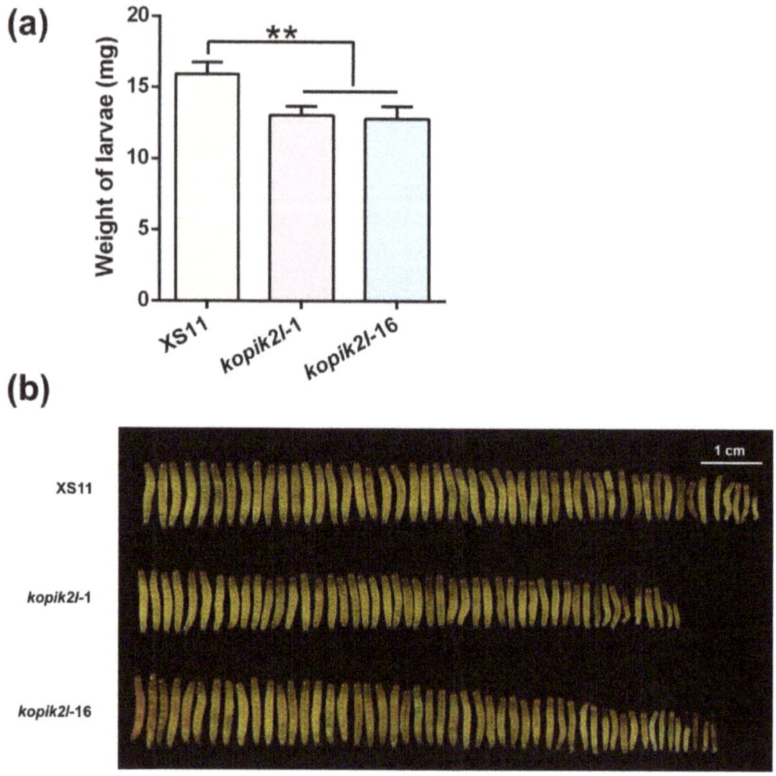

Figure 7. *OsPik-2-like* negatively regulates the resistance of rice to LF. (**a**) Mean larval mass (+SE, n = 48~51) of LF fed on WT plants and ko-*pik2l* lines for 11 days. (**b**) Phenotypes of LF larvae fed on different lines on day 11. Asterisks indicate significant differences in ko-*pik2l* lines compared with WT plants evaluated by Bayesian analysis of variance (** $p < 0.01$).

3. Discussion

Plant NLRs have widely been reported to act as resistance proteins that sense pathogen/herbivore-derived effectors, thereby triggering ETI. However, their role in plant defense responses to chewing herbivores remains poorly understood [15–25]. In this study, we revealed that a CNL protein, OsPik-2-like, plays an important role in LF larvae-induced rice defenses. First, OsPik-2-like localized to PM, and its transcript levels were significantly induced by LF larval infestation and mechanical wounding. Second, knocking out *OsPik-2-like* decreased the shoot mass, root mass leaf chlorophyll content, and stem strength of plants. Third, knocking out *OsPik-2-like* enhanced the LF-induced levels of JA and JA-Ile, as well as the activity of TrypPIs in rice, which in turn reduced the growth of LF larvae. Our findings demonstrate that *OsPik-2-like* positively regulates plant growth, but negatively mediates the resistance of rice plants to LF.

NLRs play an important role in regulating defense-related signaling pathways [3]. Two NLRs, *Sw-5b* and *Sl5R-1*, from *Solanaceae*, for instance, have been reported to activate the JA signaling pathway and then increase plant resistance to infection by *tomato spotted wilt orthotospovirus* (TSWV) [34,35]. In rice, *Bph9* and *Bph14*, two NLRs, have been reported to negatively regulate the expression of genes related to JA biosynthesis, such as genes encoding allene oxide synthase 2 (*OsAOS2*) and lipoxygenase (*OsLOX*); moreover, the levels of JA and JA-Ile in plants carrying *Bph9* or *Bph14* were significantly lower than those in WT plants in response to BPH infestation [22,25]. Here, we observed that knocking out *OsPik-2-like* significantly increased the LF-elicited levels of JA and JA-Ile in rice leaf blades (Figure 5). Moreover, LF larval infestation up-regulated the transcript levels of *OsPik-2-like*, especially late in the process. These data suggest that, like *Bph9* and *Bph14*, *OsPik-2-like* negatively modulated the biosynthesis of LF-induced JA and JA-Ile in rice. These results indicate that LF was able to suppress the JA-mediated rice defense by inducing the expression of *OsPik-2-like*. Further studies should elucidate the mechanism underlying the *OsPik-2-like*-mediated regulation of JA biosynthesis. It would also be interesting to know whether and how other defense-related signaling pathways are modulated by OsPik-2-like.

Why rice plants up-regulate the expression of *OsPik-2-like* at late stages of infestation by LF larvae, given that *OsPik-2-like* negatively regulated rice defense, remains a question. It may be that rice plants are trying to avoid the autotoxicity caused by excessive levels of defenses [36,37] and to restore growth. Given that *OsPik-2-like* suppressed the biosynthesis of LF-induced JA and JA-Ile, and that knocking out *OsPik-2-like* decreased plant growth (Figures 4 and 5), the up-regulation of *OsPik-2-like* at the late stage of LF larval infestation caused plants to maintain appropriate levels of defenses. Moreover, it made plants regrow quickly. In rice, several such genes, such as *OsMPK20-5* [38] and *OsLRR2* [39], have been reported. Alternatively, it may be related to herbivore decoy strategies [40]. Thus far, many CNLs have been reported to regulate plant disease resistance [3]. In rice, two CNLs, *Pik-1* and *Pik-2*, have been reported to enhance rice resistance to *Magnaporthe oryzae* (syn. *Pyricularia oryzae*) expressing the *AvrPik* effector [41–43]. Because many symbiotic microbe species live in herbivores [44], it may be that LF larvae regulate the expression of *OsPik-2-like* by secreting one of these microbes during their feeding, thereby suppressing the JA-mediated defense in rice. Further research should clarify this mechanism. Moreover, how OsPik-2-like influences rice growth, including which plant growth-related signaling pathways are regulated by OsPik-2-like, should also be elucidated.

The JA signaling pathway plays a central role in regulating the biosynthesis of multiple defense compounds [31,33]. For example, silencing or knocking out genes related to JA biosynthesis (such as herbivore-induced rice type 2 13-LOX gene, *OsHI-LOX*, and the allene oxide cyclase gene, *OsAOC*) or signaling (JA receptor gene, rice *CORONATINE-INSENSITIVE1* (*OsCOI1*) and the core JA-responsive transcriptional factor *OsMYC2*) in rice reduces plant resistance to LF by impairing the activity of TrypPIs and the accumulation of some phenolamines in rice plants [31,33]. Therefore, the observed results—namely, that knocking out *OsPik-2-like* enhanced the activity of LF-induced TrypPIs in plants—were related, at least in part, to the increase in the levels of LF-induced JA and JA-Ile in plants. Interestingly, we observed that knocking out *OsPik-2-like* increased the basal levels of three flavonoids and decreased the basal levels of two flavonoids; moreover, it also decreased LF-induced levels of five flavonoids (Figure 6). Although some researchers have reported that the JA signaling pathway positively modulates the production of herbivore-induced flavonoids [45], these results indicate that the regulation of flavonoid biosynthesis was complex and that other signaling pathways were also involved in this process. It has been reported that signaling pathways mediated by brassinosteroids, SA, and ET also regulate the biosynthesis of flavonoids in plants [46–48]. Future research should elucidate which other signaling pathways regulated by OsPik-2-like mediate the biosynthesis of these compounds in rice.

TrypPIs and some flavonoids are important defense compounds against herbivores. TrypPIs, for example, have been reported to suppress the growth and development of chewing herbivores, including LF, by inhibiting the activity of digestive enzymes in their midgut [30,31,49]. Some flavonoids (including prunin, carlinoside, schaftoside, and its isomers, isoschaftoside and neoschaftoside) impede the feeding, survival, and development of rice planthoppers [50–52], although no flavonoids have been reported to affect the performance of LF larvae. Moreover, two isovitexin-derived compounds, isovitexin 2′′-O-(6′′′-(*E*)-feruloyl) glucoside and isovitexin-2′-*O*-β-[6-O-E-*p*-coumaroylglucopyranoside], decrease the probing responses of *Nephotettix cincticepts* [53] and the fecundity of *Helicoverpa armigera*, respectively [54]. Hence, the enhanced resistance of ko-*pik2l* plants to LF larvae is probably related to the increase in the levels of defense compounds, such as TrypPIs and some flavonoids. Future research should clarify which compounds are responsible for the performance of LF larvae on ko-*pik2l* plants.

In summary, our results demonstrate that the PM-localized rice CNL protein, OsPik-2-like, plays an important role in regulating the interaction between rice and LF larvae. When infested by LF larvae, plants perceive signals derived from the herbivore and then initiate defense-related signaling pathways, such as the JA signaling pathway. These changes enhance the expression of defense-related genes and the level of defense compounds, such as TrypPIs, in turn decreasing the performance of LF larvae. The up-regulation of *OsPik-2-like* in rice at late stages of LF larval infestation not only reduces plant defense by suppressing the JA signaling pathway, but also promotes plant regrowth. Both consequences may be beneficial for both plants and herbivores. Plants may avoid the autotoxicity associated with excessive defenses and also experience regrowth; herbivores may improve growth due to decreased plant defense. Our study provides an interesting example of how a single gene can act as a modulator for the symbiotic relationship between plants and herbivores, balancing plant growth and defense. From an application point of view, this study shows how specific genes can be edited in plants to control insect pests.

4. Materials and Methods

4.1. Plants and Insects

The *japonica* rice variety XiuShui 11 (XS11) was used as the wild type (WT) in this study. The gene-knockout rice lines of *OsPik-2-like* (ko-*pik2l*: *kopik2l*-1 and *kopik2l*-16) were generated and screened as below. Seeds of WT and ko-*pik2l* lines were pre-germinated in plastic culture dishes (diameter 90 mm, height 15 mm) in a climate incubator under the following conditions: 28 ± 2 °C, 60% relative humidity, and 14/10 h light/dark cycle. Ten days later, rice seedlings were transplanted to 25 L hydroponic boxes (length 51 cm, width 35 cm, and height 17 cm) brimming with rice nutrient solution [55] and cultivated in a phytotron (26 ± 2 °C, 14 h light phase, 60% relative humidity). Twenty-five-day-old rice plants were individually transferred into plastic pots (diameter 7 cm, height 9.5 cm) filled with 350 mL of nutrient solution. Plants were used for experiments 4 d after transplantation.

Colonies of LF were originally obtained from a rice paddy in Hangzhou, China, and subsequently reared with wheat seedlings in a climate chamber (26 ± 2 °C, 14 h light phase, and 65 ± 10% relative humidity) for more than 30 generations.

4.2. Plant Treatment

For LF treatments, a third-instar LF larva was starved for 2 h, then placed on the youngest fully expanded leaf of a plant. Untreated plants were used as controls. For mechanical wounding in rice leaves, the youngest fully expanded leaf was punctured by rolling a fabric pattern wheel over it [45]. Untreated plants were used as controls. For mechanical wounding in rice leaf sheaths, aerial parts 2 cm high from the roots of rice plants were individually punctured 200 times using an insect needle (length 40 mm and diameter 0.45 mm; Yuxiu, Taizhou, China). Untreated plants were used as controls. For treatment with methyl jasmonate (MeJA) (Sigma-Aldrich, Darmstadt, Germany), MeJA was dissolved in 1 mL of absolute ethyl alcohol and then added to the nutrient solution to make its concentration 100 μM, as stated in previous studies [56]. Plants cultivated in the nutrient solution with an equal volume of solvent were used as controls.

4.3. RNA Extraction and Quantitative Real-Time PCR (qRT-PCR)

Total RNA was extracted from rice samples using FastPure® Universal Plant Total RNA Isolation Kit (Vazyme, Nanjing, China) following the manufacturer's instructions. One microgram of total RNA was then reverse-transcribed by using HiScript® II Q RT SuperMix for qPCR (+gDNA wiper) (Vazyme, Nanjing, China) according to the manufacturer's instructions. QRT-PCR assays were performed with Taq Pro Universal SYBR qPCR Master Mix (Vazyme, Nanjing, China) on the CFX96 Real-Time PCR System (Bio-Rad, Richmond,

CA, USA). With the *OsACTIN* (TIGR ID: LOC_Os03g50885) as an internal control, the relative expression levels of tested genes were analyzed by the $-2^{\Delta\Delta Ct}$ method [56]. All primers used in the qRT-PCR assays are listed in Table S7.

4.4. Isolating OsPik-2-like

Four micrograms of total RNA extracted from WT plants was reverse-transcribed to first-strand cDNAs using PrimeScript™ IV cDNA Synthesis Mix (TaKaRa, Dalian, China). Then, the opening read frame of *OsPik-2-like* (XM_015756755.2) was PCR-amplified from the cDNA library using specific primers OsPik-2-like-CDS-F (5′-TCTCCGCCATTTACACCCAC-3′) and OsPik-2-like-CDS-R (5′-TGCCAAAGAGGAACATCGGA-3′), and the KOD FX polymerase (TOYOBO, Shanghai, China) following the manufacturer's instructions. PCR products were subsequently cloned into the pEASY®-Blunt Zero cloning vector (TransGen, Beijing, China) and sequenced. The primers used in this study are listed in Table S8.

4.5. Structure and Phylogenetic Analysis of OsPik-2-like

The conserved domains of the deduced OsPik-2-like protein were analyzed using CD-search (https://www.ncbi.nlm.nih.gov/Structure/cdd/wrpsb.cgi; accessed on 30 August 2023). The homologs of OsPik-2-like from different plant species were identified using the BLASTP plug-in in the NCBI online database (https://blast.ncbi.nlm.nih.gov/Blast.cgi; accessed on 24 June 2024). The amino acid sequences of OsPik-2-like and their homologs from other plant species were downloaded from the NCBI website. Then, multiple sequence alignment was carried out using ClustalW program in Molecular Evolutionary Genetics Analysis (MEGA, version 11.0.13) as described previously [56]. A neighbor-joining phylogenetic tree was constructed by MEGA with 1000 bootstrap replicates using the default parameters as described [56].

4.6. Subcellular Localization Assay

For subcellular localization, the full-length coding sequence of *OsPik-2-like* without the termination codon was inserted into the pCAMBIA1301-EGFP vector [39] using ClonExpress® II One Step Cloning Kit C112 (Vazyme, Nanjing, China), yielding a *35S::OsPik-2-like-EGFP* construct. The constructed plasmid was then transformed into *A. tumefaciens* strain EHA105 by electroporation. *A. tumefaciens*—cultivated overnight and carrying the OsPik-2-like-EGFP construct—was injected into leaves of 4-week-old *N. benthamiana* plants as previously described [57]. Two days after infiltration, the EGFP fluorescence images were observed and recorded by Olympus FV3000 confocal laser scanning microscope (Olympus, Tokyo, Japan) at an excitation wavelength of 488 nm and an emission wavelength of 507 nm. The primers used in this study are listed in Table S8.

4.7. Generating and Characterizing OsPik-2-like Gene-Knockout Rice Plants

To generate *OsPik-2-like* gene-knockout rice lines, the target sequence (5′-TTCCAGGAG TCGGACATCAT-3′) was introduced into pLYsgRNAOsU6b to yield rice U6b promoter-driven single-guide RNA (sgRNA). The sgRNA expression cassette was then introduced into the plant CRISPR-Cas9 binary vector pYLCRISPR/Cas9Pubi-H [58]. The T-DNA was inserted into XS11 using *A. tumefaciens* (EHA105)-mediated transformation. Plants with mutations but lacking the T-DNA were screened by target DNA sequencing and hygromycin resistance gene identification using the method described in [39]. Two homozygous *OsPik-2-like* gene-knockout (ko-*pik2l*) lines, *kopik2l*-1 and *kopik2l*-16, were used for all experiments. Potential off-target sites were predicted by CRISPR-GE (http://skl.scau.edu.cn/home/, accessed 13 August 2023) and identified by target DNA sequencing using the specific primers listed in Table S9.

4.8. Measuring Plant Growth Parameters

Thirty-day-old WT plants and *OsPik-2-like* gene-knockout lines were used to measure plant growth parameters, which included root length, shoot height, root fresh mass, root dry mass, shoot fresh mass, shoot dry mass, stem strength, and chlorophyll content. To determine stem strength, the aerial parts (2 cm high from plant root) of rice plants were measured with a plant-stem-strength meter [59], YYD-1 (Top Cloud-Agri, Hangzhou, Zhehjiang, China). To determine chlorophyll content, the middle location of the youngest fully expanded leaves from each plant was measured with a chlorophyll meter [56], SPAD-502 Plus (Konica Minolta, Tokyo, Japan). Each experiment was replicated 20 times.

4.9. Herbivore Bioassays

To measure LF performance, freshly hatched LF neonates were allowed to feed on WT and ko-*pik2l* rice plants. The larval mass (to an accuracy of 0.1 mg) was individually measured 11 days later. Each treatment was replicated 48~51 times.

4.10. JA, JA-Ile, and IAA Analysis

For JA, JA-Ile, and IAA analysis, 30-day-old plants of the WT and ko-*pik2l* rice lines were randomly assigned to LF and control treatments as described above. The youngest fully expanded leaf of each plant was harvested at different time points (0, 1, 3, and 8 h after the start of treatment) and was ground in liquid nitrogen. JA, JA-Ile, and IAA were extracted from a sample of about 100 mg of ground rice with ethyl acetate containing labeled internal standards (D^6-JA, D^6-JA-Ile, and D^5-IAA) and then quantified by high-performance liquid chromatography–mass spectrometry–mass spectrometry (HPLC-MS-MS) (Agilent Technologies, Santa Clara, CA, USA) using the method described in [60]. Each treatment was replicated 4~6 times.

4.11. Analysis of the Activity of TrypPIs

To analyze TrypPIs, 30-day-old plants of the WT and ko-*pik2l* rice lines were randomly assigned to LF and control treatments as described above. The youngest fully expanded leaf of each plant was harvested at 0 and 2 d after the start of treatment and was ground in liquid nitrogen. Samples containing about 100 mg of ground rice were homogenized with 300 μL of cooled extraction buffer (0.1 M Tris-HC1, pH 7.6, 5% polyvinylpolypyrrolidone (Sigma-Aldrich, Darmstadt, Germany), 2 mg/mL phenylthiourea (Sigma-Aldrich, Darmstadt, Germany), 5 mg/mL diethyldithiocarbamate (Sigma-Aldrich, Darmstadt, Germany), and 0.05 M Na_2EDTA) as described previously [61]. TrypPI activity was measured by the radial diffusion method [62]. Each treatment was replicated 7 times.

4.12. Detecting Flavonoids

To detect flavonoids, 30-day-old plants of the WT and ko-*pik2l* rice lines were randomly assigned to LF and control treatments as described above. The youngest fully expanded leaf of each plant was harvested at 0 and 2 d after the start of treatment and was ground in liquid nitrogen. Flavonoids were extracted from samples of approximately 100 mg using 800 μL of 70% methanol, and then quantified by HPLC-MS-MS as previously described in [63]. The content of each compound was calculated using an external standard method [64]. Each treatment was replicated 5~6 times.

4.13. Data Analysis

Two-treatment data were analyzed using Student's t-tests (equal SDs) or t tests with Welch's correction (unequal SDs). For comparisons between data collected from WT plants and from two transgenic lines of *OsPik-2-like* (ko-*pik2l*), an a priori approach based on Bayesian analysis of variance was used. All statistical analyses were conducted with SPSS software version 26 (IBM Corp., Armonk, NY, USA).

Supplementary Materials: The following supporting information can be downloaded at: https://www.mdpi.com/article/10.3390/plants13233275/s1, Figure S1: Sequences of nucleotides and deduced amino acids of OsPik-2-like; Figure S2: Sanger sequencing results of potential off-target sites; Figure S3: Other compounds in the leaves of WT plants and ko-*pik2l* lines; Table S1: Student's t-test or t-test with Welch's correction analysis with data from Figure 3; Table S2: Bayesian analysis of variance with data from Figure 4; Table S3: Bayesian analysis of variance with data from Figure 5; Table S4: Student's *t*-test analysis with data from Figure S3; Table S5: Bayesian analysis of variance with data from Figure 6 and Figure S3; Table S6: Bayesian analysis of variance with data from Figure 7; Table S7: Primers used for real-time qPCR; Table S8: Primers used for *OsPike-2-like* cloning and subcellular localization assay; Table S9: Primers used for generation and characterization of transgenic plants.

Author Contributions: Conceptualization, Y.L.; methodology, Y.L., P.K., and C.Q.; validation, Y.L., P.K., and C.Q.; formal analysis, Y.L., P.K., and C.Q.; investigation, C.Q., X.J., and Y.Z.; resources, Y.L.; data curation, C.Q.; writing—original draft preparation, C.Q. and P.K.; writing—review and editing, Y.L.; supervision, Y.L.; project administration, Y.L.; funding acquisition, Y.L. and P.K. All authors have read and agreed to the published version of the manuscript.

Funding: This work was jointly supported by the Major Science and Technology project of Xinjiang (2023A0200904 to Y.L.), the earmarked fund for China Agriculture Research Systems (CARS-01-43 to Y.L.), and the National Natural Science Foundation of China (32202396 to P.K.).

Data Availability Statement: The data presented in this study are available on request from the corresponding author.

Acknowledgments: We thank Yie Xu for technical assistance, and Genshen Xie for the assistance in LF rearing. We thank Emily Wheeler for editorial assistance.

Conflicts of Interest: The authors declare no conflicts of interest.

References

1. Stahl, E.; Hilfiker, O.; Reymond, P. Plant-arthropod interactions: Who is the winner? *Plant J.* **2018**, *93*, 703–728. [CrossRef] [PubMed]
2. Erb, M.; Reymond, P. Molecular interactions between plants and insect herbivores. *Annu. Rev. Plant Biol.* **2019**, *70*, 527–557. [CrossRef]
3. Ngou, B.; Ding, P.; Jones, J. Thirty years of resistance: Zig-zag through the plant immune system. *Plant Cell* **2022**, *34*, 1447–1478. [CrossRef]
4. Reymond, P. Receptor kinases in plant responses to herbivory. *Curr. Opin. Biotechnol.* **2021**, *70*, 143–150. [CrossRef] [PubMed]
5. Dodds, P.N.; Rathjen, J.P. Plant immunity: Towards an integrated view of plant-pathogen interactions. *Nat. Rev. Genet.* **2010**, *11*, 539–548. [CrossRef] [PubMed]
6. Jones, J.D.G.; Vance, R.E.; Dangl, J.L. Intracellular innate immune surveillance devices in plants and animals. *Science* **2016**, *354*, aaf6395. [CrossRef]
7. Snoeck, S.; Guayazán-Palacios, N.; Steinbrenner, A.D. Molecular tug-of-war: Plant immune recognition of herbivory. *Plant Cell* **2022**, *34*, 1497–1513. [CrossRef]
8. Duxbury, Z.; Wu, C.H.; Ding, P. A comparative overview of the intracellular guardians of plants and animals: NLRs in innate immunity and beyond. *Annu. Rev. Plant Biol.* **2021**, *72*, 155–184. [CrossRef]
9. Contreras, M.P.; Ludke, D.; Pai, H.; Toghani, A.; Kamoun, S. NLR receptors in plant immunity: Making sense of the alphabet soup. *EMBO Rep.* **2023**, *24*, e57495. [CrossRef]
10. Deng, Y.; Zhai, K.; Xie, Z.; Yang, D.; Zhu, X.; Liu, J.; Wang, X.; Qin, P.; Yang, Y.; Zhang, G.; et al. Epigenetic regulation of antagonistic receptors confers rice blast resistance with yield balance. *Science* **2017**, *355*, 962–965. [CrossRef]
11. Seto, D.; Koulena, N.; Lo, T.; Menna, A.; Guttman, D.S.; Desveaux, D. Expanded type III effector recognition by the ZAR1 NLR protein using ZED1-related kinases. *Nat. Plants* **2017**, *3*, 17027. [CrossRef] [PubMed]
12. Schultink, A.; Qi, T.; Bally, J.; Staskawicz, B. Using forward genetics in *Nicotiana benthamiana* to uncover the immune signaling pathway mediating recognition of the *Xanthomonas* perforans effector XopJ4. *New Phytol.* **2019**, *221*, 1001–1009. [CrossRef] [PubMed]
13. Wang, G.; Roux, B.; Feng, F.; Guy, E.; Li, L.; Li, N.; Zhang, X.; Lautier, M.; Jardinaud, M.; Chabannes, M.; et al. The decoy substrate of a pathogen effector and a pseudokinase specify pathogen-induced modified-self recognition and immunity in plants. *Cell Host Microbe* **2015**, *18*, 285–295. [CrossRef] [PubMed]
14. Bi, G.; Su, M.; Li, N.; Liang, Y.; Dang, S.; Xu, J.; Hu, M.; Wang, J.; Zou, M.; Deng, Y.; et al. The ZAR1 resistosome is a calcium-permeable channel triggering plant immune signaling. *Cell* **2021**, *184*, 3528–3541. [CrossRef]
15. Dogimont, C.; Chovelon, V.; Pauquet, J.; Boualem, A.; Bendahmane, A. The *Vat* locus encodes for a CC-NBS-LRR protein that confers resistance to *Aphis gossypii* infestation and *A. gossypii*-mediated virus resistance. *Plant J.* **2014**, *80*, 993–1004. [CrossRef]
16. Klingler, J.P.; Nair, R.M.; Edwards, O.R.; Singh, K.B. A single gene, *AIN*, in *Medicago truncatula* mediates a hypersensitive response to both bluegreen aphid and pea aphid, but confers resistance only to bluegreen aphid. *J. Exp. Bot.* **2009**, *60*, 4115–4127. [CrossRef]
17. Wroblewski, T.; Piskurewicz, U.; Tomczak, A.; Ochoa, O.; Michelmore, R.W. Silencing of the major family of NBS-LRR-encoding genes in lettuce results in the loss of multiple resistance specificities. *Plant J.* **2007**, *51*, 803–818. [CrossRef]

18. Casteel, C.L.; Walling, L.L.; Paine, T.D. Behavior and biology of the tomato psyllid, *Bactericerca cockerelli*, in response to the *Mi-1.2* gene. *Entomol. Exp. Appl.* **2006**, *121*, 67–72. [CrossRef]
19. Nombela, G.; Williamson, V.M.; Muniz, M. The root-knot nematode resistance gene *Mi-1.2* of tomato is responsible for resistance against the whitefly *Bemisia tabaci*. *Mol. Plant Microbe Interact.* **2003**, *16*, 645–649. [CrossRef]
20. Rossi, M.U.O.C.; Goggin, F.L.; Milligan, S.B.; Kaloshian, I.; Ullman, D.E.; Williamson, V.M. The nematode resistance gene *Mi* of tomato confers resistance against the potato aphid. *Proc. Natl. Acad. Sci. USA* **1998**, *95*, 9750–9754. [CrossRef]
21. Bassetti, N.; Caarls, L.; Bouwmeester, K.; Verbaarschot, P.; van Eijden, E.; Zwaan, B.J.; Bonnema, G.; Schranz, M.E.; Fatouros, N.E. A butterfly egg-killing hypersensitive response in *Brassica nigra* is controlled by a single locus, PEK, containing a cluster of TIR-NBS-LRR receptor genes. *Plant Cell Environ.* **2024**, *47*, 1009–1022. [CrossRef] [PubMed]
22. Du, B.; Zhang, W.; Liu, B.; Hu, J.; Wei, Z.; Shi, Z.; He, R.; Zhu, L.; Chen, R.; Han, B.; et al. Identification and characterization of *Bph14*, a gene conferring resistance to brown planthopper in rice. *Proc. Natl. Acad. Sci. USA* **2009**, *106*, 22163–22168. [CrossRef] [PubMed]
23. Tamura, Y.; Hattori, M.; Yoshioka, H.; Yoshioka, M.; Takahashi, A.; Wu, J.; Sentoku, N.; Yasui, H. Map-based cloning and characterization of a brown planthopper resistance gene *BPH26* from *Oryza sativa* L. ssp. *indica* cultivar ADR52. *Sci. Rep.* **2014**, *4*, 5872. [CrossRef]
24. Ji, H.; Kim, S.; Kim, Y.; Suh, J.; Park, H.; Sreenivasulu, N.; Misra, G.; Kim, S.; Hechanova, S.L.; Kim, H.; et al. Map-based cloning and characterization of the *BPH18* gene from wild rice conferring resistance to brown planthopper (BPH) insect pest. *Sci. Rep.* **2016**, *6*, 34376. [CrossRef] [PubMed]
25. Zhao, Y.; Huang, J.; Wang, Z.; Jing, S.; Wang, Y.; Ouyang, Y.; Cai, B.; Xin, X.; Liu, X.; Zhang, C.; et al. Allelic diversity in an NLR gene *BPH9* enables rice to combat planthopper variation. *Proc. Natl. Acad. Sci. USA* **2016**, *113*, 12850–12855. [CrossRef]
26. Hu, L.; Wu, Y.; Wu, D.; Rao, W.; Guo, J.; Ma, Y.; Wang, Z.; Shangguan, X.; Wang, H.; Xu, C.; et al. The coiled-coil and nucleotide binding domains of brown planthopper resistance14 function in signaling and resistance against planthopper in rice. *Plant Cell* **2017**, *29*, 3157–3185. [CrossRef]
27. Guo, J.; Wang, H.; Guan, W.; Guo, Q.; Wang, J.; Yang, J.; Peng, Y.; Shan, J.; Gao, M.; Shi, S.; et al. A tripartite rheostat controls self-regulated host plant resistance to insects. *Nature* **2023**, *618*, 799–807. [CrossRef]
28. Jiang, L.; Zhu, Z.; Ge, L.; Yang, G.; Wu, J. Rice grain damage by combination and sequence infestations by the rice leaffolder, *Cnaphalocrocis medinalis* Guenee (Lepidoptera: Pyralidae), and the white-backed rice planthopper, *Sogatella furcifera* Horváth (Hemiptera: Delphacidae). *J. Integr. Agric.* **2014**, *13*, 2460–2470. [CrossRef]
29. Padmavathi, C.; Katti, G.; Padmakumari, A.P.; Voleti, S.R.; Subba Rao, L.V. The effect of leaffolder *Cnaphalocrocis medinalis* (Guenee) [Lepidoptera: Pyralidae] injury on the plant physiology and yield loss in rice. *J. Appl. Entomol.* **2013**, *137*, 249–256. [CrossRef]
30. Wang, X. Influence of Infestation by Herbivores with Different Feeding Habits or Treatment by β-Glucosidase on Levels of Main Defense-Related Signals in Rice Plants. Master's Thesis, Zhejiang University, Hang Zhou, China, 2006; p. 31.
31. Zhuang, Y.; Wang, X.; Llorca, L.C.; Lu, J.; Lou, Y.; Li, R. Role of jasmonate signaling in rice resistance to the leaf folder *Cnaphalocrocis medinalis*. *Plant Mol. Biol.* **2022**, *109*, 627–637. [CrossRef]
32. Kuai, P.; Lou, Y. Advances in molecular interactions between rice and insect herbivores. *Crop Health* **2024**, *2*, 6. [CrossRef]
33. Ye, M.; Luo, S.M.; Xie, J.F.; Li, Y.F.; Xu, T.; Liu, Y.; Song, Y.Y.; Zhu-Salzman, K.; Zeng, R.S. Silencing *COI1* in rice increases susceptibility to chewing insects and impairs inducible defense. *PLoS ONE* **2012**, *7*, e36214. [CrossRef] [PubMed]
34. Wu, Q.; Tong, C.; Chen, Z.; Huang, S.; Zhao, X.; Hong, H.; Li, J.; Feng, M.; Wang, H.; Xu, M.; et al. NLRs derepress MED10b- and MED7-mediated repression of jasmonate-dependent transcription to activate immunity. *Proc. Natl. Acad. Sci. USA* **2023**, *120*, e1992741176. [CrossRef]
35. Qi, S.; Shen, Y.; Wang, X.; Zhang, S.; Li, Y.; Islam, M.M.; Wang, J.; Zhao, P.; Zhan, X.; Zhang, F.; et al. A new *NLR* gene for resistance to tomato spotted wilt virus in tomato (*Solanum lycopersicum*). *Theor. Appl. Genet.* **2022**, *135*, 1493–1509. [CrossRef] [PubMed]
36. Gog, L.; Berenbaum, M.R.; DeLucia, E.H.; Zangerl, A.R. Autotoxic effects of essential oils on photosynthesis in parsley, parsnip, and rough lemon. *Chemoecology* **2005**, *15*, 115–119. [CrossRef]
37. He, Z.; Webster, S.; He, S.Y. Growth-defense trade-offs in plants. *Curr. Biol.* **2022**, *32*, R634–R639. [CrossRef]
38. Li, J.; Liu, X.; Wang, Q.; Huangfu, J.; Schuman, M.C.; Lou, Y. A group D MAPK protects plants from autotoxicity by suppressing herbivore-induced defense signaling. *Plant Physiol.* **2019**, *179*, 1386–1401. [CrossRef]
39. Kuai, P.; Lin, N.; Ye, M.; Ye, M.; Chen, L.; Chen, S.; Zu, H.; Hu, L.; Gatehouse, A.M.R.; Lou, Y. Identification and knockout of a herbivore susceptibility gene enhances planthopper resistance and increases rice yield. *Nat. Food* **2024**, *5*, 846–859. [CrossRef] [PubMed]
40. Walling, L.L. Avoiding effective defenses: Strategies employed by phloem-feeding insects. *Plant Physiol.* **2008**, *146*, 859–866. [CrossRef]
41. Kanzaki, H.; Yoshida, K.; Saitoh, H.; Fujisaki, K.; Hirabuchi, A.; Alaux, L.; Fournier, E.; Tharreau, D.; Terauchi, R. Arms race co-evolution of *Magnaporthe oryzae AVR-Pik* and rice *Pik* genes driven by their physical interactions. *Plant J.* **2012**, *72*, 894–907. [CrossRef]
42. De la Concepcion, J.C.; Franceschetti, M.; Maqbool, A.; Saitoh, H.; Terauchi, R.; Kamoun, S.; Banfield, M.J. Polymorphic residues in rice NLRs expand binding and response to effectors of the blast pathogen. *Nat. Plants* **2018**, *4*, 576–585. [CrossRef]

43. Ashikawa, I.; Hayashi, N.; Yamane, H.; Kanamori, H.; Wu, J.; Matsumoto, T.; Ono, K.; Yano, M. Two adjacent nucleotide-binding site-leucine-rich repeat class genes are required to confer *Pikm*-specific rice blast resistance. *Genetics* **2008**, *180*, 2267–2276. [CrossRef] [PubMed]
44. Pirttilä, A.M.; Brusila, V.; Koskimäki, J.J.; Wäli, P.R.; Ruotsalainen, A.L.; Mutanen, M.; Markkola, A.M.; Yount, J. Exchange of microbiomes in plant-insect herbivore interactions. *mBio* **2023**, *14*, e321022. [CrossRef]
45. Xu, J.; Wang, X.; Zu, H.; Zeng, X.; Baldwin, I.T.; Lou, Y.; Li, R. Molecular dissection of rice phytohormone signaling involved in resistance to a piercing-sucking herbivore. *New Phytol.* **2021**, *230*, 1639–1652. [CrossRef] [PubMed]
46. Chen, H.; Zhang, Q.; Lv, W.; Yu, X.; Zhang, Z. Ethylene positively regulates Cd tolerance via reactive oxygen species scavenging and apoplastic transport barrier formation in rice. *Environ. Pollut.* **2022**, *302*, 119063. [CrossRef]
47. Xu, P.; Wu, T.; Ali, A.; Wang, J.; Fang, Y.; Qiang, R.; Liu, Y.; Tian, Y.; Liu, S.; Zhang, H.; et al. Rice β-glucosidase 4 (Os1βGlu4) regulates the hull pigmentation via accumulation of salicylic acid. *Int. J. Mol. Sci.* **2022**, *23*, 10646. [CrossRef]
48. He, Y.; Zhao, Y.; Hu, J.; Wang, L.; Li, L.; Zhang, X.; Zhou, Z.; Chen, L.; Wang, H.; Wang, J.; et al. The OsBZR1-OsSPX1/2 module fine-tunes the growth-immunity trade-off in adaptation to phosphate availability in rice. *Mol. Plant* **2024**, *17*, 258–276. [CrossRef]
49. Kumar, R.; Bhutani, S.; Singh, R.; Chauhan, R.; Chowdhury, V.; Jain, K.; Rajinder, K. Enhanced resistance against the rice leaffolder (*Cnaphalocrocis medinalis*) in transgenic rice plants containing the potato proteinase inhibitor II gene. *Entomol. Gen.* **2009**, *32*, 11–22. [CrossRef]
50. Stevenson, P.C.; Kimmins, F.M.; Grayer, R.J.; Raveendranath, S. Schaftosides from rice phloem as feeding inhibitors and resistance factors to brown planthoppers, *Nilaparvata lugens*. *Entomol. Exp. Appl.* **1996**, *80*, 246–249. [CrossRef]
51. Grayer, R.J.; Harborne, J.B.; Kimmins, F.M.; Stevenson, P.C.; Wijayagunasekera, H.N.P. Phenolics in rice phloem sap as sucking deterrents to the brown planthopper, *Nilaparvata lugens*. *Acta Hortic.* **1994**, *381*, 691–694. [CrossRef]
52. Sun, K.; Kuai, P.; Lu, J.; Lou, Y. Exogenous application of paclobutrazol induces the resistance of rice to planthoppers. *Acta Entomol. Sinica* **2024**, *67*, 21–30.
53. Zhan, Z.; Matsuo, A.; Kim, C. Studies on probing stimulants for the green rice leafhopper, *Nephotettix nigropictus* (Stål) (Hemiptera: Cicadellidae), in the rice plant (*Oryza sativa* L.). *J. Pestic. Sci.* **2016**, *41*, 163–166. [CrossRef] [PubMed]
54. Caasi-Lit, M.T.; Tanner, G.J.; Nayudu, M.; Whitecross, M.I. Isovitexin-2′-*O*-β-[6-O-E-*p*-coumaroylglucopyranoside] from UV-B irradiated leaves of rice, *Oryza sativa* L. Inhibits fertility of *Helicoverpa armigera*. *Photochem. Photobiol.* **2007**, *83*, 1167–1173. [CrossRef] [PubMed]
55. Yoshida, S.; Forno, D.A.; Cock, J.H.; Gomez, K.A. *Laboratory Manual for Physiological Studies of Rice*, 3rd ed.; International Rice Research Institute: Los Baños, Philippines, 1976.
56. Ye, M.; Kuai, P.; Chen, S.; Lin, N.; Ye, M.; Hu, L.; Lou, Y. Silencing a simple extracellular leucine-rich repeat gene *OsI-BAK1* enhances the resistance of rice to brown planthopper *Nilaparvata lugens*. *Int. J. Mol. Sci.* **2021**, *22*, 12182. [CrossRef]
57. Sun, L.; Andika, I.B.; Shen, J.; Yang, D.; Chen, J. The P2 of *Wheat yellow mosaic virus* rearranges the endoplasmic reticulum and recruits other viral proteins into replication-associated inclusion bodies. *Mol. Plant Pathol.* **2014**, *15*, 466–478. [CrossRef]
58. Ma, X.; Zhang, Q.; Zhu, Q.; Liu, W.; Chen, Y.; Qiu, R.; Wang, B.; Yang, Z.; Li, H.; Lin, Y.; et al. A robust CRISPR/Cas9 system for convenient, high-efficiency multiplex genome editing in monocot and dicot plants. *Mol. Plant* **2015**, *8*, 1274–1284. [CrossRef]
59. Xie, L. Genome-Wide Association Studies and Important Gene Mining of Maize Stalk Strength Related Traits. Ph.D. Thesis, Shandong Agricultural University, Tai'an, China, 2021; p. 14.
60. Lu, J.; Robert, C.A.M.; Riemann, M.; Cosme, M.; Mène-Saffrané, L.; Massana, J.; Stout, M.J.; Lou, Y.; Gershenzon, J.; Erb, M. Induced jasmonate signaling leads to contrasting effects on root damage and herbivore performance. *Plant Physiol.* **2015**, *167*, 1100–1116. [CrossRef]
61. Jongsma, M.A.; Bakker, P.L.; Visser, B.; Stiekema, W.J. Trypsin-inhibitor activity in mature tobacco and tomato plants is mainly induced locally in response to insect attack, wounding and virus-infection. *Planta* **1994**, *195*, 29–35. [CrossRef]
62. Van Dam, N.M.; Horn, M.; Mares, M.; Baldwin, I.T. Ontogeny constrains systemic protease inhibitor response in *Nicotiana attenuata*. *J. Chem. Ecol.* **2001**, *27*, 547–568. [CrossRef]
63. Caristi, C.; Bellocco, E.; Panzera, V.; Toscano, G.; Vadalà, R.; Leuzzi, U. Flavonoids detection by HPLC-DAD-MS-MS in lemon juices from Sicilian cultivars. *J. Agric. Food Chem.* **2003**, *51*, 3528–3534. [CrossRef]
64. Wang, W.W.; Zhou, P.Y.; Mo, X.C.; Hu, L.F.; Jin, N.; Chen, X.; Yu, Z.X.; Meng, J.P.; Erb, M.; Shang, Z.C.; et al. Induction of defense in cereals by 4-fluorophenoxyacetic acid suppresses insect pest populations and increases crop yields in the field. *Proc. Natl. Acad. Sci. USA* **2020**, *117*, 12017–12028. [CrossRef] [PubMed]

plants

MDPI

Article

The Specific Impacts of Allelopathy and Resource Competition from *Artemisia frigida* on the Growth of Three Plant Species in Northern China

Qing Wang [1,2], Mengqiao Kong [1], Junwen Wang [1], Bin Gao [1] and Xiaoyan Ping [1,*]

1 School of Grassland Science, Beijing Forestry University, No. 35 Qinghua East Road, Haidian District, Beijing 100083, China; wangqingi2020@163.com (Q.W.); zhforeverj@163.com (M.K.); jwwang2022@163.com (J.W.); ibcas2011@163.com (B.G.)

2 College of Grassland Science and Technology, China Agricultural University, No. 2 Yuanmingyuan West Road, Haidian District, Beijing 100193, China

* Correspondence: pingxy@bjfu.edu.cn; Tel.: +86-(0)10-62338257

Abstract: Plant interference is a key factor influencing plant coexistence and species composition. The two primary forms of plant interference—allelopathy and resource competition—are often difficult to separate. This study conducted an outdoor pot experiment to quantify the distinct contributions of resource competition and allelopathy of *Artemisia frigida* on seedling growth of three species: *Leymus chinensis*, *Cleistogenes squarrosa*, and *Potentilla acaulis*. The index of relative neighbor effect (*RNE*) was used to quantify the overall effect of plant interference, while the inhibition rates (*IRs*) of resource competition and allelopathy were utilized to determine the specific contributions of allelopathy and resource competition from *A. frigida* on the growth of target plant species. The interference effect of *A. frigida* was found to be species-specific. The allelopathic effect of *A. frigida* played a major role in inhibiting the belowground biomass of *L. chinensis* (23.97%) and *C. squarrosa* (58.27%), while allelopathy and resource competition from *A. frigida* promoted the belowground biomass (45.12%) and aboveground biomass (46.63%) of *P. acaulis*, respectively. The combined effect of allelopathy and resource competition from *A. frigida* significantly affected the aboveground biomass of *C. squarrosa* and *P. acaulis*, as well as the belowground biomass of *L. chinensis* and *C. squarrosa*. These findings contribute to a better understanding of the patterns and mechanisms of plant species composition and its relationship with grazing intensity in this grassland ecosystem.

Keywords: *Artemisia frigida*; Eurasian Steppe; grazing intensity; outdoor pot experiment; plant interference

1. Introduction

Plant interference is a central driver for plant coexistence [1]. The negative interspecific interaction can be attributed to resource competition (competing for water, nutrients, light, or space) or allelopathy (chemical interactions among species) [2–4]. Allelopathy can alter the dynamics of resource competition by influencing the growth of recipient species [5,6], while the intensity of resource competition can also affect the release of allelochemicals [7,8]. Traditionally, resource competition has been considered the primary driver of plant community diversity and dynamics [9]. However, other researchers have found that allelopathy can significantly influence plant community composition and succession, potentially playing a role as crucial as resource competition [4,10,11].

Numerous studies have been conducted to distinguish between the two types of plant interference [7,8,12]. For example, PVC tubes have been employed to mitigate both aboveground and root competition between two species, while activated carbon (AC) has been utilized to absorb and eliminate allelochemicals leached by donor plants [12–14]. Partial least squares regression or glasshouse experiments have been employed to ascertain the relative contribution of resource competition and allelopathy on the growth of winter cereals and rice [15,16]. Density-dependent experiments have been conducted to quantify the relative contributions of allelopathy and competition to overall interference [2,3,17], while some research has focused on the impact of competition and allopathy on community composition in coastal and inland ecosystems [18]. However, nature is inherently complex, where allelopathy and resource competition often coexist and interact closely [19,20]. The specific contribution of these two types of plant interference remains unknown.

Grasslands constitute approximately 40% of the global land area, offering a plethora of benefits to over 1 billion people worldwide [21]. The most prevalent and significant use of grasslands is for livestock grazing [22,23]. Extensive research has shown that plant species composition varies considerably along grazing gradients [24,25]. *Artemisia frigida* Willd. (Prairie Sagebrush) is a resilient perennial semi-shrub found throughout the Eurasian Steppe. It exhibits remarkable resistance to drought, grazing, trampling, and regeneration and can be used as high-quality forage, especially in spring and winter [22,23,25]. Meanwhile, *Artemisia frigida* is the dominant species under moderate and heavy grazing intensities in temperate grasslands of northern China [22]. Conversely, *Leymus chinensis* prevails in the eastern Eurasian Steppe, while *Cleistogenes squarrosa* and *Potentilla acaulis* are the dominant species under light and extreme grazing intensities, respectively [25]. *Cleistogenes squarrosa* is a C_4 perennial bunchgrass noted for its robust drought tolerance, whereas *P. acaulis* is a perennial stoloniferous herb known for its pronounced allelopathy [22,24]. Both *A. frigida* and *P. acaulis* exhibit allelopathic effect, and the aqueous extracts from these two species and the volatile organic compounds (VOCs) from *A. frigida* markedly inhibit the growth of *C. squarrosa* and *L. chinensis* [22,25,26]. The ecological adaptation of *A. frigida* to prolonged continuous grazing can be attributed to its strong phenotypic plasticity [27], altered carbon allocation patterns [28], intense interspecific competitiveness [24], and allelopathy [22,25,29,30].

To assess the distinct impact of resource competition and allelopathy of *A. frigida* on the growth of common species within the Eurasian Steppe, we initiated an outdoor pot experiment in northern China. We hypothesized that the plant interference of *A. frigida* would exhibit species-specific variation among different plant species. Our study sought to address the following question: How does the interference from *A. frigida* differ across various target plant species, and how is this interference influenced by grazing intensity and specific plant species?

2. Results

2.1. Interference Between A. frigida and Three Target Plant Species

For the interference of *A. frigida* with three target plant species (AF-LC, AF-CS, and AF-PA in Table 1), *A. frigida* inhibited both the aboveground (LC 0.28, CS 0.45) and belowground biomass (LC 0.15, CS 0.64) of *L. chinensis* and *C. squarrosa* but promoted the root/shoot ratio (−0.16) of *L. chinensis*. Concurrently, *A. frigida* enhanced the aboveground biomass (−0.08), belowground biomass (−0.67), and root/shoot ratio (−0.64) of *P. acaulis*. The index of relative neighbor effect (*RNE*) of the aboveground biomass, belowground biomass, and root/shoot ratio were all significantly different among the three target species (aboveground biomass: $df = 8$, F = 19.875, $P = 0.002$; belowground biomass: $df = 8$, F = 535.398, $P < 0.001$; root/shoot ration: $df = 8$, F = 59.108, $P < 0.001$). Notably, the growth inhibitory effect of *A. frigida* on *C. squarrosa* was stronger, especially in terms of belowground biomass and root/shoot ratio, when compared to *L. chinensis*.

As for the interference of three target plant species with *A. frigida* (LC-AF, CS-AF and PA-AF in Table 1), in addition to the inhibitory effect of *P. acaulis* on the aboveground biomass (0.30) of *A. frigida*, the other two target plant species all promoted the aboveground biomass (LC −0.27, CS −0.64) and belowground biomass (LC −0.40, CS −0.52) of *A. frigida*. Meanwhile, the promoting effects of *C. squarrosa* on the aboveground and belowground biomass of *A. frigida* were both significantly greater than that of *L. chinensis* ($df = 5$, $F = 136.862$, $P = 0.001$ and $df = 5$, $F = 18.752$, $P = 0.012$).

Table 1. Interference between *Artemisia frigida* and three target plant species at the study site.

Indexes of Plant Interference	*RNE*		
	Aboveground Biomass	**Belowground Biomass**	**Root/Shoot Ratio**
AF-LC	0.28 ± 0.10 a	0.15 ± 0.04 b	−0.16 ± 0.08 b
AF-CS	0.45 ± 0.10 a	0.64 ± 0.07 a	0.30 ± 0.08 a
AF-PA	−0.08 ± 0.04 b	−0.67 ± 0.01 c	−0.64 ± 0.05 c
LC-AF	−0.27 ± 0.07 b	−0.40 ± 0.07 a	−0.18 ± 0.05 b
CS-AF	−0.64 ± 0.02 c	−0.52 ± 0.04 b	0.23 ± 0.09 a
PA-AF	0.30 ± 0.06 a	−0.59 ± 0.04 b	−0.71 ± 0.02 c

Note: Different lowercase letters in the same column indicate significant differences in the *RNE* among three target plant species. All the data are shown as mean ± SE (n = 3). AF-LC: interference of *Artemisia frigida* with *Leymus chinensis*; LC-AF: interference of *L. chinensis* with *A. frigida*; AF-CS: interference of *A. frigida* with *Cleistogenes squarrosa*; CS-AF: interference of *C. squarrosa* with *A. frigida*; AF-PA: interference of *A. frigida* with *Potentilla acaulis*; and PA-AF: interference of *P. acaulis* with *A. frigida*.

2.2. Specific Contributions of Allelopathy and Resource Competition from *A. frigida* to the Growth of Three Target Species

For *L. chinensis*, allelopathy of *A. frigida* played a predominant role in plant interference, particularly affecting the belowground biomass (Figure 1). The resource competition from *A. frigida* led to a decrease in the aboveground biomass by 1.16% but increased the belowground biomass by 17.42%. Furthermore, the allelopathy of *A. frigida* resulted in a reduction of the belowground biomass by 23.97% yet augmented the aboveground biomass by 6.13%.

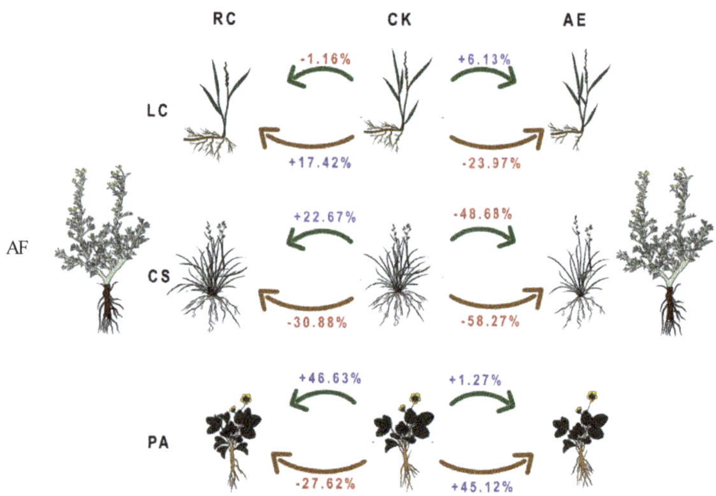

Figure 1. Specific contribution of allelopathy and resource competition from *Artemisia frigida* to the growth of three target species (LC: *Leymus chinensis*, AF: *Artemisia frigida*, CS: *Cleistogenes squarrosa*, PA: *Potentilla acaulis*; RC: resource competition; AE: allelopathy; and CK: monoculture. Red font in the figure indicates a growth-inhibiting effect, while blue font indicates a growth-promoting effect. The green arrows in the figure indicate aboveground biomass, and the brown arrows indicate belowground biomass).

For *C. squarrosa*, the allelopathy exhibited by *A. frigida* played a major role in affecting its aboveground and belowground biomass. Specifically, resource competition initiated by *A. frigida* resulted in a 22.67% increase in the aboveground biomass while simultaneously causing a 30.88% decrease in the belowground biomass. Conversely, the allelopathy of *A. frigida* resulted in a reduction in the aboveground biomass by 48.68% and the belowground biomass by 58.27%.

As for *P. acaulis*, both the aboveground and belowground biomass was affected by the resource competition from *A. frigida*. This competition led to a 46.63% increase in the aboveground biomass and a 27.62% decrease in the belowground biomass. Additionally, the allelopathy of *A. frigida* contributed to a 1.27% increase in the aboveground biomass and a 45.12% increase in the belowground biomass of *P. acaulis*.

2.3. Effects of Allelopathy and Resource Competition and Their Coupling Effect on the Plant Growth of Three Target Species

The allelopathic effect (AE vs. CK) of *A. frigida* significantly reduced the belowground biomass of *L. chinensis* (df = 17, F = 17.319, P < 0.001), while resource competition (RC vs. AC added) did not have a significant impact (Figure 2(a1,b1)). Neither factor significantly affected the aboveground biomass of *L. chinensis* (Figure 2(a1)). The combined effect of allelopathy and resource competition of *A. frigida* did not significantly influence either the aboveground or belowground biomass of *L. chinensis* (AE + RC vs. CK). However, the addition of activated carbon significantly inhibited the aboveground (df = 5, F = 18.572, P = 0.013) and belowground biomass (df = 5, F = 170.763, P < 0.001) of *L. chinensis* (AC added vs. CK).

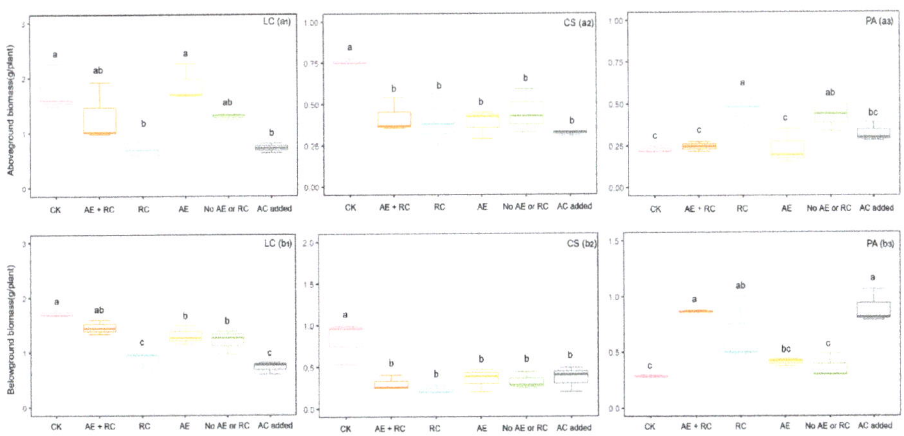

Figure 2. Effects of allelopathy and resource competition of *Artemisia frigida* on the growth of three target plant species (All the data are shown as mean ± SE (n = 3). (**a**): Effects of allelopathy and resource competition of *Artemisia frigida* on the aboveground biomass of *L. chinensis* (**a1**), *C. squarrosa* (**a2**), *P. acaulis* (**a3**). (**b**): Effects of allelopathy and resource competition of *Artemisia frigida* on the belowground biomass of *L. chinensis* (**b1**), *C. squarrosa* (**b2**), *P. acaulis* (**b3**). CK: monoculture; AE + RC: both allelopathy and resource competition; RC: resource competition; AE: allelopathy of *A. frigida*; No AE or RC: neither allelopathy nor resource competition of *A. frigida*; and AC added: activated carbon was added to evaluate the specific impact of activated carbon on plant growth of three target plant species). Different lowercase letters show the significant differences in biomass of the same plant specie between different treatments (P < 0.05).

The allelopathic effects of *A. frigida* significantly reduced both the aboveground (df = 17, F = 7.083, P = 0.003) and belowground biomass (df = 17, F = 6.529, P = 0.004) of *C. squarrosa* (Figure 2(a2,b2)). Conversely, resource competition from *A. frigida* did not have a significant effect on either the aboveground or belowground biomass of *C. squarrosa*. However, when considering the combined effects of allelopathy and resource competition from *A. frigida*, there was a significant inhibition observed in both the aboveground (df = 5, F = 30.536, P = 0.005) and belowground biomass (df = 5, F = 11.170, P = 0.029) of *C. squarrosa*. Furthermore, the introduction of activated carbon led to a significant reduction in the aboveground (df = 5, F = 134.602, P < 0.001) and belowground biomass (df = 5, F = 8.114, P = 0.046) of *C. squarrosa*.

The allelopathic effects of *A. frigida* did not significantly affect the aboveground and belowground biomass of *P. acaulis* (Figure 2(a3,b3)). However, resource competition from *A. frigida* significantly enhanced the aboveground biomass of *P. acaulis* ($df = 17$, $F = 5.870$, $P = 0.006$). The combined effect of allelopathy and resource competition from *A. frigida* significantly increased the belowground biomass of *P. acaulis* ($df = 17$, $F = 10.162$, $P = 0.001$) (Figure 2(b3)). The addition of activated carbon significantly increased the belowground biomass of *P. acaulis* ($df = 5$, $F = 45.518$, $P = 0.003$).

Our study found that, compared to monocultured *L. chinensis* and *C. squarrosa*, the interference of *A. frigida* inhibited the plant growth of these species (LC 0.22, CS 0.55). Conversely, the interference of *A. frigida* promoted the growth of *P. acaulis* compared to its monoculture condition (−0.38). However, the interference from *L. chinensis*, *C. squarrosa*, and *P. acaulis* all had a positive effect on the growth of *A. frigida* (LC −0.34, CS −0.58, PA −0.14) (Figure 3).

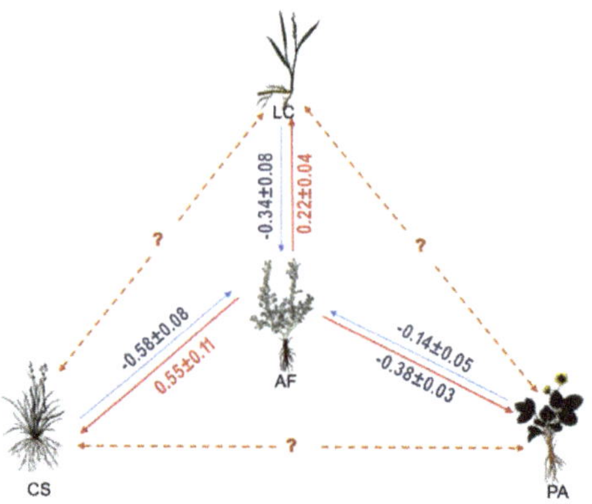

Figure 3. Interference of *Artemisia frigida* with the plant growth of three target species (We used the average *RNE* of aboveground and belowground biomass to represent the interference of *A. frigida* with three target species (red arrows). Meanwhile, the interference of three target species with *A. frigida* was also marked in the figure (blue arrows). The blue values signify a promotional effect, while the red values denote an inhibitory effect).

3. Discussion

3.1. Interference of A. frigida with Three Target Plant Species

Our results support our hypothesis that the interference of *A. frigida* was species-specific. However, this study only examined the pairwise interference of *A. frigida* on three target plant species, leaving the specific interference of *L. chinensis*, *C. squarrosa* and *P. acaulis* to each other unexplained (Figure 3). The impact of interference on phenotypic plasticity can have more profound effects on species composition than above- and belowground biomass. For instance, the inhibitory effect of *Plantago lanceolate* on the root growth of *Festuca rubra* could diminish the competitive strength of *F. rubra* against other species within the community [31]. Similarly, the inhibitory effect of *P. acaulis* on the plant growth of *A. frigida* may alter the interference of *A. frigida* with other plant species at the study site (Figure 3). Therefore, to gain a comprehensive understanding of the role of plant interference on species composition, it is necessary to consider not only pairwise interactions but also the interaction chain among coexisting plant species in future research.

The interaction between two species may be conditional, depending on environmental conditions, and herbivory may also change the balance of interference among species [32]. In a multispecies community, the species interactions are the sum of all intraspecific and interspecific interactions. Moreover, the interference among species is important in shaping community dynamics and maintaining biodiversity [33]. Concurrently, plant interference may also be influenced by plant–soil feedbacks (PSFs), which can promote the growth of neighboring species in heterogeneous environments [34]. Furthermore, competition among species within the same clades tends to favor dissociation, primarily due to asymmetrical competition within the community [35]. Species with high interspecific competition ability along with grazing intensity would result in asymmetric competition. Highly dominated communities are often characterized by strongly asymmetric competition [36]. Our findings indicate that *A. frigida* and *P. acaulis* possess a relatively high degree of interspecific competition ability in comparison to *L. chinensis* and *C. squarrosa* (Figure 3). This asymmetrical competition could potentially lead to the dominance of *A. frigida* and *P. acaulis* under conditions of heavy and extreme grazing.

3.2. Separation of Allelopathy and Resource Competition from Plant Interference

Consistent with prior research, our findings validated that the plant interference of *A. frigida* was a synergistic effect of resource competition and allelopathy [17]. Furthermore, we established that the combined effect of allelopathy and resource competition significantly impacted plant–plant interference, with the primary type and intensity of interference from *A. frigida* varying among different plant species. The allelopathic effect of *A. frigida* became important for *L. chinensis* and *C. squarrosa* under conditions of no grazing and light grazing intensity (Figure 2). Meanwhile, both the allelopathy and resource competition of *A. frigida* facilitated the dominance of *P. acaulis* under extreme grazing intensity. Therefore, this study provides a partial explanation for the mechanism of community succession in temperate grasslands along grazing gradients.

Several methods have been developed to distinguish resource competition from allelopathy [8,19,37]. However, most prior research has overlooked the combined effect of allelopathy and resource competition [38]. This study addresses two significant limitations in allelopathy research. Firstly, we implemented a field control experiment, enhancing the credibility of our results compared to those derived from greenhouse experiments. Secondly, this study not only isolates the allelopathic effects of *A. frigida* from resource competition but also examines the interactive effect of resource competition and allelopathy of *A. frigida* on the plant growth of three target plant species.

The application of activated carbon (AC) has been widely utilized in allelopathy studies to adsorb allelochemicals and mitigate soil toxicity caused by these compounds. This method effectively separates the effects of allelopathy from resource competition [39,40]. However, the use of AC can introduce controversy due to potential side effects such as altering soil nutrient availability and plant growth [40] and affecting soil microbial activity [41]. Our study confirmed that the addition of AC can influence the plant growth of target plant species, although the specific effects varied among the three target plant species (Figure 2). In contrast to the inhibitory effect of AC on the plant growth of *L. chinensis* and *C. squarrosa* (Figure 2), the addition of AC significantly promoted the plant growth of *P. acaulis* (Figure 2). This discrepancy could be attributed to the fact that AC may also absorb the allelochemicals released by *P. acaulis* itself, which has been shown to have strong allopathic effects on the seed germination of *L. chinensis* and *A. frigida* [26]. Additionally, we used AC added to *L. chinensis*, *C. squarrosa*, and *P. acaulis* as controls to assess the impact of resource competition on the plant growth of three target plant species. By incorporating controls for the effect of AC in our experimental design, our results provide a clearer reflection of the relative contributions of allelopathic interference and resource competition to plant growth.

3.3. The Impact of Allelopathy and Resource Competition and Their Coupling Effects on the Dominant Status of A. frigida in Moderate Grazing Grassland

Our findings corroborate previous research indicating that allelopathy can vary among target plant species, either stimulating or inhibiting growth [42]. For instance, the aboveground biomass of *L. chinensis* was marginally enhanced by the allelopathy of *A. frigida*, while the belowground biomass was significantly suppressed [43,44]. In this study, *C. squarrosa* demonstrated a higher sensitivity to allelopathy of *A. frigida* than *L. chinensis*, contradicting previous results [22]. This discrepancy may be attributed to *C. squarrosa*'s adaptability to environments with lower precipitation levels; thus, its interference in the community is likely to increase under conditions of aridity induced by livestock grazing [45]. The heavy rainfall during the latter stages of our experiment may have heightened *C. squarrosa*'s sensitivity to both the allelopathy and resource competition of *A. frigida*. Additionally, the belowground biomass was more significantly affected by the allelopathy of *A. frigida*, aligning with previous findings in northern China [22].

Allelopathy, whereby the growth of neighboring plants is inhibited by donor plants, can simultaneously enhance their resource competitiveness [4]. This form of intransitive competition can lead to changes in plant species composition [33]. Our study found that the combined effect of allelopathy and resource competition significantly influenced the aboveground biomass of *P. acaulis* and the above- and belowground biomass of *C. squarrosa* (Figure 2). However, we were unable to precisely quantify the contribution of this coupling effect. Our results indicated that the coupling effect varied substantially among the three target plant species. It could increase the above- and belowground biomass of *L. chinensis* and *C. squarrosa* and the belowground biomass of *P. acaulis*, and it could also decrease the aboveground biomass of *P. acaulis*. Therefore, the coupling effect of allelopathy and resource competition could either offset or amplify the actual effect, playing an important role in regulating species composition at the study site [14].

Allelopathy can facilitate the dominance of specific plants within a community, especially for alien plants [6]. Additionally, allelopathy has been identified as a key factor driving grassland succession in northern China [22,44] and forest dynamics in Mediterranean and southern China [11,46]. Resource competition is widely recognized as one of the most significant mechanisms of plant interference in shaping species composition. Concurrently, previous research has indicated that the competition among plant species in the absence of herbivores transitions to facilitation in heavily grazed pastures [47]. In stressful habitats, positive neighbor effects become increasingly important for plant interference [48], a phenomenon that may also be reflected in our study. Along with grazing gradients, the high intensity of trampling and feeding imposed by livestock can cause mechanical damage to plants of *A. frigida* and increase the release of allelochemicals [49]. A prior study revealed that grazing intensity could influence the components and concentrations of allelochemicals released by *A. frigida*, with the maximum relative contents of 1,8-cineole and β-terpineol both occurring under heavy grazing intensity [25]. These high concentrations of allelochemicals had a strong inhibitory effect on the plant growth of *C. squarrosa* and a promotional effect on the plant growth of *P. acaulis*. However, we did not take into account the concentration of allelochemicals, particularly the VOCs released by *A. frigida*, throughout the entire experiment, which should be considered in future research.

4. Materials and Methods

4.1. Study Site

This study was conducted at the Hulunbuir field observation experimental station (48°55′48″ N, 119°41′26″ E, altitude 636 m) of the Chinese Academy of Environmental Sciences in Temo Huzhu, situated in the Ewenki Autonomous Banner of Hulunbuir city, China. The study site is characterized by a mid-temperate continental monsoon climate, with an annual average temperature of −2.4 to 2.2 °C and an annual average precipitation from 330 to 350 mm (mainly occurring from June to August). The grassland at the study site is part of the Eurasian Steppe, and the soil is classified as calcic chestnuts, equivalent to Kastanozems according to the soil classification system of the Food and Agriculture Organization (FAO). In the absence of grazing, *L. chinensis* and *Stipa baicalensis* are the dominant species, while *Agropyron mongolicum*, *Achnatherum sibiricum*, and *Carex duriuscula* are common species in the study site.

4.2. Experimental Design

An outdoor pot experiment was carried out at the study site to distinguish the specific contribution of resource competition and allelopathy of *A. frigida* on the seedling growth of three target species. *Artemisia frigida* (AF) was utilized as the donor plant, while *L. chinensis* (LC), *C. squarrosa* (CS), and *P. acaulis* (PA) were selected as the recipient plants. The proportion of dry matter for each of the four species within the community is detailed in Table 2.

Table 2. The dry matter proportion of four plant species at the study site along with grazing gradients.

Plant Species	Dry Matter Proportion (%)			
	No Grazing	Light Grazing	Medium Grazing	Extreme Grazing
Leymus chinensis	35.34 ± 1.45 a	26.32 ± 1.89 b	15.53 ± 1.23 c	6.12 ± 0.78 d
Cleistogenes squarrosa	5.61 ± 0.32 d	43.64 ± 3.27 a	25.25 ± 2.32 b	13.42 ± 1.62 c
Artemisia frigida	3.26 ± 0.25 c	10.25 ± 1.67 b	38.21 ± 2.87 a	28.02 ± 2.98 a
Potentilla acaulis	0.08 ± 0.03 c	2.35 ± 0.08 bc	7.30 ± 1.12 b	32.75 ± 4.25 a

Note: Different lowercase letters in the same line indicate significant differences in the dry matter proportion of four plan species among four grazing intensities. All the data are shown as mean ± SE (n = 3).

In our pot experiment, we utilized seedlings from four plant species, all of which were collected from the same location within the study site. Seedlings of four plant species were gathered at the stem elongation stage, as defined by the BBCH scale. To ensure homogeneous growth conditions, we selected seedlings of average size, with a coefficient of variation (CV) in plant height of less than 10% for each species. We took measures to enhance the survival rate of the seedlings by preserving the integrity of their roots during excavation. Each pot was initially watered with 500 mL, and no further irrigation was provided throughout the experiment. To maintain soil consistency across the pots, we extracted 0–30 cm topsoil from the sample plot where the plants were originally collected. This soil was then sieved, meticulously mixed, and purged of any impurities. Each pot contained six seedlings: either three from both donor and recipient plants or exclusively six from recipient plants. The pots used had a capacity of 2 gallons and dimensions of 0.24 m × 0.22 m (diameter × height). We established three biological replicates for each treatment.

The experimental design is illustrated in Figure 4. Our field control experiment incorporated five treatments: allelopathy of AF(AE), resource competition of AF(RC), neither allelopathy nor resource competition of AF (No AE or RC), both allelopathy and resource competition of AF (AE + RC), and the addition of activated carbon (AC added). The allelopathic effect of *A. frigida* was neutralized by introducing activated carbon, a method frequently employed in previous studies (represented by black-dotted pots in Figure 4) [12,13,19]. We used analytical-grade activated carbon (charcoal-activated powder extra pure, Zhejiang Minxin Ecological Technology Company, Quzhou, China), added at a concentration of 20 mL/L according to Parepa et al. [50]. The activated carbon was divided into two portions: one was mixed into the soil to absorb allelochemicals released by the roots of *A. frigida*, while the other was placed on the pot surface to absorb VOCs emitted by *A. frigida*.

Resource competition was eliminated by introducing a PVC plastic sheet into the center of the pot (gray-shaded pots in Figure 4) [34]. Three individuals from each species were planted into a pot, with each species situated on opposite sides of the plastic sheet. The sheet was affixed to the sides of each pot, ensuring that each side had equal resources and space for plant growth. We conducted a standard partition study where both roots and shoots of each species were physically separated. The impact of activated carbon on seedling growth of the three target plant species was assessed by incorporating activated carbon into the pots under monoculture conditions (white spotted pots in Figure 4). The entire experiment spanned over 2 months, from June 2019 to September 2019. All 72 pots were placed outdoors throughout the experimental period.

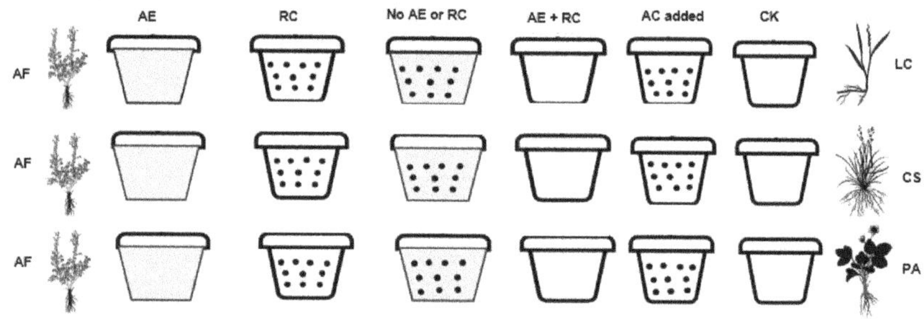

Figure 4. Design of outdoor pot experimental to differentiate between allelopathy and resource competition of *Artemisia frigida* in a temperate grassland in northern China. The allelopathic effect of *A. frigida* was eliminated by adding activated carbon to absorb allelochemicals both in the soil and air, represented by black dots on the pots. Resource competition was mitigated by introducing a PVC partition in the center of the pot, depicted as gray-shaded pots in the figure. The four plant species used in this study included LC, *Leymus chinensis;* CS, *Cleistogenes squarrosa;* AF, *Artemisia frigida;* and PA, *Potentilla acaulis*, representing the dominant species under non-grazing, light grazing, moderate grazing, and extreme grazing intensities at the study site, respectively. The experimental conditions were categorized as follows: AE (allelopathy), RC (resource competition), No AE or RC (neither allelopathy nor resource competition), AE + RC (both allelopathy and resource competition), AC added (activated carbon was added to assess its impact on plant growth), and CK (three target species planted individually).

4.3. Measurements

Biomass measurements were conducted at the end of the experiment. Seedlings were divided into aboveground and belowground parts by cutting at the root collar's apex. Roots were thoroughly washed, and both aboveground and belowground samples were all oven-dried (65 °C) for a minimum of 48 h before being weighed. We also evaluated the soil nutrient contents used for the field control experiment, and the soil description of the study site is present in Supplementary Table S1.

4.4. Data Analysis

The index of relative neighbor effect (*RNE*) was used to quantify the interference between *A. frigida* and the three target plant species, as it can overcome skewed distribution and facilitate statistical analysis compared to the commonly used index of relative competitive intensity (*RCI*) [51,52]:

$$RNE = (P_{mono} - P_{mix})/P_{max} \quad (1)$$

where P_{mono} and P_{mix} represent the performance of four plant species under monoculture and mixed culture conditions, respectively. The maximum value between P_{mono} and P_{mix} is denoted as P_{max}. For this research, we utilized aboveground biomass, belowground biomass, and root/shoot ratio as performance parameters. The *RNE* value ranged from −1 to 1, where positive and negative *RNE* values signify competitive and facilitatory effects, respectively.

Meanwhile, the index of inhibitory rate (*IR*) was used to distinguish the specific impacts of resource competition and allelopathy from *A. frigida* on the three target plant species:

$$IR = (1 - T/C) \times 100\% \quad (2)$$

where T represents the performance of four plant species under polyculture conditions, while C represents the performance of four plant species under monoculture conditions. The positive and negative values of *IR* indicate inhibitory and facilitatory effects, respectively.

We calculated the indices of *IR* separately for resource competition and allelopathy, noting that the values of *T* and *C* differed between resource competition and allelopathy. To calculate the indices of *IR* under resource competition for *A. frigida*, *T* represents the performance of the recipient species in polyculture conditions, influenced solely by interspecific competition, with allelopathy eliminated by the addition of activated carbon (RC). Conversely, *C* represents the performance of the recipient species in monoculture conditions with the inclusion of activated carbon (AC added), ensuring that the effect of activated carbon on plant growth was counterbalanced [40].

As for calculating the indices of *IR* under allelopathy of *A. frigida*, *T* represents the performance of the recipient species in polyculture conditions influenced solely by allelopathy. To eliminate resource competition, a PVC partition was introduced in the center of each pot (AE). Conversely, *C* denotes the performance of recipient species in monoculture conditions (CK).

4.5. Statistical Analyses

This study tried to ascertain the significance of *RNE* and *IR* among target plant species and the significance of plant growth of target plant species among five treatments (AE, RC, AE + RC, No AE or RC, AC added). ANOVA is used for significance, and the LSD is used for grouping when data show significant differences among treatments. All experimental data are reported as the mean ± standard error (SE). A significance level of $\alpha = 0.05$ was adopted for all analytical determinations. Statistical analyses were performed using R statistical software version 3.6.1 (R Project for Statistical Computing, Auckland, NZ). All the graphs were generated with ggplot2.

5. Conclusions

Overall, our findings indicate that the type and intensity of plant interference significantly vary among different species. The growth of *L. chinensis*, *C. squarrosa*, and *P. acaulis* is primarily affected by resource competition, allelopathy, and the combined effect of allelopathy and resource competition from *A. frigida*, respectively. The inhibitory effect of *A. frigida* on *L. chinensis* and *C. squarrosa*, coupled with its promotional effect on *P. acaulis*, could partially elucidate the community succession observed at the study site in relation to grazing gradients. Both inhibitory and facilitation models make a substantial contribution to the process of community succession. Our results may aid in distinguishing the mechanisms involved in plant interference and community succession within grassland under long-term grazing. Future work should focus on mutual interference among *L. chinensis*, *C. squarrosa*, and *P. acaulis* in plant interactions.

Supplementary Materials: The following supporting information can be downloaded at: https://www.mdpi.com/article/10.3390/plants13233286/s1, Table S1: Soil description of the study site used for pot experiment.

Author Contributions: Q.W.: Writing-original draft, visualization, data curation, formal analysis, investigation, resources, software; M.K.: Visualization, formal analysis, investigation, software; J.W.: Visualization, data curation, formal analysis, investigation, software; B.G.: Conceptualization, data curation, formal analysis, investigation, methodology, resources; X.P.: Conceptualization, data curation, methodology, supervision, validation, writing—original draft, writing—review & editing, funding acquisition. All authors have read and agreed to the published version of the manuscript.

Funding: This research was funded by National Natural Science Foundation of China (31700388) and Open bidding for selecting the best candidates for National Forestry and Grassland Administration (202401-06-04).

Data Availability Statement: Data is contained within the article.

Acknowledgments: We thank Xiaofan He for the technical assistance in the preparation of the manuscript. In particular, the authors would like to express their gratitude to three anonymous reviewers for their valuable comments and suggestions, which have significantly contributed to the improvement of our manuscript.

Conflicts of Interest: The authors declare that they have no known competing financial interests or personal relationships that could have appeared to influence the work reported in this paper.

References

1. Kong, C.H.; Li, Z.; Li, F.L.; Xia, X.X.; Wang, P. Chemically mediated plant-plant interactions: Allelopathy and allelobiosis. *Plants* **2024**, *13*, 626. [CrossRef] [PubMed]
2. Liu, D.L.; An, M.; Johnson, I.R.; Lovett, J.V. Mathematical modelling of allelopathy: IV. Assessment of contributions of competition and allelopathy to interference by barley. *Nonlinearity Biol. Toxicol. Med.* **2005**, *3*, 213–224. [CrossRef] [PubMed]
3. San Emeterio, L.; Damgaard, C.; Canals, R.M. Modelling the combined effect of chemical interference and resource competition on the individual growth of two herbaceous populations. *Plant Soil* **2007**, *292*, 95–103. [CrossRef]
4. Hierro, J.L.; Callaway, R.M. The ecological importance of allelopathy. *Annu. Rev. Ecol. Evol. Syst.* **2021**, *52*, 25–45. [CrossRef]
5. Meiners, S.J.; Kong, C.H.; Ladwig, L.M.; Pisula, N.L.; Lang, K.A. Developing an ecological context for allelopathy. *Plant Ecol.* **2012**, *213*, 1221–1227. [CrossRef]
6. Zhang, Z.; Liu, Y.; Yuan, L.; Weber, E.; van Kleunen, M. Effect of allelopathy on plant performance: A meta-analysis. *Ecol. Lett.* **2021**, *24*, 348–362. [CrossRef]
7. Inderjit; del Moral, R. Is separating resource competition from allelopathy realistic? *Bot. Rev.* **1997**, *63*, 221–230. [CrossRef]
8. He, H.B.; Wang, H.B.; Fang, C.X.; Lin, Z.H.; Yu, Z.M.; Lin, W.X. Separation of allelopathy from resource competition using rice/barnyardgrass mixed-cultures. *PLoS ONE* **2012**, *7*, e37201. [CrossRef]
9. Tilman, D. *Resource Competition and Community Structure*; Princeton University Press: Princeton, NJ, USA, 1982.
10. Mallik, A.U. Allelopathy: Advances, Challenges and Opportunities. In *Allelopathy in Sustainable Agriculture and Forestry*; Zeng, R.S., Mallik, A.U., Luo, S.M., Eds.; Springer: New York, NY, USA, 2008; pp. 25–38.
11. Fernandez, C.; Monnier, Y.; Santonja, M.; Gallet, C.; Weaton, L.A.; Prevosto, B.; Saunier, A.; Baldy, V.; Bousquet-Melou, A. The impact of competition and allelopathy on the trade-off between plant defense and growth in two contrasting tree species. *Front. Plant Sci.* **2016**, *7*, 594. [CrossRef]
12. Qasem, J.R. A new technology separating allelopathy from competition in pot experiments. *Int. J. Agric. Sci. Food Technol.* **2017**, *3*, 19–25. [CrossRef]
13. Nilsson, M.C. Separation of allelopathy and resource competition by the boreal dwarf shrub *Empetrum hermaphroditum* Hagerup. *Oecologia* **1994**, *98*, 1–7. [CrossRef] [PubMed]
14. Bieberich, J.; Lauerer, M.; Drachsler, M.; Heinrichs, J.; Müller, S.; Feldhaar, H. Species- and developmental stage-specific effects of allelopathy and competition of invasive *Impatiens glandulifera* on co-occurring plants. *PLoS ONE* **2018**, *13*, e0205843. [CrossRef] [PubMed]
15. Reiss, A.; Fomsgaard, I.S.; Mathiassen, S.K.; Stuart, R.M.; Kudsk, P. Weed suppression by winter cereals: Relative contribution of competition for resources and allelopathy. *Chemoecology* **2018**, *28*, 109–121. [CrossRef]
16. Masum, S.M.; Hossain, M.A.; Akamine, H.; Sakagami, J.; Ishii, T.; Nakamura, I.; Asaduzzaman, M.; Bhowmik, P.C. Performance of Bangladesh indigenous rice in a weed infested field and separation of allelopathy from resource competition. *Weed Biol. Manag.* **2019**, *19*, 39–50. [CrossRef]
17. Uddin, M.N.; Robinson, R.W. Allelopathy and resource competition: The effects of *Phragmites australis* invasion in plant communities. *Bot. Stud.* **2017**, *58*, 29. [CrossRef]
18. Martines, I.P.; Kojouharov, H.V.; Grover, J.P. A chemostat model of resource competition and allelopathy. *Appl. Math Comput.* **2009**, *215*, 573–582. [CrossRef]
19. Liu, C.; Bu, Z.J.; Mallik, A.; Rochefort, L.; Hu, X.F.; Yu, Z. Resource competition and allelopathy in two peat mosses: Implication for niche differentiation. *Plant Soil* **2020**, *446*, 229–242. [CrossRef]
20. Mougi, A. Allelopathic adaptation can cause competitive coexistence. *Theor. Ecol.* **2013**, *6*, 165–171. [CrossRef]
21. Bai, Y.; Cotrufo, M.F. Grassland soil carbon sequestration: Current understanding, challenges, and solutions. *Science* **2022**, *377*, 603–608. [CrossRef]
22. Li, X.F.; Wang, J.; Huang, D.; Wang, L.X.; Wang, K. Allelopathic potential of *Artemisia frigida* and successional changes of plant communities in the northern China steppe. *Plant Soil* **2011**, *341*, 383–398. [CrossRef]
23. Schönbach, P.; Wan, H.; Gierus, M.; Bai, Y.F.; Müller, K.; Lin, L.J.; Susenbeth, A.; Taube, F. Grassland responses to grazing: Effects of grazing intensity and management system in an Inner Mongolian steppe ecosystem. *Plant Soil* **2011**, *340*, 103–115. [CrossRef]
24. Gao, Y.Z.; Wang, S.P.; Han, X.G.; Patton, B.D.; Nyren, P.E. Competition between *Artemisia frigida* and *Cleistogenes squarrosa* under different clipping intensities in replacement series mixtures at different nitrogen levels. *Grass Forage Sci.* **2005**, *60*, 119–127. [CrossRef]
25. Wang, Q.; Zhang, H.; Yang, Q.; Wang, T.M.; Zhang, Z.W.; Liu, J.L.; Shi, M.M.; Ping, X.Y. The impact of grazing intensity on the allelopathic effect of *Artemisia frigida* in a temperate grassland in northern China. *Flora* **2022**, *288*, 152005. [CrossRef]
26. Zhang, Y.; Tang, S.; Liu, K.; Li, X.; Huang, D.; Wang, K. The allelopathic effect of *Potentilla acaulis* on the changes of plant community in grassland, northern China. *Ecol. Res.* **2015**, *30*, 41–47. [CrossRef]
27. Li, J.H.; Li, Z.; Ren, J.Z. Effect of grazing intensity on clonal morphological plasticity and biomass allocation patterns of *Artemisia frigida* and *Potentilla acaulis* in the Inner Mongolia steppe. *New Zealand J. Agric. Res.* **2005**, *48*, 57–61.

28. Li, F.R.; Zhao, A.F.; Zhou, H.Y.; Zhang, T.H.; Zhao, X. Effects of simulated grazing on growth and persistence of *Artemisia frigida* in a semiarid sandy rangeland. *Grass Forage Sci.* **2002**, *57*, 239–246. [CrossRef]
29. Zhang, R.M.; Zuo, Z.J.; Gao, P.J.; Hou, P.; Wen, G.S.; Gao, Y. Allelopathic effects of VOCs of *Artemisia frigida* Willd. on the regeneration of pasture grasses in Inner Mongolia. *J. Arid Environ.* **2012**, *87*, 212–218. [CrossRef]
30. Zhang, R.; Zhang, W.; Zuo, Z.J.; Li, R.; Wu, J.; Gao, Y. Inhibition effects of volatile organic compounds from *Artemisia frigida* Willd: On the pasture grass intake by lambs. *Small Rumin. Res.* **2014**, *121*, 248–254. [CrossRef]
31. Padilla, F.M.; Mommer, L.; de Caluwe, H.; Smit-Tiekstra, A.E.; Wagemaker, C.A.M.; Ouborg, N.J.; de Kroon, H. Early root overproduction not triggered by nutrients decisive for competitive success belowground. *PLoS ONE* **2013**, *8*, e55805. [CrossRef]
32. Callaway, R.M.; Pennings, S.C.; Richards, C.L. Phenotypic plasticity and interactions among plants. *Ecology* **2003**, *84*, 1115–1128. [CrossRef]
33. Aschehoug, E.T.; Brooker, R.; Atwater, D.Z.; Maron, J.L.; Callaway, R.M. The mechanisms and consequences of interspecific competition among plants. *Annu. Rev. Ecol. Evol. Syst.* **2016**, *47*, 263–281. [CrossRef]
34. Wang, S.; Callaway, R.M. Plasticity in response to plant-plant interactions and water availability. *Ecology* **2021**, *102*, e03361. [CrossRef] [PubMed]
35. Freckleton, R.P.; Watkinson, A.R. Asymmetric competition between plant species. *Funct. Ecol.* **2001**, *15*, 615–623. [CrossRef]
36. Wang, L.; Delgado-Baquerizo, M.; Wang, D.; Isbell, F.; Liu, J.; Feng, C.; Liu, J.; Zhong, Z.; Zhu, H.; Yuan, X.; et al. Diversifying livestock promotes multidiversity and multifunctionality in managed grasslands. *Proc. Natl. Acad. Sci. USA* **2019**, *116*, 6187. [CrossRef]
37. Falquet, B.; Roux, D.; Henriet, L.; Tschuy, F.; Wirth, J. Simple method to separate resource competition from allelopathic root interactions. *Allelopath. J.* **2014**, *34*, 227–240.
38. Inderjit, K.M.; Foy, C.L. On the significance of field studies in allelopathy. *Weed Technol.* **2001**, *15*, 792–797. [CrossRef]
39. Ridenour, W.M.; Callaway, R.M. The relative importance of allelopathy in interference: The effects of an invasive weed on a native bunchgrass. *Oecologia* **2001**, *126*, 444–450. [CrossRef]
40. Lau, J.A.; Puliafico, K.P.; Kopshever, J.A.; Setltzer, H.; Jarvis, E.P.; Schwarzländer, M.; Strauss, S.Y.; Hufbauer, R.A. Inference of allelopathy is complicated by effects of activated carbon on plant growth. *New Phytol.* **2008**, *178*, 412–423. [CrossRef]
41. Meynet, P.; Hale, S.E.; Davenport, R.J.; Cornelissen, G.; Breedveld, G.D.; Werner, D. Effect of activated carbon amendment on bacterial community structure and functions in a PAH impacted urban soil. *Environ. Sci. Technol.* **2012**, *46*, 5057–5066. [CrossRef]
42. Wu, C.; Chen, Y.; Grenier-Héon, D. Effects of allelopathy and availability of nutrients and water resources on the survival and growth of plant species in a natural Dacrydium forest. *Can. J. For. Res.* **2021**, *52*, 100–108. [CrossRef]
43. Ninkovic, V. Volatile communication between barley plants affects biomass allocation. *J. Exp. Bot.* **2003**, *54*, 1931–1939. [CrossRef] [PubMed]
44. Zuo, Z.J.; Zhang, R.M.; Gao, P.J.; Wen, G.S.; Gao, Y. Allelopathic effects of *Artemisia frigida* Willd. on growth of pasture grasses in Inner Mongolia, China. *Biochem. Syst. Ecol.* **2011**, *39*, 377–383.
45. Xu, Y.; Hao, Y. A review of ecological research on *Cleistogenes squarrosa*. *Acta Agrestia Sin.* **2016**, *24*, 726–730. (In Chinese)
46. Qin, F.; Liu, S.; Yu, S. Effects of allelopathy and competition for water and nutrients on survival and growth of tree species in *Eucalyptus urophylla* plantations. *For. Ecol. Manag.* **2018**, *424*, 387–395. [CrossRef]
47. Callaway, R.M.; Kikodze, D.; Chiboshvili, M.; Khetsuriani, L. Unpalatable plants protect neighbors from grazing and increase plant community diversity. *Ecology* **2005**, *86*, 1856–1862. [CrossRef]
48. Maestre, F.; Bautista, S.; Cortina, J. Positive, negative, and net effects in grass-shrub interactions in Mediterranean semiarid grasslands. *Ecology* **2003**, *84*, 3186–3197. [CrossRef]
49. Weston, L.A.; Barney, J.N.; DiTommaso, A. A review of the biology and ecology of three invasive perennials in New York State: Japanese knotweed (*Polygonum cuspidatum*), mugwort (*Artemisia vulgaris*) and pale swallow-wort (*Vincetoxicum rossicum*). *Plant Soil* **2005**, *277*, 53–69. [CrossRef]
50. Parepa, M.; Bossdorf, O. Testing for allelopathy in invasive plants: It all depends on the substrate! *Biol. Invasions* **2016**, *18*, 2975–2982. [CrossRef]
51. Callaway, R.M.; Brooker, R.W.; Choler, P.; Kikvidze, Z.; Lortie, C.; Michalet, R.; Paolini, L.; Pugnaire, F.; Newingham, B.; Aschehoug, E.T.; et al. Positive interactions among alpine plants increase with stress. *Nature* **2002**, *417*, 844–848. [CrossRef]
52. Weigelt, A.; Jolliffe, P. Indices of plant competition. *J. Ecol.* **2003**, *91*, 707–720. [CrossRef]

plants

MDPI

Article

Assessment of 3-Cyanobenzoic Acid as a Possible Herbicide Candidate: Effects on Maize Growth and Photosynthesis

Luiz Henryque Escher Grizza, Isabela de Carvalho Contesoto, Ana Paula da Silva Mendonça, Amanda Castro Comar, Ana Paula Boromelo, Ana Paula Ferro, Rodrigo Polimeni Constantin, Wanderley Dantas dos Santos, Rogério Marchiosi * and Osvaldo Ferrarese-Filho *

Laboratory of Plant Biochemistry, Department of Biochemistry, State University of Maringá, Maringá 87020-900, Brazil; luizheg@gmail.com (L.H.E.G.); isabelacontesoto@gmail.com (I.d.C.C.); silvamendoncaanapaula@gmail.com (A.P.d.S.M.); amandacomar9@gmail.com (A.C.C.); paulaboromelo123@gmail.com (A.P.B.); apferro2@uem.br (A.P.F.); rpconstantin@uem.br (R.P.C.); wdsantos@uem.br (W.D.d.S.)

* Correspondence: rmarchiosi@uem.br (R.M.); osferrarese@gmail.com (O.F.-F.)

Abstract: Chemical weed control is a significant agricultural concern, and reliance on a limited range of herbicide action modes has increased resistant weed species, many of which use C4 metabolism. As a result, the identification of novel herbicidal agents with low toxicity targeting C4 plants becomes imperative. An assessment was conducted on the impact of 3-cyanobenzoic acid on the growth and photosynthetic processes of maize (*Zea mays*), a representative C4 plant, cultivated hydroponically over 14 days. The results showed a significant reduction in plant growth and notable disruptions in gas exchange and chlorophyll *a* fluorescence due to the application of 3-cyanobenzoic acid, indicating compromised photosynthetic activity. Parameters such as the chlorophyll index, net assimilation (A), stomatal conductance (g_s), intercellular CO_2 concentration (C_i), maximum effective photochemical efficiency ($F_{v'}/F_{m'}$), photochemical quenching coefficient (q_P), quantum yield of photosystem II photochemistry (ϕ_{PSII}), and electron transport rate through PSII (ETR) all decreased. The A/PAR curve revealed reductions in the maximum net assimilation rate (A_{max}) and apparent quantum yield (ϕ), alongside an increased light compensation point (LCP). Moreover, 3-cyanobenzoic acid significantly decreased the carboxylation rates of RuBisCo (V_{cmax}) and PEPCase (V_{pmax}), electron transport rate (J), and mesophilic conductance (g_m). Overall, 3-cyanobenzoic acid induced substantial changes in plant growth, carboxylative processes, and photochemical activities. The treated plants also exhibited heightened susceptibility to intense light conditions, indicating a significant and potentially adverse impact on their physiological functions. These findings suggest that 3-cyanobenzoic acid or its analogs could be promising for future research targeting photosynthesis.

Keywords: chlorophyll *a* fluorescence; enzyme inhibitor; gas exchange; phenolics; phytotoxicity; pyruvate *O*-phosphate dikinase; weeds

1. Introduction

C4 plants, such as maize (*Zea mays*), sugarcane (*Saccharum officinarum*), millet (*Pennisetum glaucum*), and sorghum (*Sorghum bicolor*), have evolved highly specialized photosynthetic pathways that enable them the competitive ability to capture and use atmospheric CO_2. This efficiency is facilitated by the actions mesophyll and bundle sheath cells working in the C4 carbon concentration mechanism [1]. More specifically, the assimilation and reduction of CO_2 occur in two distinct steps, implemented by different enzymes. Initially, CO_2 is assimilated in the cytoplasm of mesophyll cells, where it is converted into bicarbonate, then incorporated into phosphoenolpyruvate to form oxaloacetate by the action of phosphoenolpyruvate carboxylase (PEPCase). The oxaloacetate is then converted into malate or aspartate, which are transported to the bundle sheath cells. In these cells, malate or aspartate is decarboxylated, releasing CO_2, which is in turn captured by RuBisCo for further assimilation [1]. Because the high-CO_2 environment, the carboxylative activity of RuBisCo increases and reduces oxygen interference, minimizing photorespiration [2,3]. As a result, C4 plants grow faster and accumulate more biomass, especially in areas with high temperature and high light. The ability of C4 plants to concentrate CO_2 provides them with a significant advantage in challenging climates, making them essential for food, feed, and also bioenergy production [4]. In addition, C4 plants contribute to about 18% of global plant productivity, highlighting their essential role in both ecosystems and agriculture [5].

Agroecosystems, much like any other ecosystem, are greatly influenced by their surrounding environment and climate, which in turn impacts the various living organisms within them. Since the primary goal of an agroecosystem is to meet specific human needs, not all organisms within it are desirable. For instance, pests, diseases, and weeds are typically unwelcome. Weeds are common in all agricultural settings and can significantly reduce both productivity and crop quality. Climate change, with higher CO_2 levels, rising temperatures, and changing rainfall patterns, directly impacts agriculture, favoring the growth of competitive C4 weeds [6]. Weeds present a significant challenge due to their rapid growth, the ability to absorb resources, and the interception of sunlight, all of which can disrupt the agroecosystem. To address these challenges, integrated weed management strategies—such as crop rotation, selective herbicide use, and other agronomic practices—are essential [7,8].

Herbicides are a primary tool for weed control, but their repeated use often leads to the emergence of resistant weed biotypes. These resilient weeds not only survive but thrive, dominating plant populations and creating new challenges. Atrazine, a pre-emergent herbicide widely used in maize cultivation, exemplifies this challenge. It inhibits photosystem II, disrupting photosynthesis after being absorbed primarily by plant roots [9]. However, its extensive use has raised significant concerns due to its persistence in the environment, leaching into groundwater and surface water, and its long half-life. These characteristics, combined with evidence of adverse health effects, including carcinogenicity and endocrine disruption, led to its ban in the European Union [10]. Despite this, atrazine remains one of the most commonly used herbicides in countries like Brazil [11]. The environmental and health concerns associated with herbicides such as atrazine, coupled with the rise of resistant weed biotypes, emphasize the urgent need for novel herbicides with unique mechanisms of action [12]. These alternatives must balance efficacy and environmental safety to ensure sustainable weed management practices. Currently, there are 533 documented cases of herbicide-resistant weeds worldwide, involving 273 species (156 dicots and 117 monocots). These resistant weeds have evolved to survive against 21 out of 31 known herbicide sites of action and 168 different herbicides, impacting 101 crops in 72 countries. Some species have developed resistance to up to seven herbicide sites of action, highlighting the significant challenge posed by resistant weeds in the *Poaceae* family, which contains 92 resistant species—double that of the second most affected family, *Asteraceae* [9]. Many of the most problematic weed species are C4 plants [13], with 14 of the 18 most noxious species belonging to this group [14]. The continuous rise of herbicide-resistant weeds, coupled with the lack of new herbicides with novel modes of action since 1980 [12], highlights the pressing need for more effective, selective, and environmentally friendly herbicide solutions.

Our research team is dedicated to identifying and evaluating novel herbicides with innovative modes of action. We utilize a combination of in silico, in vitro, and in vivo approaches to assess the effectiveness of enzyme inhibitors with potential herbicidal activity [15,16]. Targeting enzymes essential for C4 photosynthesis, such as pyruvate *O*-phosphate dikinase (PPDK), could be one promising strategy to develop herbicides [17,18]. Inhibiting PPDK, a crucial enzyme in mesophyll cells, has been shown to reduce photosynthetic activity, potentially providing a method for controlling C4 weeds [19–21]. Enzymatic inhibitors, which downregulate target enzymes, play a critical role in controlling plant metabolism. Because PPDK is essential for C4 plants to assimilate CO_2 through photosynthesis, we evaluate selective inhibitors of this enzyme as potential candidates for new herbicides. We have already tested a few compounds, selected from a virtual library and anchored to the PPDK active site of *Digitaria sanguinalis*, to develop specific PPDK inhibitors. Using virtual screening and drug repositioning approaches, we found five promising chemicals (bromoacetic, bromopyruvic, 2-bromobutyric, 2-bromopropionic, and 3-cyanobenzoic acids) as potential PPDK inhibitors. Among these, bromoacetic acid showed a dose-dependent inhibition of PPDK activity in maize, significantly affecting gas exchange and reducing the photosynthetic rate [22]. These findings laid the groundwork for further exploration of PPDK inhibitors. More recently, we found that 3-cyanobenzoic acid not only inhibited maize seedling growth but also triggered oxidative stress, increasing reactive oxygen species (ROS) production [23]. Despite the lack of literature on this specific molecule, 3-cyanobenzoic acid is a simple phenolic compound, part of a class of molecules derived from benzoic and cinnamic acids, which are known for their extensive biological activities [24]. These include allelopathic interactions between plants, where these simple phenols can affect growth by disrupting electrical conductivity in membranes, hormonal balance, and stomatal closure, and have been shown to impact structural components of mitochondria and chloroplasts as primary targets [24]. Notably, this compound can also be synthesized by bacteria, presenting a promising opportunity for its bio-sourced production, which could enhance its sustainability and commercial viability [25]. Furthermore, 3-cyanobenzoic acid is like certain O-group herbicides, such as chloramben, dicamba, naptalam, and dimethyl-2,3,5,6-tetrachloroterephthalate, which are herbicides known for mimicking auxin and inhibiting microtubule assembly [9,26].

Based on our earlier findings, we hypothesized that the adverse impacts of 3-cyanobenzoic acid extend beyond root damage and that this compound could potentially serve as a targeted herbicide for C4 plants. Thus, this study aimed to evaluate the phytotoxic effects of 3-cyanobenzoic acid on maize (*Zea mays*), focusing on its impact on growth, photosynthetic processes, and biochemical parameters. Maize was selected as a model plant due to its C4 metabolism and ease of cultivation. To evaluate enzyme activity and photochemical efficiency, we measured intercellular CO_2 concentration (C_i) and constructed irradiance light curves. Chlorophyll *a* fluorescence emission, a noninvasive and quantitative method, was used to monitor changes in photosynthetic processes. From the fluorescence data, we calculated key parameters such as photochemical yield, electron transport ability, and energy dissipation. We also generated OJIP curves and conducted the JIP fluorescence test to analyze energy flow through reaction centers and photosystems, providing a comprehensive assessment of the compound's effects on plant photosynthetic performance. By assessing these effects, we aim to explore the potential of 3-cyanobenzoic acid as a candidate herbicide targeting C4 plants.

2. Materials and Methods

The maize (*Zea mays* L. cv. IPR-164) seeds were cleaned by soaking in a 2% sodium hypochlorite for three minutes, followed by thorough rinsing with deionized water to ensure cleanliness. The seeds were then evenly spread on pre-moistened germination paper (Germitest® CEL-060, Netlab, São Paulo, Brazil) and covered with another sheet, rolled up, and kept in the dark at 25 °C for three days to promote germination. Germinated seedlings with radicle lengths between 5 and 6 cm were randomly selected and grouped into sets of 12, then transferred to hydroponic systems. These seedlings were placed in Styrofoam trays and placed in glass containers filled with 200 mL of $^1/_4$ strength nutrient solution at pH 6.0 [27], which contained either 0 (control), 0.5, or 1.0 mM of 3-cyanobenzoic acid (Sigma–Aldrich, St. Louis, MO, USA). The seedlings were grown under controlled conditions with a 12 h light/12 h dark photoperiod at 25 °C and an irradiance of 300 μmol photons m^{-2} s^{-1}. The nutrient solution was replaced every other day to maintain optimal growth conditions. On the 14th day (V2 stage) entire plants were carefully harvested for biometric measurements, while the second fully expanded leaf was sampled for further biochemical and physiological analyses.

2.1. *Biometric Measurements*

Plant height was determined by measuring the distance from the point of seed attachment to the culm up to the base of the last fully developed leaf and the main root. Stem diameter was precisely measured at the first internode using a digital caliper. Fresh biomass of shoots and roots was assessed through gravimetric analysis. Leaf area estimation was performed using the Easy Leaf Area Free 1.02 software for mobile devices [28]. These measurements ensured consistency and accuracy, with a total of 18 plants evaluated per treatment.

2.2. *SPAD Index, Chlorophyll a Fluorescence, and Gas Exchange*

Chlorophyll content was measured using a SPAD-502 m (Konica Minolta®, Ramsey, NJ, USA). Chlorophyll *a* fluorescence parameters were assessed with a pulse-amplitude modulated fluorometer (PAM) integrated into an LI-6800 portable photosynthesis system (Li-Cor®, Lincoln, NE, USA). Measurements were conducted in the middle region of the last fully expanded leaf, previously dark-adapted for 10 h to fully oxidize photosystem II (PSII), ensuring optimal data accuracy [29]. Key fluorescence parameters were recorded, including minimal fluorescence intensity (F_0), observed in the absence of light when PSII reaction centers were fully open, and maximal fluorescence intensity (F_m), elicited under a high-intensity multiphase flash of 8000 μmol m^{-2} s^{-1} when PSII centers were fully closed. After that, the maximum photochemical efficiency ($F_m - F_0/F_m$) of the plants was calculated [30]. A total of eighteen plants were evaluated for each treatment.

Next, measurements focused on light-adapted leaf regions under controlled conditions (irradiance of 1400 μmol m^{-2} s^{-1} and CO_2 concentration of 400 μmol mol^{-1}) to achieve complete photosynthetic induction. Using the PAM fluorometer and the LI-6800 infrared gas analyzer (IRGA, Li-Cor®), fluorescence and gas exchange parameters were recorded. Variables such as net assimilation (A), stomatal conductance (g_s), maximum quantum yield of PSII photochemistry ($F_{v'}/F_{m'}$), effective quantum yield of PSII photochemistry (ϕ_{PSII}), electron transport rate (ETR), photochemical quenching coefficient (q_P), and non-photochemical quenching (NPQ) were recorded. The measurements were performed at 27 °C; relative humidity of 60%; air flow of 700 μmol^{-1}; CO_2 concentration of 400 μmol^{-1}; photosynthetically active radiation (PAR) of 1400 μmol m^{-2} s^{-1}; and multiphase saturating light flash of 8000 μmol m^{-2} s^{-1}. Values were recorded from duplicated points obtained when constructing light curves and A/C_i curves, which are explained in detail in the following sections. The values were recorded point by point with adaptation times ranging from 60 to 120 s after changing the parameters associated with the curve (light or CO_2), indicating the steady state of the leaf area when the instantaneous rate of change of water,

CO_2 concentration, and chlorophyll *a* fluorescence was varied in the recorded values less than one in the device sample chamber. A total of 22 plants were evaluated per treatment, ensuring high reproducibility and reliability of the results.

2.3. Net Assimilation Curves in Response to Changes in Photosynthetically Active Radiation (A/PAR)

Plants were light-adapted under controlled conditions (1400 µmol m^{-2} s^{-1} irradiance and a fixed CO_2 concentration of 400 µmol mol^{-1} within the sample chamber) until full photosynthetic induction was achieved. Other environmental parameters were maintained as previously described to ensure consistency. The net assimilation rate (A) curve was generated by systematically recording photosynthetically active radiation (PAR) values at a series of pre-defined intervals using the LI-6800 IRGA. Measurements at each PAR level were taken over 60 to 120 s, allowing the plants to reach a stable photosynthetic state before data collection. PAR levels ranged from 2000 to 0 µmol photons m^{-2} s^{-1}, following a descending sequence of 2000, 1800, 1600, 1400, 1200, 1000, 800, 600, 400, 200, 175, 150, 125, 100, 75, 50, 25, and 0 µmol photons m^{-2} s^{-1}. This comprehensive range ensured an accurate representation of the light response curve. Measurements were taken from four plants per treatment.

The resulting A/PAR curve was well-suited to the light range of 0 to 2000 µmol m^{-2} s^{-1}, with particular emphasis on the linear phase observed between 0 and 175 µmol m^{-2} s^{-1}. Key parameters derived from the curve included the maximum net assimilation rate (A_{max}), apparent quantum yield (ϕ), light compensation point (LCP), and dark respiration rate (R_D). These parameters were calculated using a Microsoft® Excel-based tool developed by Lobo et al. [31], ensuring precision in data analysis and parameter estimation.

2.4. Net Assimilation Curves in Response to Changes in Intercellular CO_2 Concentration (A/C_i)

Like the A/PAR curve measurements, acclimatization of the plants to light intensity at 1400 µmol m^{-2} s^{-1} was ensured prior to data collection. The recording time was standardized to capture the steady-state photosynthetic responses of the plants. During the experiment, the light intensity was maintained at a constant PAR of 1400 µmol m^{-2} s^{-1}, while the concentration of available CO_2 was systematically varied. The measurements began at a CO_2 concentration of 400 µmol mol^{-1}, progressively decreasing in steps without reaching zero. Following this, CO_2 levels were increased in the sequence: 400, 200, 25, 50, 75, 100, 125, 150, 175, 200, 225, 250, 270, 300, 325, 350, 400, 400, 500, 600, 700, 800, 1000, 1200, 1400, 1600, 1800, and 2000 µmol mol^{-1}. This method ensured comprehensive profiling of photosynthetic responses at varying CO_2 concentrations. Measurements were performed on six plants per treatment, providing robust and replicable data [2,32]. To analyze the results, spreadsheets developed by Zhou et al. [2] were used to fit assimilation curves based on the concentration of intercellular CO_2, tailored specifically for C4 plants. From these fitted curves, key photosynthetic parameters were derived, including the maximum RuBisCo carboxylation rate (V_{cmax}), maximum PEPCase carboxylation rate (V_{pmax}), photosynthetic electron transport rate (J), and mesophilic conductance (g_m).

2.5. Transient Responses of the Chlorophyll a Fluorescence (OJIP Curve) and Derived Parameters (JIP Test) Parte Superior do Formulário

Fast fluorescence kinetics were measured using the inductive flash feature of the LI-6800 IRGA PAM fluorometer (Li-Cor®, Lincoln, NE, USA) in plants that had been dark-adapted for 10 h to ensure fully open PSII reaction centers. Fluorescence emissions were recorded at millisecond intervals for one second, capturing the rapid changes associated with the OJIP transient. Data from 18 plants per treatment were collected to produce OJIP curves, with time converted to a logarithmic scale, showcasing different stages between O (F_0) and P (F_M) phases. The photochemical phase (O–J) reflects the reduction of the primary quinone acceptor (Q_A), while the thermal phase (J–I) corresponds to electron transfer between Q_A and the secondary quinone acceptor (Q_B). The I–P phase represents fluorescence emissions driven by the pool of oxidized plastoquinone molecules [33–35]. This detailed division of phases enables a comprehensive understanding of electron transport dynamics within the photosynthetic apparatus.

Fluorescence data were analyzed using the JIP test, which translates the OJIP curves into specific energy fluxes and parameters indicative of photosynthetic efficiency. Key energy fluxes included absorption per reaction center (ABS/RC), trapped energy per reaction center (TR_0/RC), electron transport flux per reaction center (ET_0/RC), dissipated energy per reaction center (DI_0/RC), and the density of active PSII reaction centers per cross-section (RC/CS). These metrics provided insights into energy dynamics and the operational efficiency of PSII. The maximum quantum yield of primary photochemistry (TR_0/ABS), quantum yield for electron transport from Q_A to Q_B (ET_0/ABS), and the probability that a trapped exciton moves an electron into the electron transport chain beyond Q_A^- (ET_0/TR_0) were also assessed. Two critical performance indices, PI_{ABS} and PI_{TOTAL}, were derived. PI_{ABS} quantifies the conservation of absorbed energy from PSII up to Q_B reduction, while PI_{TOTAL} is an indicator of the conservation of energy absorbed by the PSII until the decline in the final PSI acceptors) [33–35]. Together, these indices provide an integrated evaluation of the energy conservation ability within the photosynthetic electron transport chain, offering a robust framework for assessing plant photosynthetic performance under varying treatments.

2.6. Statistical Analysis

The experimental design used was completely randomized, and the data were submitted to one-way analysis of variance (ANOVA). Significant differences were then compared using Dunnett's test at both 5% and 1% levels of significance. Data from the A/PAR and A/C_i curves were adjusted using Microsoft® Excel spreadsheet provided by Lobo et al. [31] and Zhou et al. [2]. Furthermore, parameters derived from the JIP test were calculated using data from the chlorophyll *a* transient curve, normalized, and subjected to the same tests. All statistical analyses were performed using GraphPad Prism® program, version 8 (La Jolla, CA, USA).

3. Results

3.1. Effects on Biometric Measurements

Treatment with 3-cyanobenzoic acid significantly inhibited maize growth (Table 1). At 1.0 mM, plant height decreased by 20.3%. The leaf area was reduced by 21.4% at 0.5 mM and by 42% at 1.0 mM, while the culm diameter decreased by 26.5% at 0.5 mM and 34.9% at 1.0 mM. Shoot fresh biomass saw a 45.9% decrease at 1.0 mM compared to the control. Root length also showed significant reductions, with a 45.5% decrease at 0.5 mM and a striking 65.7% decrease at 1.0 mM. Moreover, fresh root biomass decreased by 62.9% at 1.0 mM. These decreases are illustrated in Figure 1. These substantial decreases underscore the significant impact of 3-cyanobenzoic acid on various growth parameters of maize.

Table 1. Effects of 3-cyanobenzoic acid on maize plant growth during a 14-day hydroponic cultivation period.

	3-Cyanobenzoic Acid (mM)		
	0	**0.5**	**1.0**
Plant height (cm)	9.51 ± 0.27	8.77 ± 0.26	7.57 ± 0.20 **
Shoot fresh weight (g)	1.11 ± 0.12	0.88 ± 0.06	0.60 ± 0.05 **
Leaf area (cm^2)	38.0 ± 2.84	29.9 ± 1.71 *	22.0 ± 1.38 **
Culm diameter (cm)	3.34 ± 0.19	2.46 ± 0.08 **	2.17 ± 0.08 **
Root length (cm)	28.7 ± 1.77	15.4 ± 0.81 **	9.84 ± 0.47 **
Root fresh weight (g)	0.33 ± 0.04	0.25 ± 0.02	0.12 ± 0.01 **

Means values (n = 18 ± SEM) significantly different from the control are marked with * $p \leq 0.05$, ** $p \leq 0.01$, according to Dunnett's test.

Figure 1. Hydroponically grown maize plants treated with 3-cyanobenzoic acid for 14 days: 0 mM (**A**), 0.5 mM (**B**), and 1.0 mM (**C**). Scale bars represent 10 cm.

3.2. *Effects on SPAD Index, Chlorophyll a Fluorescence and Gas Exchange*

Maize plants grown with 3-cyanobenzoic acid for 14 d exhibited mild chlorosis. The total chlorophyll content, as indicated by the SPAD index in the second fully expanded leaf, decreased by 19.3% at 1.0 mM (Figure 2A). F_v/F_m values remained unchanged (Figure 2B).

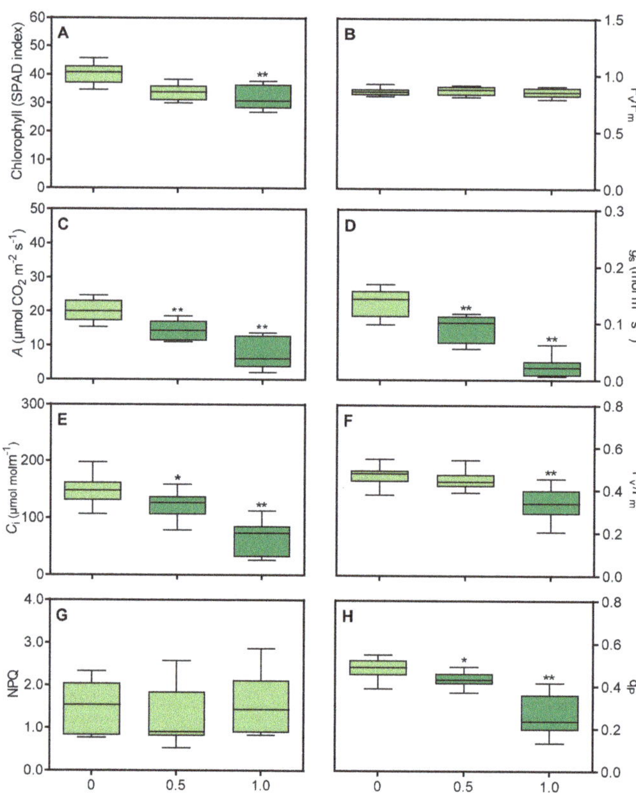

Figure 2. Effects of 3-cyanobenzoic acid on hydroponically grown maize plants for 14 days. Parameters measured include (**A**) chlorophyll content (SPAD index), (**B**) maximum quantum efficiency of PSII photochemistry (F_v/F_m), (**C**) net assimilation (*A*), (**D**) stomatal conductance (g_s), (**E**) intercellular CO_2 concentration (C_i), (**F**) maximum efficiency of PSII ($F_{v'}/F_{m'}$), (**G**) non-photochemical quenching (NPQ), and (**H**) photochemical quenching (q_P). Means values (n = 16–22 ± SEM) significantly different from the control are marked with * $p \leq 0.05$, ** $p \leq 0.01$, according to Dunnett's test.

Under full photosynthetic conditions (1400 μmol m^{-2} s^{-1} irradiance and 400 μmol mol^{-1} CO_2), several key parameters were significantly reduced compared to the controls. The *A* (Figure 2C) decreased by 23% to 0.5 mM and 56.7% to 1.0 mM. Similarly, g_s showed a marked reduction, decreasing by 32.8% at 0.5 mM and 68.2% at 1.0 mM (Figure 2D). The C_i also decreased by 21.4% at 0.5 mM and 44% at 1.0 mM (Figure 2E). In addition, the $F_{v'}/F_{m'}$ decreased by 24.8% at 1.0 mM (Figure 2F). While NPQ (Figure 2G) remained stable, q_P (Figure 2H) decreased by 10.9% at 0.5 mM and 39.3% at 1.0 mM.

The ϕ_{PSII} (Figure 3A) and ETR (Figure 3B) decreased by 15.3% (at 0.5 mM) and 51.4% (at 1.0 mM).

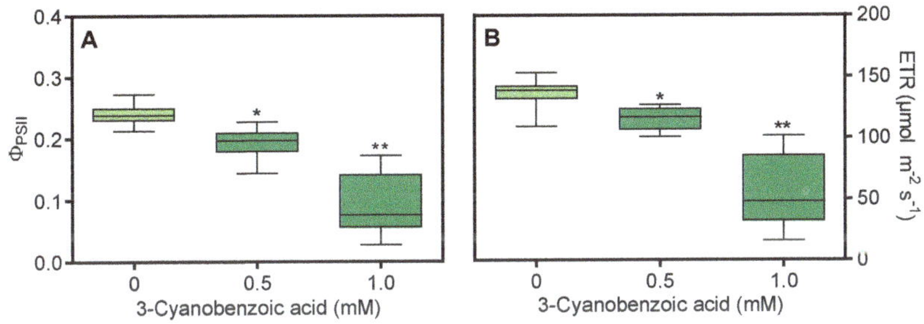

Figure 3. Effects of 3-cyanobenzoic acid on hydroponically grown maize plants for 14 days on quantum yield of photosystem II photochemistry (ϕ_{PSII}) (**A**) and electron transport rate through PSII (ETR) (**B**). Means (n = 22 ± SEM) marked with * or ** are statistically different from the control according to Dunnett's test at 5% and 1% significance levels, respectively.

3.3. Effects on A/PAR Curve and Derived Parameters

Significant changes were seen in the *A* versus PAR (Figure 4). At 0.5 mM, *A* levels were already lower than those of the control. At 1.0 mM, the decline in *A* was more pronounced at higher PAR levels and in the slope of the first linear phase.

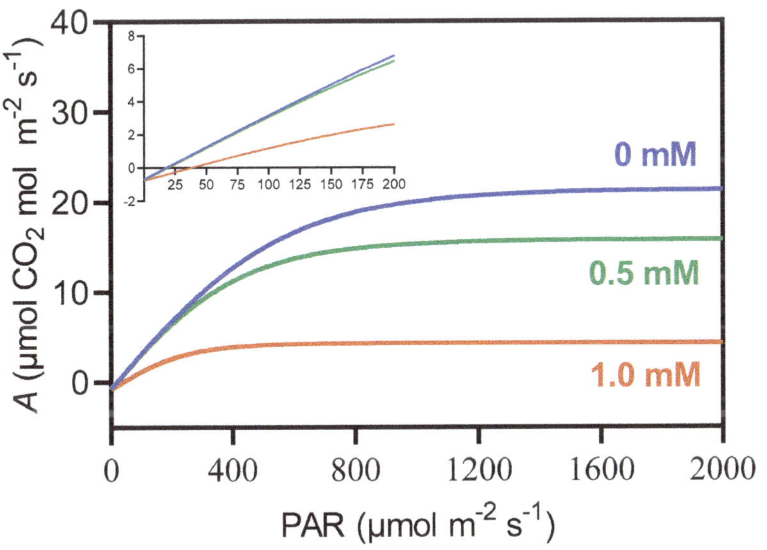

Figure 4. Average net assimilation (*A*) curves in response to varying photosynthetically active radiation (PAR) for maize plants grown hydroponically with 3-cyanobenzoic acid for 14 days. The initial linear region of the graph is magnified for clarity. Data are presented as mean values (n = 4).

The A_{max} decreased by a striking 82% at 1.0 mM (Figure 5A). The ϕ decreased by 57% at 1.0 mM (Figure 5B). Additionally, the LCP increased by 175% at 1.0 mM (Figure 5C), while R_D remained unaffected (Figure 5D).

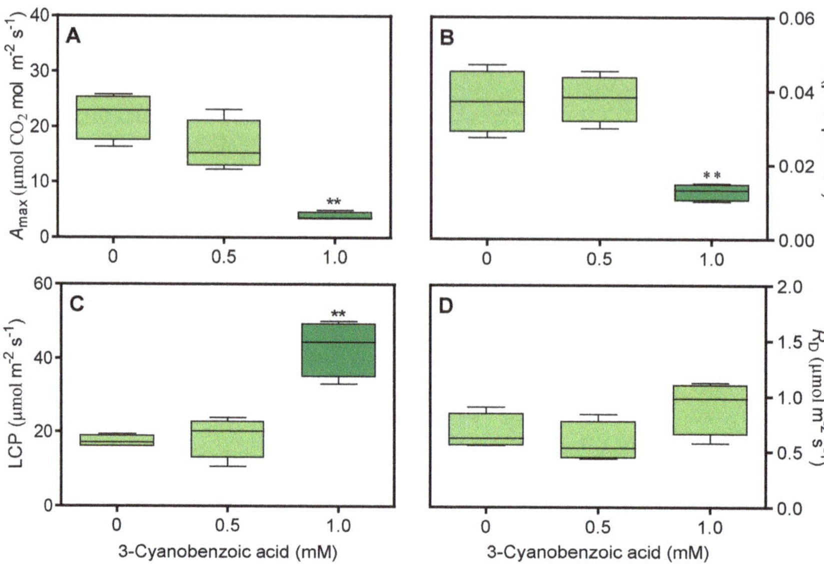

Figure 5. Effects of 3-cyanobenzoic acid on hydroponically grown maize plants after 14 days, focusing on parameters derived from the *A*/PAR curve: (**A**) net assimilation (A_{max}), (**B**) apparent quantum yield (ϕ), (**C**) light compensation point (LCP), and (**D**) dark respiration rate (R_D). Means values (n = 3–4 ± SEM) significantly different from the control are marked with, ** $p \leq 0.01$, according to Dunnett's test.

3.4. Effects on A/C_i Curve and Derived Parameters

To differentiate the effects of CO_2 diffusion through the stomata, the *A* versus C_i curve was analyzed to derive key parameters, including V_{cmax}, V_{pmax}, *J*, and g_m. The impact of 3-cyanobenzoic acid was clear at even low CO_2 concentrations and became more pronounced at higher concentrations (Figure 6).

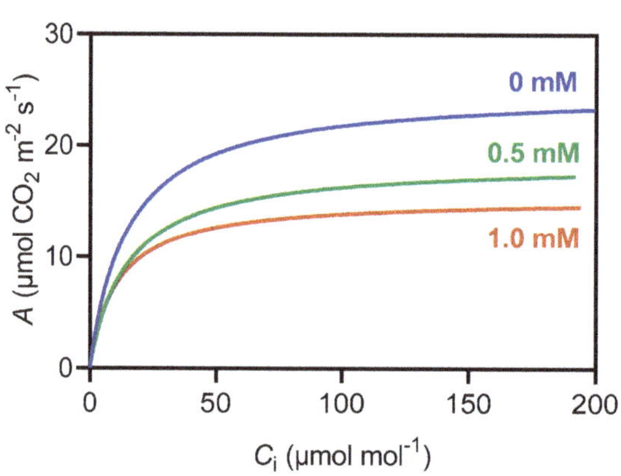

Figure 6. Average net assimilation (*A*) curves in response to varying intercellular CO_2 concentration (C_i) forma maize plants grown hydroponically with 3-cyanobenzoic acid for 14 days. Data are presented as mean values (n = 4–6).

At 1.0 mM, V_{cmax} decreased by 32% (Figure 7A), V_{pmax} by 42% (Figure 7B), J by 34.8% (Figure 7C), and g_m by 28% (Figure 7D).

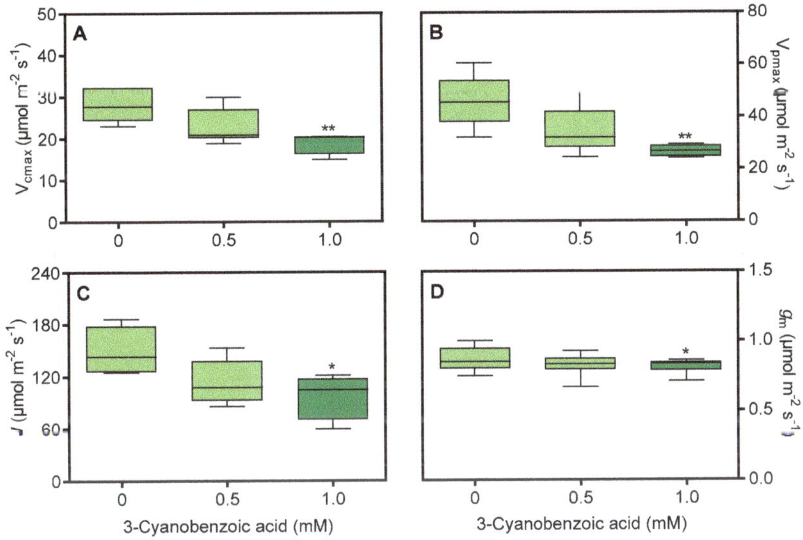

Figure 7. Effects of 3-cyanobenzoic acid on maize plants grown hydroponically for 14 days, focusing on parameters derived from the A/C_i curve: (**A**) maximum carboxylation rate of RuBisCo (V_{cmax}), (**B**) maximum carboxylation rate of PEPCase (V_{pmax}), (**C**) rate of photosynthetic electron transport (J), and (**D**) mesophyll conductance (g_m). Means values (n = 4–6 ± SEM) significantly different from the control are marked with * $p \leq 0.05$, ** $p \leq 0.01$, according to Dunnett's test.

3.5. Effects on Chlorophyll a Transient Curve and Derived Parameters

No significant changes were seen in the OJIP curve (Figure 8).

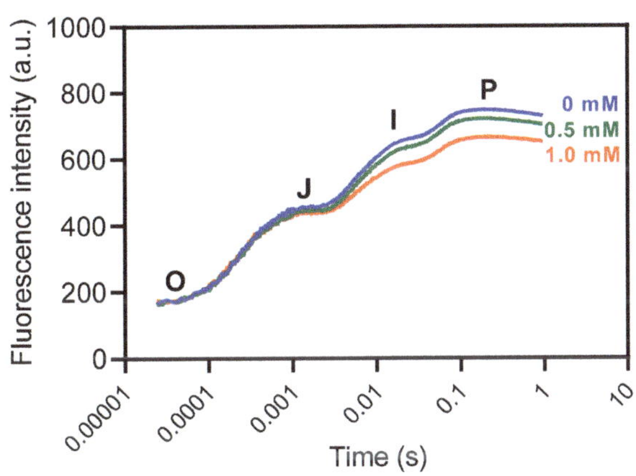

Figure 8. Chlorophyll *a* fluorescence OJIP transient curves in maize plants grown hydroponically with 3-cyanobenzoic acid for 14 days. The OJIP curve represents key fluorescence intensities: the minimal fluorescence when all PSII reaction centers are open (O step), the intensity at 0.002 s (J step), the intensity at 0.03 s (I step), and the maximal fluorescence when all PSII reaction centers are closed (P step, at 0.3 s). Data are presented as means (n = 18 ± SEM).

Finally, a noteworthy reduction was seen in RC/CS, which decreased by 25.8% at 0.5 mM (Figure 9).

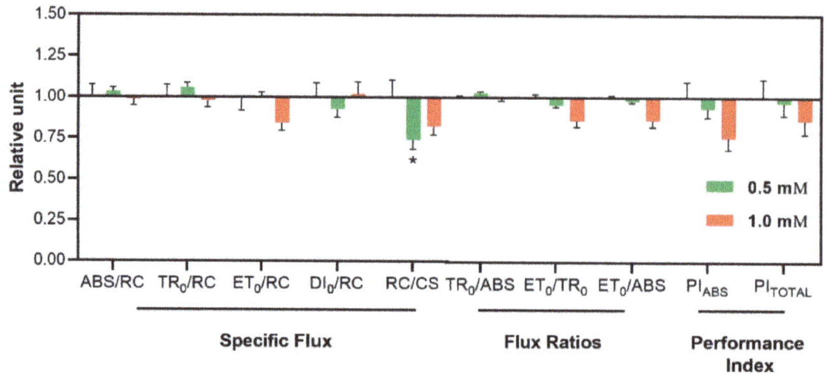

Figure 9. Effects of 3-cyanobenzoic acid on specific energy flux parameters in hydroponically grown maize plants after 14 days of treatment. Parameters include: absorption flux per reaction center (ABS/RC), energy trapping per reaction center (TR_0/RC), electron transport per reaction center (ET_0/RC), energy dissipation per reaction center (DI_0/RC), reaction center density per cross-sectional area (RC/CS), quantum yield of primary PSII photochemistry (TR_0/ABS), efficiency with which a trapped electron is transferred from Q_A to Q_B (ET_0/TR_0), quantum yield of electron transport from Q_A to Q_B (ET_0/ABS), and performance indices (PI_{ABS} and PI_{Total}). Data are presented as means (n = 18 ± SEM). Mean value significantly different from the control is marked with * $p \leq 0.05$, according to Dunnett's test.

4. Discussion

Treatment with 3-cyanobenzoic acid resulted in a significant reduction in both foliar and root growth, as well as in key photosynthetic, photochemical, and carboxylative parameters in maize plants grown hydroponically for 14 days. These findings suggest that the compound impairs multiple aspects of plant development and function. Further steps aim to understand the reasons behind these declines and identify primary targets.

Our earlier study demonstrated significant damage to root growth and development in maize seedlings exposed to 3-cyanobenzoic acid for four days. This included reduced root length and biomass, a decrease in secondary roots, increased root respiration, heightened antioxidant activity, and elevated ROS production. While the mitochondrial respiratory chain was not directly affected, the observed damage was linked to membrane destabilization, deficiencies in nutrient and water absorption, and hormonal imbalances [23]. These effects are consistent with those observed in other plants exposed to monophenols, such as cinnamic and benzoic acids. Upon absorption, these acids undergo deprotonation, leading to the dissipation of the proton gradient and membrane depolarization. This cascade of events triggers an increase in ROS production and antioxidant activity, disrupting ion transport, nutrient absorption, and water balance [24,36–38]. Additionally, these disturbances extend to hormonal imbalances, stomatal closure, reduced chlorophyll content, diminished RuBisCo activity, and damage to photosynthetic reaction centers and electron transport chains [24].

Maize plants grown for 14 days exhibited damage like that observed in seedlings [23], suggesting an acclimatization response to unfavorable conditions. However, it is essential to determine whether the observed effects are solely due to continuous root damage or if there are also direct impacts on the aerial parts of the plant, including photosynthesis. Regardless of light intensity or CO_2 supply, a clear reduction in net photosynthesis was observed. The negative correlation between compound concentration and photosynthetic performance—reflecting a trade-off between defense mechanisms and growth—is expected [39,40]. Consequently, the decrease in photosynthesis cannot be attributed solely to energy partitioning. A more comprehensive understanding of the underlying mechanisms requires the identification and integration of additional relevant parameters.

Various simple phenolics have been shown to impact photosynthetic parameters, with stomatal closure being one of the most common effects [37,41–43]. Typically, an increase in C_i is accompanied by a reduction in g_s, signaling a decrease in the efficiency of carboxylation cycles. This has been demonstrated in soybean plants exposed to L-DOPA [44] and in poplar seedlings exposed to a phenolic mixture [43]. However, alternative patterns can also emerge. In cases of photosynthesis limitation due to stomatal closure, a decrease in g_s is often paired with a reduction in C_i [24]. Similar findings were observed in cucumber treated with cinnamic and benzoic acid derivatives [41]. As evidenced in the present study, this pattern suggests that the damage to diffusion processes is more severe than the impairment of the carboxylative systems (such as the C4 cycle, Hatch–Slack pathway, and C3 Calvin–Benson cycle). This observation, per se, provides a strong rationale for the noted decrease in C_i, emphasizing the greater impact on stomatal function and CO_2 diffusion.

Stomatal closure likely explains much of the observed reductions in photosynthetic parameters. By limiting CO_2 availability within the leaf, this closure, combined with a decrease in g_m, results in a substrate shortage for key enzymes such as PEPCase (V_{pmax}) and RuBisCo (V_{cmax}). This shortage compromises their carboxylative activities, which accounts for the observed declines in photosynthetic efficiency. Alternatively, 3-cyanobenzoic acid may have exerted a direct inhibitory effect on these enzymes. For instance, Gao et al. [38] showed that salicylic acid reduced RuBisCo activity and disrupted photochemical structures in *Chlamydomonas reinhardtii*. Similarly, Huang and Bie [45] reported that cinnamic acid decreased RuBisCo activity in *Vigna unguiculata* without affecting C_i or ϕ_{PSII}, suggesting that direct enzyme inhibition could be a contributing factor in the present study.

Maize, a C4 plant, benefits from an efficient CO_2 concentration mechanism that provides a distinct advantage over C3 plants. This mechanism is driven by PEPCase, which has a higher carboxylation rate than RuBisCo, enabling more effective CO_2 fixation [1–3]. However, the observed decline in V_{pmax} was more pronounced than the reduction in V_{cmax}, suggesting a compromised CO_2 concentration mechanism. This impairment is further worsened by the lack of ribulose-1,5-bisphosphate (RuBP), the substrate for RuBisCo, which weakens the carboxylation process. The decreased plateau in the A/C_i curve shows a limitation in RuBP regeneration, a critical process for maintaining efficient photosynthesis [2,46,47]. RuBP regeneration relies heavily on ATP and NADPH, which are generated through the electron transport chain, as evidenced by the reduced ETR and the decrease in J. These findings underscore the impact of 3-cyanobenzoic acid on maize's photosynthetic machinery, particularly on its ability to regenerate RuBP and sustain efficient carbon fixation. Although our results indicate significant disruptions in photosynthetic parameters, it is important to acknowledge that concluding 3-cyanobenzoic acid directly targets key photosynthetic pathways at this stage may be premature. The observed effects on photosynthesis, such as reductions in ETR, V_{cmax}, V_{pmax}, and chlorophyll content, could also be indirect consequences of oxidative stress, which is known to impair photosynthetic machinery through ROS accumulation and membrane destabilization. To confirm whether the primary target of 3-cyanobenzoic acid is within photosynthetic pathways, further studies are required. A combination of omics approaches, such as transcriptomics and proteomics, would allow us to identify changes in the expression and abundance of proteins involved in photosynthesis and oxidative stress responses [48]. For example, decreased levels of PPDK, RuBisCo, or other Calvin cycle enzymes could point to direct inhibition. Additionally, metabolomics could reveal disruptions in ATP and NADPH production, as well as in the pools of photosynthetic intermediates. Complementary in silico molecular docking studies could help predict the binding affinity of 3-cyanobenzoic acid to target enzymes, such as RuBisCo, PPDK, or components of the electron transport chain. Such simulations would provide insights into the likelihood of direct enzymatic inhibition. Moreover, targeted enzymatic assays would be essential to confirm the direct inhibition of photosynthetic enzymes, distinguishing them from generalized oxidative stress effects. These combined approaches would clarify the mode of action of 3-cyanobenzoic acid and determine whether it is a direct or indirect disruptor of photosynthesis.

Photosynthesis involves the capture of light energy and its conversion into chemical energy, processes that are reflected in key parameters such as q_P, $F_{v'}/F_{m'}$, ϕ_{PSII}, and ETR. The observed changes in these parameters show a significant reduction in ATP and NADPH production, suggesting an energy deficit that hampers the plant's ability to conduct essential metabolic functions. Additionally, the insufficient dissipation of energy impacts the NPQ mechanism, leading to lower ROS production in leaves compared to roots [23]. Several phenolic compounds, including *O*-methoxybenzoic, benzoic, usnic, salicylic, cinnamic, syringic, and phthalic acids, are known to interfere with antenna complexes, disrupting the function of reaction centers and blocking electron transport between Q_A and Q_B. These disruptions reduce photochemical efficiency and increase heat dissipation, as shown in earlier studies [24,38,42,43]. Similarly, photochemical quenching parameters change similarly to *A* in plants under growth stress, where ATP and NADPH consumption are critical. Variations in carboxylation efficiency, CO_2 availability, or carbohydrate export from cells can further influence this consumption, highlighting the complex interdependencies within the photosynthetic process [49].

To evaluate potential structural damage to the photochemical process, chlorophyll *a* fluorescence is measured in the dark, when the carboxylation cycle is inactive. Key parameters such as F_v/F_m, SPAD index, OJIP curve, and JIP test are critical for this assessment. F_v/F_m reflects the maximum efficiency of photosystem II, indicating the proportion of open reaction centers. The lack of change in F_v/F_m suggests that there was no severe damage to photochemical structures, as the parameter remains stable even with reduced chlorophyll content [29,38]. However, while the unaltered F_v/F_m indicates that 3-cyanobenzoic acid did not severely damage the photochemical machinery, the observed decrease in reaction centers per cross-section (RC/CS) and total chlorophyll levels suggests that some degree of damage may have occurred, potentially affecting overall photosynthetic efficiency.

In their study on the effects of various phenolic compounds on the photosynthesis of *C. reinhardtii*, Gao et al. [38] found that while benzoic acid did not affect F_v/F_m, it significantly altered chlorophyll and carotenoid levels, as well as the OJIP curve and JIP-test parameters. Specifically, benzoic acid reduced the RC/CS ratio, like 3-cyanobenzoic acid, and affected PI_{ABS}, an important indicator of PSII photoinhibition under abiotic stress, while also enhancing the J phase in the OJIP curve. These changes closely resemble those caused by DCMU, a PSII inhibitor that disrupts electron transport between plastoquinones, thereby reducing PSII quantum efficiency by disconnecting smaller antenna molecules and blocking electron flow beyond Q_A. Based on these observations, it is likely that 3-cyanobenzoic acid induces photochemical damage to PSII reaction centers by interfering with antenna molecules, leading to a reduction in chlorophyll content. This decrease in chlorophyll is a primary effect of phenolic compounds on photosynthesis, likely due to either decreased biosynthesis or increased degradation of chlorophyll [24]. Additionally, carotenoids, which function as accessory pigments in antenna complexes, may also be reduced because of hormonal imbalances, potentially linked to an increase in abscisic acid, a carotenoid-derived hormone [50].

The effects of 3-cyanobenzoic acid on maize seedling roots suggest a disruption in hormonal balance [23]. Abscisic acid (ABA) is known to induce stomatal closure [40,51], but certain phenolic compounds, such as hydroxybenzoic and hydroxycinnamic acids, can counteract this effect. Similarly, 3-cyanobenzoic acid, like 3-hydroxybenzoic acid, exhibits diverse and complex impacts on plant physiology [24,52]. Monosubstituted benzoic acids, like 3-cyanobenzoic acid, can inhibit ethylene synthesis at high concentrations, while some monophenols stimulate auxin degradation [24]. Given these effects, it is plausible to hypothesize that 3-cyanobenzoic acid may contribute to root damage and hinder maize growth and photosynthesis. This could be linked to elevated ABA levels relative to other hormones, which may lead to reduced stomatal conductance and altered carotenoid synthesis. Such changes in hormonal regulation could help explain the observed damage to reaction centers, further impairing the plant's ability to efficiently capture light and perform photosynthesis.

Lastly, the LCP, which is the minimum irradiance needed to sustain basal metabolism, showed the most significant change [34]. The increase in LCP can be attributed to a combination of reduced carbon assimilation, impaired quantum efficiency, and elevated root respiration [23]. Despite the challenges imposed by 3-cyanobenzoic acid, maize plants exhibited a degree of adaptability, adjusting their metabolic processes to cope with the altered growth conditions. However, in environments with insufficient light, maize plants remain vulnerable, as they are unable to meet the high demands of basal metabolism. At high irradiance, NPQ would likely be inadequate to prevent the formation of ROS in the aerial parts of the maize plants, further exacerbating the stress.

Future research on 3-cyanobenzoic acid should prioritize elucidating its biochemical targets, particularly PPDK, through comprehensive enzymatic assays, and determining its selectivity between C4 weeds and C3 crops. Field trials are essential to assess its efficacy, stability, and interactions within the environment, including its soil adsorption and bioavailability. Exploring chemical modifications could improve both its potency and safety profile, while in-depth toxicity evaluations and advanced formulation strategies will be critical for optimizing its use as a preemergence herbicide. Despite its promising potential, several challenges remain, including environmental persistence, high development costs, stringent regulatory hurdles, and the risk of resistance development. A strategic, multidisciplinary approach is vital to addressing these challenges effectively. These factors will play a crucial role in shaping future studies and advancing the compound's development toward practical application.

5. Conclusions

The presence of 3-cyanobenzoic acid significantly hampers maize growth by disrupting photosynthetic, photochemical, and carboxylative processes. The primary driver of reduced photosynthesis is root damage, which diminishes both stomatal and mesophyll conductance, thereby limiting CO_2 availability for RuBisCo and PEPCase activity. This disruption in CO_2 assimilation is compounded by damage to the antenna complexes and reaction centers of photosystem II, likely caused by reduced chlorophyll content and a possible diversion of carotenoids for ABA synthesis, which triggers stomatal closure. While maize plants exhibit some resilience to 3-cyanobenzoic acid, their vulnerability becomes evident under low-light conditions due to the high metabolic costs associated with maintaining basic physiological functions. Under high light conditions, the inadequacy of the photosynthetic machinery leads to increased oxidative stress, further exacerbating growth limitations. These findings highlight the potential of 3-cyanobenzoic acid and its derivatives as herbicide candidates, targeting key photosynthetic pathways. By leveraging advanced screening technologies and analyzing plant-specific responses, such as those observed in maize, we can develop selective and environmentally sustainable herbicides. Evaluating the phytotoxicity of potential plant protection products is essential to ensuring their effectiveness against target weeds while minimizing adverse effects on non-target species and ecosystems. These assessments provide valuable insights into the compounds' mechanisms of action, selectivity, and safety, facilitating the development of effective and sustainable agricultural solutions.

Author Contributions: Conceptualization, funding acquisition, writing—review and editing, O.F.-F.; conceptualization, funding acquisition, supervision, writing—review and editing, R.M.; investigation, methodology, validation, visualization, writing—original draft, L.H.E.G.; investigation, validation, I.d.C.C., A.P.d.S.M., A.C.C., A.P.B. and A.P.F.; data curation, R.P.C. and W.D.d.S. All authors have read and agreed to the published version of the manuscript.

Funding: This research was funded by the Coordination of Higher Education Personnel Improvement–Brazil (CAPES, Finance Code 001), and Araucaria Foundation [grant numbers 013/2017 (PRONEX) and 002/2016].

Data Availability Statement: Data are contained within the article.

Acknowledgments: L.H.E.G., I.d.C.C., A.P.d.S.M., A.C.C. and A.P.B. are receiving a scholarship from the CAPES (Finance Code 001). W.D.d.S., R.M. and O.F.-F. are research fellows of the National Council for Scientific and Technological Development (CNPq).

Conflicts of Interest: The authors declare no conflicts of interest.

Abbreviations

The following abbreviations are used in this manuscript:

A	net assimilation
ABS/RC	absorption flux per reaction center at t = 0 or a measure for an average antenna size
A_{max}	net assimilation rate
C_i	intercellular CO_2 concentration
DI_0/RC	energy dissipation per reaction center
E	transpiration rate
ETR	electron transport rate through PSII
ET_0/ABS	quantum yield for electron transport (at t = 0) from Q_A to Q_B
ET_0/RC	electron transport flux per reaction center at t = 0
ET_0/TR_0	probality (at t = 0) of a trapped exciton transfer an electron beyond Q_A^- into the ETR
F_0	minimal fluorescence intensity when all the reaction centers of PSII are open
F_m	maximum fluorescence intensity when all reaction centers of PSII are closed
F_v	variable fluorescence
F_v/F_m	maximum quantum yield of PSII photochemistry
$F_{v'}/F_{m'}$	maximum quantum yield of PSII photochemistry in light
g_m	mesophilic conductance
g_s	stomatal conductance
J	photosynthetic electron transport rate
LCP	light compensation point
NPQ	non-photochemical quenching
OJIP	chlorophyll *a* fluorescence transient with O-J-I-P phases
PAR	photosynthetically active radiation
PEPCase	phosphoenolpyruvate carboxylase
PI_{ABS}	indicator of the conservation of energy absorbed by PSII until the decrease of the Q_B
PI_{TOTAL}	indicator of energy conservation in PSII up to the reduction of final PSI acceptors.
PPDK	pyruvate *O*-phosphate dikinase
Q_A	primary quinone acceptor of PSII
Q_B	secondary quinone acceptor of PSII
q_P	photochemical quenching coefficient
RC/CS	amount of active PSII reaction centers per cross-section
R_D	dark respiration rate
RuBisCo	ribulose-1,5-bisphosphate carboxylase/oxygenase

TR_0/ABS	maximum quantum yield of primary photochemistry at t = 0
TR_0/ABS	quantum efficiency of primary PSII photochemistry
TR_0/RC	trapped energy flux per reaction center at t = 0
V_{cmax}	maximum RuBisCo carboxylation rate
V_{pmax}	maximum PEPCase carboxylation rate
ϕ	apparent quantum yield
ϕ_{PSII}	effective quantum yield of PSII photochemistry

References

1. Wasilewska-Dębowska, W.; Zienkiewicz, M.; Drozak, A. How Light Reactions of Photosynthesis in C4 Plants Are Optimized and Protected under High Light Conditions. *Int. J. Mol. Sci.* **2022**, *23*, 3626. [CrossRef] [PubMed]
2. Zhou, H.; Akçay, E.; Helliker, B.R. Estimating C4 photosynthesis parameters by fitting intensive A/Ci curves. *Photosynth. Res.* **2019**, *141*, 181–194. [CrossRef] [PubMed]
3. Alvarez, C.E.; Maurino, V.G. Adaptive diversity in structure and function of C4 photosynthetic components. *Biochem. Soc. Trans.* **2023**, *51*, 1157–1168. [CrossRef] [PubMed]
4. Aggarwal, P.R.; Muthamilarasan, M.; Choudhary, P. Millet as a promising C4 model crop for sustainable biofuel production. *J. Biotechnol.* **2024**, *395*, 110–121. [CrossRef]
5. Talukder, P.; Sinha, B.; Biswas, S.; Ghosh, A.; Banerjee, A.; Paul, S. A study on the prospect of converting C3 plants into C4 plants. *Biocatal. Agric. Biotechnol.* **2024**, *58*, 103191. [CrossRef]
6. Roy, D.; Ghosh, S.; Datta, D.; Sreekanth, D.; Pawar, D.; Mukherjee, P.K.; Ghosh, D.; Santra, S.C.; Moulick, D. Climate Resilient Weed Management for Crop Production. In *Climate–Resilient Agriculture*; Hasanuzzaman, M., Ed.; Springer Nature: Cham, Switzerland, 2023; Volume 2, pp. 125–156. [CrossRef]
7. Perotti, V.E.; Larran, A.S.; Palmieri, V.E.; Martinatto, A.K.; Permingeat, H.R. Herbicide resistant weeds: A call to integrate conventional agricultural practices, molecular biology knowledge and new technologies. *Plant Sci.* **2020**, *290*, 10255. [CrossRef]
8. Riechers, D.E.; Soltani, N.; Chauhan, B.S.; Concepcion, J.C.T.; Geddes, C.M.; Jugulam, M.; Kaundun, S.S.; Preston, C.; Wuerrfel, R.J.; Sikkema, P.H. Herbicide resistance is complex: A global review of cross-resistance in weeds within herbicide groups. *Weed Sci.* **2024**, 1–22. [CrossRef]
9. Heap, I. International Herbicide-Resistant Weed Database. Available online: https://www.weedscience.org/Home.aspx (accessed on 10 November 2024).
10. Gupta, P.K. Toxicity of Herbicides. In *Veterinary Toxicology—Basic and Clinical Principles*, 3rd ed.; Gupta, R.C., Ed.; Academic Press: Cambridge, MA, USA, 2018; pp. 553–567. [CrossRef]
11. Brovini, E.M.; de Deus, B.C.T.; Vilas-Boas, J.A.; Quadra, G.R.; Carvalho, L.; Mendonça, R.F.; Pereira, R.d.O.; Cardoso, S.J. Three-bestseller pesticides in Brazil: Freshwater concentrations and potential environmental risks. *Sci. Total Environ.* **2021**, *771*, 144754. [CrossRef]
12. Dayan, F.E.; Duke, S.O. Discovery for New Herbicide Sites of Action by Quantification of Plant Primary Metabolite and Enzyme Pools. *Engineering* **2020**, *6*, 509–514. [CrossRef]
13. Elmore, C.; Paul, R. Composite List of C4 Weeds. *Weed Sci.* **1983**, *31*, 686–692. [CrossRef]
14. Rao, V.S. *Principles of Weed Science*, 2nd ed.; CRC Press: Boca Raton, FL, USA, 2000. [CrossRef]
15. Almeida, A.M.; Abrahão, J.; Seixas, F.A.V.; Bueno, P.S.A.; Oliveira, M.A.S.d.; Tomazini, L.F.; Constantin, R.P.; dos Santos, W.D.; Marchiosi, R.; Ferrarese-Filho, O. Unraveling Shikimate Dehydrogenase Inhibition by 6-Nitroquinazoline-2,4-diol and Its Impact on Soybean and Maize Growth. *Agronomy* **2024**, *14*, 930. [CrossRef]
16. Martarello, D.C.I.; Grizza, L.H.E.; Foletto-Felipe, M.d.P.; Mendonça, A.P.d.S.; Constantin, R.P.; Ferro, A.P.; dos Santos, W.D.; Constantin, R.P.; Marchiosi, R.; Ferrarese-Filho, O. *S*-Benzyl-L-cysteine Inhibits Growth and Photosynthesis, and Triggers Oxidative Stress in Ipomoea grandifolia. *Agronomy* **2024**, *14*, 1633. [CrossRef]
17. Doyle, J.; Burnell, J.N.; Haines, D.S.; Llewellyn, L.E.; Motti, C.A.; Tapiolas, D.M. A Rapid Screening Method to Detect Specific Inhibitors of Pyruvate Orthophosphate Dikinase as Leads for C_4 Plant-Selective Herbicides. *J. Biomol. Screen.* **2005**, *10*, 67–75. [CrossRef] [PubMed]
18. Nguyen, G.T.T.; Erlenkamp, G.; Jäck, O.; Küberl, A.; Bott, M.; Fiorani, F.; Gohlke, H.; Groth, G. Chalcone-based selective inhibitors of a C4 plant key enzyme as novel potential herbicides. *Sci. Rep.* **2016**, *6*, 27333. [CrossRef] [PubMed]
19. Haines, D.S.; Burnell, J.N.; Doyle, J.; Llewellyn, L.E.; Motti, C.A.; Tapiolas, D.M. Translation of in vitro inhibition by marine natural products of the C4 acid cycle enzyme pyruvate P(i) dikinase to in vivo C4 plant tissue death. *J. Agric. Food Chem.* **2005**, *53*, 3856–3862. [CrossRef]

20. Motti, C.A.; Bourne, D.G.; Burnell, J.N.; Doyle, J.R.; Haines, D.S.; Liptrot, C.H.; Llewellyn, L.E.; Ludke, S.; Muirhead, A.; Tapiolas, D.M. Screening marine fungi for inhibitors of the C4 plant enzyme pyruvate phosphate dikinase: Unguinol as a potential novel herbicide candidate. *Appl. Environ. Microbiol.* **2007**, *73*, 1921–1927. [CrossRef]
21. Minges, A.; Groth, G. Small-molecule inhibition of pyruvate phosphate dikinase targeting the nucleotide binding site. *PLoS ONE* **2017**, *12*, e181139. [CrossRef]
22. Constantin, R.P.; Melo, G.S.R.; Martarello, D.C.I.; Abrahão, J.; dos Santos, W.D.; Ferrarese-Filho, O.; Marchiosi, R. Bromoacetic acid inhibits pyruvate orthophosphate dikinase and reduces photosynthetic activity in maize. *Acta Physiol. Plant.* **2021**, *43*, 54. [CrossRef]
23. Souza, M.C.; Constantin, R.P.; Abrahão, J.; Foletto-Felipe, M.P.; Grizza, L.H.E.; Constantin, R.P.; dos Santos, W.D.; Ferrarese-Filho, O.; Marchiosi, R. Inhibitory Effect of 3-Cyanobenzoic Acid on Initial Growth of Maize Seedlings and Its Biochemical Impacts on Antioxidant and Energy Metabolisms. *J. Plant Growth Regul.* **2024**, *43*, 458–472. [CrossRef]
24. Marchiosi, R.; dos Santos, W.D.; Constantin, R.P.; Lima, R.B.; Soares, A.R.; Finger-Teixeira, A.; Mota, T.R.; Oliveira, D.M.; Foletto-Felipe, M.P.; Abrahão, J.; et al. Biosynthesis and Metabolic Actions of Simple Phenolic Acids in Plants. *Phytochem. Rev.* **2020**, *19*, 865–906. [CrossRef]
25. Bengis-Garber, C.; Gutman, A.L. Bacteria in organic synthesis: Selective conversion of 1,3-dicyanobenezene into 3-cyanobenzoic acid. *Tetrahedron Lett.* **1988**, *29*, 2589–2590. [CrossRef]
26. Délye, C.; Jasieniuk, M.; Le Corre, V. Deciphering the evolution of herbicide resistance in weeds. *Trends Genet.* **2013**, *29*, 649–658. [CrossRef] [PubMed]
27. Dong, J.; Wu, F.; Zhang, G. Influence of cadmium on antioxidant capacity and four microelement concentrations in tomato seedlings (*Lycopersicon esculentum*). *Chemosphere* **2006**, *64*, 1659–1666. [CrossRef]
28. Easlon, H.M.; Bloom, A.J. Easy Leaf Area: Automated Digital Image Analysis for Rapid and Accurate Measurement of Leaf Area. *Appl. Plant Sci.* **2014**, *2*, 1400033. [CrossRef]
29. Murchie, E.H.; Lawson, T. Chlorophyll fluorescence analysis: A guide to good practice and understanding some new applications. *J. Exp. Bot.* **2013**, *64*, 3983–3998. [CrossRef]
30. Loriaux, S.D.; Avenson, T.J.; Welles, J.M.; Mcdermitt, D.K.; Eckles, R.D.; Riensche, B.; Genty, B. Closing in on maximum yield of chlorophyll fluorescence using a single multiphase flash of sub-saturating intensity. *Plant Cell Environ.* **2013**, *36*, 1755–1770. [CrossRef]
31. Lobo, F.A.; De Barros, M.P.; Dalmagro, H.J.; Dalmolin, A.C.; Pereira, W.E.; De Souza, E.C.; Vourlitis, G.L.; Rodríguez Ortíz, C.E. Fitting net photosynthetic light-response curves with Microsoft Excel—A critical look at the models. *Photosynthetica* **2013**, *51*, 445–456. [CrossRef]
32. Bellasio, C.; Beerling, D.J.; Griffiths, H. An Excel tool for deriving key photosynthetic parameters from combined gas exchange and chlorophyll fluorescence: Theory and practice: Descriptive modelling of gas exchange data. *Plant Cell Environ.* **2016**, *39*, 1180–1197. [CrossRef]
33. Strasser, R.J.; Srivastava, A.; Tsimilli-Michael, M. The fluorescence transient as a tool to characterize and screen photosynthetic samples. In *Probing Photosynthesis Mechanisms, Regulation and Adaptation*; Yunus, M., Pathre, U., Mohanty, P., Eds.; CRC Press: Boca Raton, FL, USA, 2000; pp. 445–483.
34. Avarseji, Z.; Mohassel, M.H.R.; Nezami, A.; Abbaspoor, M.; Mahallati, N. Dicamba^{+2}, 4-D affects the shape of the Kautsky curves in wild mustard (*Sinapis arvensis*). *Plant Knowl. J.* **2012**, *1*, 41–45. Available online: https://profdoc.um.ac.ir/articles/a/1032111.pdf (accessed on 25 November 2024).
35. Banks, J.M. Continuous excitation chlorophyll fluorescence parameters: A review for practitioners. *Tree Physiol.* **2017**, *37*, 1128–1136. [CrossRef]
36. Chipley, J.R. Sodium Benzoate and Benzoic Acid. In *Antimicrobials in Food*; Davidson, P.M., Taylor, T.M., David, J.R.D., Eds.; CRC Press: Boca Raton, FL, USA, 2020. [CrossRef]
37. Zhou, Y.; Yu, J. Allelochemicals and photosynthesis. In *Allelopathy*; Reigosa, M., Pedrol, N., González, L., Eds.; Springer: Dordrecht, The Netherlands, 2006; pp. 127–139. [CrossRef]
38. Gao, Y.; Liu, W.; Wang, X.; Yang, L.; Han, S.; Chen, S.; Strasser, R.J.; Valverde, B.E.; Qiang, S. Comparative phytotoxicity of usnic acid, salicylic acid, cinnamic acid and benzoic acid on photosynthetic apparatus of *Chlamydomonas reinhardtii*. *Plant Physiol. Biochem.* **2018**, *128*, 1–12. [CrossRef] [PubMed]
39. Sumbele, S.; Fotelli, M.N.; Nikolopoulos, D.; Tooulakou, G.; Liakoura, V.; Liakopoulos, G.; Bresta, P.; Dotsika, E.; Adams, M.A.; Karabourniotis, G. Photosynthetic capacity is negatively correlated with the concentration of leaf phenolic compounds across a range of different species. *AoB Plants* **2012**, *2012*, pls025. [CrossRef] [PubMed]
40. Kostidis, G.; Karabourniotis, G. The intraspecific trade-off between photosynthesis and phenolics in selected crop species: A meta-analysis study. *Discov. Plants* **2024**, *1*, 29. [CrossRef]
41. Yu, J.Q.; Ye, S.F.; Zhang, M.F.; Hu, W.H. Effects of root exudates and aqueous root extracts of cucumber (*Cucumis sativus*) and allelochemicals, on photosynthesis and antioxidant enzymes in cucumber. *Biochem. Syst. Ecol.* **2003**, *31*, 129–139. [CrossRef]

42. Lu, X.F.; Zhang, H.; Lyu, S.S.; Du, G.D.; Wang, X.Q.; Wu, C.H.; Lyu, D.G. Effects of exogenous phenolic acids on photosystem functions and photosynthetic electron transport rate in strawberry leaves. *Photosynthetica* **2018**, *56*, 616–622. [CrossRef]
43. Li, K.; Zhang, T.; Li, H.; Zhang, L.D.; Li, F. Phenolic acids inhibit the photosynthetic productivity of poplar. *Photosynthetica* **2020**, *58*, 1210–1216. [CrossRef]
44. Marchiosi, R.; Bido, G.S.; Böhm, P.A.F.; Soares, A.R.; Silva, H.A.; Ferro, A.P.; Ferrarese, M.L.L.; Ferrarese-Filho, O. Photosynthetic response of soybean to L-DOPA and aqueous extracts of velvet bean. *Plant Growth Regul.* **2016**, *80*, 171–182. [CrossRef]
45. Huang, X.; Bie, Z. Cinnamic acid-inhibited ribulose-1,5-bisphosphate carboxylase activity is mediated through decreased spermine and changes in the ratio of polyamines in cowpea. *J. Plant Physiol.* **2010**, *167*, 47–53. [CrossRef]
46. Evans, J.R.; von Caemmerer, S. Carbon Dioxide Diffusion inside Leaves. *Plant Physiol.* **1996**, *110*, 339–346. [CrossRef]
47. Long, S.P.; Bernacchi, C.J. Gas exchange measurements, what can they tell us about the underlying limitations to photosynthesis? Procedures and sources of error. *J. Exp. Bot.* **2003**, *54*, 2393–2401. [CrossRef]
48. Foletto-Felipe, M.P.; Abrahão, J.; Contesoto, I.C.; Ferro, A.P.; Grizza, L.H.E.; Menezes, P.V.M.C.; Wagner, A.L.S.; Seixas, F.A.V.; Oliveira, M.A.S.; Tomazini, L.F.; et al. Inhibition of sulfur assimilation by S-benzyl-L-cysteine: Impacts on growth, photosynthesis, and leaf proteome of maize plants. *Plant Physiol. Biochem.* **2024**, *216*, 109173. [CrossRef] [PubMed]
49. Baker, N.R. Chlorophyll Fluorescence: A Probe of Photosynthesis In Vivo. *Annu. Rev. Plant Biol.* **2008**, *59*, 89–113. [CrossRef] [PubMed]
50. Chauhan, J.; Prathibha, M.; Singh, P.; Choyal, P.; Mishra, U.N.; Saha, D.; Kumar, R.; Anuragi, H.; Pandey, S.; Bose, B.; et al. Plant photosynthesis under abiotic stresses: Damages, adaptive, and signaling mechanisms. *Plant Stress* **2023**, *10*, 100296. [CrossRef]
51. Holappa, L.D.; Blum, U. Effects of exogenously applied ferulic acid, a potential allelopathic compound, on leaf growth, water utilization, and endogenous abscisic acid levels of tomato, cucumber, and bean. *J. Chem. Ecol.* **1991**, *17*, 865–886. [CrossRef]
52. Purohit, S.; Laloraya, M.M.; Bharti, S. Effect of phenolic compounds on abscisic acid-induced stomatal movement: Structure—Activity relationship. *Physiol. Plant.* **1991**, *81*, 79–82. [CrossRef]

plants

MDPI

Article

Chemical Control of the Invasive Weed *Trianthema portulacastrum*: Nethouse Studies

Yaakov Goldwasser [1,*], Onn Rabinowitz [2], Jackline Abu-Nasser [3], Evgeny Smirnov [3], Guy Achdary [3] and Hanan Eizenberg [3]

1 Valley Farmers Center Ltd., Migdal Haemek 2310201, Israel
2 Northern Research and Development, Kiryat Shmona 1101600, Israel; onnrab@gmail.com
3 Newe Ya'ar Research Center, Ramat Yishay 3009500, Israel; jackline@volcani.agri.gov.il (J.A.-N.); evegenysm30@gmail.com (E.S.); achdarig@volcani.agri.gov.il (G.A.); eizenber@volcani.agri.gov.il (H.E.)
* Correspondence: yaakov.goldwasser@mail.huji.ac.il

Abstract: *Trianthema portulacastrum* L. (Aizoaceae), commonly known as desert horse purslane or black pigweed, is a C4 dicot succulent invasive annual plant that is widespread in agricultural fields in Southeast Asia, tropical America, Africa, and Australia. In Israel, *Trianthema portulacastrum* is an invasive weed of increasing importance in agricultural fields, including mainly corn, tomato, alfalfa watermelon, and groundnut crops. The significance of this weed in crops has been recently reported in neighboring countries of Jordan and Egypt. In previous studies, we have examined and described the spread, biology, and germination requirements of *Trianthema portulacastrum* in Israel. The present study aimed to investigate the efficiency of single pre- and post-emergence herbicides and the combination of pre-applied herbicides for the control of this invasive weed in pots in a nethouse. We conducted three sequential experiments in a nethouse: (1) screening of pre-emergence herbicides, (2) screening of post-emergence herbicides, and (3) assessment of residual activity of combined pre-emergence herbicides in three distinct Hula Valley soil types. Efficacy was evaluated through weekly assessments of seedling emergence and vigor, with the final shoot fresh weight determined upon the experiment's completion. In all experiments, weekly counts and vigor estimation of *T. portulacastrum* seedlings were conducted, and shoot fresh weights were determined at the end of the experiments. The results of pre-emergence herbicide screening showed that Fomesafen, Terbutryne, Flurochloridon, Sulfosulfuron, Cyrosulfamid + Izoxaflutole, and Dimethenamid were the most effective herbicides, leading to complete eradication of *T. portulacastrum* plants. Results of the post-emergence screening revealed that Saflufenacil, Foramsulfuron, Tembotrione + Isoxdifen-ethyl, and Rimsulfurom Methyl completely controlled the weed. In the soil residual study, three herbicide combinations (Fomesafen + Terbutryn, Sulfosulfuron + Fomesafen, and Dimethenamid + Flurochloridon) provided effective control across all soil types. These findings provide a foundation for future field trials investigating integrated pre- and post-emergence herbicide programs for *T. portulacastrum* management in various crops.

Keywords: controlled environment studies; desert horse purslane; herbicides; Hula Valley Israel; invasive plants; pre-emergence; post-emergence

1. Introduction

Trianthema portulacastrum L. (Aizoaceae), commonly known as desert horse purslane or black pigweed, is a widespread weed in Southeast Asia, America, Africa, Australia, and Asia, with no known center of origin [1] (Figure 1). It is a branched annual C4

https://doi.org/10.3390/plants14010019

dicot plant with ovate, simple leaves, and pink, five-petal bisexual flowers that are insect-pollinated. The flowers produce capsules that release many small, kidney-shaped, hard-coated seeds throughout the warm spring–autumn period. Seed dispersal occurs through multiple vectors, including soil movement, water flow, and agricultural equipment, with seeds maintaining viability in the soil for several years [1–3]. The agricultural impact of *T. portulacastrum* spans diverse cropping systems worldwide, with recent research documenting significant infestations in groundnuts [4], rice [5–12], pearl millet [13], and greengram [14–16]. *T. portulacastrum* has documented medicinal properties, including antimicrobial, analgesic, anti-inflammatory, anti-hyperglycemic, and hepatoprotective activities [17–20].

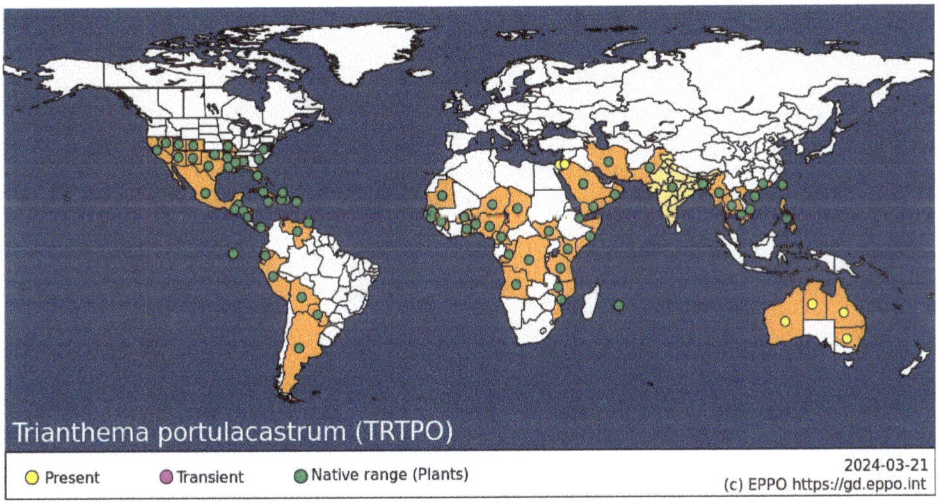

Figure 1. Global distribution of *T. portulacastrum*, updated March 2024, according to the European and Mediterranean Plant Protection Organization (EPPO) Global Database. https://gd.eppo.int.

There have been numerous reports of this weed as an invasive species in the Middle East and the Indian subcontinent. In Egypt, *T. portulacastrum* was reported to be an invasive weed in field crops and orchards, with high phenotypic differences resulting from variable soil moisture, soil salinity, and temperature regimes in diverse geographical regions [21]. Controlled-environment seed germination studies have revealed impressive reproductive capacity, with individual plants producing an average of 1931 seeds, each weighing an average of 1.08 g in dry weight. A different study revealed that high temperatures of 30–45 °C stimulate maximum seed germination [22]. Reports from India and Pakistan have also described this plant as an invasive and noxious weed in many crops, with few effective control measures. A study of chemical and cultural control measures indicated that efficient weed control was achieved when these measures were applied in the early *T. portulacastrum* growth stages, up to 40 days after emergence [23]. Attempts at cultural and biological measures to control this troublesome weed have reported only limited success [24–26].

Recent reports on the invasion of this weed into Jordan and Israel led to the inclusion of *T. portulacastrum* on the 2019 invasive weed alert list of the European and Mediterranean Plant Protection Organization (EPPO) [27]. In Israel, *T. portulacastrum* field infestation throughout agricultural regions has been reported in recent years. This invasive weed thrives in warm, moist soil environments, and it is acclimated to the agricultural conditions of irrigated crops in Israel, especially in the hot and moist conditions characteristic of the Hula Valley region.

In previous studies, we surveyed and mapped *T. portulacastrum* populations in the Hula Valley in Israel, examined the biology of the plant, collected seeds for laboratory and greenhouse studies, determined seedling emergence depths, and studied the effects of temperature and light on germination, leading to the development of a temperature-based seed germination model [3].

The aim of the present study was to screen in pots under nethouse conditions single pre- and post-emergence herbicides and pre-emergence herbicide combinations in different soils of the Hula Valley in Israel for sustainable control of *T. portulacastrum*. This research provides a foundation for developing practical field-scale management strategies for this increasingly important invasive weed in Israel and in other global infested regions.

2. Materials and Methods

2.1. Screening Herbicides for T. portulacastrum Control

Screening for effective herbicides for the control of *T. portulacastrum* was conducted in 0.25 L pots containing Newe Ya'ar heavy soil placed in a nethouse of 50 mesh at the Newe Ya'ar Research Center. *T. portulacastrum* seeds used in this study were collected in the Hula Valley in 2019–2020, immersed at the beginning of each experiment in 97% sulfuric acid for 30 min, and then thoroughly washed with sterile water to ensure the maximum germination rate.

Herbicide applications were conducted with a "Resses" table sprayer delivering 200 L/ha. Following application, pots were placed in a nethouse and irrigated daily using mini sprinklers.

2.2. Pre-Emergence Herbicide Screening

For the pre-emergence experiment, *T. portulacastrum* seeds were treated for 30 min with 97% sulfuric acid and sown in pots at 2 cm soil depth in Newe Ya'ar soil. Seeding was performed with 10 seeds per pot with five replication pots per treatment. Herbicides were then applied to the soil in the pots, followed by the placement of pots in a nethouse (Figure 2). In the treatment in which Dimethenamid was applied, the herbicide was sprayed on the soil surface, followed by its incorporation throughout the soil. *T. portulacastrum* plant vitality assessments were conducted weekly, with final shoot fresh weights determined 32 days after application (DAA) by harvesting plants at the soil level.

Figure 2. Pots of the pre-emergence trial at 18 DAA in the nethouse, 5 replications = 5 pots per treatment. 0 = control, 1 = Dimethenamid 2520 g/ha, 2 = Imazapic 480 g/ha, 3 = imazamox 32 g/ha, 4 = Saflufenacil 35 g/ha, 5 = Fomesafen 625 g/ha, 6 = Terbutryne 1000 g/h, 7 = Flurochloridon 625 g/ha, 8 = Sulfosulfuron 38 g/ha, 9 = Cyprosulfamid 72 g/ha + Izoxaflutole 72 g/ha.

2.3. Post-Emergence Herbicide Screening

For the post-emergence experiment, *T. portulacastrum* seeds were sown as described in Experiment 1, but with six replicate pots per treatment. Herbicide applications were made when plants reached the four-true-leaf stage (13 days after sowing). Weekly assessments documented plant vigor and survival rates, with final shoot fresh weights determined 32 DAA. See Figure 3.

Figure 3. Pots of the post-emergence trial at 8 DAA in the nethouse, 6 replications with 6 pots per treatment. Herbicide and application rate: 0 = control, 1 = Imazapic 480, 2 = Saflufenacil 35 g/ha, 3 = Foramsulfuron 45 g/ha, 4 = Tembotrione 44 g/ha + Isoxdifen-ethyl 110 g/ha, 5 = Imazamox 32 g/ha, 6 = Bentazon 1920 g/ha, 7 = Pyraflufen 10 g/ha, 8 = Rimsulfuron Methyl 37.5 g/ha, 9 = Aclonifen 1200 g/ha.

2.4. Pre-Emergence Herbicide Residual Effects on Three Hula Valley Soils

This experiment evaluated herbicide combination treatments across three representative Hula Valley soil types Menara, Shamir, and Lehavot soils (see Table 1) in 800 cm^3 pots with four replicates per treatment. Seeds were sown at 2 cm depth at three intervals (0, 25, and 50 DAA) to assess treatment longevity and placed at the Newe-Ya'ar nethouse. Weekly monitoring of plant vigor and survival continued until 50 days after treatment (DAT) for each seeding date, concluding with fresh weight determination. Pre-emergence application of herbicides at each application date was performed using the Newe Ya'Newear table sprayer delivering 200 L/ha, followed by the return of the pots to the nethouse. The pots were irrigated daily using a mini-sprinkler system. Assessment of *T. portulacastrum* plant vigor and the number in each pot was conducted weekly, and, at the end of the experiment, 50 DAT for each seeding date, plants from each pot were cut at soil height and their fresh weight determined.

Table 1. Soil properties of the Newe Ya'ar soil and the three Hula Valley soils used in the pre-emergence herbicide residual effect experiment.

Soil Location	Location Coordinates	Total Lime (%)	Active Lime (%)	Sand (%)	Silt (%)	Clay (%)	pH	Organic Matter (%)
Newe Ya'ar	32°42′13.6″ N 35°11′05.7″ E	-	12.5	15.2	26.3	58.6	7.5	2.0

Table 1. *Cont.*

Soil Location	Location Coordinates	Total Lime (%)	Active Lime (%)	Sand (%)	Silt (%)	Clay (%)	pH	Organic Matter (%)
Lehavot	33°8′13.46″ N 35°38′25.14″ E	22.1	9.6	11.0	38.4	50.6	7.5	1.94
Shamir	33°4′42.87″ N 35°37′15.24″ E	67.3	18.4	15.0	46.4	38.6	7.4	3.45
Menara	33°6′20.14″ N 35°35′0.5″ E	40.0	13.0	23.0	34.4	42.6	7.4	7.68

3. Results

3.1. Pre-Emergence Herbicide Screening Assay

The weekly estimations of *T. portulacastrum* plant vigor conducted for the pre-emergence pot trial reveal that at the end of the experiment, 32 DAA, Fomesafen, Terbutryne, Flurochloridon, Sulfosulfuron, Cyprosulfamid + Izoxaflutole, and Dimethenamid controlled 95–100% of the weed infestation throughout the experiment, Saflufenacil application resulted in partial control, and Imazamox was the least effective (Figure 5). At the end of the experiment, 32 DAA, all herbicide treatments excluding Imazamox drastically reduced the *T. portulacastrum* final shoot fresh weight in comparison to the non-treated control, whereas Fomesafen and Cyprosulfamid + Izoxaflutole applications led to 100% eradication of the plants (Figure 4).

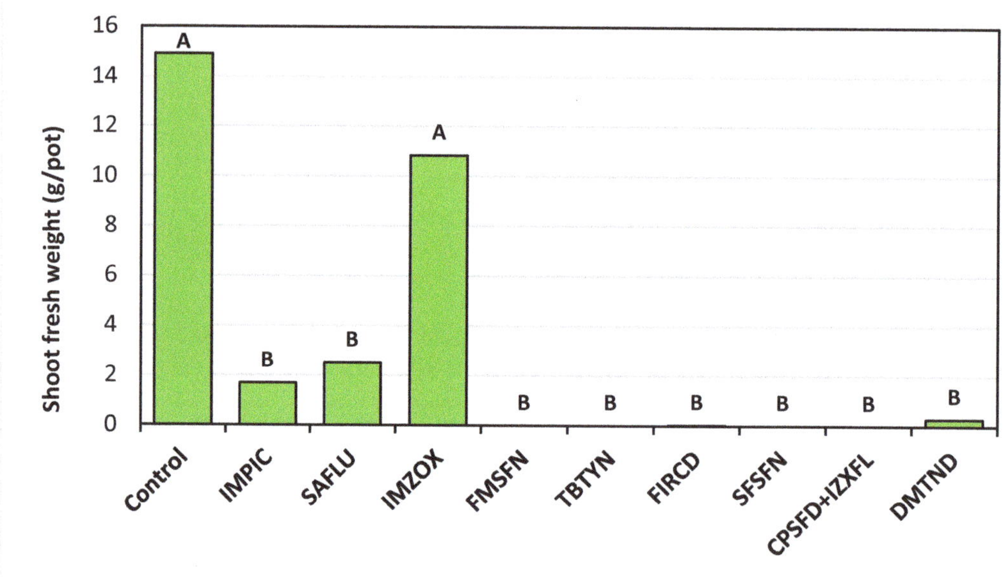

Figure 4. *T. portulacastrum* final shoot fresh weight at the termination of the pre-emergence pot experiment, 32 DAA. See Table 2 for full herbicide chemical names and application rates. Bars topped with different letters are statistically different according to Tukey–Kramer HSD, $p = 0.01$.

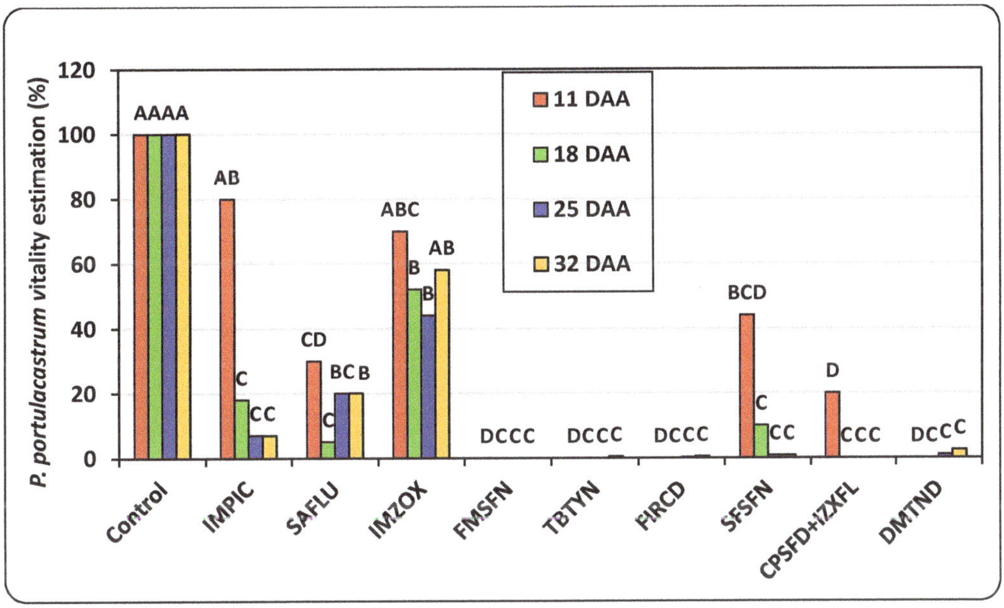

Figure 5. Weekly vitality estimations of *T. portulacastrum* plants in the pre-emergence pot experiment. Bars at each DAA topped with different letters are statistically different according to Tukey–Kramer HSD, $p = 0.01$. See Table 2 for full herbicide chemical names and application rates. DAA: days after application.

Table 2. Herbicides tested in the pre- and post-emergence nethouse screening studies.

Herbicide Chemical Name	Abbreviation	Herbicide Trade Name [a]	Active Ingredient	Application Rate g/ha
Aclonifen	ACNFN	Challenge	600 g/L	1200
Bentazon	BNTON	Bazagran	480 g/L	1920
Cyprosulfamid+ Izoxaflutole	CPSFD+IZXFL	Balance Smart	240 g/L+ 240 g/L	72+ 72
Dimethenamid	DMTND	Frontier Optima	720 g/L	2520
Flurochloridon	FIRCD	Racer	250 g/L	625
Fomesafen	FMSFN	Flex, Relax	250 g/L	625
Foramsulfuron	FRSFN	Equip	22.5 g/L	45
Imazamox	IMZOX	Pulsar	40 g/L	32
Imazapic	IMPIC	Cadre	240 g/L	480
Metribuzin	MTBZN	Sencor	70%	21
S-Metolachlor	MTLCR	Dual S Gold	915 g/L	119
Pyraflufen	PYFFN	Ecopart	20 g/L	10
Rimsulfuron Methyl	RMSFM	Titus	25%	37.5
Saflufenacil	SAFLU	Heat	70%	35
Sulfosulfuron	SFSFN	Monitor	75%	38
Tembotrione+ Isoxdifen-ethyl	TMBON+ ISXFE	Laudis	44 g/L+ 22 g/L	110+ 55
Terbutryne	TBTYN	Terbutrex	500 g/L	1000

[a] Herbicide trade names, as registered in Israel.

3.2. Post-Emergence Herbicide Assay

The weekly assessments of plant vigor conducted for the post-emergence pot trial revealed that the herbicides Saflufenacil, Foramsulfuron, Tembotrione + Isoxdifen-ethyl, Pyraflufen, and Rimsulfuron Methyl were the most effective in *T. portulacastrum* control throughout the trial, while Imazamox and Aclonifen were not effective (Figure 6). At the end of the experiment, 32 DAA, all herbicide treatments statistically reduced the *T. portulacastrum* final shoot fresh weight in comparison to the non-treated control, whereas Saflufenacil, Foramsulfuron, and Tembotrione + Isoxdifen-ethyl were the most effective, leading to 100% eradication of the plants (Figure 7).

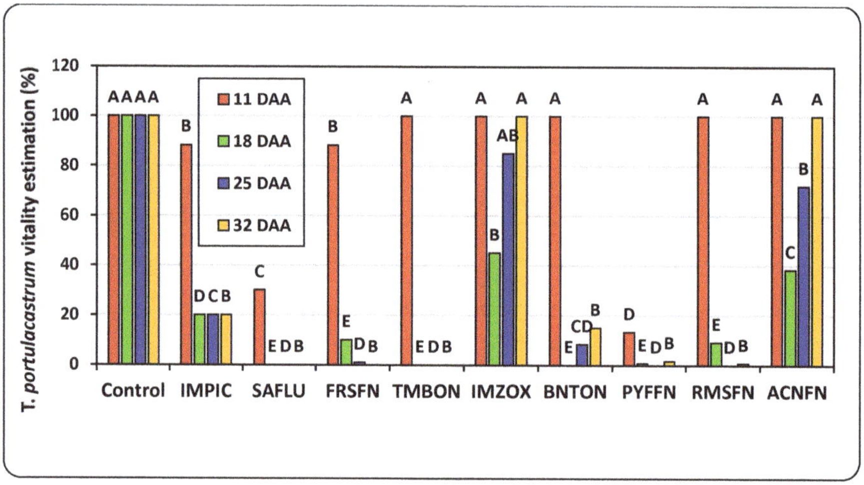

Figure 6. Weekly vitality estimations of *T. portulacastrum* plants in the post-emergence pot experiment. Herbicides were applied on *T. portulacastrum* at the 4-leaf stage on August 23rd, 13 days after seeding. See Table 2 for full herbicide chemical names and application rates. Bars at each DAA topped with different letters are statistically different according to Tukey–Kramer HSD, $p = 0.01$.

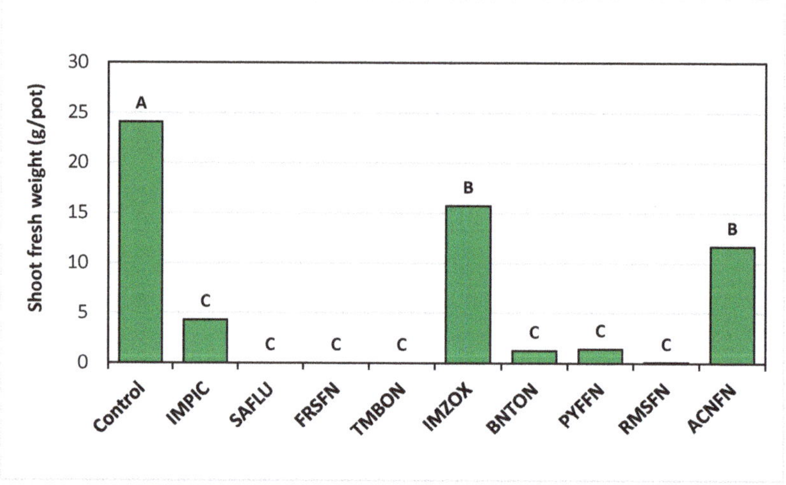

Figure 7. *T. portulacastrum* shoot fresh weights at the termination of the post-emergence applied herbicides experiment, 32 DAA. See Table 2 for full herbicide chemical names and application rates. Bars topped with different letters are statistically different according to Tukey–Kramer HSD, $p = 0.01$.

3.3. Pre-Emergence Herbicide Residual Effect on Three Hula Valley Soils

The final shoot fresh weight of *T. portulacastrum* shoots in the three Hula Valley soils 0 DAA of pre-emergence applied herbicides is exhibited in Figures 8–10, while the final shoot fresh weight in the three soils 50 DAA is shown in Figures 11–13. In the Lehavot soil, at 0 DAA, all herbicide combinations effectively and significantly provided complete control of the weed, whereas the non-treated control yielded an average of 14.3 g F.W of *T. portulacastrum* per pot (Figure 8). In the Menara soil, at 0 DAA, all herbicide combinations effectively and significantly controlled the weed, excluding the Saflufenacil + Terbutryne treatment, which yielded 25.8 g per pot F.W. of *T. portulacastrum* compared to 38.4 g in the control treatment (Figure 9).

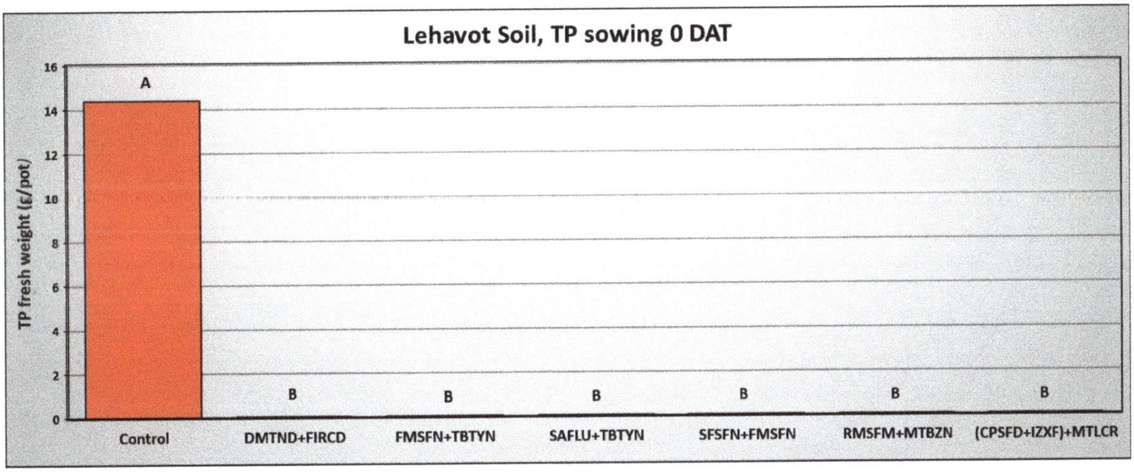

Figure 8. *T. portulacastrum* final shoot fresh weight at the termination of the pre-emergence pot experiment in Lehavot soil seeded 0 days after herbicide application. See Table 2 for herbicide abbreviations and rates. Bars topped with different letters are statistically different according to Tukey–Kramer HD, $p = 0.01$.

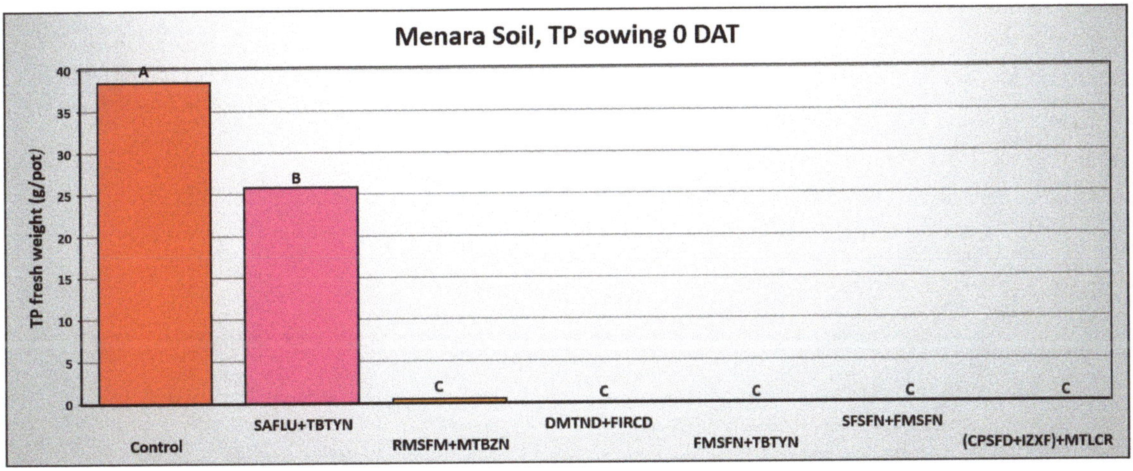

Figure 9. *T. portulacastrum* final shoot fresh weight at the termination of the pre-emergence pot experiment in Menara soil seeded 0 days after herbicide application. See Table 2 for herbicide abbreviations and rates. Bars topped with different letters are statistically different according to Tukey–Kramer HD, $p = 0.01$.

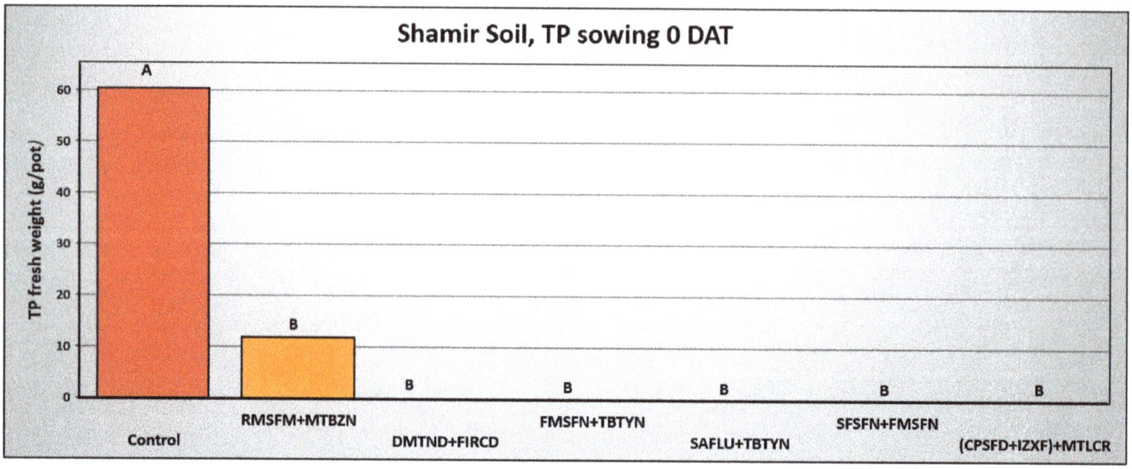

Figure 10. *T. portulacastrum* final shoot fresh weight at the termination of the pre-emergence pot experiment in Shamir soil seeded 0 days after herbicide application. See Table 2 for herbicide abbreviations and rates. Bars topped with different letters are statistically different according to Tukey–Kramer HD, $p = 0.01$.

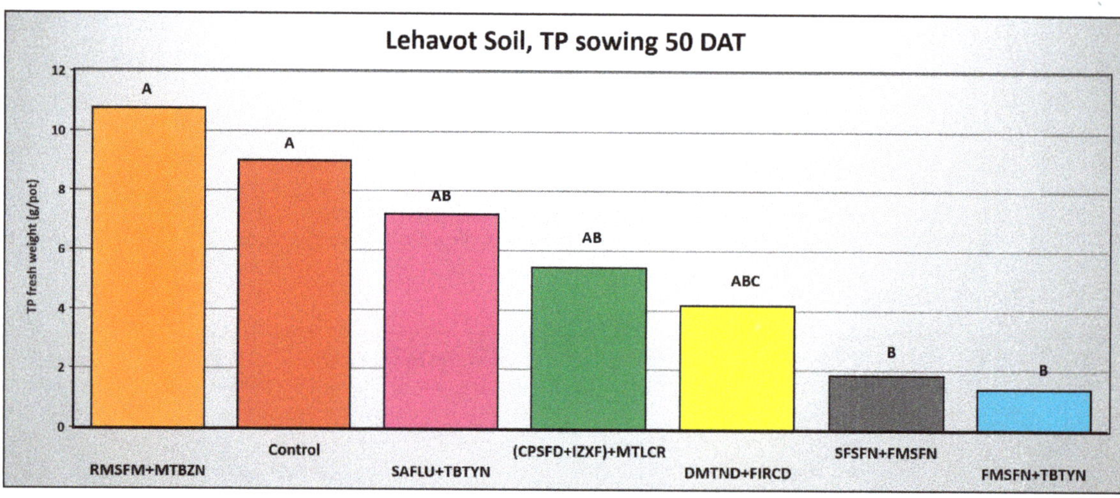

Figure 11. *T. portulacastrum* final shoot fresh weight at the termination of the pre-emergence pot experiment in Lehavot soil seeded 50 days after herbicide application. See Table 2 for herbicide abbreviations and rates. Bars topped with different letters are statistically different according to Tukey–Kramer HD, $p = 0.01$.

In the Shamir soil, all herbicide combinations effectively and significantly controlled the weed at 0 DAA, excluding the Rimsulfuron Methyl + Metribuzin treatment, which yielded 11.8 g per pot F.W. of *T. portulacastrum* compared to 60.4 g per pot F.W. of *T. portulacastrum* in the control treatment. (Figure 10).

Overall, these results demonstrate that most pre-emergence herbicides tested resulted in excellent control of *T. portulacastrum* at 0 DAA, excluding Saflufenacil + Terbutryn in the Menara soil and rimsulfuron Methyl + Metribuzin in the Shamir soil.

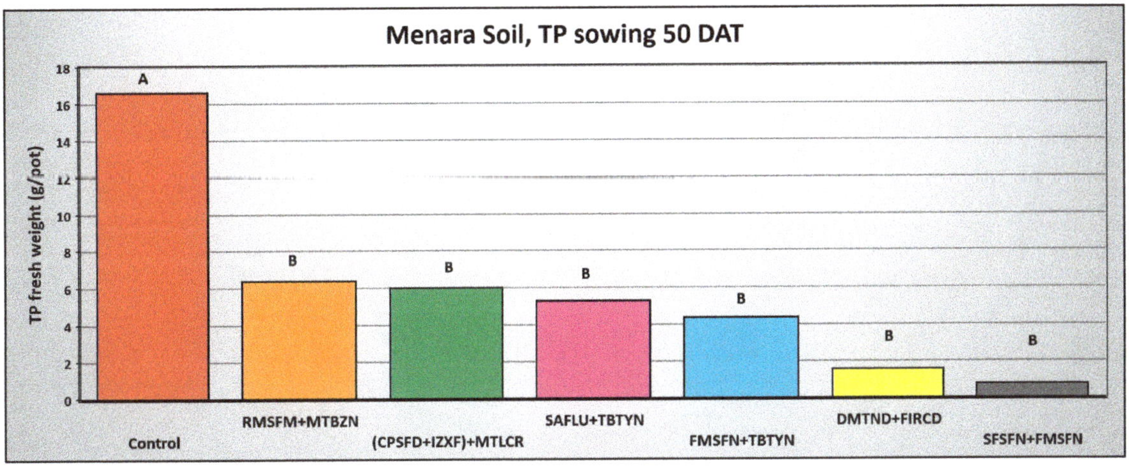

Figure 12. *T. portulacastrum* final shoot fresh weight at the termination of the pre-emergence pot experiment in Menara soil seeded 50 days after herbicide application. See Table 2 for herbicide abbreviations and rates. Bars topped with different letters are statistically different according to Tukey–Kramer HSD, $p = 0.01$.

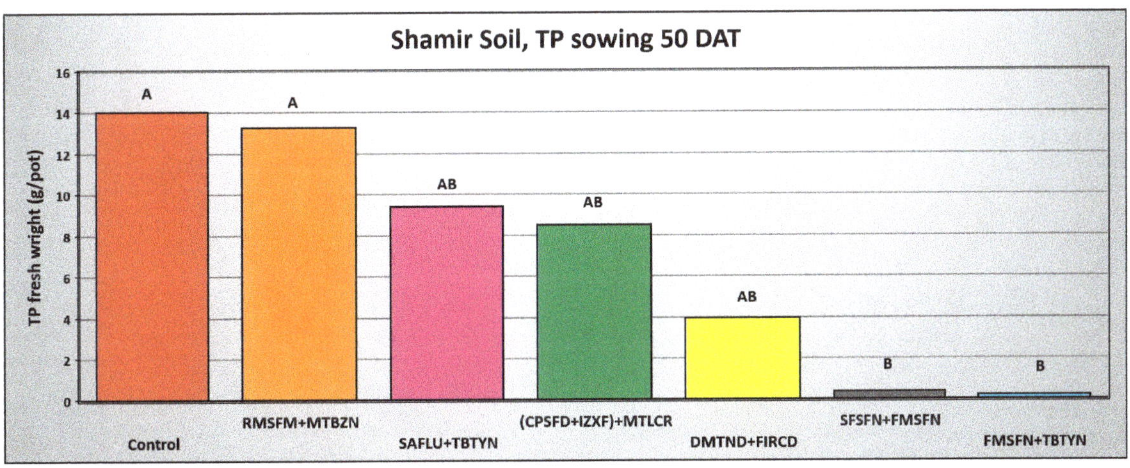

Figure 13. *T. portulacastrum* final shoot fresh weight at the termination of the pre-emergence pot experiment in Shamir soil seeded 50 days after herbicide application. See Table 2 for herbicide abbreviations and rates. Bars topped with different letters are statistically different according to Tukey–Kramer HSD, $p = 0.01$.

The results of *T. portulacastrum* control by the herbicide combinations at 50 DAA indicate a decrease in control efficacy. However, the combinations of Sulfosulfuron + Fomesafen and Fomesafen + Terbutryne effectively and significantly controlled the weed in the Lehavot and the Shamir soils (Figures 11 and 13), while in the Menara soil all herbicide combinations significantly controlled the weed (Figure 12).

A table summarizing the different herbicides used in this study, including their chemical names, trade names, and registered crops in Israel, is presented in Table 3. The most effective herbicides identified in this study are outlined in blue, while the Israeli-registered crops for each herbicide are outlined in yellow. This table shows that effective and selective herbicides are available for *T. portulacastrum* control across various crops. Specifically for the crops in the Hula Valley, the selective and efficient registered herbicides are for alfalfa, tomato, maize, and groundnuts.

Table 3. Herbicide chemical and trade names and their registered crops in Israel.

Herbicide Chemical Name	Herbicide Trade Name	Registered Crops in Israel
Aclonifen	Challenge	Parsley, tomato [b], onion, chickpea, sweet potato, dill, coriander
Bentazon	Bazagran	Wheat, pea, groundnuts, onion, green bean, vetch
Cyprosulfamid+ Izoxaflutole [a]	Balance Smart	Maize, chickpea
Dimethenamid	Frontier Optima	Groundnut's, potato, maize
Flurochloridon	Racer	Carrot, chickpea, sunflower, parsley, coriander, dill, cotton, sweet potato
Fomesafen	Flex, Relax	Peas, chickpea, green beans,
Foramsulfuron	Equip	Maize
Imazamox	Pulsar	Alfalfa, green beans, peas, clover, vetch, groundnuts
Imazapic	Cadre	Groundnuts, tomato
Metribuzin	Sencor	Carrot, potato, tomato
S- Metolachlor	Dual S Gold	Maize, green beans, sunflower, groundnuts, paprika, chickpea
Pyraflufen	Ecopart	Wheat, vetch, onion, groundnuts, orchards
Rimsulfuron Methyl	Titus	Tomato, maize, potato
Saflufenacil	Heat	Wheat, maize, orchards, maize
Sulfosulfuron	Monitor	Wheat, tomato, lawn
Tembotrione + Isoxdifen-ethyl	Laudis	Maize
Terbutryn	Terbutrex	Groundnuts, chickpea, sweet pea, sunflower

[a] Outlined in blue are the outstanding herbicides for *T. portulacastrum* control, as observed in the present nethouse study. [b] Outlined in yellow are the common crops grown in the Hula Valley for which further field trials will be conducted.

4. Discussion

Based on our research on the biology and germination traits of the rapidly spreading invasive weed *T. portulacastrum* in Israel [3], the present study aimed to conduct nethouse studies for the screening of effective pre- and post-emergence herbicides aimed at controlling this troublesome weed in crops. The effective pre-emergence herbicides that were tested in this study for controlling *T. portulacastrum* were Fomesafen, Terbutryne, Flurochloridon, Sulfosulfuron, Cyrosulfamid + Izoxaflutole, and Dimethenamid. The most effective post-emergence herbicides were Saflufenacil, Foramsulfuron, Tembotrione + Isoxdifen-ethyl, and Rimsulfuron Methyl. In the final nethouse experiment using three representative Hula Valley soils, three pre-emergence herbicide combinations demonstrated superior control of *Trianthema portulacastrum* across all soil types: Fomesafen + Terbutryne, Sulfosulfuron + Fomesafen, and Dimethenamid + Flurochloridon. Of these, the first two combinations are approved for weed management in processing tomatoes, while the third is authorized for use in groundnut cultivation.

Chemical control trials of *T. portulacastrum* have been reported in various crops by multiple researchers, but only two of these crops are grown in the Hula Valley and Israel: maize and groundnuts. In groundnuts, a recent comprehensive review summarizes the literature on the overall herbicide efficacy against *T. portulacastrum* infestations in groundnut experiments over the last 20 years [4]. However, the available literature demonstrates a substantial range in herbicide efficacy due to differences in environmental conditions and fluctuations in weed germination and infestation across years and locations. Grichar [2,28] reports experiments in groundnuts in the US and provides the following results for pre-emergence regimes:

Dimethenamid (51%), Pendimethalin (67%), and Flumioxazin (64–73%). For post-emergence regimes, the efficacy is as follows: Acifluorfen (17–77%), Lactofen (68–100%), 2,4-DB (13–84%), Imazethapyr (65%), Diclosulam (7–73%), Imazapic (7–62%), and Imazethapyr (13–27%). While several herbicides are available for weed control in groundnut, no single herbicide can provide season-long weed control due to limited application timing, lack of extended residual activity, variability in the weed control spectrum, and rotational restrictions. Consequently, effective weed management in groundnuts necessitates the use of herbicide mixtures and/or sequential application of pre-plant-incorporated, pre-emergence, and/or post-emergence herbicides, as well as residual herbicides. Groundnut has a long growing season (140 to 160 days) for development and maturity [29], and residual herbicides allow weeds to be more competitive for light, particularly during the early stage of crop growth [30].

5. Conclusions

This study provides valuable insights into the chemical control of the invasive weed *Trianthema portulacastrum* through a systematic evaluation of pre-emergence, post-emergence, and residual herbicide treatments under controlled nethouse conditions. Based on global observations across multiple crops, including our findings, effective management of *T. portulacastrum* requires a combination of herbicide treatments due to its continuous germination throughout the growing season and declining herbicide efficacy over time. This management approach necessitates both herbicide mixtures and sequential applications, including pre-plant incorporated (PPI), preemergence, and postemergence. This challenge is particularly pronounced in slow-growing crops with sparse canopy coverage, which create favorable conditions for both continuous *T. portulacastrum* germination and rapid establishment of this aggressive C4 weed. Under nethouse conditions, pot experiments demonstrated effective *T. portulacastrum* control using several herbicides. Preemergence treatments, including Fomesafen, Terbutryne, Flurochloridon, Sulfosulfuron, Cyrosulfamid + Izoxaflutole, and Dimethenamid, provided successful control, as did postemergence applications of Saflufenacil, Foramsulfuron, Tembotrione + Isoxdifen-ethyl, and Rimsulfuron Methyl. Furthermore, herbicide combinations of Fomesafen + Terbutryne and Sulfosulfuron + Fomesafen demonstrated high efficacy at 50 days after application (DAA) when tested across three representative soil types from the Hula Valley. In recent years, *T. portulacastrum* infestations have increased significantly and are spreading to new regions. Thus, following our studies and our understanding of the biological traits of this invasive weed, the major importance of the present research was the investigation of control measures for pre- and post-applied selective herbicides and herbicide combinations that could be employed safely in different crops. By combining our understanding of this invasive weed's biological characteristics with extensive herbicide trials, we identified effective selective herbicides and herbicide combinations that can be safely implemented across diverse cropping systems. Nevertheless, the risks arising from chemical residues of plant protection products in edible plant tissues, particularly those resulting from complex agrochemical "cocktails", are limited by chemical regulation standards, thus warranting comprehensive examination.

Our findings establish a foundation for developing comprehensive management strategies, although field-scale validation remains essential. Future research priorities include optimizing application rates and timing, conducting extensive field trials, and integrating herbicide applications with cultural control practices. Based on the findings of the present study, detailed elements of the *T. portulacastrum* control strategy will be studied under field conditions.

Author Contributions: Validation, G.A.; Formal analysis, E.S.; Investigation, J.A.-N.; Data curation, O.R.; Writing – original draft, Y.G.; Supervision, H.E. All authors have read and agreed to the published version of the manuscript.

Funding: This research was funded by the Chief Scientist, Israel Ministry of Agriculture as support for Upper Valley agriculture.

Data Availability Statement: The original contributions presented in this study are included in the article. Further inquiries can be directed to the corresponding author.

Conflicts of Interest: Author Yaakov Goldwasser was employed by the company Valley Farmers Center Ltd. The remaining authors declare that the research was conducted in the absence of any commercial or financial relationships that could be construed as a potential conflict of interest.

References

1. Kaur, M.; Agarwal, N.K. *Trianthema portulacastrum* L.—The noxious weed and its control. *Adv. Plants Agric. Res.* **2017**, *6*, 62–64. [CrossRef]
2. Grichar, W. Horse purslane (Trianthema portulacastrum) control in peanut (Arachis hypogaea). *Weed Technol.* **1993**, *7*, 570–572. [CrossRef]
3. Goldwasser, Y.; Rabinowitz, O.; Achdary, G.; Kapiluto, O.; Abu-Nasser, J.; Smirnov, E.; Eizenberg, H. The Invasive Weed *Trianthema portulacastrum* in Israel. *Plants* **2024**, *13*, 518. [CrossRef]
4. Daramola, O.S.; Iboyi, J.E.; MacDonald, G.E.; Maupin, B.; Myhre, E.; Griffin-LaHue, D. A systematic review of chemical weed management in peanut (*Arachis hypogaea*) in the United States: Challenges and opportunities. *Weed Sci.* **2024**, *72*, 5–29. [CrossRef]
5. Bishnoi, U.; Menon, S. Weed management in the direct seeded rice: A review. *Pharma Innov. J.* **2022**, *11*, 1416–1420.
6. Hussain, I.; Shahzad, H.; Ullah, S.; Nazir, M.J.; Rizwan, M. Herbicidal eradication of weeds in maize crop. *Pak. J. Agric. Res.* **2022**, *35*, 250–258. [CrossRef]
7. Sharma, N.; Rayamajhi, M. Different aspects of weed management in maize (*Zea mays* L.): A brief review. *Adv. Agric.* **2022**, *2022*, 7960175. [CrossRef]
8. Hirwe, O.R.; Kumar, S.; Sri, K.H.; Reddy, P.M.; Kumar, N.; Nandana, S.K.; Sharma, S.; Fayaz, S. Different weed management techniques in maize (*Zea mays* L.): A review. *Int. J. Plant Soil Sci.* **2023**, *35*, 179–191. [CrossRef]
9. Yogananda, S.B.; Thimmegowda, P.; Shruthi, G.K. Performance of sequential application of pre and post-emergence herbicides for management of weeds in aerobic rice (*Oryza sativa*). *Indian J. Agron.* **2022**, *67*, 12–19. [CrossRef]
10. Jannu, N.; Narender, G. Weed management in direct seeded rice in Indian perspective: A review. *Pharma Innov.* **2023**, *12*, 2522–2532.
11. Pleasant, J.M.; Burt, R.F.; Frisch, J.C. Integrating mechanical and chemical weed management in corn (*Zea mays*). *Weed Technol.* **1994**, *8*, 217–223. [CrossRef]
12. Habib, R.; Mubeen, K.; Matloob, A.; Baqir Hussain, M.B. Interference of horse purslane (*Trianthema portulacastrum* L.) on growth and yield of soybean (*Glycine max* (L.) Merr.). *Saudi J. Biol. Sci.* **2020**, *26*, 61. [CrossRef]
13. Dubey, R.P.; Chethan, C.R.; Choudhary, V.K.; Mishra, J.S. A review on weed management in millets. *Indian J. Weed Sci.* **2023**, *55*, 141–148. [CrossRef]
14. Ramesh, T.; Rathika, S. Management of *Trianthema portulacastrum* through herbicides in greengram. *Indian J.* **2020**, *52*, 286–289.
15. Easwari, R.; Perumal, M.S.; Suseendran, K.; Ganesa, B.A. Effect of pre and post-emergence herbicides on weed parameters and growth of irrigated green gram. *Pharma Innov. J.* **2023**, *12*, 2803–2806.
16. Rathika, S.; Udhaya, A.; Ramesh, T.; Shanmugapriya, P. Weed management strategies in green gram: A review. *Pharma Innov. J.* **2023**, *12*, 5574–5580.
17. Balyan, R.S.; Bhan, V.M. Germination of horse purslane (*Trianthema portulacastrum*) in relation to temperature, storage conditions, and seeding depths. *Weed Sci.* **1986**, *34*, 513–515. [CrossRef]
18. Kumar, V.; Aneja, K.R. Management of horse purslane (*Trianthema portulacastrum* L.): An overview. *Res. J. Bot.* **2016**, *11*, 25–32. [CrossRef]
19. Prakash, A.; Janmeda, P.; Sharma, V. Bioactivity and pharmacological potential of *Trianthema portulacastrum* L. (Angiosperms: Aizoaceae): An overview. *Plant Sci. Today* **2019**, *6*, 590–599. [CrossRef]
20. Das, U.; Saha, T.; Ghosh, R.; Das, S.K. *Trianthema portulacastrum* L.: Traditional medicine in healthcare and biology. *Indian J. Biochem. Biophys.* **2020**, *57*, 127–145.
21. Fakhr, M.; Mazrou, Y.S.; Ellmouni, F.Y.; ElSaied, A.; Elhady, M.E.; Elkelish, A.A.; Nour, I.H. Investigating the phenotypic plasticity of the invasive weed *Trianthema portulacastrum* L. *Plants* **2022**, *11*, 77. [CrossRef]
22. Fahmy, G.M.; Moussa, S.; Farrag, H.F.; Rehem, R.A.A.E. Seed and germination traits for the summer weed *Trianthema portulacastrum* L. *Egypt. J. Exp. Biol. (Bot.)* **2019**, *15*, 235–242. [CrossRef]
23. Mubeen, K.; Yonas, M.W.; Khalofah, A.; Ikram, R.M.; Sarvwar, N.; Shehzad, M.; Wasaya, A.; Rehman, H.U.; Yasir, T.A.; Aziz, M.; et al. Interference of horse purslane (*Trianthema portulacastrum* L.) and other weeds affect yield of autumn planted maize (*Zea mays* L.). *Saudi J. Biol. Sci.* **2021**, *28*, 2291–2300. [CrossRef] [PubMed]

24. Aneja, K.R. Biological control of horse purslane (*Trianathema portulacastrum* L.) by fungal pathogens. *J. Mycopathalogical Res.* **2010**, *48*, 181–185.
25. Khalik, A.; Matloob, A.; Farook, M.; Mushtaq, M.N.; Kahan, M.B. Effect of crop residues applied isolated or in combination on the germination and seedling growth of horse purslane (*Trianthema portulacastrum*). *Planta Daninha* **2011**, *29*, 121–128. [CrossRef]
26. Saeed, M.U.; Marwat, K.B.; Hassan, G.; Khan, A.Z.; Khan, I.A. Interference of horse purslane (*Trianthema portulacastrum* L.) with maize (*Zea mays* L.) at different densities. *Pak. J. Bot.* **2010**, *42*, 173–179.
27. EPPO Alert List: *Trianthema portulacastrum* L. Available online: https://gd.eppo.int/taxon/TRTPO (accessed on 20 December 2024).
28. Grichar, W. Herbicide systems for control of horse purslane (*Trianthema portulacastrum* L.), smell melon (*Cucumis melo* L.), and palmer amaranth (*Amaranthus palmeri* S. Wats) in peanut. *Peanut Sci.* **2008**, *35*, 38–42. [CrossRef]
29. Chaudhari, S.; Jordan, D.L.; Grey, T.L.; Prostko, E.P.; Jennings, K.M. Weed control and peanut (*Arachis hypogaea* L.) response to acetochlor alone and in combination with various herbicides. *Peanut Sci.* **2018**, *45*, 45–55. [CrossRef]
30. Wilcut, J.W.; York, A.C.; Grichar, W.J.; Wehtje, G.R. The biology and management of weeds in peanut (*Arachis hypogaea*). In *Advances in Peanut Science*; Pattee, H.E., Stalker, H.T., Eds.; American Peanut Research and Education Society: Stillwater, OK, USA, 1995; pp. 207–244.

plants

MDPI

Article

Electrophysiological and Behavioral Responses of *Orchestes steppensis* (Coleoptera: Curculionidae) to *Ulmus* Plant Volatiles

Meng Yang [1,2,†], Qin Li [1,2,*,†], Guoshuai Zhao [1,2], Yalin Liu [1,2] and Yonggen Lou [3]

1 College of Life Science and Technology, Xinjiang University, Urumqi 830046, China; 15172242757@163.com (M.Y.); 15688998265@163.com (G.Z.); m18749236916@163.com (Y.L.)
2 Xinjiang Key Laboratory of Biological Resources and Genetic Engineering, Urumqi 830046, China
3 Key Laboratory Biological Crop Pathogens & Insects Zhejiang Pro, State Key Laboratory Rice Biology, Institute Insect Science, Zhejiang University, Hangzhou 310058, China; yglou@zju.edu.cn
* Correspondence: liqin@xju.edu.cn
† These authors contributed equally to this work.

Abstract: The flea-weevil *Orchestes steppensis* Korotyaev (Coleoptera: Curculionidae) is an Eastern Palaearctic Steppe species, and a serious pest of elm trees (*Ulmus* spp., Ulmaceae) by feeding on the leaves (adults) or mining them heavily (larvae) in Xinjiang, China. In order to search for chemical and ecological pest management practices, the olfactory preferences of *O. steppensis* for leaves of three elm species were investigated. The results revealed that *O. steppensis* has different host preferences for the three elm species: *U. pumila* L. first, followed by *Ulmus laevis* Pall. and *Ulmus densa* Litw. last. Volatile organic compounds from the leaves of the three *Ulmus* species were collected using dynamic headspace adsorption and analyzed through gas chromatography–mass spectrometry (GC-MS). A total of 94 volatile components in the healthy leaves and the infested leaves by *O. steppensis* of the three elm species were identified and analyzed, and 13 active compounds were identified using coupled gas chromatography–electroantennographic (GC-EAD) recording and GC-MS analysis. The response of *O. steppensis* to different concentrations of active compounds was determined using EAG, and the behavioral response to the highest EAG concentration of each active compound was determined. The results demonstrated that 3-hexen-1-ol, 3,7-dimethyl-1,3,6-Octatriene, methyl salicylate, 1-hexanol, and 3-hexen-1-ol, acetate were attractive to *O. steppensis*, while nonanal and 2-hexenal were repellent.

Keywords: *O. steppensis*; chemical ecology; host preference; volatile compound; GC-EAD; EAG; *Ulmus*

1. Introduction

Ulmus spp. (Ulmaceae) are drought resistant, cold tolerant, and adaptable, the characteristics that play a very important role in windbreak plantations and sand fixations in desert and semi-desert environments and urban landscapes in Xinjiang (Xinjiang Uyghur Autonomous Region of China) [1,2]. Therefore, they are widely distributed and have also been planted extensively in many urban and rural areas of this region. However, they have been seriously damaged by *Orchestes steppensis* Korotyaev, 2016 (Coleoptera: Curculionidae) in recent years. In 2022, the occurrence area of *O. steppensis* in Hutubi County and Changji City reached 2000 hm^2 [3]. A large number of pesticides are used to control this pest, but there is little research on chemical ecological control [4]. It is reported that host volatile organic compounds (VOCs) play an important role in monitoring and trapping pest populations, which have been used as chemical and ecological pest management practices for control of Thripidae and Scarabaeidae pests [5,6].

O. steppensis is an oligophagous insect with a body length of 2.6~3.2 mm and feeds on the leaves of *Ulmus* spp. (adults) or mines them heavily (larvae), thus making the trees incapable of vigorous growth, and even causing defoliation and death of the lower, small branches [2]. It is widely spread across China, Kazakhstan, Mongolia, North America, and Russia [2,7–9]. In China, this pest is distributed in Anhui, Beijing, Gansu, Hebei, Jiangsu, Jilin, Liaoning, Nei Mongol, Ningxia, Shaanxi, Shandong, Shanghai, Tianjin, and Xinjiang [2,10,11]. In Xinjiang, *O. steppensis* completes one generation per year. The overwintering adults emerge from hibernation from late March to early April. After mating, they lay eggs along the main veins of elm leaves. The eggs develop until mid-May when the new generation begins to emerge as adults [12]. The degree of damage of *O. steppensis* to the three elm species is different; the most severe damage occurs to *Ulmus pumila* L., followed by *Ulmus densa* Litw., and finally *Ulmus laevis* Pall. [2].

Insect antennae have tactile and olfactory functions used for searching for host plants and avoiding natural enemies [13]. Herbivorous insects can search for their host for feeding or egg-laying by sensing the VOCs of their host plants [14,15]. GC-EAD, EAG, and olfactory behavioral assays are commonly used to determine the effects of chemicals on the biological activity of insects; GC-EAD and EAG initially screen for active substances and their optimal concentrations, and olfactometers are used to evaluate the biological effects of these active substances and concentrations [16,17].

The aims of our study were to detect whether the host preference of *O. steppensis* to the three elm species was affected by the VOCs of the elm leaves firstly and then to study the VOCs in the healthy leaves and the herbivore-induced plant volatiles (HIPVs) infested by *O. steppensis* to make clear the specific VOCs in the elm leaves attractive or repellent to this weevil pest. Finally, the biological activity of the identified compounds was tested in the indoor bioassay.

2. Materials and Methods

2.1. Insects and Plants

Adults of *O. steppensis* were collected in the field by net sweeping from April to October 2022~2023 in Urumqi City, Xinjiang, China. Live weevils were placed in an intelligent artificial climate chamber (25 ± 1 °C, 45% ± 5%RH, photoperiod 13L:11D) (BD-PRX-150B; Nanjing Beidi Experimental Instrument Co., Ltd., Nanjing, China). They were fed with fresh *U. pumila* leaves, and the leaves were replaced every 7 d. Three elm (*Ulmus*) species, *U. pumila*, *U. densa*, and *U. laevis*, planted at Xinjiang University were the test plants.

2.2. Tested Chemicals

The chemicals used in the experiments are listed in Table 1. The glass materials used during the experiments were heated 1 h at 99.9 °C before and after being used for experimental purposes.

Table 1. List of the chemicals used in the experiments tested.

No.	Compounds	CAS Number	Purity	Test	Company
1	2-hexenal	6728-26-3	98%	EAG, Y-tube experiments	Macklin [1]
2	3,7-dimethyl-1,3,6-octatriene	13877-91-3	≥98%	EAG, Y-tube experiments	Macklin [1]
3	3-hexen-1-ol, acetate	3681-71-8	98%	EAG, Y-tube experiments	Macklin [1]
4	1-hexanol	111-27-3	99%	EAG, Y-tube experiments	Macklin [1]
5	3-hexen-1-ol	928-96-1	98%	EAG, Y-tube experiments	Macklin [1]
6	nonanal	124-19-6	96%	EAG, Y-tube experiments	Macklin [1]
7	decanoic acid, propyl ester	30673-60-0	98%	EAG, Y-tube experiments	Adamas [2]
8	methyl salicylate	119-36-8	AR, ≥99%	EAG, Y-tube experiments	Macklin [1]
9	benzoic acid, 2-methylpropyl ester	120-50-3	99%	EAG, Y-tube experiments	Macklin [1]
10	guaiacol	90-05-1	GC, >99%	EAG, Y-tube experiments	Macklin [1]

Table 1. *Cont.*

No.	Compounds	CAS Number	Purity	Test	Company
11	benzyl alcohol	100-51-6	AR, ≥99%	EAG, Y-tube experiments	Macklin [1]
12	phenylethyl alcohol	60-12-8	AR	EAG, Y-tube experiments	Macklin [1]
13	2,4-di-tert-butylphenol	96-76-4	97%	EAG, Y-tube experiments	Macklin [1]
14	paraffin liquid	8012-95-1	AR	EAG, Y-tube experiments	Zhi Yuan [3]
15	ethanol absolute	64-17-5	AR, 99.7%	EAG, Y-tube experiments	Hu Shi [4]
16	n-Hexane	110-54-3	HPLC	Collection of plant volatiles	Da Mao [5]
17	dichloromethane	75-09-2	HPLC	Collection of plant volatiles	Da Mao [5]
18	methyl benzoate	93-58-3	HPLC	Detection of plant volatiles	Xi Ya [6]

[1] Macklin: Shanghai, China; [2] Adamas: Shanghai, China; [3] Zhi Yuan: Tianjin, China; [4] Hu Shi: Shanghai, China; [5] Da Mao: Tianjin, China; [6] Xi Ya: Shangdong, China.

2.3. Selection Preferences of the *O. steppensis* to Leaves of Three Elm Species

A four-arm olfactometer (XLM4-150; Nanjing Xuelai Biotechnology Co., Ltd., Nanjing, China) was used to test the host preference of *O. steppensis* for leaves of three elm species [18–20]. Adults were starved for 24 h prior to the experiment, and each weevil adult was tested only once. A female or male adult was chosen at random for the test, and the sexes can be differentiated using morphological characteristics of the genitalia [8,21]. The selection preferences of *O. steppensis* for air (control) and leaves (experimental) of *U. pumila*, *U. laevis*, and *U. densa*, respectively, were first tested. Two upper taste channels of the four-armed olfactometer were closed, and two lower taste channels were open for the test. The experimental channel was connected to 3.0 g fresh, healthy elm leaves as the odor source; the control channel was filled with clean air. Fifty adults of *O. steppensis* per group were tested, and for accuracy, the tests were repeated three times. Secondly, the selection preference of *O. steppensis* of three elm leaves was further tested. Three channels were connected to 3.0 g three elm leaves as the odor source, and one channel was connected to the clean air; 70 adults were tested in each group, and similarly, the tests were conducted in triplicate.

The diameter of the four-armed olfactometer was 150 mm, the length of each odorant channel was 15 cm, and the angle between the two channels was 90°. The gas flow rate of each odorant channel was adjusted to 400 mL/min before the measurement, and after 5 min of aeration, the adult insects were introduced into the insect hole at the top of the four-armed olfactometer and timed when they started to crawl. If the insect crawled through 2/3 of any odorant channel (10 cm from the insect hole) and stayed for more than 30 s or passed through the odorant channel directly within 10 min, it was regarded that the individual chose the substance connected to this odorant channel, and the time when the individual reached the bottle of the odorant source was recorded subsequently; otherwise, it was regarded as unchosen and was not included in the statistical analysis. We found that *O. steppensis* has upward convergence, light convergence, and strong aggregation; therefore, in order to avoid the influence on the experimental results, the following measures need to be taken: firstly, the four-armed olfactometer was placed horizontally, and the activity room and the odorant source bottle were darkened by using a visor, in order to avoid the influence of the light angle and the color of the plant in the bottle on the results of the experiments; secondly, for every 5 insects tested, the four-armed olfactometer needed to be rotated clockwise by 90° to avoid the influence of positional effects; for every 20 insects tested, the testing area was cleaned with anhydrous ethanol and distilled water to avoid the influence of insect source pheromones.

2.4. Elm Leaf Volatiles Collection and Analysis

The headspace dynamic technique of Wu et al. (2022) was used to collect elm leaf volatiles with minor modifications [20,22,23]. The microwave oven bags (oven bags,

43 cm × 55 cm, Reynolds, ND, USA) were used to enclose healthy or insect-infested elm branches and sealed. The leaves were not removed from the plants. The original air in the bag was evacuated with an atmospheric sampler, and fresh air was pumped into the bag through an activated carbon tube (5 mm × 8 mm, 200 mm glass tube filled with 1.5 g of 20~40 mesh activated carbon adsorbent) until the bag was full of air. A glass adsorption column with 50 mg of Porapak Q (80–100 mesh) between glass wool plugs at both ends was then attached at the outlet. After 4 h of continuous collection by airflow (1 L/min) through the activated carbon tube, volatiles adsorbed on the Porapak Q adsorbent were eluted with 1 mL of dichloromethane and then 1 mL of n-hexane, respectively. After the elution, it was transferred into a 2 mL Agilent vial and stored in a refrigerator at −20 °C for the use of GC-MS and GC-EAD. The methyl benzoate (100 ng/μL) was used as an internal standard. HIPVs by *O. steppensis* were collected by inoculating 100 *O. steppensis* individuals in the healthy volatiles experiment, allowing them to feed for 24 h, removing the insects, and collecting them again in bags using the method described above. A blank control was set up to prevent the gas components in the bag or in the air from influencing the experiment. The collection of volatiles was repeated three times.

Healthy volatile organic compounds of the three elm species and the HIPVs infested by *O. steppensis* were analyzed using a gas chromatography–mass spectrometer (Agilent, Santa Clara, CA, USA, GC 7890B/MS 5977A). The volatiles were transferred in splitless mode to the analytical column (DB-WAX: 60 m × 0.25 mm ID, 0.25 μm film) placed in the GC oven. The GC oven temperature was initially held at 40 °C for 0.5 min and raised at 10 °C/min to 150 °C and held for 8 min, then 20 °C/min to 240 °C and held for 8 min. At 70 eV, EI-mass spectra were acquired while scanning from m/z 45 to 650 at a rate of 2.40 scans s^{-1}. The MS transfer line and ion source were set to 250 °C and 230 °C, respectively. Identification of healthy VOCs and HIPVs of the three elm species was based on a comparison of mass spectra with those reported in the Agilent MassHunter NIST 14.0 library. Categories of the compounds were determined by referring to [24] and the key functional groups in the compounds. The retention index (RI) of the compounds was calculated by I = 100n + 100 × {[logt'R(X) − logt'R(n)]\[logt'R(n + 1) − logt'R(n)]}. The retention times of normal alkanes of C7-C40 as the standard substance were conducted using the same procedure as the compounds for calculation of the RI of the compounds. Quantitative analysis was performed using the internal standard method to calculate the relative content of each compound. The relative amount of internal standard in each sample could be calculated from the volume of the eluate and its concentration. Each compound in the sample was finally expressed as ng/μL.

Relative content of the tested compound = Chromatographic peak area of the tested compound × Relative content of the internal standard/Chromatographic peak area of the internal standard.

2.5. *Electrophysiological Responses of O. steppensis Antennae to Elm Leaf Volatiles by GC-EAD*

GC-EAD was used to test the electrophysiological responses of *O. steppensis* antennae to elm leaf volatiles [25,26]. The conditions for gas chromatography were the same as those described above for GC-MS. EAG (SYNTECH, Christinenthal, Germany) consists of the stimulus controller CS-55, the MP-15 micromanipulator, the 2-channel programmable signal acquisition controller IDAC-2, the GC/EAD interface temperature controller TC-02, and EC-03 gas phase shunt material heating system. The outlet of the GC column was split at a 1:1 ratio with a universal Y-splitter (Agilent) between the FID and EAD. Healthy *O. steppensis* females and males were taken after 24 h of starvation, and the antennae were removed with a scalpel along the length of about 50 μm from the base and the end of the antennae under the stereomicroscope. The treated antennae were then connected to two saline-filled glass electrodes connected to silver wires at both ends of the tentacle potentiostat. The base of the antenna is connected to the reference electrode, and the end of the antenna is connected to the recording electrode; 2 μL of collected elute of elm leaf volatile compounds were manually injected, and the split method was used for non-split injecting, respectively, using antennae taken from different individual insects.

2.6. EAG Responses

The preparation of antennae in the EAG experiment was the same as in the GC-EAD experiment described above. We used standards of compounds that can trigger electrophysiological phenomena in antennae for the GC-EAD experiments. The compounds 2-hexenal, 3,7-dimethyl-1,3,6-octatriene, 3-hexen-1-ol-acetate, 1-hexanol, 3-hexen-1-ol, nonanal, decanoic acid propyl ester, methyl salicylate, 2-methylpropyl ester benzoic acid, guaiacol, benzyl alcohol, phenylethyl alcohol, and 2,4-di-tert-butylphenol were diluted in paraffin liquid to five different concentrations (0.01, 0.1, 1, 10, and 100 mg/mL) for EAG and Y-tube olfactometer behavioral assays. All compounds were obtained from commercial suppliers (Table 1). We did not use all standard compounds because some were not available at the time of our study; 10 μL of the sample were taken to be tested and were dropped evenly on a 1 cm × 3 cm strip of filter paper, and then the solvent was allowed to evaporate for 10 s before it was placed in a glass Pasteur tube as a source of stimulation. The EAG values of the samples were corrected using solvent liquid paraffin as a blank control. The samples were tested in order from low to high concentration. After the baseline had stabilized, the test was started. Each stimulation time was 0.5 s, the stimulation interval was 60 s, and the flow rate of moist air was 1 L/min. Four repetitions were made in the case of males and females of *O. steppensis*, respectively, using antennae taken from different individual insects.

2.7. Y-Tube Olfactometer Behavioral Assays

The behavioral responses of *O. steppensis* to the composition of volatile compounds from the host elm leaves were investigated in a Y-tube olfactometer (stem and arms: 15 cm long, respectively, 1 cm internal diam., and 30°Y angle) [25,27]. Behavioral responses of adult *O. steppensis* individuals to 13 GC-EAD experiments with biologically active compounds and concentrations at the highest EAG response values were examined. Filter paper strips dripped with 10 μL of volatile solution were placed at the end of one side of the flavor source, and filter paper strips infused with the same volume of liquid paraffin were placed at the end of the other side of the flavor source as a control. A single healthy adult *O. steppensis*, starved for 24 h, was inserted into the mouth of the main arm tube and left to crawl to start timing. Entering 2/3 of the side arm of the odor source within 5 min and staying for more than 30 s was regarded as selecting the odor source of that side arm, and entering 2/3 of the control side arm and staying for more than 30 s was regarded as selecting the control; if the test bug stayed in the main arm or did not exceed 2/3 of the side wall within 5 min, it was recorded as not selecting for the volatile. Fifty *O. steppensis* were tested for each test solution.

2.8. Statistical Analysis

All statistical analyses of data were performed using SPSS 20.0 (IBM). In the selection preferences of the *O. steppensis* to leaves of the three elm species experiment, differences in the number of choices of *O. steppensis* for the same elm leaf and the time to reach the same flavor source bottle between male and female adults of *O. steppensis* were determined by *t*-tests, whereas differences in the number of choices of male and female adults of *O. steppensis* between the experimental groups of *U. pumila*, *U. densa*, and *U. laevis* and the control group were determined by chi-square tests; differences in selection number and selection time of the same sex of *O. steppensis* adults to three elm species and also to the clean air were conducted by one-way ANOVA and Duncan's multiple comparisons. In the experiment of EAG responses of female and male adults of *O. steppensis* to 13 compounds, differences in the EAG relative values to the same compound at the same concentration between male and female adults of this weevil were determined by *t*-tests, whereas differences in the EAG relative values of male or female adults of *O. steppensis* to different concentration of the same compound were conducted by one-way ANOVA and Duncan's multiple comparisons. The behavioral response of *O. steppensis* was analyzed using the chi-square test.

3. Results

3.1. Selection Preferences of O. steppensis for Leaves of Its Three Elm Species Host

The results of the study found that the number of choices of *O. steppensis* for the experimental groups of *U. pumila* ($x^2 = 42.32$, $p < 0.01$), *U. densa* ($x^2 = 23.92$, $p < 0.01$), and *U. laevis* ($x^2 = 29.04$, $p < 0.01$) were all extremely significantly higher than those of the control group (Figure 1), and that the leaves of three species of elm, *U. pumila, U. densa,* and *U. laevis,* had an attraction effect on the *O. steppensis*.

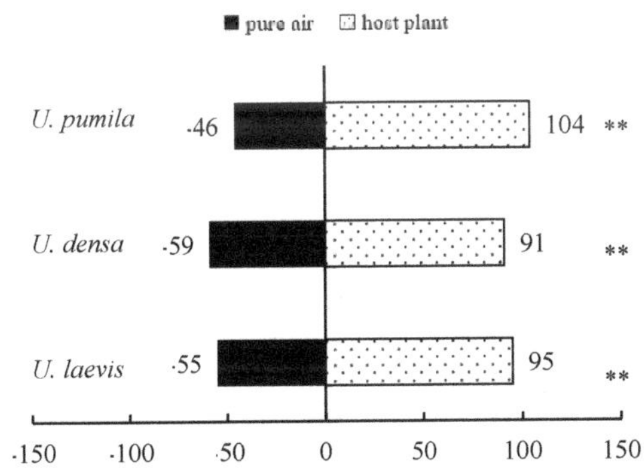

Figure 1. Olfactory responses of *O. steppensis* to three species host plant leaves. Note: ** indicates a significant difference in the level of $p < 0.01$ between the experiment and the control determined by the x^2 test.

There was no significant difference in the reaction time required by adult *O. steppensis* to reach the flavor source between the experimental group and its control group in the three experiments ($p = 0.52 > 0.05$; $p = 0.42 > 0.05$; $p = 0.75 > 0.05$). There was no significant difference ($p = 0.28 > 0.05$; $p = 0.13 > 0.05$; $p = 0.83 > 0.05$; $p = 0.83 > 0.05$; $p = 0.05 = 0.05$; $p = 0.51 > 0.05$) in the time taken by male and female adults to reach the same flavor source bottle (Table 2).

Table 2. The average time for *O. steppensis* to arrive at the flavor bottles of three species host plant leaves: mean ± SEM (s).

	Adult Sex	*U. pumila*	*U. densa*	*U. laevis*
test	female	159.62 ± 15.18	207.35 ± 17.34	215.11 ± 3.86
	male	147.13 ± 11.95	194.89 ± 15.71	177.07 ± 11.00
control	female	144.04 ± 10.33	187.24 ± 9.54	203.51 ± 7.93
	male	184.61 ± 23.79	203.33 ± 25.74	182.04 ± 19.33

The results of the olfactory behavioral responses of *O. steppensis* to three species of elm are shown in Figure 2A, indicating that there were differences in the number of choices made by *O. steppensis* to the leaves of different species. *O. steppensis* females showed the highest number of selections for *U. pumila,* followed by *U. laevis* and finally, *U. densa* ($F = 196.00$, $p = 0.00 < 0.01$); males showed the highest number of selections for *U. pumila* followed by *U. laevis* and *U. densa* ($F = 359.56$, $p = 0.00 < 0.01$). There was no significant difference in the number of selections made by *O. steppensis* males and females for the same flavor source ($p = 0.35 > 0.05$). The results indicate that *O. steppensis* preferences for leaves of different elm species are different, and the preferences are, in descending order, *U. pumila* > *U. laevis* > *U. densa* > blank.

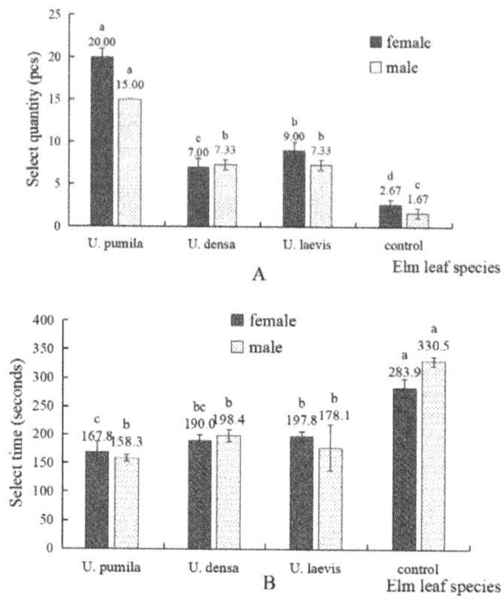

Figure 2. Number of selections of *O. steppensis* adults (**A**) and selection time (**B**) towards the three species of *Ulmus* and the control group. Note: Different small letters indicate that the number or time of selection of different flavor source bottles by the weevil adults of the same sex is significantly different at the $p < 0.05$ level and the difference was determined by one-way ANOVA and Duncan's multiple comparisons.

The selection time of *O. steppensis* to three *Ulmus* species is shown in Figure 2B. Among them, *O. steppensis* had the shortest selection time for *U. pumila*, followed by *U. laevis* and finally *U. densa* ($F = 104.33$, $p = 0.00 < 0.01$). The results showed that there was no significant difference between the selection time of male and female *O. steppensis* for the same flavor source ($p = 0.73 > 0.05$).

3.2. GC-MS of Elm Leaf Volatiles

A total of 94 compounds were detected by GC-MS in leaf volatiles of three elm species before and after damage by the *O. steppensis*, as shown in Table 3. A total of 10 categories, including 4 benzenes, 21 alcohols, 3 phenols, 4 ethers, 10 aldehydes, 12 acids, 8 terpenes, 3 ketones, 6 alkanes, and 23 ester compounds, were detected. After *U. pumila* was fed on by *O. steppensis*, the percentage of aldehydes increased from 10.36% to 14.67%; the percentage of acids increased from 25.5% to 40.2%; the percentage of terpenes increased from 3.57% to 3.83%; and the percentage of ketones increased from 0.34% to 0.73%. After *U. densa* elm was fed on by *O. steppensis*, the percentage of ether content increased from 0% to 0.47%; the percentage of acid content increased from 31.71% to 38.54%; the percentage of alkane content increased from 4.84% to 5.93%; and the percentage of ester content increased from 13.90% to 15.73%. After *U. laevis* was harmed by *O. steppensis*, the proportion of benzene content increased from 1.64% to 2.51%; the proportion of alcohols and hydrocarbons increased from 12.45% to 21.15%; the proportion of ethers increased from 0.69% to 0.71%; the proportion of aldehydes increased from 1.50% to 12.94%; the proportion of terpenoids increased from 1.38% to 6.97%; the proportion of ketones increased from 0% to 0.36%; and alkane content increased from 1.87% to 3.00%.

Table 3. Identified volatile components released from two leaf states of three elm species and their Quantities (ng/μL) relative to the internal standards: mean ± SEM.

No.	Retention Time (min)	Retention Index (RI)	Compound	Categories	*U. pumila* Healthy	Insect-Infested	*U. densa* Healthy	Insect-Infested	*U. laevis* Healthy	Insect-Infested
1	7.595	974	glyceric acid	Acid	-	-	-	0.43 ± 0.04 a*	0.03 ± 0 A*	-
2	7.976	1000	glycolaldehyde dimethyl acetal	Aldehyde	-	-	-	-	-	0.63 ± 0.05 a*
3	8.202	1014	toluene	Benzene	0.29 ± 0.06 A	0.33 ± 0.02 b	-	-	-	0.68 ± 0.12 a*
4	8.547	1036	3-methylcyclopentanol	Alcohol	0.2 ± 0.06 A	0.13 ± 0.01 b	-	-	0.07 ± 0.01 A	0.68 ± 0.04 a*
5	9.722	1105	p-xylene	Benzene	0.08 ± 0.01 A	0.35 ± 0.06 a*	-	-	-	0.27 ± 0.04 a*
6	9.903	1116	bicyclo[2.1.1]hexan-2-ol, 2-ethenyl-	Alcohol	0.21 ± 0.06 A	0.3 ± 0.02 c	-	1.86 ± 0.21 a*	0.02 ± 0 B	0.97 ± 0.14 b*
7	9.948	1119	trans-1(7),8-p-menthadien-2-ol	Alcohol	-	-	-	-	0.02 ± 0 A*	-
8	10.432	1147	heptanal	Aldehyde	0.7 ± 0.06 A	1.33 ± 0.24 a	0.75 ± 0.1 A	1.65 ± 0.01 a*	-	1.22 ± 0.09 a*
9	10.528	1152	cyclopentanone, 2-(1-methylpropyl)-	Ketone	0.16 ± 0.02 A	0.27 ± 0.1 a	-	-	-	-
10	10.722	1163	2-hexanol	Alcohol	0.21 ± 0 A	0.28 ± 0.04 b	-	-	-	1.21 ± 0.04 a*
11	10.727	1164	(+)-dipentene	Terpenoid	-	-	-	-	0.18 ± 0.01 A*	-
12	11.003	1179	2-hexenal	Aldehyde	1.37 ± 0.18 A	4.36 ± 0.76 a*	2.12 ± 0.2 A	5.94 ± 0.21 a*	-	1.77 ± 0.02 b*
13	11.172	1188	3,7,11-trimethyl-1-dodecanol	Alcohol	0.14 ± 0.04 A	0.19 ± 0.01 b	-	-	-	1.57 ± 0.02 a*
14	11.261	1193	1,3,6-octatriene, 3,7-dimethyl-	Terpenoid	0.25 ± 0.07 A	1.77 ± 0.02 a*	-	-	-	-
15	11.776	1223	tetradecane	Alkane	0.17 ± 0.03 A*	-	-	-	-	-
16	12.075	1241	2,3,5,8-tetramethyldecane	Alkane	0.19 ± 0.04 A*	-	-	-	-	-
17	12.079	1241	vanillin	Aldehyde	-	-	-	-	-	1.21 ± 0.11 a*
18	12.105	1242	cis-2-penten-1-ol	Alcohols	-	1.57 ± 0.02 a*	-	-	-	-
19	12.153	1245	3-hexen-1-ol, acetate(z)	Esters	1.52 ± 0.16 A*	1.06 ± 0.02 a	-	-	3.25 ± 0.75 A*	-
20	12.461	1263	(z)-hex-2-enyl acetate	Esters	0.26 ± 0.05 A	0.49 ± 0.04 a*	-	-	-	-
21	12.673	1275	1-hexanol	Alcohols	0.4 ± 0.06 A	1.99 ± 0.11 a*	-	-	-	-
22	13.217	1305	3-hexen-1-ol, (z)-	Alcohols	4.49 ± 0.54 B	4.31 ± 0.31 c	10.87 ± 1.09 A	23.93 ± 0.62 a*	2.56 ± 0.56 B	16.82 ± 0.36 b*
23	13.413	1317	nonanal	Aldehyde	0.39 ± 0.11 B	1.88 ± 0.37 c*	6.17 ± 1.15 A	13.38 ± 0.54 a*	0.31 ± 0.05 B	7.9 ± 0.95 b*
24	13.650	1331	citronellol	Terpenoid	-	-	-	-	0.3 ± 0.08 A	6.63 ± 0.49 a*
25	14.205	1363	2,6,10-trimethyldodecane	Alkane	0.69 ± 0.04 A*	0.38 ± 0.04 b	-	-	0.26 ± 0.04 B	1.91 ± 0.44 a*
26	14.282	1368	2,6,10-trimethyltridecane	Alkane	-	-	-	-	-	2.27 ± 0.14 a*
27	14.546	1382	trans-2-ethyl-2-hexen-1-ol	Alcohol	-	0.25 ± 0.04 a*	-	-	-	-
28	14.733	1393	1,2-15,16-diepoxyhexadecane	Ether	-	-	-	-	-	0.38 ± 0.05 a*
29	14.763	1394	2,3-dimethylcyclohexanol	Alcohol	0.28 ± 0.07 A*	-	-	-	-	-
30	15.840	1452	linalool	Terpenoid	-	0.2 ± 0.01 a*	-	-	-	-
31	16.171	1469	benzaldehyde	Aldehyde	0.62 ± 0.04 A	2.09 ± 0.1 a*	-	-	-	2.31 ± 0.2 a*
32	18.162	1558	geranyl isovalerate	Ester	0.73 ± 0.06 A*	0.3 ± 0.05 b	-	-	-	2.74 ± 0.44 a*
33	18.554	1574	nonanoic acid, propyl ester	Ester	0.57 ± 0.11 A	0.64 ± 0.01 b	-	2.01 ± 0.33 a*	0.38 ± 0.08 A	2.61 ± 0.01 a*
34	19.081	1595	benzeneacetaldehyde	Aldehyde	1.28 ± 0.07 B	7.45 ± 0.54 a*	2.74 ± 0.06 A*	0.44 ± 0.13 c	0.21 ± 0.01 C	2.08 ± 0.23 b*
35	19.353	1605	(e)-2-hexenyl benzoate	Ester	0.41 ± 0.03 A	-	-	-	0.08 ± 0.01 B	1.62 ± 0.09 a*
36	19.714	1617	benzoicacid, 1-methylethylester	Ester	-	-	-	3.27 ± 0.47 a*	-	-
37	19.887	1623	benzoic acid, ethyl ester	Ester	0.96 ± 0.1 B	2.29 ± 0.19 b*	3.69 ± 0.59 A	7 ± 0.87 a*	0.85 ± 0.03 B	3.73 ± 0.19 b*

Table 3. *Cont.*

No.	Retention Time (min)	Retention Index (RI)	Compound	Categories	*U. pumila* Healthy	Insect-Infested	*U. densa* Healthy	Insect-Infested	*U. laevis* Healthy	Insect-Infested
38	20.068	1628	radicinin	Ketone	-	-	-	-	-	0.31 ± 0.01 a*
39	20.521	1643	patchouli alcohol	Terpenoid	-	0.42 ± 0.04 a*	-	-	-	-
40	21.060	1660	tridecanedial	Aldehyde	0.54 ± 0.04 A*	-	-	-	-	-
41	21.080	1660	palmitic acid	Acid	-	-	-	-	0.21 ± 0.01 A*	-
42	21.160	1663	decanoic acid propyl ester	Ester	0.62 ± 0.05 A*	0.46 ± 0.02 c	-	1.17 ± 0.11 b*	0.24 ± 0.05 B	1.78 ± 0.14 a*
43	21.212	1664	7-epi-cis-sesquisabinene hydrate	Alcohol	-	-	-	-	-	1.38 ± 0.32 a*
44	21.392	1670	β-pinene	Terpenoid	-	0.28 ± 0.07 a*	-	-	-	-
45	21.460	1672	e-3-pentadecen-2-ol	Alcohol	-	-	-	-	-	2.5 ± 0.79 a*
46	22.404	1700	methyl salicylate	Ester	0.38 ± 0.14 AB	0.41 ± 0.07 c	0.67 ± 0.05 A	1.59 ± 0.07 b*	0.1 ± 0.01 C	3.25 ± 0.52 a*
47	22.542	1703	benzoicacid,2-methylpropylester	Ester	0.55 ± 0.07 B	2 ± 0.03 b	1.92 ± 0.08 A	4.03 ± 0.24 a	0.52 ± 0.03 B	2.17 ± 0.02 b*
48	22.615	1705	4-ethyl-4-methyl-2-pentadecyl-1,3-dioxolane	Ether	-	-	-	-	-	1.38 ± 0.11 a*
49	22.620	1705	2-pentadecyl-1,3-dioxocane	Ether	0.2 ± 0.05 A*	-	-	0.9 ± 0.33 a*	0.24 ± 0.01 A*	-
50	22.957	1714	o-toluylic acid, 2-butyl ester	Esters	0.37 ± 0.04 A	0.84 ± 0.04 a*	-	-	0.16 ± 0.01 B	1.35 ± 0.23 a*
51	23.117	1718	dipropyl succinate	Esters	0.56 ± 0.05 B	1.08 ± 0.03 b*	1.67 ± 0.16 A*	-	0.41 ± 0.06 B	1.66 ± 0.1 a*
52	23.402	1725	guaiacol	Phenol	0.31 ± 0.16 A	1.53 ± 0.2 a*	-	-	-	1.72 ± 0.28 a*
53	23.485	1727	benzoic acid, 4-methyl-, propyl ester	Ester	-	1.12 ± 0.28 a*	-	-	-	-
54	23.550	1728	phenylacetic acid propyl ester	Ester	3.01 ± 0.35 A	7.03 ± 0.73 b*	4.7 ± 0.66 A	9.34 ± 0.73 b*	3.36 ± 0.55 A	42.54 ± 3.37 a*
55	23.610	1730	m-toluylic acid, 2-butyl ester	Ester	0.24 ± 0.08 B	0.84 ± 0.24 a	-	1.89 ± 0.29 a*	0.55 ± 0.06 A	4.38 ± 1.8 a
56	23.687	1732	benzyl alcohol	Alcohol	2.12 ± 0.32 B	2.71 ± 0.18 b	13.47 ± 1.18 A	11.38 ± 0.59 a	1.66 ± 0.26 B	11.6 ± 2.06 a*
57	23.774	1734	pentadecanal	Aldehyde	-	-	-	-	-	0.19 ± 0.05 a*
58	23.847	1736	3-methylene-1-oxa-spiro[4.5]decan-2-one	Ketone	-	0.58 ± 0.02 a*	-	-	-	0.59 ± 0.21 a*
59	24.056	1741	phenylethyl alcohol	Alcohol	0.65 ± 0.06 A	2.59 ± 0.47 b*	-	-	-	8.91 ± 0.46 a*
60	24.370	1748	4-phenyltoluene	Benzene	1.24 ± 0.08 B	2.73 ± 0.17 b*	3.21 ± 0.32 A	5.25 ± 0.1 a*	0.57 ± 0.04 C	5.25 ± 0.86 a*
61	24.545	1752	2-hexadecanol	Alcohol	0.32 ± 0.02 A	1.47 ± 0.05 b*	-	3.11 ± 0.32 a*	-	1.78 ± 0.36 b*
62	24.598	1754	myristic acid	Acid	-	-	-	-	0.48 ± 0.05 A*	-
63	24.644	1755	17-octadecynoic acid	Acid	-	0.63 ± 0.13 a*	-	-	-	-
64	24.654	1755	13-tetradecynoic acid methyl ester	Ester	0.19 ± 0.01 A*	-	-	-	-	-
65	24.909	1761	fumaric acid, ethyl 2-methylallyl ester	Ester	-	-	-	-	-	4.54 ± 0.3 a*
66	24.995	1763	octadecanoic acid	Acid	-	-	-	7.5 ± 1.22 a*	-	-
67	25.233	1768	methyl myristate	Ester	0.36 ± 0.02 A	1.04 ± 0.07 b*	-	-	-	2.77 ± 0.24 a*
68	25.290	1770	2-pentyl-3-methyl-2-cyclopenten-1-one	Terpenoid	0.92 ± 0.07 A	1.02 ± 0.08 a	-	-	-	1.85 ± 0.36 a*
69	25.540	1775	1-hexadecanol	Alcohol	0.27 ± 0.07 A	0.62 ± 0.15 a	-	-	-	-
70	25.576	1776	2-pentadecanone, 6,10,14-trimethyl-	Terpenoid	0.52 ± 0.05 A	0.78 ± 0.08 b	-	-	-	8.72 ± 0.72 a*
71	25.654	1778	13-heptadecyn-1-ol	Alcohol	-	-	-	-	-	1.33 ± 0.41 a*
72	25.660	1778	elaidic acid	Acid	-	-	-	-	0.24 ± 0.03 A*	-
73	25.763	1780	trans-13-octadecenoic acid	Acid	-	-	-	-	0.45 ± 0.01 A*	-

Table 3. *Cont.*

No.	Retention Time (min)	Retention Index (RI)	Compound	Categories	*U. pumila* Healthy	Insect-Infested	*U. densa* Healthy	Insect-Infested	*U. laevis* Healthy	Insect-Infested
74	26.189	1790	7-hexadecyn-1-ol	Alcohol	-	0.19 ± 0.01 a*	-	-	-	-
75	26.551	1798	biphenyl	Benzene	-	-	0.7 ± 0.06 A*	-	-	-
76	26.693	1802	5-octadecenal	Aldehyde	-	-	-	-	-	14.61 ± 3.7 a*
77	26.713	1802	cis-10-heptadecenoic acid	Acid	-	-	-	-	1.16 ± 0.14 A*	-
78	26.819	1807	cis-3-hexenyl benzoate	Ester	-	-	-	-	0.76 ± 0.15 A*	-
79	26.887	1809	1-hexadecanol,2-methyl-	Alcohol	-	0.19 ± 0.01 b*	-	0.9 ± 0.07 b*	-	3.41 ± 1.18 a*
80	26.909	1810	9,12,15-octadecatrienoic acid, (z,z,z)-	Acid	-	-	3.38 ± 0.49 A*	-	-	-
81	27.148	1820	oleic acid	Acid	0.56 ± 0.04 A	0.94 ± 0.07 b*	-	-	0.57 ± 0.11 A	5.12 ± 0.04 a*
82	27.280	1825	2-methyl-4-(2,6,6-trimethylcyclohex-1-enyl)but-2-en-1-ol	Alcohol	-	0.18 ± 0.02 a*	-	-	-	-
83	27.755	1844	hexadecanoic acid, methyl ester	Ester	1.22 ± 0.17 A*	0.34 ± 0.03 b	0.64 ± 0.09 B*	-	1.13 ± 0.15 A*	0.5 ± 0.01 a
84	28.121	1858	2-methoxy-5-[(e)-1-propenyl]phenol	Phenol	0.22 ± 0.03 A	0.42 ± 0.02 a*	-	-	-	-
85	28.164	1860	(z)-9-hexadecenoic acid methyl ester	Ester	-	-	-	-	0.05 ± 0.01 A*	-
86	28.299	1865	2,6,10-trimethyltetradecane	Alkane	0.47 ± 0.08 A*	-	-	-	0.21 ± 0.06 A	2.53 ± 0.52 a*
87	28.480	1872	3-ethoxy-3,7-dimethyl-1,6-octadiene	Ether	-	0.18 ± 0.02 a*	-	-	-	-
88	28.650	1879	2,4-di-tert-butylphenol	Phenol	1.89 ± 0.05 B	3.04 ± 1.06 b	7.33 ± 0.17 A	7.97 ± 0.51 a	1.43 ± 0.33 B	8.24 ± 0.08 a*
89	28.895	1888	methyl 14-methylhexadecanoate	Ester	0.37 ± 0.06 A*	-	-	-	-	-
90	29.046	1893	1,2-propnediol,3-(tetradecyloxy)	Alcohol	-	0.43 ± 0.07 a*	-	-	-	-
91	30.016	1946	linoleic acid	Acid	-	-	1.04 ± 0.01 A	3.44 ± 0.4 a*	1.07 ± 0.03 A*	-
92	30.296	1962	methyl stearate	Ester	0.11 ± 0.01 A	0.15 ± 0.01 b	-	-	-	1.43 ± 0.1 a*
93	30.486	1973	benzoic acid	Acid	11.5 ± 4.37 B	45.32 ± 7.19 b*	25.89 ± 2.78 A	62.88 ± 3.65 a*	10.51 ± 4.51 B	36.97 ± 2.95 b*
94	30.961	2000	n-eicosane	Alkane	1.02 ± 0.25 B	0.88 ± 0.05 b	4.63 ± 0.4 A	11.42 ± 2.46 a	0.18 ± 0.02 B	0.69 ± 0.17 b*

Note: "-" indicates that the compound was not detected. Different capital letters indicate significant differences in the volatile content of three healthy leaves. Lower case letters indicate significant differences in the volatile content of three leaves damaged by *O. steppensis*, and * indicates significant differences in the volatile content of leaves before and after damage to the same species of elm.

The experimental results showed that three elm compound compositions changed significantly in healthy VOCs and HIPVs infested by *O. steppensis* in the leaves of these three elm species. A total of 53 healthy VOCs and 57 HIPVs in the leaves of *U. pumila* were detected. A total of 19 and 25 compounds were detected in *U. densa* leaves before and after feeding by the *O. steppensis*, respectively. Ten new compounds were added to the HIPVs of *O. steppensis* leaves induced by *O. steppensis*; 38 and 54 compounds were detected in *U. laevis* leaves before and after feeding by *O. steppensis*. Twenty-nine new compounds in *U. laevis* were found after feeding by the *O. steppensis*.

For the healthy leaves of three elm species, there were 25 compounds specific to *U. pumila*, 2 compounds specific to *U. densa*, and 11 compounds specific to *U. laevis*, as shown in Figure 3A. There were 16 compounds common to *U. pumila* and *U. densa*, 26 compounds common to *U. pumila* and *U. laevis*, and 15 compounds common to *U. densa* and *U. laevis*. The three species of elm shared 14 compounds. In the state of feeding by *O. steppensis* on elm, as shown in Figure 3B, there were 18 compounds specific to *U. pumila*, 5 compounds specific to *U. densa*, and 15 compounds specific to *U. laevis*. There were 20 compounds common to *U. pumila* and *U. densa*, 39 compounds common to *U. pumila* and *U. laevis*, and 20 compounds common to *U. densa* and *U. laevis*. The three species of elm shared 20 compounds.

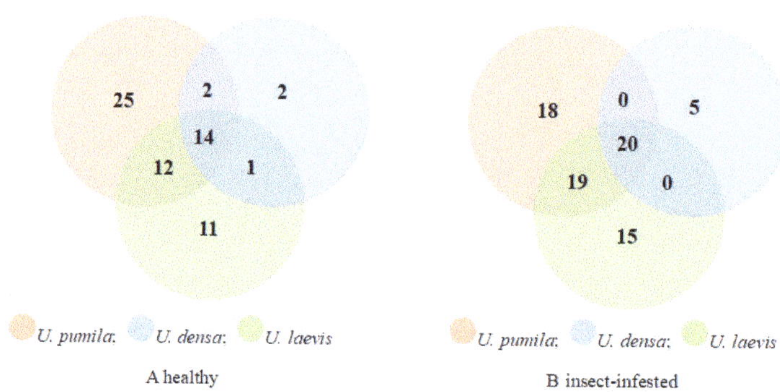

Figure 3. Venn set of VOCs from two leaf states of three elm species. The digits in the figure indicate unique (nonoverlapping color parts) and shared (overlapping color parts) VOC numbers among three elm species.

3.3. GC-EAD Analysis

GC-EAD analyses with two leaf states of three elm species volatile samples identified volatile compounds that elicited antennal response in *O. steppensis* (Figure 4). The results of the GC-EAD experiments screened a total of 13 effective compounds and their distribution in two leaf states of three elms (Table 4).

The results of the GC-EAD experiment screened a total of 13 effective compounds and their distribution in the two leaf states of the three elms, as shown in Table 4. *U. pumila* healthy and damaged leaves were screened for 13 and 13 effective compounds, respectively; *U. densa* healthy and damaged leaves were screened for 7 and 8 effective compounds, respectively; and *U. laevis* healthy and damaged leaves were screened for 8 and 10 effective compounds, respectively.

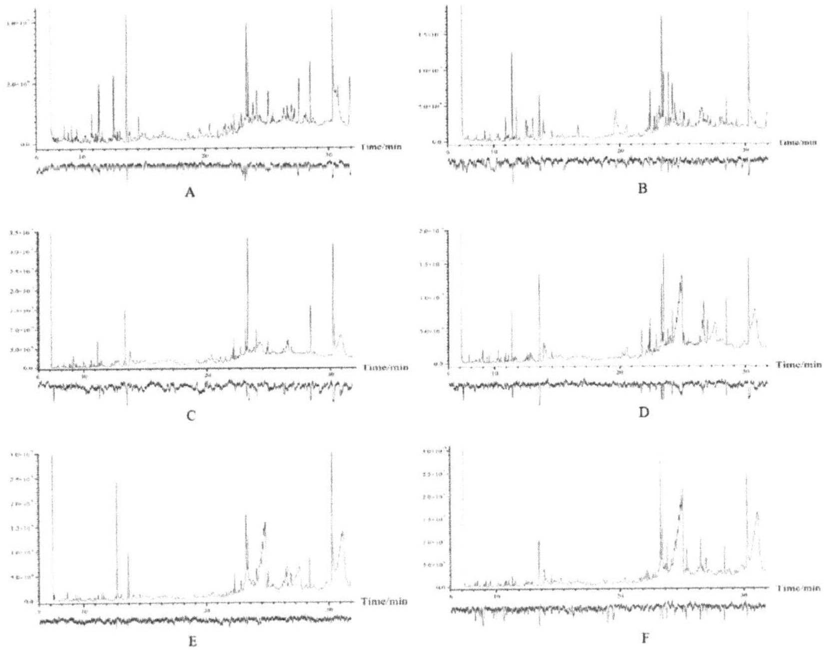

Figure 4. GC-EAD of volatiles from elm leaves by *O. steppensis*. (**A**): Healthy *U. pumila*; (**B**): Infested *U. pumila*; (**C**): Healthy *U. densa*; (**D**): Infested *U. densa*; (**E**): Healthy *U. laevis*; (**F**): Infested *U. laevis*.

Table 4. Distribution of 13 active compounds in the volatiles of two leaf states of three elms by GC-EAD experiments.

No.	Retention Time (min)	Compound	*U. pumila*		*U. densa*		*U. laevis*	
			Healthy	Insect-Infested	Healthy	Insect-Infested	Healthy	Insect-Infested
1	11.003	2-hexenal	+	+	+	+	+	+
2	11.261	3,7-dimethyl-1,3,6-octatriene	+	+	-	-	-	-
3	12.153	3-hexen-1-ol,acetate	+	+	-	-	-	-
4	12.673	1-hexanol	+	+	-	-	-	-
5	13.217	3-hexen-1-ol	+	+	+	+	+	+
6	13.413	nonanal	+	+	+	+	+	+
7	21.16	decanoic acid, propyl ester	+	+	-	+	+	+
8	22.404	methyl salicylate	+	+	+	+	+	+
9	22.542	benzoic acid, 2-methylpropyl ester	+	+	+	+	+	+
10	23.402	guaiacol	+	+	-	-	-	+
11	23.687	benzyl alcohol	+	+	+	+	+	+
12	24.056	phenylethyl alcohol	+	+	-	-	-	+
13	28.65	2,4-di-tert-butylphenol	+	+	+	+	+	+

Note: "+" in the table means that an antennae response was detected, and "-" means that no antennae response was detected.

3.4. EAG Responses of Female and Male Adults of *O. steppensis* to 13 Single Compounds

The normalized EAG responses of female and male adults of *O. steppensis* to 2-hexenal, 3,7-dimethyl-1,3,6-octatriene, 3-hexen-1-ol, acetate, 1-hexanol, 3-hexen-1-ol, nonanal, decanoic acid, propyl ester, methyl salicylate, Benzoic acid, 2-methylpropyl ester, Guaiacol, Benzyl alcohol, Phenylethyl Alcohol, and 2,4-Di-tert-butylphenol at five concentrations (0.01, 0.1, 1, 10, and 100 mg/mL diluted in paraffin liquid) were measured, as shown in Figure 5.

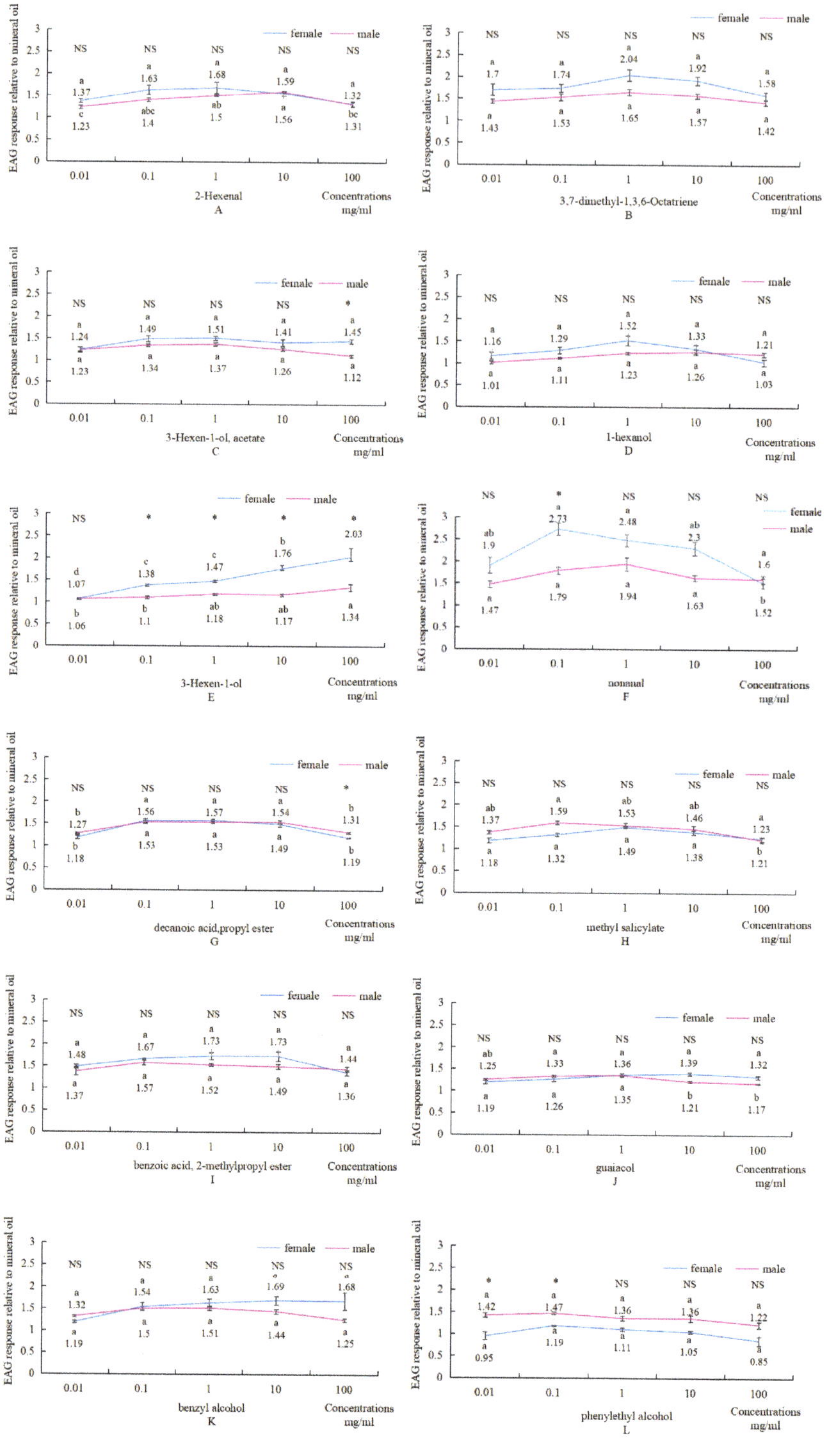

Figure 5. *Cont.*

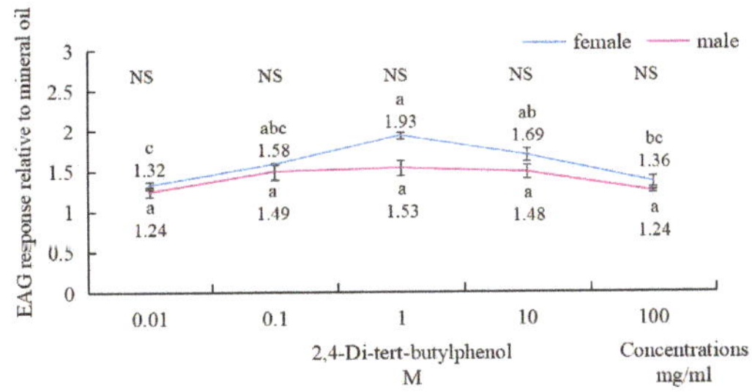

Figure 5. Normalized EAG responses of male and female adults of *O. steppensis* to 13 volatile compounds. (**A**): 2-hexenal; (**B**): 3,7-dimethyl-1,3,6-octatriene; (**C**): 3-hexen-1-ol,acetate; (**D**): 1-hexanol; (**E**): 3-hexen-1-ol; (**F**): nonanal; (**G**): decanoic acid, propyl ester; (**H**): methyl salicylate; (**I**): benzoic acid, 2-methylpropyl ester; (**J**): guaiacol; (**K**): benzyl alcohol; (**L**): phenylethyl alcohol; (**M**): 2,4-di-tert-butylphenol. Notes. Different small letters indicate that EAG responses of female and male adults of *O. steppensis* to different concentrations of the same compound are significantly different at the $p < 0.05$ level, which was determined by one-way ANOVA and Duncan's multiple comparisons; difference in EAG responses of *O. steppensis* to the same concentrations of the same compound between male and female adults of this weevil pest was determined by *t*-test (NS: $p > 0.05$; *: $0.01 < p < 0.05$).

O. steppensis showed EAG responses to all 13 compounds. *O. steppensis* showed the highest EAG relative response values to 1 mg/mL 2-hexenal, 1 mg/mL 1,3,7-dimethyl-3,6-Octatriene, 1 mg/mL 3-hexen-1-ol, acetate, 1 mg/mL 1-hexanol, 100 mg/mL 3-hexen-1-ol, 0.1 mg/mL nonanal, 1 mg/mL decanoic acid, propyl ester, 1 mg/mL methyl salicylate, 1 mg/mL Benzoic acid, 2-methylpropyl ester, 1 mg/mL guaiacol, 1 mg/mL benzyl alcohol, 0.1 mg/mL Phenylethyl Alcohol, and 1 mg/mL 2,4-di-tert-butylphenol had the largest relative EAG response values.

3.5. Y-Tube Tests of O. steppensis Choice to Compounds

Different volatile compounds showed significantly different effects on the selection behavior of *O. steppensis*. Behavioral experiments using the optimal concentrations of effective compounds demonstrated that *O. steppensis* prefer to choose 100 mg/mL of 3-hexen-1-ol, 1 mg/mL of 1,3,7-dimethyl-3,6-Octatriene, and 1 mg/mL of methyl salicylate more than in the control group and their selection difference are extremely significant, as shown in Figure 6.

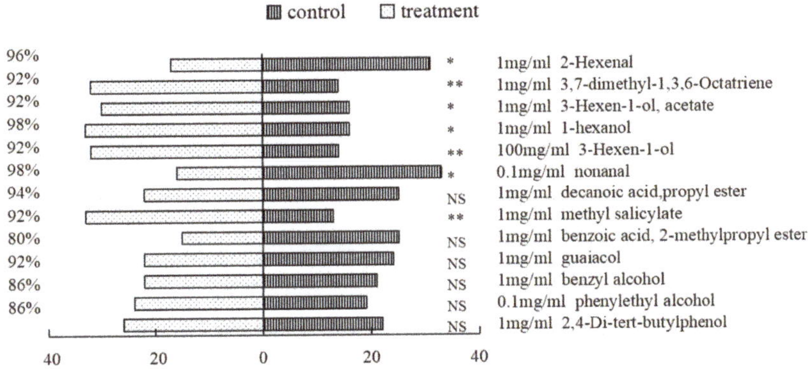

Figure 6. Normalized EAG responses of *O. steppensis* adults to volatile compounds. Note: Y-tube behavior choices of *O. steppensis* towards different compounds at different concentrations determined by the χ^2 test. (NS: $p > 0.05$; *: $0.01 < p < 0.05$; **: $p < 0.01$).

4. Discussion

The results of the volatile bioassay showed that when *O. steppensis* searches for host plants, olfactory detection plays a leading role, and the selection rate varies for different host plants. Both male and female *O. steppensis* showed selective preference for *U. pumila*, *U. densa*, and *U. laevis*. The results revealed that *O. steppensis* has different host preferences for the three elm species: *U. pumila* first, followed by *Ulmus laevis* and *Ulmus densa* last. The results were inconsistent with the reports of Zhang (2000) [12] and Li (2018) [2], and they reported that the damage severity in descending order of *O. steppensis* on these three elm species was *U. pumila*, *U. densa*, and *U. laevis*. In this study, we drew a conclusion that the reason why the indoor olfactory selection behavior is not completely consistent with the field damage may be because the current study is only based on the healthy volatiles results of three elm species' attraction to *O. steppensis*, while the effect of leaf damage on the attraction of *O. steppensis* remains unknown at present. Further investigation is needed to determine whether the behaviors of phytophagous insects–such as feeding, mating, and laying eggs on elm leaves–are influenced by the physical properties (such as size and shape) and chemical composition (such as nutrients) of the leaves [28].

Different tree species have different chemical compositions and relative contents before and after infestation, which constitute different chemical fingerprints of the host plant [29]. In our study, the highest compounds in healthy and post-injury leaves of *U. pumila* were 3-hexen-1-ol and phenylacetaldehyde, and for *U. densa*, HIPVs were benzyl alcohol and 3-hexen-1-ol, respectively, and benzoic acid and phenylacetic acid propyl ester for *U. laevis*. Plant volatiles are mostly mixtures with few individual compounds, and various compounds in certain proportions make up the chemical fingerprints of different plant species. Different plant–insect systems release different volatiles, and the composition of compounds released from the damaged parts of plants is related to their species and phytophagous insect species [30]. At present, there have been some studies on elm volatiles both domestically and internationally, and the results of volatile composition tests differ among different elm species and different parts of the plant. The results of this experiment showed that there were significant differences in the volatiles of the leaves of three elm species. *U. pumila* leaves had the most and richest variety of volatiles before and after damage. Changes in volatiles have an effect on the behavior of phytophagous insects after a plant is damaged [31].

EAG bioassay of volatile compounds showed that adult males and females differed significantly only at partial concentrations of 3-hexen-1-ol, acetate, 3-hexen-1-ol, nonanal, decanoic acid, propyl ester, and phenylethyl alcohol. Different concentrations of the same compound display different physiological activity for herbivorous insects, indicating that the specific concentrations of plant volatiles are important factors that influence insect responses [32]. The antennae potential response of insects only responds to the insect's sensitivity to volatiles; behavioral experiments are required to determine the specific role of volatiles [33]. In the EAG experiments, the EAG values of females were generally higher than those of males, except for phenylethanol, and the results indicated that *O. steppensis* females were more sensitive to host plant volatiles than males in their antennae potential responses. It has been reported that female insects localize, feed, and lay eggs through host plant volatiles and, therefore, have a greater EAG response to host plant volatiles [34,35]. It is hypothesized that *O. steppensis* females need to search for hosts for oviposition and, therefore, need to be more sensitive to plant volatiles in the environment [36].

Electrophysiological and behavioral results indicated that 100 mg/mL of 3-hexen-1-ol, 1 mg/mL of 3,7-dimethyl-1,3,6-octatriene, methyl salicylate, 1-hexanol, and 3-hexen-1-ol, acetate had an attractive effect on *O. steppensis*; 0.1 mg/mL of nonanal and 1 mg/mL of 2-hexenal had a repellent effect on *O. steppensis*. In previous studies, the results of antennae potentials and behavioral responses were inconsistent, and it was believed that the behavioral selection of insects to volatiles was also influenced by environmental factors, their own physiological conditions, and other aspects [37–39]. The electrophysiological and behavioral results of this experiment were also inconsistent, and it was believed that the olfactory localization of *O. steppensis* did not only rely on the antennae sensors but was also affected by other olfactory sensors on the body surface. Although the antennae had an electrophysiological response, there was no associated behavior. Healthy leaf volatiles of *U. pumila*, *U. densa*, and *U. laevis* contained five, two, and three eliciting components, respectively, and the relative content of the two repellent components (2-hexenal and nonenal) in the healthy leaves of *U. pumila* was lower than in that of *U. densa*. There was only one repellent component, nonenal, in the healthy leaves of *U. laevis* compared to the former two elm species. Consequently, *U. pumila* was the most attractive host for *O. steppensis* elm, followed by *U. laevis* and finally *U. densa*. This result is consistent with the results of laboratory experiments on behavioral responses of *O. steppensis* to the leaves of three elm species. It was reported that phytophagous insects rely on a series of compounds in specific ratios for host plant selection [40]. In the current study, the experiment was only conducted on a single component, and the behavioral experiments were not conducted on a mixture of the components. In addition, the test insects were collected in the field, and their age was not clear, which may have interfered with the experimental results to some extent. Whether the optimal concentration of the screened compounds specifically has the effect of trapping *O. steppensis* in the forest needs to be followed up with wind tunnel tests and forest experiments.

Author Contributions: Conceptualisation, M.Y. and Q.L.; methodology, M.Y., Q.L. and Y.L. (Yonggen Lou); software, M.Y. and Q.L.; validation, M.Y. and Q.L.; formal analysis, M.Y.; investigation, M.Y. and Q.L.; resources, Q.L.; data curation, M.Y. and Q.L.; writing—original draft preparation, M.Y., Q.L. and Y.L. (Yonggen Lou); writing—review and editing, G.Z., Y.L. (Yalin Liu) and Y.L. (Yonggen Lou); visualization, Q.L.; supervision, Q.L.; project administration, Q.L.; funding acquisition, Q.L. All authors have read and agreed to the published version of the manuscript.

Funding: This work was funded by the Natural Science Foundation of Xinjiang Autonomous Region (2023D01C14), the National Natural Science Foundation of China (grant no. 31900349), the Third Xinjiang Scientific Expedition Program (grant no. 2022xjkk1505), the Youth Talent Promotion Project of Autonomous Region Association for Science and Technology of China (grant no. RCTJ60) to Qin Li, and Postdoctoral program of Beet pest survey project (grant no. TCT-JSYF-20221122) to Qin Li.

Data Availability Statement: Data is contained within the article.

Acknowledgments: We are very thankful to Andrew Polaszek (Natural History Museum, London, United Kingdom) for his great help in improving the English language and its clarity throughout the manuscript. We also thank the Hongying Hu research group of Xinjiang University, China, for helping collect part of the weevils specimens; Jinhui Yang (Analysis and Testing Center of Xinjiang University, China) for her great help in elm leaf volatiles collection and analysis; Xiaoyu Li (College of Life Science and Technology of Xinjiang University, China) in calculation of the retention time of the compounds.

Conflicts of Interest: The authors declare no conflict of interest.

References

1. Abduwaha, A.; Adili, S. Survey on selecting elm types and spatial distribution of larvae of *Rnynchaenus dlni* Linnaeus. *J. Xinjiang Agric. Univ.* **2016**, *39*, 127–131.
2. Li, Q.; Triapitsyn, S.V.; Wang, C.; Zhong, W.; Hu, H. Biological traits and the complex of parasitoids of the elm pest *Orchestes steppensis* (Coleoptera: Curculionidae) in Xinjiang, China. *Bull. Entomol. Res.* **2018**, *108*, 48–57. [CrossRef] [PubMed]

3. Cheng, X. Monitoring Investigation and Control Technology of *Rhynchaenus alini* in Nursery Garden. *Mod. Agric. Sci. Technol.* **2023**, *16*, 87–90.
4. Gao, Z.; Ye, X.; Yang, F.; Xia, Y. The key technology of aircraft prevention and control of *Rhynchaenus alni* in Changji City. *Rural. Sci. Technol.* **2017**, *3*, 33–35.
5. Wolde-Hawariat, Y.; Seyoum, E.; Jembere, B.; Negash, M.; Hansson, B.S.; Hillbur, Y. Behavioral and electrophysiological response of sorghum chafer *Pachnoda interrupta* (Coleoptera: Scarabaeidae) to plant compounds. *Int. J. Trop. Insect Sci.* **2007**, *27*, 53–61. [CrossRef]
6. Li, Y.; Chen, L.; Wei, H. Host preference and EAG response of 4 scarab beetles to head space volatiles of host and non-host plants. *J. Environ. Entomol.* **2018**, *40*, 1328–1334.
7. Korotyaev, B.A. On the distribution dynamics of some weevils (Coleoptera: Coccinellidae, Bruchidae, Curculionidae) in the plains of the northwestern Caucasus, Russia. *Entomol. Rev.* **2015**, *95*, 137–140. [CrossRef]
8. Korotyaev, B.A. New data on the changes in the abundance and distribution of several species of beetles (Coleoptera) in European Russia and the Caucasus. *Entomol. Rev.* **2016**, *96*, 620–630. [CrossRef]
9. Vlasov, D.V. Finding of the Flea-Weevil *Orchestes steppensis* Korotyaev, 2016 (Coleoptera, Curculionidae: Rhamphini) in Yaroslavl Province. *Entomol. Rev.* **2019**, *99*, 1389–1390. [CrossRef]
10. Li, Y.; Liu, H. Research on the occurrence and prevention of *Rhynchaenus alni*. *Liaoning Agric. Sci.* **2004**, 100.
11. Li, X.; Feng, Y.; An, X.; Yang, G. Spatial distribution pattern of the larvae of *Orchestes alni* and influence factors in Helan Mountains. *Chin. J. Appl. Entomol.* **2020**, *57*, 450–459.
12. Zhang, B.; Xu, Z.; Shen, L. A new pest in Yili area-*Rhynchaenus alni* Linnaeus. *For. Xinjiang* **2000**, *1*, 45.
13. Gotzek, D.; Robertson, H.M.; Wurm, Y.; Shoemaker, D. Odorant binding proteins of the red imported fire ant, *Solenopsis invicta*: An example of the problems facing the analysis of widely divergent proteins. *PLoS ONE* **2011**, *6*, e16289. [CrossRef] [PubMed]
14. Myrick, A.J.; Baker, T.C. Locating a compact odor source using a four-channel insect electroantennogram sensor. *Bioinspiration Biomim.* **2011**, *6*, 016002. [CrossRef] [PubMed]
15. Antonatos, S.; Anastasaki, E.; Balayiannis, G.; Michaelakis, A.; Magiatis, P.; Milonas, P.; Papadopoulos, N.T.; Papachristos, D.P. Identification of volatile compounds from fruits aroma and citrus essential oils and their effect on oviposition of *Ceratitis capitata* (Diptera: Tephritidae). *Environ. Entomol.* **2023**, *52*, 327–340. [CrossRef] [PubMed]
16. Reinecke, A.; Ruther, J.; Hilker, M. The scent of food and defence: Green leaf volatiles and toluquinone as sex attractant mediate mate finding in the European cockchafer *Melolontha melolontha*. *Ecol. Lett.* **2002**, *5*, 257–263. [CrossRef]
17. Sun, Y.; Shao, K.; Lu, Y.; Shi, Q.; Wang, W.; Chen, L. Electrophysiological and alarm behavioral responses of *Solenopsis invicta* Buren (Hymenoptera: Formicidae) to alkoxypyrazines. *J. Asia-Pac. Entomol.* **2017**, *20*, 541–546. [CrossRef]
18. Wang, X.; Jin, Y.; Wang, Y.; Zhong, Y.; Li, L.; Chen, F. Responses of brown planthopper, *Nilaparvata lugens* (Stäl) to specific rice plant volatile organic compounds (VOCs). *Chin. J. Appl. Entomol.* **2022**, *59*, 50–57.
19. Gao, M.; Xie, P.; He, Q.; Ye, Z.; Han, Q. Behavioral responses of *Diaphorina citri* Kuwayama to buds of two host plant species and their volatiles. *J. Environ. Entomol.* **2023**, *46*, 1–17.
20. Yan, X.; Wang, Z.; Ma, L.; Song, C.; Zhao, J.; Wang, C.; Hao, C. Electroantennogram and behavioural responses of *Plutella xylostella* to volatiles from the non-host plant *Mentha spicata*. *J. Appl. Entomol.* **2023**, *147*, 511–519. [CrossRef]
21. Korotyaev, B.A.; Sofronova, E.V. New data on the distribution and host plants of weevils (Coleoptera, Curculionoidea: Apionidae, Curculionidae) in the south of Baikal Siberia and in Mongolia. *Entomol. Rev.* **2016**, *96*, 1289–1296. [CrossRef]
22. Wu, J.; Zheng, K.; Gu, Y.; Fan, J.; Wu, L. Electrophysiological and behavioral responses of *Anoplophora chinensis* (Coleoptera: Cerambycidae) to the volatiles from Salix ohsidare. *Acta Entomol. Sin.* **2022**, *65*, 1477–1487.
23. Zhang, Y.; Sun, H.; Feng, J.; Xu, C.; Liu, S.; Guo, K.; Wei, H.; Xu, R.; Qiao, H.; Lu, P. Oviposition preferences of *Gephyraulus lycantha* (Diptera: Cecidomyiidae) females with respect to three *Lycium* spp. host plants. *Chin. J. Appl. Entomol.* **2023**, *60*, 629–640.
24. Silva, D.B.; Weldegergis, B.T.; Van Loon, J.J.; Bueno, V.H. Qualitative and Quantitative Differences in Herbivore-Induced Plant Volatile Blends from Tomato Plants Infested by Either *Tuta absoluta* or *Bemisia tabaci*. *J. Chem. Ecol.* **2017**, *43*, 53–65. [CrossRef]
25. Liu, J.; Zhang, J.; Shen, J.; Zhao, H.; Ma, W.; Jiang, Y. Differences in EAG Response and Behavioral Choices between Honey Bee and Bumble Bee to Tomato Flower Volatiles. *Insects* **2022**, *13*, 987. [CrossRef] [PubMed]
26. Bozsik, G.; Molnár, B.P.; Szőcs, G. Comparison of antennal responses of *Ovalisia festiva* and *Phloeosinus aubei* to volatile compounds of their common host, *Thuja occidentalis*. *Physiol. Entomol.* **2022**, *47*, 136–146. [CrossRef]
27. Magsi, F.H.; Luo, Z.; Zhao, Y.; Li, Z.; Cai, X.; Bian, L.; Chen, Z. Electrophysiological and Behavioral Responses of *Dasychira baibarana* (Lepidoptera: Lymantriidae) to Tea Plant Volatiles. *Environ. Entomol.* **2021**, *50*, 589–598. [CrossRef] [PubMed]
28. Pang, B.; Cheng, J. Relationships between the Plant Surface and Insects. *Chin. J. Ecol.* **1998**, *4*, 53–59.
29. Kristensen, T.G.; Nielsen, J.; Pedersen, C.N.S. A tree-based method for the rapid screening of chemical fingerprints. *Algorithms Mol. Biol.* **2010**, *5*, 1–10. [CrossRef] [PubMed]
30. Ji, X. Ecological Effects of Volatile Semiochemicals Induced by Wheat Aphid and Species Diversity on *Macrosiphum avenae*. Master's Thesis, Shandong Agricultural University, Shangdong, China, 2009.

31. Hu, X.; Su, S.; Liu, Q.; Jiao, Y.; Peng, Y.; Li, Y.; Turlings, T.C. Caterpillar-induced rice volatiles provide enemy-free space for the offspring of the brown planthopper. *Elife* **2020**, *9*, e55421. [CrossRef] [PubMed]
32. Chi, D.; Li, X.; Yu, J.; Xie, X.; Wang, G. The EAG response and behavior of the *Saperda populnea* L. to volatiles from poplar branches. *Acta Ecol. Sin.* **2011**, *31*, 334–340. [CrossRef]
33. Fan, N.; Huang, J.; Liu, S.; Zhang, K.; Zhang, X.; Zheng, H. Electrophysiological and olfactory behavioral responses of adult *Callosobruchus chinensis* (Coleoptera: Chrysomelidae) to the volatiles from *Vigna angularis* and *V. radiata* seeds. *Acta Entomol. Sin.* **2024**, *67*, 58–67.
34. Das, P.D.; Raina, R.; Prasad, A.R.; Sen, A. Electroantennogram responses of the potato tuber moth, *Phthorimaea operculella* (Lepidoptera; Gelichiidae) to plant volatiles. *J. Biosci.* **2007**, *32*, 339–349. [CrossRef] [PubMed]
35. Li, G.; Chen, Y.; Yan, R.; Lei, Y.; Chen, X.; Li, B. EAG and behavioral responses of adults of *Adoxophyes orana* (Lepidoptera:Tortricidae) to volatiles of its major host plants. *Acta Entomol. Sin.* **2022**, *65*, 1026–1037.
36. Palma, R.; Mutis, A.; Manosalva, L.; Ceballos, R.; Quiroz, A. Behavioral and electrophysiological responses of *Hylastinus obscurus* to volatiles released from the roots of *Trifolium pratense* L. *J. Soil Sci. Plant Nutr.* **2012**, *12*, 183–193. [CrossRef]
37. Canale, A.; Benelli, G.; Germinara, G.S.; Fusini, G.; Romano, D.; Rapalini, F.; Desneux, N.; Rotundo, G.; Raspi, A.; Carpita, A. Behavioural and electrophysiological responses to overlooked female pheromone components in the olive fruit fly, *Bactrocera oleae* (Diptera: Tephritidae). *Chemoecology* **2015**, *25*, 147–157. [CrossRef]
38. Bai, R.; Chen, L.; Wang, W. Electrophysiological responses of four scarab beetles to the volatiles produced by different plants. *J. Environ. Entomol.* **2021**, *43*, 492–499.
39. Li, M.; Wang, S. Study on Electroantennogram and Behavioral Response of Potosia brevitarsis Lewis to 9 Volatiles in Sugar-Vinegar Liquid. *Xinjiang Agric. Sci.* **2022**, *59*, 466–473.
40. Visser, J. Host odor perception in phytophagous insects. *Annu. Rev. Entomol.* **1986**, *31*, 121–144. [CrossRef]

MDPI AG
Grosspeteranlage 5
4052 Basel
Switzerland
Tel.: +41 61 683 77 34

Plants Editorial Office
E-mail: plants@mdpi.com
www.mdpi.com/journal/plants

www.ingramcontent.com/pod-product-compliance
Lightning Source LLC
Chambersburg PA
CBHW061848210926
51850CB00006B/1008
9783725879472